Unity 2022 by Example

A project-based guide to building 2D and 3D games, enhanced for AR, VR, and MR experiences

Scott H. Cameron

‹packt›

Unity 2022 by Example

Group Product Manager: Rohit Rajkumar

Publishing Product Manager: Vaideeshwari Muralikrishnan

Book Project Manager: Sonam Pandey

Senior Editor: Anuradha Joglekar

Technical Editor: Reenish Kulshrestha

Copy Editor: Safis Editing

Proofreader: Anuradha Joglekar

Indexer: Manju Arasan

Production Designer: Vijay Kamble and Prafulla Nikalje

DevRel Marketing Coordinators: Anamika Singh and Nivedita Pandey

Publication date: June 2024

Production reference: 1030524

Published by Packt Publishing Ltd.

Grosvenor House

11 St Paul's Square

Birmingham

B3 1RB, UK

ISBN 978-1-80323-459-5

www.packtpub.com

To my luvluv Bheng, my little buddy "Arckie," and my peanut gang, Yzhen, Zhyon, and Xhymn – I love you all dearly. Sincere thanks for all your loving support, encouragement, and inspiration.

– Daddy Scott

In memory of our beloved daughter, Zhyon, who loved to draw.

Foreword

There aren't many industries that cost you nothing to enter, yet still have the potential to earn you millions of dollars in just a few years. Typically, you would start a career by earning an expensive tertiary diploma, then perhaps complete an unpaid internship, and then start in your chosen industry with an entry-level salary.

But of course, who really wants typical?

Game development is an entirely open industry. Literally anyone with a halfway-decent PC and a desire to experiment, learn, and iterate can develop a simple new game in a few weeks. Of course, the larger the game you wish to produce, the more time you'll need to dedicate to its production. The point is that your success is only a question of your own time input, not whether you have an expensive degree or even any industry experience.

And by "you," I mean *you*, the reader. In a very real sense, your path between dreamer and success story lies within these pages. The first step, you've already taken: deciding to use a game development engine, rather than create everything from scratch.

Unity was first released 20 years ago and so has become a mature, polished product. In the world of game engines, Unity's user base is unmatched, in part because the full-featured engine is completely free for individuals and independent studios, and in part because it's not difficult to develop a working familiarity and build up from there.

That's exactly what this book does. In this hefty tome, Scott Cameron will take you step by step through the complete project-creation process, from installing the Unity Hub to lighting, even to machine learning, and finally to publishing your polished game on Steam, all painstakingly compiled by an expert on the Unity engine.

An award-winning graphic artist, Unity Certified Expert, and good friend of mine, Scott's broad experience in software and game development spans three decades. Even as an experienced developer myself, Scott remains my go-to guy for any technical hurdle I might have in Unity. His methodical, no-nonsense approach to coding and design makes him the perfect author for an all-encompassing book such as this.

Through each chapter, Scott introduces new technical concepts exactly when the reader needs them. He provides enough context and tips to proceed, plus cross-references and online links if the reader wants to dive deeper into a particular topic.

Almost everyone on the planet has played some kind of digital game on console, PC, mobile, or VR. If you've played *Pokémon Go, Beat Saber, Monument Valley, Among Us, Overcooked*, or a host of other popular titles, then you have already experienced Unity as a consumer. Scott's guidance in *Unity 2022 By Example* will lead you from a novice to a competent, confident game creator.

Global revenue for games is approaching US$400 billion, having long since eclipsed the combined global revenue of film and music, and it's growing fast. There is room in the industry for you and the game that's in your head right now. Get started reading and absorbing this book!

The rest is up to you.

Edward Falzon MBUS MINFTECH
Unity-certified Expert Programmer, President, Pluck Games

Contributors

About the author

Scott H. Cameron has over 25 years of experience in information technology, software development, and design. He's had the pleasure of crafting innovative solutions across diverse industries, from entertainment to manufacturing and financial services to architecture.

For the past decade, his latest adventure has immersed him in game development as a Unity Certified Expert Programmer, senior engineer team lead, and expert course mentor at Zenva Academy. Whether creating captivating VR experiences, making remarkable indie games, or revolutionizing foreign language learning through gamification, he's driven by a love for pushing creative boundaries and sharing his experiences.

I wrote this book amid a global pandemic, a super typhoon, and the loss of our daughter. I couldn't have done it without my wife's unwavering strength and support and my children's inspiration. Love ya!

Thanks to friends, Edward, Kendrick, Sean, and Alvin, for influencing and motivating me. My gratitude also goes to Nica and Miguel, the artists.

Lastly, I'm thankful to my exceptional partners at Packt, the technical reviewer, and everyone who helped me write this book.

About the reviewer

John Estes is a long-time user of the Unity game engine in his work toward gamification in healthcare. He has served as an expert adviser to Jintronix, a start-up company using Microsoft Kinect in Unity for rehabilitation gamification. He has also used his technical, artistic, and healthcare skills as a research manager in a university simulation lab focused on VR research in neurological rehabilitation. Today, John focuses on freelance work in real-time cinematography as well as designing and programming independent games and educational materials for vestibular rehabilitation in Unity.

Almost everyone on the planet has played some kind of digital game on console, PC, mobile, or VR. If you've played *Pokémon Go*, *Beat Saber*, *Monument Valley*, *Among Us*, *Overcooked*, or a host of other popular titles, then you have already experienced Unity as a consumer. Scott's guidance in *Unity 2022 By Example* will lead you from a novice to a competent, confident game creator.

Global revenue for games is approaching US$400 billion, having long since eclipsed the combined global revenue of film and music, and it's growing fast. There is room in the industry for you and the game that's in your head right now. Get started reading and absorbing this book!

The rest is up to you.

Edward Falzon MBUS MINFTECH
Unity-certified Expert Programmer, President, Pluck Games

Contributors

About the author

Scott H. Cameron has over 25 years of experience in information technology, software development, and design. He's had the pleasure of crafting innovative solutions across diverse industries, from entertainment to manufacturing and financial services to architecture.

For the past decade, his latest adventure has immersed him in game development as a Unity Certified Expert Programmer, senior engineer team lead, and expert course mentor at Zenva Academy. Whether creating captivating VR experiences, making remarkable indie games, or revolutionizing foreign language learning through gamification, he's driven by a love for pushing creative boundaries and sharing his experiences.

I wrote this book amid a global pandemic, a super typhoon, and the loss of our daughter. I couldn't have done it without my wife's unwavering strength and support and my children's inspiration. Love ya!

Thanks to friends, Edward, Kendrick, Sean, and Alvin, for influencing and motivating me. My gratitude also goes to Nica and Miguel, the artists.

Lastly, I'm thankful to my exceptional partners at Packt, the technical reviewer, and everyone who helped me write this book.

About the reviewer

John Estes is a long-time user of the Unity game engine in his work toward gamification in healthcare. He has served as an expert adviser to Jintronix, a start-up company using Microsoft Kinect in Unity for rehabilitation gamification. He has also used his technical, artistic, and healthcare skills as a research manager in a university simulation lab focused on VR research in neurological rehabilitation. Today, John focuses on freelance work in real-time cinematography as well as designing and programming independent games and educational materials for vestibular rehabilitation in Unity.

Table of Contents

Part 2: 2D Game Design

2

Creating a 2D Collection Game

3

Completing the Collection Game

Part 3: 2D Game Design Continued

4

5

9

Completing the Adventure Game   263

Part 4: 3D Game Design

10

Creating a 3D First Person Shooter (FPS)   313

Part 5: Enhancing and Finishing Games

13

14

15

Finishing Games with Commercial Viability | 507

Preface

Hello! The Unity game engine presents an exciting opportunity for aspiring game developers to bring their creative visions to life. This book is designed for those committed to making finished games while following best practices and wanting to release games commercially. By the end of this comprehensive journey, you'll have not only gained the knowledge and skills in Unity 2022 but also a sense of accomplishment, knowing that you're one step closer to achieving your goal of creating and publishing games.

This book aims to guide you through structuring your game projects effectively, covering everything from code organization to asset management and scene hierarchy. I delve into fundamental game design principles, including crafting core gameplay loops, defining player actions, implementing mechanics, and establishing win/lose conditions.

Throughout the chapters, we tackle vital challenges newer game developers may encounter, such as gaps in game design concepts, familiarity with Unity's features and Scripting API, and proficiency in C# programming. Unity's distinctive methodology for constructing game elements is central to our approach, emphasizing a component-based approach driven by C# scripts.

This book takes a practical approach to learning as I guide you through a series of projects. We'll start with a simple 2D object collection game and gradually progress to immersive 3D **first person shooter (FPS)** and **mixed reality (MR)** experiences. Each chapter builds on the previous one, introducing new concepts and techniques and providing practical examples, hands-on exercises, and the occasional challenge or bonus activity. This way, you'll not only understand the theory but also gain valuable practical skills.

Moreover, I address essential aspects often overlooked in traditional learning materials, such as publishing games for commercialization, project continuance and version control, and supporting games throughout their lifetime. You'll also gain insights into **Unity Gaming Services (UGS)** and the process of operating a game as a service. This comprehensive approach ensures that you'll be well-prepared for all stages of game development.

By the end of this book, you'll possess solid foundational knowledge of Unity 2022 and be equipped to structure projects in a maintainable and extensible way for commercially releasing games. So, let's embark on this journey together and unleash your potential as a game developer in the vibrant ecosystem of Unity!

Who this book is for

This book caters to individuals aspiring to become game developers, specifically those dedicated to crafting complete games within Unity while adhering to coding best practices and having ambition for a commercial game release. The ideal reader already has Unity installed and possesses a foundational understanding of navigating the Unity Editor interface. They also have sound file organizational skills, beginner-level art skills, and some experience implementing basic functionality through C# scripting.

What this book covers

Chapter 1, Foundational Knowledge of Unity 2022, provides the introductory steps to learn how to use Unity 2022 to create a game easily! It covers installing Unity Hub and exploring the available templates to kickstart game projects. It also covers how the Package Manager provides additional functionality while keeping project footprints small and avoiding unwanted tooling. This chapter concludes by beginning a new game project, learning about the GameObject, and creating a simple 2D character.

Chapter 2, Creating a 2D Collection Game, provides an in-depth exploration of expanding the scene with additional GameObjects to craft a 2D top-down setting for the start of a collection game. It covers the utilization of Unity's Tilemap feature for efficient environment creation. This chapter also introduces the fundamental concepts of the C# programming language and will guide you through creating custom scripts to enhance functionality within the game environment. The chapter ends by outlining the integration of player input using Unity's latest event-based input system.

Chapter 3, Completing the Collection Game, provides an introduction to utilizing Unity's **Cinemachine** (**CM**), a robust camera control system, to follow the player around within the game environment. It will also introduce integrating a **user interface** (**UI**) into the game, leveraging Unity's UI system (uGUI) and custom-coded C# components, including features such as a timer and score tracking.

Chapter 4, Creating a 2D Adventure Game, provides a guide to creating a 2D side-on adventure game using Unity 2D tooling such as Sprite Shape. The chapter covers importing artwork, incorporating dynamic moving platforms, and optimizing game performance. By the end of the chapter, you will have a refined understanding of game design documentation and the ability to design interactive 2D environments, including triggers that provide secondary actions to create more engaging player experiences.

Chapter 5, Continuing the Adventure Game, provides an overview of how to create a rigged and animated 2D player character using the Unity 2D Animation package. The chapter will cover setting up the character avatar using the PSD Importer, implementing an input action map for controlling player movement with a controller script, and animating the character using Mecanim.

Chapter 6, Introduction to Object Pooling in Unity 2022, provides an introduction to object pooling, a crucial optimization pattern used in game development to maintain performance and avoid lags in gameplay. This chapter utilizes this pattern to maintain a performant player shooting mechanic by using Unity's object pooling API.

Chapter 7, Polishing the Player's Actions and Enemy Behavior, provides a focus on learning how to improve the visual aesthetics of the player character artwork by introducing Unity's graph node-based shader authoring tool Shader Graph and the Trail Renderer component. It then presents the creation of diverse **non-player character** (**NPC**) enemies using **Scriptable Objects** (**SOs**) to configure the different enemy types. The chapter culminates with introducing enemy behavior management through a state pattern.

Chapter 8, Extending the Adventure Game, provides the steps to establish health management and damage infliction mechanisms to equip the player, enemy, and any destructible object with a simple yet effective health system. The chapter also introduces the creation of an enemy wave spawner that enables dynamic spawning.

Chapter 9, Completing the Adventure Game, provides an introduction to creating a global event system that facilitates communication across diverse C# classes and promotes modularity and scalability. Utilizing the newly created event system, the following section covers developing a quest system suitable for any number of custom missions. Additionally, this chapter covers integrating and customizing a puzzle system from the Unity Asset Store to suit a game developer's specific needs.

Chapter 10, Creating a 3D First Person Shooter (FPS), provides the skills to design and construct a gray-box 3D environment utilizing modular components within the Unity Editor using ProBuilder and Prefabs. It will also cover how to quickly integrate an FPS character controller with the Unity Starter Asset and adapt and optimize existing 2D code for environment interactions into their 3D API methods.

Chapter 11, Continuing the FPS Game, provides instruction on updating and enhancing the previous chapter's 3D environment by replacing Prefabs and materials, polishing assets, scattering objects using Polybrush, and improving lighting with light baking and light probes. It then introduces the application of wear-and-tear effects to add realism and decorate the environment using decals.

Chapter 12, Enhancing the FPS Game with Audio, provides a focus on improving the player experience by adding audio to a Unity game project. Throughout the chapter, you are provided with the skills and knowledge necessary to create an audio manager and related reusable audio player components to play music, ambient sounds, and **sound effects** (**SFX**). The chapter finishes with examples of implementing footstep sounds for the player character and a reverb zone within the game level.

Chapter 13, Implementing AI with Sensors, Behavior Trees, and ML-Agents, provides an introduction to essential **artificial intelligence** (**AI**) behavior and NPC navigation using Unity's AI Navigation package and refactored 2D components from the previous chapters. The chapter also covers enhancing enemy NPC dynamics by introducing sensors and behavior trees. It explores integrating **machine learning** (**ML**) tools with Unity ML-Agents to add advanced AI capabilities to games.

Chapter 14, Entering Mixed Reality with the XR Interaction Toolkit, provides the knowledge to develop **mixed reality** (**MR**) games and experiences that use a player's physical space to create immersive and novel gameplay experiences. This chapter uses the Unity XR Interaction Toolkit to make a final boss room encounter in MR. It explores the design process, working with AR Foundation planes, placing interactable objects, and implementing mechanics.

Chapter 15, Finishing Games with Commercial Viability, provides essential knowledge for effective game development project management by exploring **Games as a Service (GaaS)**, Unity DevOps and LiveOps resources, source code management via Unity Version Control, in-game economies, platform distribution, and UGS implementation. It covers managing and securing a project development lifecycle, distributing commercially viable games, and reaching an intended player audience.

To get the most out of this book

You will need Unity Hub and a version of the Unity 2022 Editor installed on your computer. *Chapter 1* will walk you through the installation of both if you don't already have Unity installed. All code examples will work with the latest LTS release version of Unity 2022.3. All code examples and downloadable project files have been confirmed with the latest Unity 2022.3 LTS release version available at the time of publication.

Software/hardware covered in the book	Operating system requirements
Unity Hub	Windows, macOS, or Linux
Unity 2022 Editor	Windows, macOS, or Linux
C# version 9	
Unity Gaming Services (UGS)	

To complete the examples provided throughout the chapters, you need only a Unity ID account and a Unity Personal license (free).

If you are using the digital version of this book, we advise you to type the code yourself or access the code from the book's GitHub repository (a link is available in the next section). Doing so will help you avoid any potential errors related to the copying and pasting of code.

Download the example code files

You can download the example code files for this book from GitHub at `https://github.com/PacktPublishing/Unity-2022-by-Example`. If there's an update to the code, it will be updated in the GitHub repository.

We also have other code bundles from our rich catalog of books and videos available at `https://github.com/PacktPublishing/`. Check them out!

Conventions used

There are a number of text conventions used throughout this book.

`Code in text`: Indicates code words in text, database table names, folder names, filenames, file extensions, pathnames, dummy URLs, user input, and Twitter handles. Here is an example: Create a new folder in the `Assets/Sprites/Character` directory."

A block of code is set as follows:

```
internal class Tags
{
    // Ensure all tags are spelled correctly!
    public const string Player = "Player";
}
```

Bold: Indicates a new term, an important word, or words that you see onscreen. For instance, words in menus or dialog boxes appear in **bold**. Here is an example: "In the **Geometry** panel in the window's bottom-right section, ensure **Weights** is enabled "

> Tip | Maximize Unity window
>
> To maximize the currently active window in Unity, as seen with the scene view window in *Figure 10.3*, you can use the keyboard shortcut *Shift* + spacebar.

Get in touch

Feedback from our readers is always welcome.

General feedback: If you have questions about any aspect of this book, email us at `customercare@packtpub.com` and mention the book title in the subject of your message.

Errata: Although we have taken every care to ensure the accuracy of our content, mistakes do happen. If you have found a mistake in this book, we would be grateful if you would report this to us. Please visit `www.packtpub.com/support/errata` and fill in the form.

Piracy: If you come across any illegal copies of our works in any form on the internet, we would be grateful if you would provide us with the location address or website name. Please contact us at `copyright@packt.com` with a link to the material.

If you are interested in becoming an author: If there is a topic that you have expertise in and you are interested in either writing or contributing to a book, please visit `authors.packtpub.com`.

Share Your Thoughts

Once you've read *Unity 2022 by Example*, we'd love to hear your thoughts! Scan the QR code below to go straight to the Amazon review page for this book and share your feedback.

`https://packt.link/r/1-803-23459-8`

Your review is important to us and the tech community and will help us make sure we're delivering excellent quality content.

Download a free PDF copy of this book

Thanks for purchasing this book!

Do you like to read on the go but are unable to carry your print books everywhere?

Is your e-book purchase not compatible with the device of your choice?

Don't worry! Now with every Packt book, you get a DRM-free PDF version of that book at no cost.

Read anywhere, any place, on any device. Search, copy, and paste code from your favorite technical books directly into your application.

The perks don't stop there, you can get exclusive access to discounts, newsletters, and great free content in your inbox daily

Follow these simple steps to get the benefits:

1. Scan the QR code or visit the following link:

https://packt.link/free-ebook/9781803234595

2. Submit your proof of purchase.
3. That's it! We'll send your free PDF and other benefits to your email directly.

Part 1:
Introduction to Unity

The tasks outlined in this part will provide you with the necessary knowledge to navigate and utilize the **Unity Editor** effectively while creating new projects for your 2D games. Setting up the first 2D project of the book – a simple 2D top-down object collection game – will walk you through the key steps of project creation in the Unity Hub, introduce the Unity Editor windows and toolbars, create 2D objects in the Hierarchy and Scene windows, and set object properties such as **Sorting Layers** in the **Inspector** window.

This part includes the following chapter:

- *Chapter 1, Foundational Knowledge of Unity 2022*

1

Foundational Knowledge of Unity 2022

Getting started in Unity 2022 is easy with **Unity Hub**. Unity Hub serves a few very useful purposes, and we'll be going through installing it and learning about its features. In this chapter, we'll not only install Unity Hub and the **Unity Editor**, but we'll also break down the different templates that are available to kickstart your game and AR/VR projects.

In addition to templates that provide a starting foundation for your new project, Unity also provides added functionality through the Package Manager. The Package Manager allows Unity to give a small project size footprint and not bloat the Editor with a default installation of unneeded or unwanted tooling (referring to relatively simple add-on programs that combine to accomplish a task). Being familiar with and understanding what packages are available will surely help save time and increase the quality of your project.

Finding your way around the Unity Editor is only half an introduction. The second half of this chapter teaches you how to create content and make things interactable. We'll do this by first creating a simple 2D character right in the Editor using built-in tooling. This all starts with the **GameObject** – Unity's building block.

The book takes a project-based approach to learning, so we'll walk through designing a game, creating a game, and solving problems along the way.

In this chapter, we're going to cover the following main topics.

- Unity Hub – choosing the **2D Universal Render Pipeline (URP) template**

- Getting to know the Unity Editor and installing packages

- Introducing the GameObject! All about the **Transform** and **components**

- **2D Sprites** with **Sprite Creator** – understanding the **Sprite Renderer** and draw ordering

- **Game Design Document (GDD)** – introducing the 2D collection game

By the end of this chapter, you'll be able to create a new Unity project, be comfortable finding your way around the Unity Editor, understand the initial criteria for a game design document, and be prepared to create the **2D Sprite**-based character that is the first element of our game.

Technical requirements

To follow along in this chapter, you'll need a computer with Windows 7+ 64-bit, Mac OS X 10.12+, or Linux (Ubuntu 16.04, 18.04, and CentOS 7) running. You'll need sufficient free hard drive space for not only the Unity Editor installation but also the project files. We recommend 25 GB for the Unity install folder, with 3 GB of free space for the installation temp files (the temp files are usually located on your OS installation drive), and 10 GB should be sufficient for the project files.

Unity Hub – Choosing the 2D URP template

The Unity Hub makes it simple to manage the installed Unity Editor versions and add or remove modules for installed Editors, and it helps manage your different projects. If you're entirely new to Unity, then that previous sentence might be a bit confusing. Why would we have to manage different installed Editors? Simply put, software changes. As the Unity Editor evolves, it introduces new features and changes to its scripting API.

> Tip
>
> As a general rule of thumb, you should not upgrade the Unity Editor version for your project once you've started production. Doing so can have undesirable effects, such as broken renderings or code that no longer compiles. We'll discuss this more in the following sections when installing the Unity Editor and selecting our project template.

Installing Unity Hub

Let's get started on our journey by first getting Unity Hub installed. We'll be using the **Unity Personal** license throughout this book, which is the free version of Unity. Free here doesn't mean we will be limited in features or capabilities in building our games. It just means you are only allowed to use this version for free if you fall under the criteria for requiring a paid license (if you earned less than $100K of revenue or funds raised in the last 12 months).

> Important note
>
> If you are an eligible student, you may want to check out the **Unity Student plan** at https://unity.com/products/unity-student. It provides access to **Unity Pro**, a selection of quality assets, and **Unity Gaming Services**, such as **Cloud Build**, that professionals and studios building games use on Unity.

Okay, let's do this. Perform the following steps to install the Unity Hub:

1. Go to `https://unity.com/download` and select the download link for your OS. This will download `UnityHubSetup.exe` to your `Downloads` folder.

2. Click on the executable from your web browser, navigate to your `Downloads` folder, and double-click the executable to launch the installation.

3. Keep the defaults or change the installation path if you want to install to a different hard drive location (only use local drives and not network drives, as this could cause problems).

4. After clicking **Install** and letting the installation process complete, click **Finish** to run Unity Hub.

When opening Unity Hub for the first time, you will be prompted to **Sign in** or **Create account** if you have not created one already. Your Unity account, known as **Unity ID**, will be used for licensing. Unity requires an active license to install the Unity Editor. The Personal Edition license is free, and Unity Hub will generate one for you.

With Unity Hub now installed, let's continue installing the Unity Editor.

Installing the Unity Editor – What version?

We have a new installation of Unity Hub and are now ready to install a Unity Editor version. As previously mentioned, Unity Hub allows you to have multiple versions of the Unity Editor installed to manage the different projects you'll create over time. Opening Unity Hub without any Editor version installed will default to prompting the latest Unity Editor version in the **LTS** stream to be installed. LTS simply means **long-term support**. This is usually the best version choice to lock in and base a new project on because it will be the most stable version available and is guaranteed to be supported for the next two years. To ensure stability with the LTS version, no new features are added to the tooling or scripting API. If you want to create with the latest engine features, you'll have to choose one of the newer **tech streams** since no additional tools or technologies will be introduced in the LTS stream.

Okay, simply stated, what does this mean?

- If you're starting production today or about to ship, and you desire stability and support over the lifetime of your development and release cycle, choose the LTS stream. At the time of writing, this is 2020.3 LTS.

- If you're starting production today but want to leverage newer tools and technology, and the latest official release version is close to becoming the new LTS stream, choose the latest official release version. At the time of writing, this is 2021.2 (becoming the latest LTS stream when 2021.3 is officially released).

- If you want to create on the cutting edge of the tools and technology available and don't mind managing potential crashes and bugs, then choose the latest pre-release (beta) version.

> **Important note**
>
> This book is written specifically for Unity 2022. If you have an earlier version of Unity already installed, then the instructions or features may differ, so it may be difficult – or impossible – to follow along. We recommend installing the latest version of Unity 2022 to complete the projects we'll create in the following chapters.

To gauge your production schedule against the Unity LTS releases, refer to the following timeline:

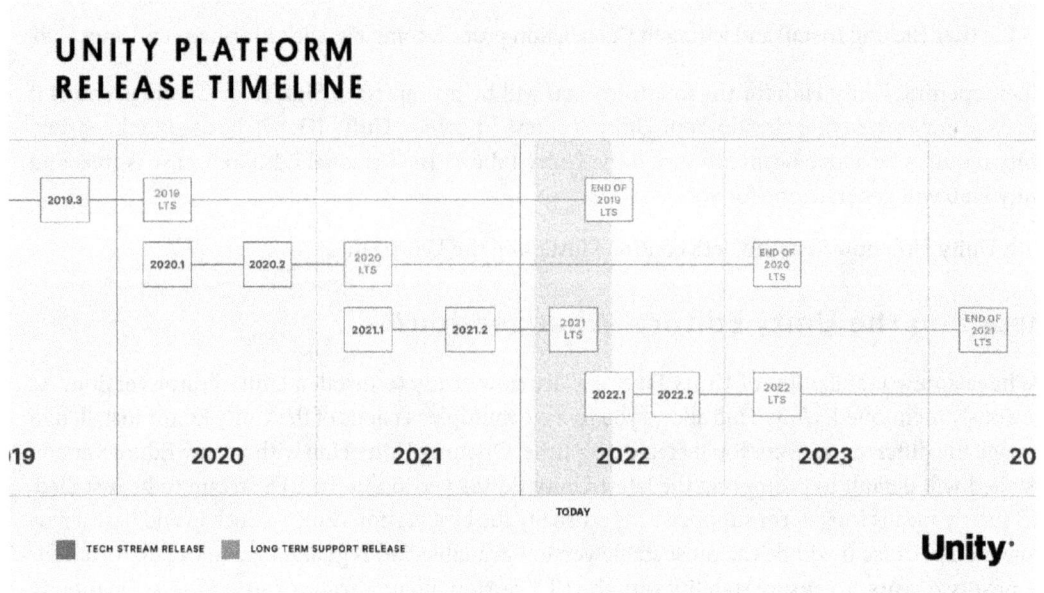

Figure 1.1 – Unity platform release timeline

Unity provides detailed LTS release information on the **Unity QA** resources page at `https://unity3d.com/unity/qa/lts-releases`.

Proceed to install the Unity 2022 Editor with the following steps:

1. If you've just finished installing Unity Hub, you'll be prompted to install the latest LTS version of the Unity Editor. We want to install a Unity 2022 release specifically, so proceed by clicking **Skip installation** in the lower-right corner of the dialog, unless, at the time you're reading this, the 2022 LTS version has been released, in which case you can just install it and skip the remaining steps!

2. In the main Unity Hub window, select **Installs** in the left-side pane.

3. Click the **Install Editor** button in the window's top-right corner.

4. The **Official releases** tab should be selected by default (the latest beta versions can be found under the **Pre-releases** tab).

5. In the **OTHER VERSIONS** section below the **LONG TERM SUPPORT (LTS)** section, find the latest 2022 release version listed and click **Install**.

6. The next screen you'll be presented with is selecting the dev tools, platforms, and documentation modules. Since we're assuming you're installing Unity for the first time, and we don't currently have any specific requirements for our first project, we'll just keep the defaults. Click on **Continue**.

Unity Hub will now download and install the selected version. Depending on your internet connection and hard drive speed, this will take a bit of time – an average base installation requires roughly 3 GB of downloads (temporary files) and approximately 7 GB of free hard drive space for installation.

In this section, we learned about the different Unity Editor versions available and how to install them. In the next section, we'll learn about render pipelines to determine how to proceed with a new project.

What is a render pipeline?

In the next section, when we create our project, we'll select a template to base our project on. This requires a bit of explanation to understand the templates' options fully. We'll create a 2D game (a two-dimensional game represented by planar images), so selecting a 2D template makes sense, but these are the available template names for 2D: **2D**, **2D (URP)**, and **2D Mobile**. We won't be creating a mobile game, so we can rule out **2D Mobile**, but what does **(URP)** mean?

If you're unfamiliar with how video games work under the hood, **rendering** refers to how the 2D graphics or 3D models are drawn to the screen to generate an image. Over time, as Unity evolved its rendering technology to better suit the types of games creators were making, they understood the need to make changes and improve the rendering technology. A performant and customizable render pipeline architecture was introduced to serve creators best called a **Scriptable Render Pipeline** (**SRP**). The template named **2D** will use Unity's **built-in renderer**, whereas **2D (URP)** will use the **Universal Render Pipeline** (**Universal RP**, or **URP**). The Universal RP is a default SRP Unity provides as a starting basis for making performant games on the broadest device platforms – this will eventually replace the built-in legacy renderer as the default.

> **Important note**
>
> We'll discuss the URP feature set relevant to adding renderer features to the projects throughout the book but not compare it directly with the built-in renderer features. A feature comparison table between URP and the built-in renderer can be found at `https://docs.unity3d.com/Packages/com.unity.render-pipelines.universal%407.1/manual/universalrp-builtin-feature-comparison.html`

This section taught you what a render pipeline is and how to select the correct 2D project template. Now you'll use what you learned to create our first project!

Creating a project

We'll be creating our project from scratch using one of the templates Unity Hub provides to make sure everything is set up correctly for our game's rendering requirements. Since we'll start by creating a 2D game, we'll select the **2D (URP) Core** template. Core means that the template won't provide any example assets, samples, or learning content in this context. It will give an empty 2D project with a pre-configured URP 2D renderer setup – perfect, just what we need!

Proceed with the following steps to create a new 2D URP project in Unity Hub:

1. With **Projects** selected in the left-side pane, click the **New project** button in the window's top-right corner.
2. Verify that the Editor version at the top of the window is set to the installed 2022 version.
3. Find the **2D (URP)** template in the list and click to select it.
4. Next, in the right-side pane, give your project a name by entering it in the **Project name** field.
5. And finally, verify the installation path in the **Location** field, then click on **Create project**.

Now you can create your project with the preceding steps while referring to the following screenshot (making sure to have the correct **Editor version** selected at the top in case you have multiple versions installed already):

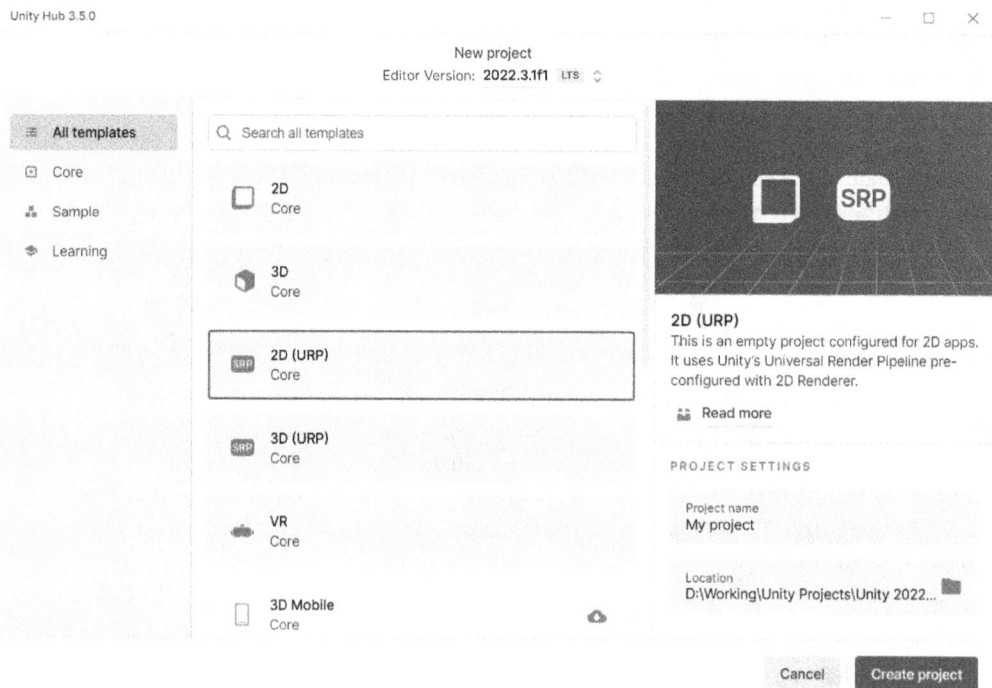

Figure 1.2 – Unity Hub project templates

In this section, you learned how to install the Unity Hub and installed a specific version of the Unity Editor. Then, you learned what a render pipeline is and how that relates to creating a new project. We'll continue discussing the Unity Editor now by introducing its main features.

Getting to know the Unity Editor and installing packages

Having just created a new 2D URP project and opened up the Editor, let's take a tour of the Unity 2022 interface! We'll cover only the most common features in this section and dig deeper into the features of the specific windows and toolbars as we work through our projects in the later chapters, providing the information in the context of the required task.

When we open the Unity Editor for the first time, it will use the default Windows layout, as shown in the following screenshot:

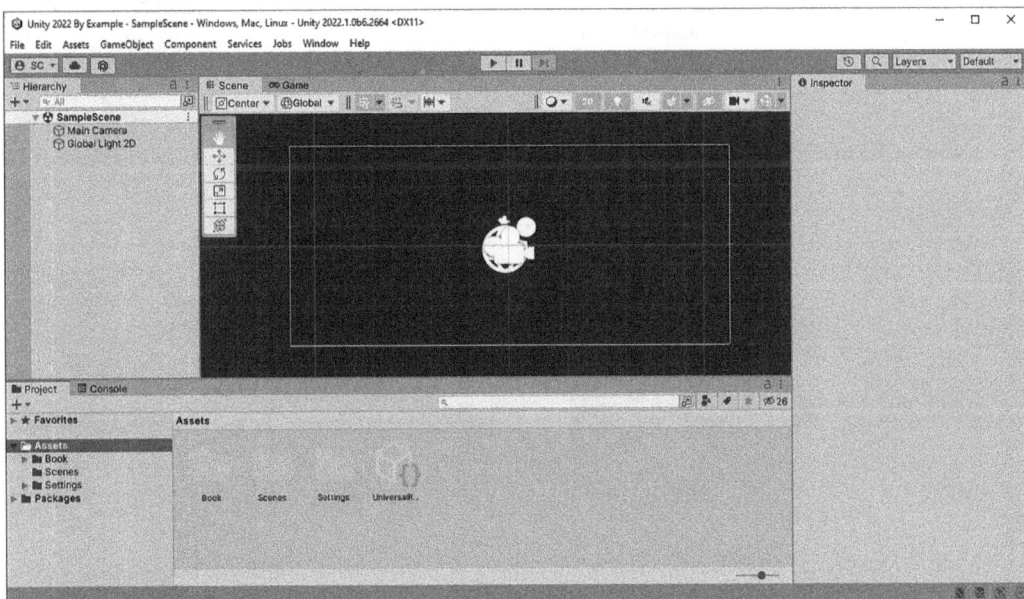

Figure 1.3 – The Unity 2022 Editor default layout

The common Editor windows that you'll be using most of the time are the following:

- **Scene** – this is the window where we build our content visually. Objects added to the **Scene Hierarchy** with renderer components will be visible in the scene and game views.

- **Game** – the simulation you see in the game view represents what players will see rendered in your final playable distribution builds. You will also playtest your game in this window when entering **Play Mode**.

- **Project** – the **Project** window is comparable to your OS's file manager. It's where you'll import and organize the files that make up the **Assets** for the project you're creating (such as 3D models, 2D images, sounds & music, plugins, and so on).

- **Hierarchy** – this is the scene's **GameObject Hierarchy** (more on GameObjects in the next section). Understanding how to organize your scene's objects in the **Hierarchy** window (as in parent-child relationships) will be essential to work on your projects effectively, so this will be a topic of discussion in the coming chapters.

- **Inspector** – in the natural order of things in Unity, the **Inspector** window comes next as it's tied directly to the GameObjects' in the **Hierarchy** window. When an object is selected in the **Scene Hierarchy**, the **Inspector** shows all its details (**Transform** and components).

- **Console** – information, warnings, errors, and any relevant trace information are displayed in a list view that can be sorted and filtered. Debugging any problems occurring in your project will be performed from the **Console** information displayed.

Note that in the default layout, clicking on a tab, such as **Game** or **Console** in *Figure 1.3*, will bring it to the "front" for interaction. Tabs can also be dragged and docked to other windows to provide a fully customized layout.

In addition to these windows, Unity also has some toolbars, such as the following:

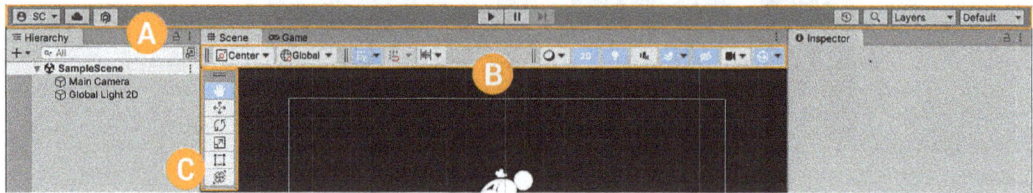

Figure 1.4 – The Unity 2022 Editor toolbars

- **Main Toolbar** (*A*) – the buttons on the upper left provide access to your **Unity Account** (Unity ID), **Unity Cloud Services**, and the current **version control system** (**VCS**) (we'll tackle version control in a later chapter). The centered buttons are the play (for entering **Play Mode**), pause, and step controls. On the right side, the buttons are **Undo History**, **Global Search**, the **Layers Visibility** dropdown, and finally, the **Editor Layout** dropdown (as noted earlier, we're looking at the default window layout; you can change the layout using one of the presets or save your own).

- **Scene Toolbar** (*B*) – the tools, starting on the left side, are:

  - **Tool Handle Position** (Center, Pivot) – when moving objects in the scene, the action will be based on this position, either the Center of the object or the object's Pivot.

> **Tip**
>
> If the **Anchor** position of your GameObject in the scene View doesn't look correct, don't forget to check this setting!

- **Tool Handle Rotation** (Global, Local) – when rotating objects in the scene, they rotate relative to their Transform in **Global** or **Local** space.

> **Tip**
>
> You may need to change between Global or Local space settings to rotate an object correctly. If your rotations don't look correct, don't forget to check this setting!

- The remainder of the tools include grid visibility and snapping settings, **Draw Mode**, 2D or 3D scene View (we are currently in 2D), toggles for scene lighting, audio, effects, hidden objects, scene View camera settings, and Gizmos. There's a lot to unpack here, but don't worry, we'll touch on these as we work through the projects we'll be creating in the coming chapters.

- **Manipulation Tools Toolbar** (*C*) – a floating toolbar, also known as **Overlays**, within the **Scene** window. This toolbar provides the essential tools for working with GameObjects within the scene View. The tools include View, Move, Rotate, Scale, Rect, and Transform.

One more "toolbar" along the bottom of the Editor window is called the **Status bar** (not pictured). The Status bar mostly provides the current status of specific processes, such as the last **Console** warning or error message (left-side), the progress of lighting generation (right-side), and the code compilation spinner (right-corner).

> **Additional reading | Unity documentation**
>
> You can find more information on Unity's interface at `https://docs.unity3d.com/2022.3/Documentation/Manual/UsingTheEditor.html`

In this section, you learned about the familiar Editor windows and toolbars and how they can manipulate objects in the scene View. Let's see how we can extend the features and tooling in the Editor now with Packages.

The Unity Package Manager

Without knowing, you were already introduced to packages earlier when we discussed the Universal RP. Since we started with a URP template, we didn't have to do anything special, but Unity provides **Scriptable Render Pipeline** support through packages! Packages provide a way for Unity to offer multiple versions of an engine feature or service without requiring a new installation of the Editor. You can even try out the latest pre-release version of a package to stay on the cutting edge of the technology and quickly revert to a stable or alternate version should you encounter any problems. The Package Manager is accessible from the top menu: **Window | Package Manager**.

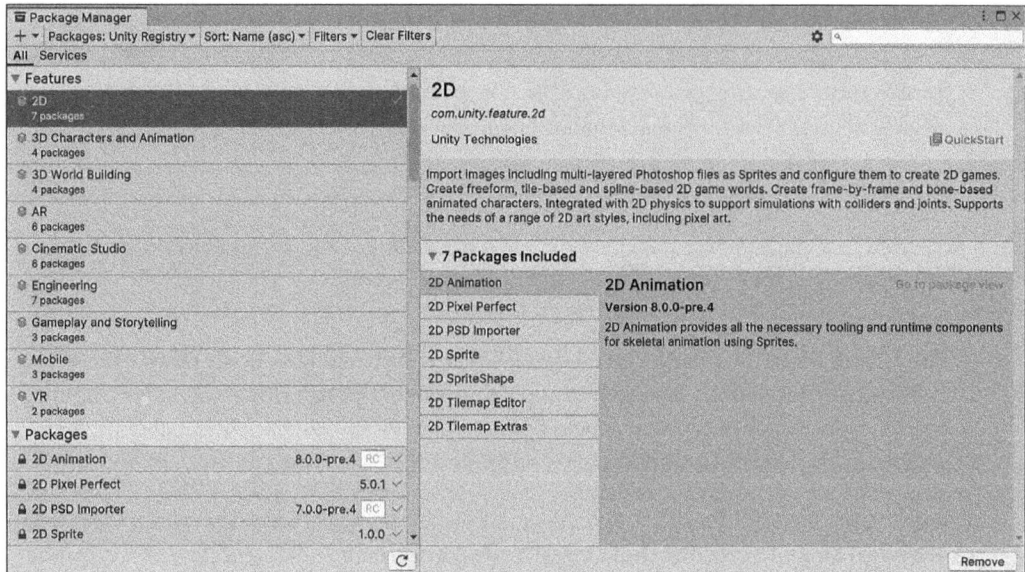

Figure 1.5 – The Unity Package Manager

Feature sets in the **Package Manager** are bundles of common tooling that provide a simpler and more streamlined install experience. The **2D** feature set (selected in *Figure 1.4*) is for creators working with 2D projects. In our case, again, having started from the **2D URP** template, the **2D** feature set has already been imported to our project (indicated by the green checkmark). We're good to go!

Should you need to make any changes to the packages in your project, from the dropdown in the **Package Manager** top menu, you can see what packages are available from what is already in your project via **In Project**, or either the **Unity Registry** or **My Assets** (which are assets you've purchased in the Unity Asset Store) selections. Managing a package is as simple as selecting it in the list and choosing an available function from the buttons displayed in the bottom-right window. For example, **Download**, **Install**, **Remove**, or **Update**.

> Tip
> Packages are project-specific, so you will need to ensure you have the desired packages installed for each new project you create!

> **New to Unity 2022**
>
> In the **Package Manager** window, you can now multi-select packages in the list for adding, updating, or removing in a single operation.

In this section, you learned about the Editor windows and toolbars and how to add/remove features and tooling with packages that extend the Editor's capabilities. Next, we'll start learning about the GameObject.

Introducing the GameObject – All about Transform and components

Simply put, anything you want to add to your scene will be added as a GameObject. The GameObject is the base building block for everything that exists in a scene. It also acts as a container for adding functionality with components. When a GameObject is added to the **Hierarchy**, is it active by default but can be deactivated – "turning off" all components added to it – with the checkbox to the left of the name field (to the left of **Main Camera** in *Figure 1.6*). A component can be for either visual or functional purposes, or in some cases, both! Functional components implement Unity's scripting API, and Unity provides components for supporting the engine's many features. You'll create custom component scripts using the C# language, which will be covered in *Chapter 2*.

In this section, you will learn how GameObjects are added to a scene, then you'll be introduced to the Transform component, and you'll learn how to work with components.

Adding GameObjects to the scene

New empty GameObjects and objects of a specific type are easily added to the **Hierarchy** with the **Create** menu. The **Create** menu is easily accessible from Unity's top menu under GameObject, from the **Hierarchy** window directly with the + (plus sign) icon dropdown menu at the top, or by right-clicking anywhere in the **Hierarchy** window. GameObjects in the form of **Prefabs** and other supported types can be added to the scene View directly by dragging and dropping from the **Project** window – this is often the quickest way to accomplish some tasks, and we'll be exploiting this feature while building the book's projects.

The Transform component

GameObjects added to the scene have a default Transform component that dictates its **Position**, **Rotation**, and **Scale** in 3D space – a Cartesian coordinate system using three mutually perpendicular coordinate axes with Y-Up, namely X-axis, Y-axis, and Z-axis. Positioning graphics in the scene can be performed by either manipulating the Transform manually (by typing in values), with the **Manipulation Tools** (clicking and dragging in the **Scene** window), or through code by using the Unity scripting API to access the GameObject's **Transform** properties and methods.

In the following screenshot, we can see the **Main Camera** GameObject's **Transform** in the **Inspector** and how the **Position** value (`Vector3`) is represented in 3D space:

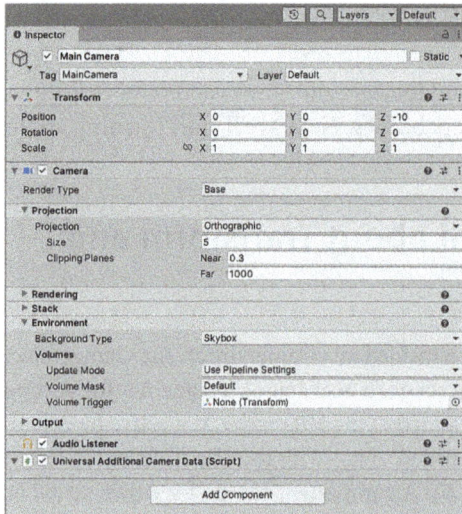

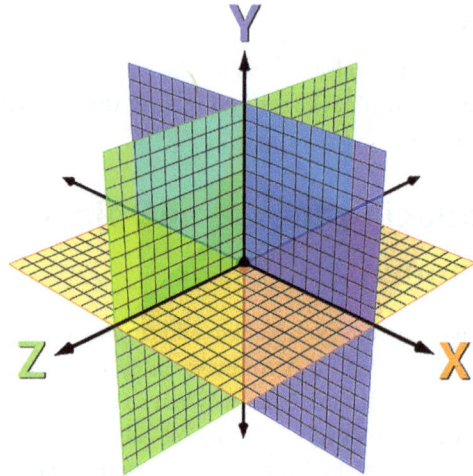

Figure 1.6 – Inspector Transform and related 3D coordinate system

In the preceding figure, we can see that the **Main Camera** (a GameObject) in the scene was selected in the **Hierarchy** because its name is stated at the top of the **Inspector** window. The following section under that is the GameObject's **Transform** values indicating that **Main Camera** is sitting -10 units on the Z-axis from the origin (0, 0, 0 where all axes intersect). The **Rotation** values on all axes are at 0, meaning no rotation is applied to this GameObject. Similarly, the **Scale** value on all axes is at 1, meaning no scaling is applied to the default scale of this GameObject. We'll modify these values in context while creating our player character graphics in the *2D sprites with Sprite Creator – Understanding the Sprite Renderer and draw ordering* section.

Before creating our player, let's discuss components in more detail since we'll be directly working with them.

Components

Below the **Transform** section, we can see several components already added that provide functionality for the GameObject's purpose as our scene's main camera (cameras in our **Scene Hierarchy** determine what the player sees rendered in the game view). To add more functionality to this camera, click the **Add Component** button at the bottom of the **Inspector** window. You'll be presented with a filtered and searchable list of available components to add. The components will include not only the ones provided by Unity, such as the **Camera** component in *Figure 1.6*, but also any scripts you've already created and added to the project. In the next chapter, we'll tackle creating our scripts and adding them as components.

Components can be added, removed, and copied/pasted, and their values can be copied/pasted and even saved as presets! This is all accomplished via the **Component** header section. Click and hold the header to manually reposition a component up/down in the **Inspector** window or right-click for moving and other functions. All of the aforementioned functions are accessible either through the right-click dialog popup or the icons on the right side of the component header; those are **Reference**, **Presets**, and a vertical ellipsis equivalent to right-clicking in the title.

Components add powerful features to your projects, but things come together even more when components combine and work together with other components. Unity is based on this **component architecture**. We will be diving into how to structure your projects best to leverage Unity components in a well-structured single-responsibility design pattern that is easy to work with and friendly for both designers and developers to use.

In this section, you learned about the importance of components and how to work with them in **Inspector**. We'll be working with components throughout the book, starting with the **Sprite Renderer** in the next section.

2D sprites with Sprite Creator – Understanding the Sprite Renderer and draw ordering

Let's dig right in and put into practice what we just learned about GameObjects by creating a simple sprite-based character we'll be using as the player in a collection game. This will be our first project in the book! We'll make the player character from scratch using Unity's built-in **Sprite Creator** graphics.

In this section, we'll create a new scene, add sprites, and learn how to manipulate and layer sprites together to make our player character.

Creating a new scene

First, let's create a new scene by going to **File** | **New Scene** (or using the *Ctrl/Cmd + N* shortcut). This will open the **New Scene** dialog and prompt us to select a **Scene Template**. We're going to be using the **Lit 2D (URP)** template since we'll be working with 2D while also using the Universal RP, and we want to take advantage of all the advanced lighting features URP offers.

> **Important note**
>
> When the scene opens, it's good practice to save it immediately! Save it now by going to **File** | **Save** (or by pressing *Ctrl/Cmd + S*), selecting a folder in your project (usually Assets/ Scenes), and giving it a descriptive name. Now, anytime you make changes and want to save your progress, simply use the *Ctrl/Cmd + S* shortcut to save. You'll want to do this regularly and make it a habit – crashes occur when you least expect it, and you don't want to lose any significant progress that will have to be recreated.

With our new scene being created, let's add some sprites!

Creating a sprite using Sprite Creator

Next, create a sprite in our scene using the **Create** menu by right-clicking within the **Hierarchy** window (or using the + dropdown at the top of the window) and then going to **2D Objects | Sprites | Circle**. You should now have a circle visible in your scene View placed at the origin position of (0, 0, 0). Yay!

Let's parent this to a new empty GameObject to create our player object with a good structure – separating the graphics from the functionality we'll add as components to the root GameObject. We'll do this by performing the following steps:

1. Create a new empty GameObject by, again, right-clicking within the **Hierarchy** window (or using the + dropdown at the top of the window) and then selecting **Create Empty**. Newly created GameObjects will be positioned at the origin position of (0, 0, 0).

2. The default name of a new GameObject, as you guessed it, is **GameObject**. It is highlighted for editing by default, so you can easily rename it without additional steps. We'll use the default name for now, so hit the *Enter* key.

3. Create another new GameObject, but this time name it Graphics.

4. Now, we'll set up the player character's GameObject structure by drag and drop. First, click and drag the **Circle** object to **Graphics**, then click and drag the **Graphics** object to **GameObject**.

> **Important note**
>
> We could have saved a few steps by first selecting the **Circle** GameObject in the **Scene Hierarchy**, right-clicking on it to open the **Create** menu, then selecting **Create Empty Parent**. The previous process was essential to demonstrate how GameObjects in the **Hierarchy** window can be parented and moved to other positions in the **Hierarchy**. To give this a try now, undo the parenting by going to **Edit | Undo** (or *Ctrl/Cmd + Z*) and then redo the parenting of the GameObjects.

5. Finally, rename the root object from GameObject to Player. First, select it in the **Hierarchy** window and press the *F2* (Windows) or *Enter* (Mac) key. Alternatively, with the chosen item in the **Hierarchy**, use the name field at the top of the **Inspector** window to rename it. You should end up with the following sprite and GameObject setup in your scene:

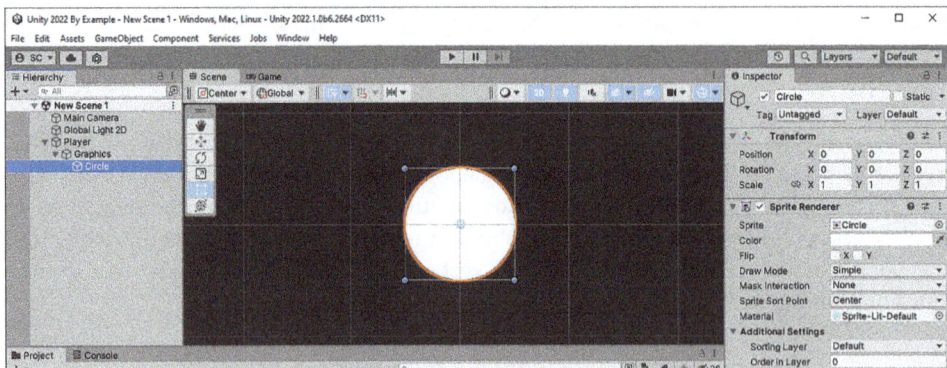

Figure 1.7 – 2D Sprite player character GameObject Hierarchy

You learned how to add a sprite shape to your scene and understand the parenting of GameObjects to create a good structure. Before tapping fully into your arcane artistic talents to create our player character, knowing how to get around in the scene View first will surely be beneficial.

Navigating the scene View

Moving around in the scene **View** will help your drawing efforts by zooming in/out on details and focusing on the part you're working on. While in **2D Mode**, you'll exclusively be using both panning and zooming:

- **Pan** the scene View around by pressing the right-mouse button, using the **View Tool** ("hand" icon in the **Toolbar Overlay**) by clicking and dragging, or you can also use the keyboard by pressing the arrow keys.

- **Zoom** in/out of the scene View by scrolling the mouse wheel.

- Additionally, you can bring focus on an object in the scene View by double-clicking it in the **Hierarchy** window.

> **Additional reading | Unity documentation**
>
> You can read more about scene View navigation at `https://docs.unity3d.com/2022.3/Documentation/Manual/SceneViewNavigation.html`

We're all set to start building our character in the next section.

Creating our player character

We will create a ladybug for the player character in our collection game project. We'll use **Sprite Creator** sprites to design our character right inside the Unity Editor! In a later chapter, we'll be importing original art assets to use as the different sprites in the game. For now, we'll be limited to using some basic shapes to build out our character design, but with some creativity, the results can look quite lovely. The Transform values for **Position**, **Rotation**, and **Scale** and their corresponding **Manipulation Tools (Toolbar Overlay)** will be used extensively to draw our character.

> **Important note**
>
> Placeholder graphics created by a programmer during early development – and in some cases, before an artist has finalized artwork – are commonly referred to as "programmer art." This term is sometimes used negatively to indicate mediocre artwork, but don't let that stop you from being creative! Games such as Geometry Dash, 140, and VVVVVV all use simple graphics to great effect.

Let's start by selecting the **Circle** object (that we previously created) in the **Hierarchy** window (or double-clicking to bring it into focus in the scene View) – note that we'll want to keep all our new sprites as children of the **Graphics** object in the **Hierarchy** (as seen in *Figure 1.8*). This will be the ladybug's body, so let's give it a nice red color.

The **Sprite Renderer** component has a field for color just below the **Sprite** field that indicates we're using the **Circle** sprite shape. Clicking the color will pop up the **Color Picker** dialog (see *Figure 1.8*). When you have a nice shade of red selected – as indicated in the top-right corner of the dialog – simply click the close button in the dialog's title bar (the **x**).

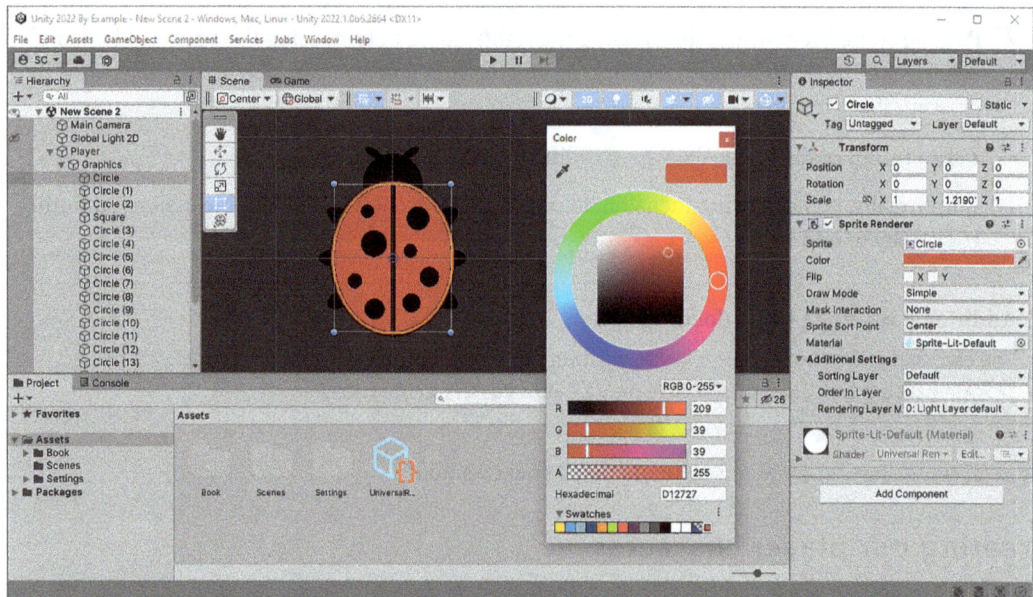

Figure 1.8 – The Ladybug Sprite Renderer component and the Color Picker dialog

With our body sprite having a nice shade of red, we can now shape it by manipulating its **Transform** in the next section.

Using Manipulation Tools

In *Figure 1.8*, you can see that we've already gone ahead and finished our ladybug character design. Let's walk through the process of creating it now. You should have a red circle in your scene View, but it needs to be… less circular. Let's turn it into an ellipse by scaling it on its Y-axis. This can be accomplished by either typing in a value in the **Transform | Scale | Y** field or adjusting manually using the **Rect Tool** (selected in the **Toolbar Overlay** in *Figure 1.8*, and **Scale Tool** may also be used).

> **Important note**
>
> The 2D scene View is represented by the X and Y axes for horizontal and vertical values, respectively. The Z-axis would represent depth, but in **2D Mode**, we won't be manipulating the Z-axis value and will be using **Sprite Sorting** and **Ordering in Layers**.

While manipulating manually with the **Rect Tool**, click and drag on the edge (1 axis) or corner (2 axes) of the box surrounding the **Circle** object to resize it.

Two modifier keys that can help create shapes more easily are the *Shift* and *Alt* keys – but keep the keys pressed while dragging, do not press them before doing so.

- To maintain the current aspect ratio while scaling, hold down the *Shift* key while dragging.
- To scale an object equally on both sides from the center pivot, hold down the *Alt* key while dragging.

> **Tip**
>
> You can hold both the *Shift* and *Alt* keys down while manipulating an object.

When you have a body shape that looks good, let's move on by creating an outline for it. An outline will help provide a good separation of our player character from the background environment. With the **Circle** object still selected (you can simply click on shapes in the scene View to select them), press *Ctrl/Cmd + D* to duplicate it. This will create a copy of the **Circle** sprite and append a number incrementally to the name for every duplicate made. Set the new shape's color to black and scale it up uniformly to be slightly larger than the red circle. While performing this action, you'll likely realize that we now have a problem – the black shape hides the red body. Let's fix it.

Sprite Layers and Ordering

In **Inspector**, **Additional Settings** is a section within the **Sprite Renderer** component. If the **Sorting Layer** and **Order in Layer** fields are not visible directly underneath it, click on **Additional Settings** to expand it (refer to the bottom of the **Inspector** window in *Figure 1.7*). We can change the drawing order of sprites in two ways: 1) by specifying **Sorting Layer** or 2) by a value specified in the **Order in Layer** field.

Sorting Layers is a topic we'll be diving into in the coming chapters since we only need a single Layer for our current purpose – think of Layers as pages in a book where the order in which they are read front to back can be rearranged. Now, working with the **Order in Layer** field, we enter positive and negative integer values to set the sprites drawing order front to back on its **Sorting Layer** (currently Default). The default value is 0, putting every shape at the same depth. Since we want the black shape to be in the back of the red shape, set its **Order in Layer** value to -1. You can now readjust its size to give a nice outline thickness to the red shape.

> **Tip**
>
> You can select multiple shapes and change the **Order in Layer** value for all at once by holding the *Ctrl* or *Shift* key when selecting.

Creating the rest of the ladybug character is just a matter of duplicating a sprite shape already in the scene or creating a new one. The **Sprite Creator** shapes available are **square**, **circle**, **capsule**, **diamond**, and **hexagon**.

Move the shapes into position by simply clicking and dragging on the sprite in the scene View and placing it in position. Hovering the mouse pointer near the corner point (blue dot) will show the rotation cursor. While this cursor is showing, clicking and dragging will rotate it.

> **Additional reading | Unity documentation**
>
> You can read more about positioning GameObjects at `https://docs.unity3d.com/2022.3/Documentation/Manual/PositioningGameObjects.html`

Go ahead and create your own ladybug player character now. Use all of the **Transform** fields, **Rect Tool** with the modifier keys, and the **Order in Layer** field to draw. You will perform these actions repeatedly throughout creating the projects in the book for not only sprites but also UI elements. Have fun with it!

In this section, you learned how to create a new **Scene**, add and duplicate sprites, and manipulate those sprites to make our player character with the **Order in Layer** value. Next, before wrapping up this chapter, we'll discuss game design.

Game Design Document (GDD) – Introducing the 2D collection game

Throughout the book, the game we'll be creating will be defined and structured using what is known as **Game Design Document** (**GDD**). This document will serve as our point of reference as we decide how to develop the game's core aspects. Our game is going to be pretty simple but packed with features. Most of the information written in the GDD is self-explanatory, but a few concepts might be new, so let's start by reviewing them:

- What is the name of the game?

  - This one is self-explanatory, so just don't stress over this now! Anything as a working title will work just fine – have fun with it.

- What is the game's **core loop**?

  - The core loop is what makes your game an enjoyable and satisfying experience for players. It is the series of actions the player repeatedly performs to accomplish the objective.

Let's see an example GDD filled out for the 2D collection game we'll make throughout the book:

Name of Game	Outer World
What is the theme, setting, or genre?	2D Sci-fi Platformer
Summary **What's the big picture?**	An adventure game that takes the player on a journey from peaceful farming to battling robotic systems infected by an evil alien plant entity. The game takes place on an alien planet where the player's race has established a habitat on the planet's surface. The habitat is fully automated and maintained by robots managed by a central control system. An evil alien plant entity has infiltrated the control systems and taken over the robots individually and the central system. The goal of the alien entity is not known, but it must be stopped if the player's race is to survive on this planet.
What is the game's unique feature?	Multiple game modes provide a novel and exciting approach to gameplay: simulation, adventure, and shooting.
What games inspired you and why?	Metroid, Mega Man, and Stardew Valley.
Describe the gameplay, the core loop, and the progression.	Collect energy shards to clean seedlings in a space station habitat as the timer ticks down!

Table 1.1 – GDD for the game

As needed, we'll be adding to the GDD over time, but this will serve us well as a starting point. Yay!

This section introduced you to a simple GDD template and you learned the basic but essential questions to answer when making a remarkable game.

Summary

This chapter was a quick introduction to installing Unity Hub, installing the Unity Editor, and why we chose the 2D Universal RP template for creating our new 2D project. In this chapter, you learned the importance of **Package Manager** for adding tooling and features to the Unity Editor specific to our project's needs. You then learned how to make your way around the different windows and toolbars, and you used them to create and manipulate GameObjects in our new 2D scene. We also got a headstart on building the collection game by drawing our player character within the Unity Editor – learning the importance of Layer sorting order in the process.

Finally, we started defining what the game we're creating will be by deciding on a name for the game, writing our summary, and specifying some gameplay in the GDD – this will provide the necessary direction for the production of the game.

In the next chapter, we'll dive into creating the "Outer World" collection game environment, learning how to implement game mechanics through scripts, and adding a basic UI.

Image sources

Figure 1.1 – Unity platform release timeline

1. **Webpage reference**: `https://blog.unity.com/engine-platform/unity-20221-tech-stream-is-now-available`

2. **Image URL**: `https://blog-api.unity.com/sites/default/files/2022-05/image2.jpg`

Part 2: 2D Game Design

In this part, you will continue game design for a 2D collection game, where you will create an environment and address level design in the GDD for the gameplay's win/lose conditions. You will be introduce controlling the player character with the new Input System, the defined mechanics' functionality for interacting with collectibles and hazards, and adding C# scripts (components) to GameObjects. You will learn how to create your own physics-based interactivity for a 2D game.

This part includes the following chapters:

2
Creating a 2D Collection Game

In *Chapter 1*, you were introduced to the Unity Editor and its common windows and toolbars. We also created our first player character – the ladybug! Sadly, the ladybug is currently sitting in a vast nothingness without any goals providing value for its existence.

In this chapter, we'll start putting the previous knowledge gained to good use by first adding more GameObjects to our Scene and creating a 2D, top-down environment for the *Outer World* collection game (our ladybugs will be happy indeed).

Creating a 2D top-down environment and designing a level is a trivial task when using the **Tilemap** feature. It allows you to create a palette of **Tiles** and then simply draw the level right within the Unity scene View.

There is only so much you can do with GameObjects and Unity's built-in components alone, so you'll be introduced to the **C#** language. Writing your own scripts in C# allows you to create the functionality your games and experiences require.

We'll discuss how to create scripts and understand the best-practice approach to structuring code, principles, and patterns to follow that will keep your code easy to work with, maintain, and extend. These approaches to writing code will benefit you individually but also in a team environment, working with other professionals, since the structure and practices are widely adopted in the industry.

The first C# script we'll be tackling is a **controller** to solve the problem of our ladybug character not having the ability to move. Throughout the book, we will use a *problem-solving* approach to define the process because it frames the coding requirements with an analytical mindset, breaking down a problem into smaller tasks while considering the overall solution. We'll also be working with player input, so you'll learn how to set up and connect your code to Unity's new event-based **Input System**.

In this chapter, we're going to cover the following main topics:

- Creating a 2D, top-down game environment with **Tilemap**
- An introduction to creating scripts in C# – IDE, SOLID principles, and design patterns
- Coding a simple player controller with the new Input System

By the end of this chapter, you'll be able to quickly create a 2D, top-down, tile-based game environment, design a level, and understand how to create and edit C# scripts while considering best practice principles and design patterns. Our ladybug character will also be able to explore a cool new environment by responding to player input.

Technical requirements

To follow along in this chapter, you'll need to have **Visual Studio Community 2022** installed; we will be using this version throughout the chapters and projects in the book. This should have been installed with the Unity Editor in *Chapter 1*.

You can download the complete project on GitHub at `https://github.com/PacktPublishing/ Unity-2022-by-Example`.

Creating a 2D, top-down game environment with Tilemap

The first thing we will accomplish to create the collection game environment is some level design. The design of the level will affect how the game plays – how fun and how challenging it is. This will not be a comprehensive dive into level design; after all, we're making a very simple collection game.

We'll be focusing on just one principle of level design right now – guiding the player. Since this is a 2D, top-down view, the easiest way for us to guide the player is by using shapes in the level and introducing hazards. We can only do so much with such a simple game, but the principles can be applied to larger and more complex games.

Visualizing design isn't everyone's strong suit – we all need creative help and inspiration sometimes. And while a **Game Design Document** (**GDD**) is good at describing things, it can only convey so much in so many words. As they say, a picture is worth a thousand words, so have a look at my initial sketch of what I was planning for the collection game's level design:

Figure 2.1 – The collection game-level sketch

If you're lacking inspiration or just prefer to follow along (no one here will judge you for it), then use my sketch for your own ends. I'll probably sound like a broken record by the time you're well into this book, but remember this – if you're not having fun, you're doing something wrong.

Next, let's explore some examples of guiding the player.

Level design – Guiding the player

By creating shapes in the level (also known as **leading lines**), we encourage the player to move in desirable directions. For example, from the position the player is spawned in, let's start by giving them an open area directly in front of them.

The sides of this open area will begin to angle inward as it gets farther from the player in the direction they face. This will create the appearance of an arrow shape. This shape will invite the player subconsciously to move forward in the direction the "arrow" is pointing.

> **Spawn/respawn/despawn player**
>
> **Spawning** refers to the creation of the player in the game. This typically occurs at the beginning of gameplay when starting the game or when starting a level.
>
> A **respawn** is when the player is spawned after death or some game event that requires the player to restart at a specific position.
>
> When the player is **despawned**, they are removed from the game world.

Shapes are one of the tools we have in level design to guide the player through the environment, while another is hazards. Let's have a look at using hazards in the next section.

Hazards

The level design in a video game can influence the player's movement through visual cues, and one of those cues is hazards. These elements can guide the player toward their next objective and progress gameplay forward while also changing the level's difficulty by placing hazards in areas that the player must visit to complete the game's objectives.

> **Additional reading | 2D level design**
>
> Level Design Patterns in 2D Games by *Ahmed Khalifa*: `https://www.gamedeveloper.com/design/level-design-patterns-in-2d-games`

In this section, you learned that shapes and hazards are simple concepts to guide the player, and they can be used with great effect if done well. It takes practice to make guidance subtle and blended/hidden into an environment design.

Let's start practicing by creating the level for the 2D collection game. In the next section, we'll be using the **Tilemap** feature that is a part of Unity's 2D toolset, and the first step in this workflow is to create a **Tile Palette**.

Creating Tile Palettes

You may have closed Unity since the last time we worked on the player character in *Chapter 1*. To return where we left off in the project, open Unity Hub. The default view will list all of the projects you previously created. The list also includes some additional information about the project, such as the last time it was modified and the Editor version being used for it. You can see an example of this in the following screenshot:

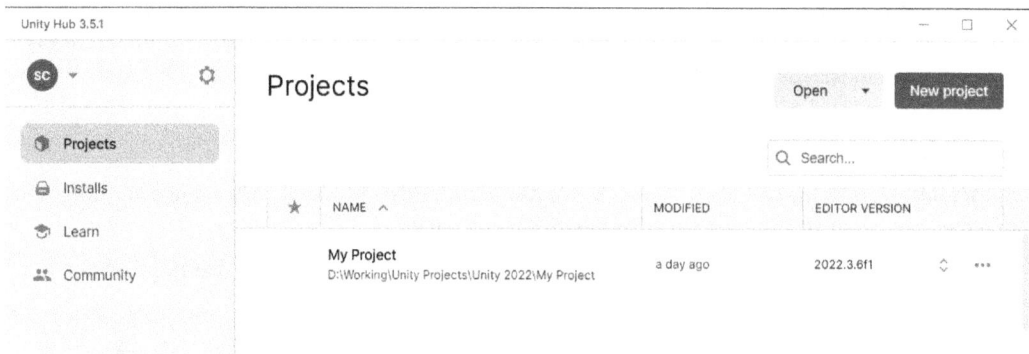

Figure 2.2 – The Unity Hub Projects list

> **Adding and removing projects from Unity Hub**
>
> You can add and remove projects from Unity Hub to keep your project list relevant. Add a project from disk by using the **Open** button at the top right of the window. Use the horizontal ellipses on the right side of each entry in the list to remove the project. Please note that removing a project from Hub will not delete it from the disk and will have to be cleaned up manually outside of Hub.

Open the project you previously created from the **2D URP Core** template (in the screenshot shown in *Figure 2.2*, it is the **My Project** project). Do that by simply clicking on the project name in the list.

Before actually creating our tilemap, we need sprite images to work with. The following section will address adding sprites to the project that will form the basis of our first tilemap, used to construct the game's level.

2D game assets

To create the environment for our level, we'll be using some freely available game assets. We'll use original art assets in the upcoming chapters with different art pipelines and workflows, so you will gain a more rounded knowledge when dealing with 2D art in Unity.

> **Free game assets**
>
> **Kenney** provides readymade game assets with no strings attached! Thousands of sprites are available – for many different themes and genres – for you to use in your own projects, with any kind of use allowed, even commercial. You can find all of the Kenney assets listed here: `https://kenney.nl/assets`

Kenney is a website that provides many pre-made sprite sheets and tilemap images. The main difference between these two graphics terms can be broken down like this:

- A **sprite sheet image** is an image made up of several smaller images (the left image in *Figure 2.5*)
- A **tilemap image** is an image made up of several smaller images arranged in a two-dimensional grid (the right image in *Figure 2.5*)

> **Important note**
>
> The terms *sprite sheet* and *tilemap* are often used interchangeably without much difference in meaning. I've attempted to clarify the terms practically and provide examples in the following sections.

2D art is generally provided in sprite sheets since a single image is a more optimized way of working with images in a game engine instead of using individual images for each asset. Even if the 2D art is provided in single images, Unity has tooling that combines the individual sprite images into a single image for optimization (in this case, the combined or packed image in Unity is referred to as a **Sprite Atlas**). As mentioned previously, Kenney provides game assets in both sprite sheets and tilemap images.

> **Optimization note**
>
> Sprite sheets are *more optimized* because they reduce the number of draw calls the renderer (graphics card) has to perform. All of the sprites included on the sheet can be drawn simultaneously, treating the entire sprite sheet as a single draw, whereas if they were individual sprites, they would need to be drawn separately. This reduction in draw calls can improve game performance (increasing the frame rate).

Now that we understand what 2D game asset images are, let's import some assets from Kenney in the next section to understand importing and working with the images in practice.

Importing sprites

Importing images into Unity is easy, but let's first start with defining some organization in our project by creating some folders. Perform the following steps to create the `Sprites`, `Tile Palettes`, and `Tile Palettes/Tiles` folders:

1. In the **Project** window, select the `Assets` folder.
2. Right-click within the **Project** window to bring up the **Create** menu (or click on the + button at the top just under the **Project** tab) and select **Create | Folder**.
3. Name the folder `Sprites` and hit *Enter*.
4. Repeat *step 2* to create the `Tile Palettes` folder.
5. Now, after selecting the `Tile Palettes` folder, repeat *step 2* to create the `Tiles` folder (this folder is a subfolder of the `Tile Palettes` folder).

We'll be using two sprite sheets from Kenney in the collection game project to create the game environment – the **Top-down Tanks Redux** and **Tower Defense (top-down)** game assets. You can get the sprite sheets by downloading them directly from the `https://kenney.nl/assets` website, or they are also provided in the book's project files available in the GitHub project (refer to the link in the *Technical requirements* section).

Once the folders are created and the sprite sheets have downloaded, we are now ready to import sprites. Yay!

You can import sprites into Unity by either dragging them from the File Manager (Windows)/Finder (Mac) and dropping them into the Unity **Project** window or by right-clicking within the **Project** window and selecting **Import New Asset…**.

First, we will be working with the *Tower Defense (top-down)* game asset's `towerDefense_tilesheet.png` sprite sheet. Import the image into the `Assets/Sprites` folder, and once the asset has finished importing, selecting it in the **Project** window will display the sprite import settings in the **Inspector** window.

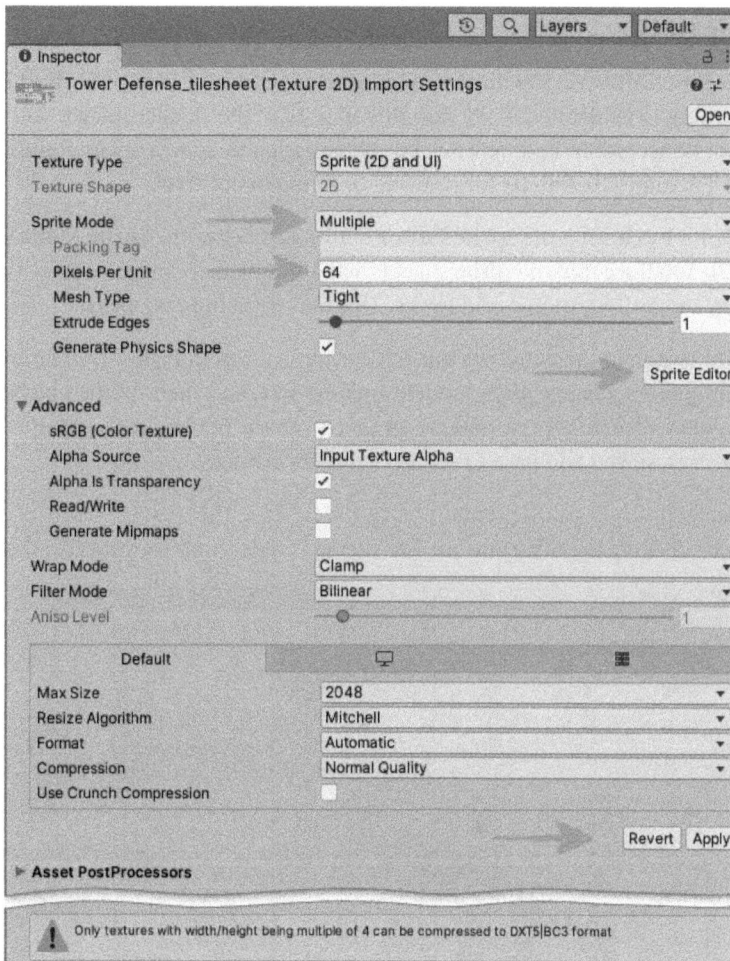

Figure 2.3 – Sprite import settings

Since the sprite sheet comprises many images, we first want to change **Sprite Mode** to **Multiple**. Any time you change the import settings, you'll have to click the **Apply** button at the bottom of the list of fields. Once clicked, we'll move on to slicing the sprite sheet to create the individually selectable sprites.

> **Power of two textures**
>
> It is best to size textures with dimensions that are powers of two on all sides. The sizes are 2, 4, 8, 16, 32, 64, 128, 256, 512, 1,024, or 2,048 pixels (px). Note that the textures do not have to be square (for example, 256 x 1,024). Unity will use the non-power of two textures, although they will not be compressed (taking up more video memory) and not optimized (in some cases, slower to load and render). The **Inspector** window will display a warning when they cannot be compressed, as seen at the bottom of *Figure 2.3*.

Slicing the sprite sheet

With **Sprite Mode** set to *Multiple*, the imported image is now treated as a sprite sheet, with several smaller images needing to be defined. If we skip this step, all of the smaller images within the sprite sheet image will not be accessible, and we won't be able to assign them to a **Sprite Renderer** or create tiles for use in the **Tilemap Tile Palette** for drawing or filling in our level.

Open the Sprite Editor by clicking the **Sprite Editor** button just below the **Sprite Mode** section fields (refer to *Figure 2.3*). You'll see that the sprite sheet is actually a tilemap image because the sprites are already arranged in a two-dimensional grid for us. This makes slicing easy-peasy!

Next, click the **Slice** button in the window's top-left corner (as seen in *Figure 2.4*). From the options presented in the dropdown, change **Type** to **Grid by Cell Size**, and then set the pixel size to **X**: 64 and **Y**: 64. This Kenney tilemap's sprite cells are 64 pixels square, but you may end up working with sprites of different sizes, so you will have to set this value accordingly.

We can keep the default values for the remainder of the fields in the dialog, so proceed to slice the image into sprites by clicking the **Slice** button. The slicing results should appear like the following:

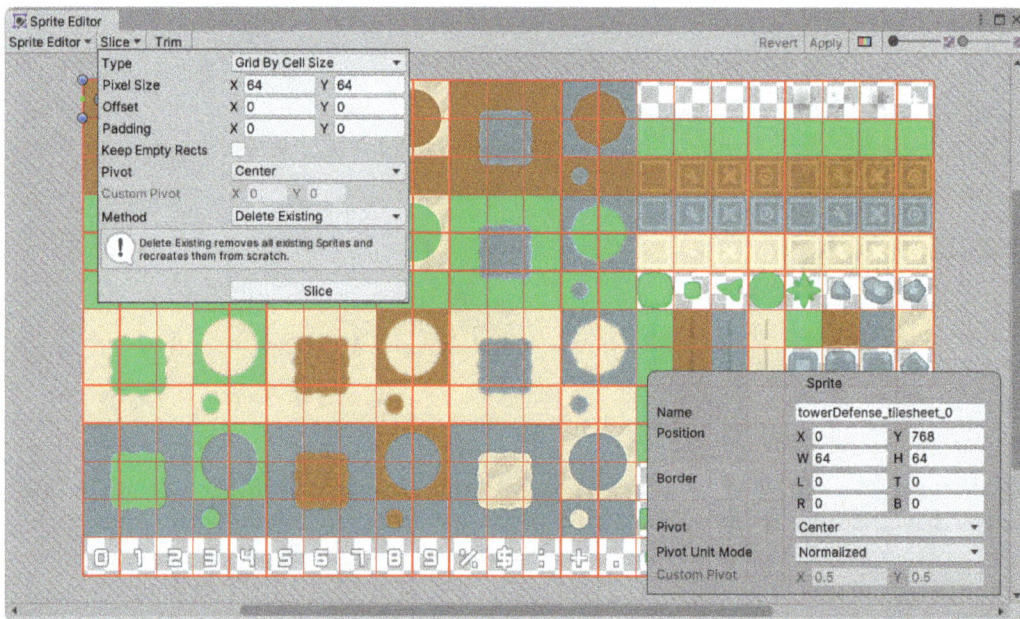

Figure 2.4 – The Sprite Editor Tower Defense (top-down) tilemap

Trimming sprites

For some sprites that don't completely fill the grid cell, you may get some bleeding from neighboring sprites if some of the spacing was not adjusted just right. This is easy to fix by clicking on the offending sprite in the Sprite Editor and then clicking the **Trim** button (directly to the right of the **Slice** button).

Let's proceed to slice the other Kenney asset we'll be using for the objects in the environment of the collection game now – the *Top-down Tanks Redux* game asset's `Assets/Sprites/topdowntanks_onlyObjects_default.png` sprite sheet.

Again, after importing, select the image in the **Project** window to display the import settings in the **Inspector** window, set **Sprite Mode** to **Multiple**, click the **Apply** button (at the bottom), and finally, click the **Sprite Editor** button.

You'll likely notice right away that something is different with this image – it's not a tilemap. Kenney provides game assets as sprite sheets for more game engines than just Unity. While this sprite sheet may work well in some other game engines, it, unfortunately, does not in Unity.

Suppose the smaller images had more padding (blank space represented by transparent pixels surrounding the image). In that case, Unity's **Automatic** slice type setting may work as intended to slice into individual sprites – it does with sufficient padding – but in this case, it did not work and treated the whole sprite sheet image as a single Sprite.

So, to use the images from a sprite sheet that does not have sufficient padding to work well with Unity's automatic slicing, you'll have to open your favorite image editing software (for example, Adobe Photoshop, Gimp, or Krita) and rearrange the images with padding, or arrange them in a grid for slicing by cell size.

As you can see in the following screenshot, I chose to copy the images from the original Kenney sprite sheet (the left side) and create a new tilemap image with a consistent cell size layout (the right side – this image is included in the book's project files).

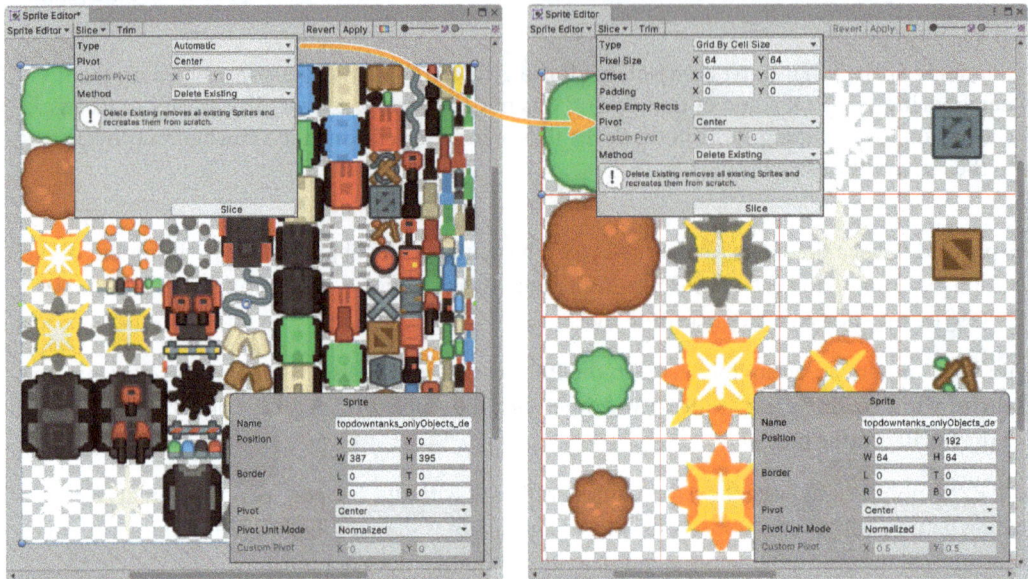

Figure 2.5 – Sprite Editor – automatic versus grid slicing

Optimization note

To reduce draw calls using a rendering technique called batching, a Sprite Atlas can be added to the project. Assigning individual sprites, sprite sheet images, and even folders to the Sprite Atlas will pack them into a single sprite sheet image to render from. This atlas will be the only asset being drawn from instead of multiple sprite images, where each additional sprite image would add a draw call to the renderer.

As a bonus activity, create a Sprite Atlas in the **Project** window from the **Create** menu by choosing **Create | 2D | Sprite Atlas**. You can then assign both imported and sliced sprite sheets to the **Objects for Packing** list or the `Sprites` folder itself. You may need to verify that the feature is enabled (default) in **Project Settings | Editor | Sprite Packer | Mode = Sprite Atlas v2 – Enabled**.

Now that we have our tilemaps sliced into individual Sprites, we're ready to create a Tile Palette. The Sprites will be added to a Tile Palette to create Tiles used to draw out the level. Let's make our first Tile Palette in the next section.

Additional reading | Unity documentation

Sprite Editor: `https://docs.unity3d.com/2022.3/Documentation/Manual/SpriteEditor.html`

Creating Tile Palettes

To get started creating Tile Palettes, perform the following steps:

1. First, open the **Tile Palette** window by going to **Windows | 2D | Tile Palette**.

2. We'll use the *Tower Defense (top-down)* game asset as our main environment map, so create a new **Tile Palette** by clicking the **Create New Palette** drop-down menu and selecting *New Palette*. This drop-down list is where we will be selecting the current Palette to use to draw the tilemap in the scene View.

3. Name it `Environment Main`; keep the defaults since we're using a rectangle grid.

4. Click the **Create** button, as shown in *Figure 2.6*.

5. A Tile Palette is an asset saved in the `Project` folders, so you will be prompted to save it. Select the folder we previously created – `Assets/Tile Palettes`.

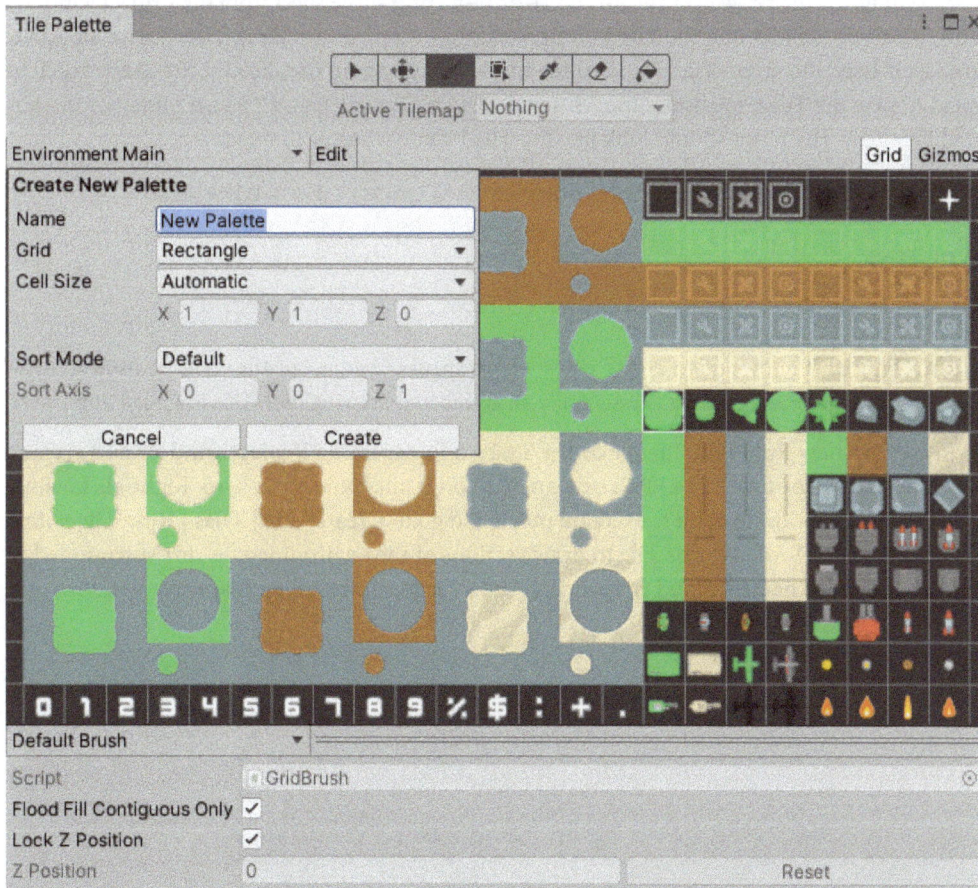

Figure 2.6 – Creating a Tile Palette and Tiles

> **Additional reading | Unity documentation**
> Creating a Tile Palette: `https://docs.unity3d.com/2022.3/Documentation/Manual/Tilemap-Palette.html`

With the palette created, we can move on to the last step required before we can start drawing our level, and we'll tackle that in the next section by adding Tiles to the Palette.

Creating Tiles

Tiles are what the Tile Palette uses to draw sprites into the Scene. You may be wondering why we need to create Tiles if we already have our sprites sliced and ready to use as individual images. That's because Tiles can provide additional functionality when drawing your tilemaps. Other types of Tiles are the **Rule Tile** and **Animated Tile**, but we'll start by using the default Tile type.

Let's start by dragging the `towerDefense_tilesheet.png` sprite sheet from the **Project** window's `Assets/Sprites` folder into the **Tile Palette** window – where it says **Drag Tile, Sprite, or Sprite Texture assets here** – to create Tiles for all of the individual sprites in the sliced sprite sheet. You'll be prompted to save the Tiles, so choose the `Assets/Tile Palettes/Tiles` folder we already prepared for them. The result should look like the Tile Palette window in *Figure 2.6*.

Before we start drawing our level, let's have a quick look at two very useful types of Tiles.

Types of Tiles

The common types of tiles include:

- **Rule Tile**: This allows us to create rules where specified adjacent tiles will be used to draw complex shapes more easily. It is essentially an automated drawing tool. This is a huge time-saver!

  To create a new Rule Tile, change to the `Tiles` folder in the **Project** window and create a subfolder at `Assets/Tile Palettes/Tiles` named `Rule Tiles`. Right-click within the **Project** window to open the **Create** menu and go to **Create | 2D | Tiles | Rule Tile** within the new folder. We'll use this Tile to draw sections of the environment as dirt surrounded by grass sprites, so name it `Environment Area 1`, as shown in the **Inspector** window:

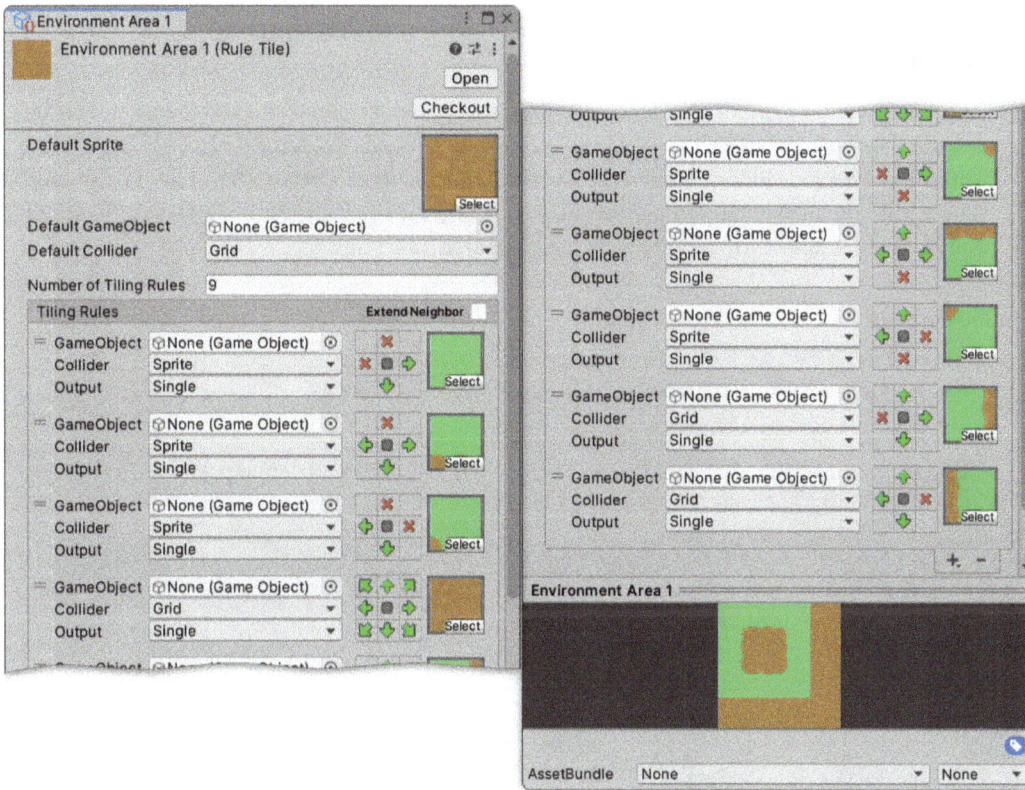

Figure 2.7 – Rule Tile setup

We'll add Tiling Rules to create a filled square with appropriate corner sprites, adjacent to the horizontal and vertical side sprites. That makes nine sprites in total – four on the sides, four in the corners, and one in the center.

Set the **Number of Tiling Rules** setting to 9, and proceed to assign all the sprites from the environment sprite sheet by using *Figure 2.7* as a guide. Once the sprites are assigned, we can set the "3 x 3" boxes that visualize the behavior of the Rule. Again, using *Figure 2.7*, set the box rules accordingly.

Additional reading | Unity documentation

Rule Tile: `https://docs.unity3d.com/Packages/com.unity.2d.tilemap.extras%403.0/manual/RuleTile.html`

- **Animated Tile**: This allows us to create a sprite animation by assigning several sprites to replace each other at a specified speed.

  To create a new Animated Tile, change to the `Tiles` folder in the **Project** window and create a subfolder at `Assets/Tile Palettes/Tiles`, named `Animated Tiles`. Right-click within the **Project** window to open the **Create** menu and go to **Create | 2D | Tiles | Animated Tile** within the new folder. To use this tile, drag a sequence of sprites to the **Drag a Sprite or Sprite Texture assets to start creating an Animated Tile** section.

Additional reading | Unity documentation

Animated Tile: `https://docs.unity3d.com/Packages/com.unity.2d.tilemap.extras%403.0/manual/AnimatedTile.html`

For a quick recap concerning all the folders we've created to organize and contain all of our art assets, our **Project** window should look similar to the following:

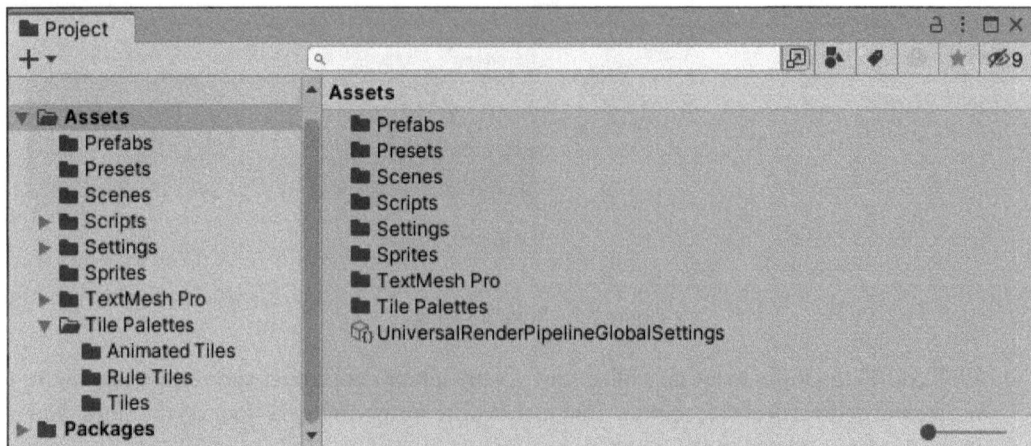

Figure 2.8 – Project folder structure

In this section, you learned how to create Tile Palettes, understood how to import and slice sprite sheets to create individual sprites, and explored the advantages of using Rule Tiles to make drawing easier.

This is all the prerequisite setup we need to draw the environment for our collection game level, so let's get started doing just that. In the next section, we'll add several tilemaps for the different components that make up the level design.

Building the collection game environment with Tilemap

We'll need a new scene for the game level, so go to **File | New Scene** (*Ctrl/Cmd + N*). This will bring up the **New Scene** window, where we'll select the **Lit 2D (URP)** Scene Template and then click the **Create** button in the window's lower-right corner, as shown in the following screenshot:

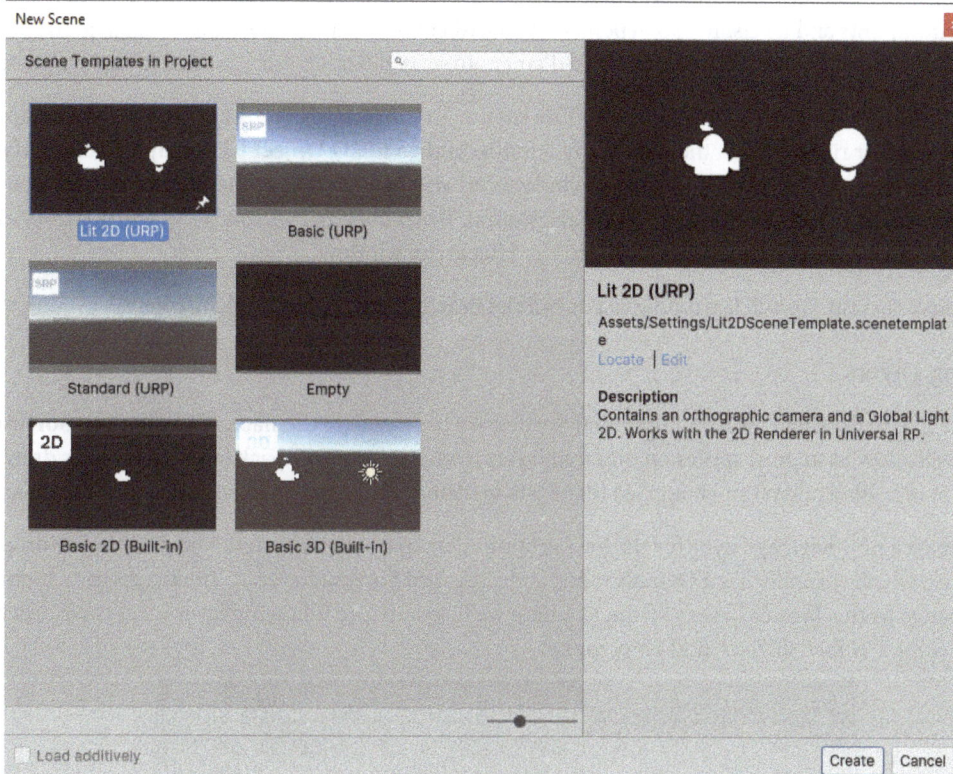

Figure 2.9 – New Scene Templates

> **Scene Templates – Unity documentation**
>
> Unity provides a set of built-in Scene Templates, but you can add your own user-defined templates to create new scenes containing all of your starting content. For additional information on creating Scene Templates, refer to the Unity documentation here: `https://docs.unity3d.com/2022.3/Documentation/Manual/scene-templates.html`

A new **Lit 2D (URP)** scene will not prompt to save when created (some do prompt, depending on what the template contains), so go ahead and save the scene as your first step (*Ctrl/Cmd + S*) – get in the habit of saving frequently. A good name to start with here could be `Game` or `Level`, since our game will consist entirely of just a single scene (that is, for a single game level).

With our new game scene saved, we're ready to start adding tilemaps in the next section!

Adding tilemaps to a Scene

A **Tilemap** is a component added to a GameObject that adds functionality just like any other component we previously discussed. The **Tilemap** component stores and manages the **Tile** assets used for creating 2D environments and levels. It also depends on a **Grid** component that provides a visual guide and aligns the painted tiles. We will make multiple tilemaps to serve different visual and functional purposes.

To create a tilemap in the scene, from the **File** menu, go to **GameObject** | **2D Object** | **Tilemap** | **Rectangular**, or right-click in the **Hierarchy** window and go to **2D Object** | **Tilemap** | **Rectangular** (the + button at the top of the **Hierarchy** window can also be used). Give the tilemap a descriptive name for the layer we'll be drawing – the background. I've named it `Tilemap - Background`. Refer to *Figure 2.11* for an example of this in the **Hierarchy** window.

In the next section, we will learn how to control the drawing order of multiple tilemaps.

Sorting Layers

Like we have a Sorting Layer on individual Sprites, we have one on tilemaps. The **Tilemap Sorting Layer** will allow us to draw sprites on different layers to control front-to-back layering (objects drawn on top of or behind others) and is applied to the whole tilemap (all the sprites that make up the tilemap).

Let's create a new **Sorting Layer** for the background tilemap by clicking on the **Sorting Layer** drop-down list, which currently has a **Default** value, and click **Add Sorting Layer…**. The **Inspector** window will change to the **Tags & Layers** settings, where we'll specify the following layers and their order: *Background*, *Collider*, *Objects*, and *Foreground*.

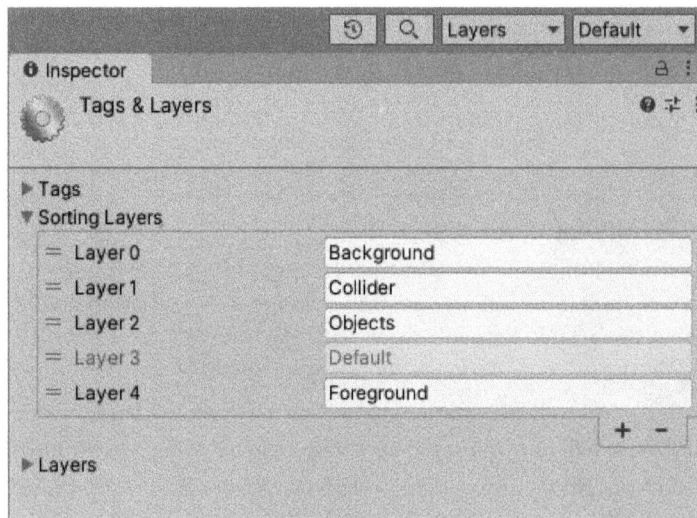

Figure 2.10 – Sorting Layers

The renderer will draw the layers in the order they appear here, with **Layer 0** (**Background**) at the back, **Layer 4** (**Foreground**) at the front, and the layers in between drawn in their respective locations.

Clicking the + button will add a new layer to the list while clicking the - button with a layer selected will remove it. Dragging = to the left of the field (for example, **Layer 0**) up/down will reorder. Once we've finished adding and setting the layer orders, click on **Tilemap - Background** in the **Hierarchy** window again and select *Background* in the **Tilemap Renderer Sorting** Layer drop-down field.

We still have the option of creating a second Tilemap background on the same Sorting Layer, and we can control which background draws above/below the other background by using the **Order in Layer** field – this is useful if you have two different style environments for your background and would like to manage them separately.

We are finally ready to start drawing with our Tiles in the next section!

Drawing the level in the scene view

If you've ever previously used a drawing program, then the painting tools in the Tile Palette's toolbar will seem familiar, except that you will now be limited to painting on a grid. Paintbrush, flood fill, box fill, and eraser tools are all standard painting tools in drawing programs.

We won't cover all the painting tools in detail, but I encourage you to explore them on your own as you draw out your environment and level design – don't be afraid to make mistakes! Like most things in Unity – and pretty much every program – you can quickly undo things by using *Ctrl*/*Cmd* + *Z*.

Remember to save your progress as you go to keep what you like. Have fun with your drawing, and take the time to experiment because you never know where this type of creativity will lead.

Scene navigation

As a reminder for getting around in the scene View while drawing out your level, refer back to *Navigating the scene View* section in *Chapter 1*.

Follow these steps every time you want to draw in the scene View using the painting tools:

1. Start by selecting the tilemap you want to paint on, either in the **Hierarchy** window or within the **Tile Palette** window, using the **Active Tilemap** drop-down list (just below the toolbar).

2. Select the **Palette** for the **Tile** you want to paint with using the **Tile Palettes** drop-down list located just below the **Active Tilemap** section on the left. For instance, we will choose the **Environment Main** palette that we created earlier.

3. Choose the tile to paint with by clicking on it in the tiles section grid. Remember that we're starting with the background of our level design, so pick a solid square tile that we can use to fill a large area.

4. To paint individual tiles, by clicking on a grid cell in the scene View or painting continuous lines of sprites by clicking and dragging, select the **Paintbrush** tool (the paintbrush icon).

5. To draw a rectangular shape filled with the selected tile, select the **Box Fill** tool (the box icon).

6. You can also fill a larger contiguous area of grid cells with the selected tile by using the **Flood Fill** tool (the bucket icon).

7. The **Eraser** tool can erase tiles from the cell grid by clicking on them or clicking and dragging.

The following figure illustrates the background tilemap selected in the **Hierarchy** window (*A*), a view of its components in the **Inspector** window (*B*), the environment palette (**Environment Main**) that we're using to draw the background with (*C*), and both the selected tile (**grass**) (*D*) and painting tool (**Paintbrush**) currently being used (*E*).

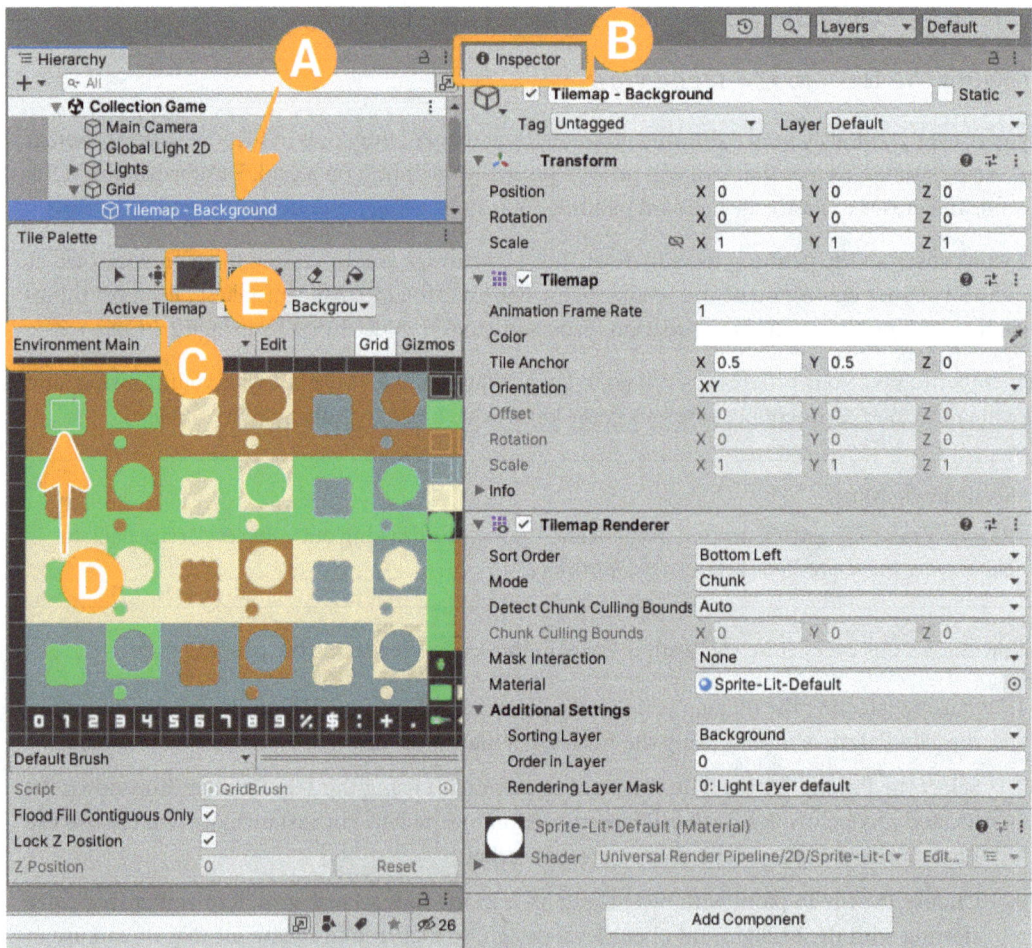

Figure 2.11 – The Tilemap in the Hierarchy and Tile Palette windows

> **Additional reading | Unity documentation**
> **Painting on Tilemaps**: `https://docs.unity3d.com/2022.3/Documentation/Manual/Tilemap-Painting.html`

As a reminder of what we're drawing here, we're creating the background for the level of our collection game – defining the general play area. To visualize the overall level design plan, it may help to start drawing some of your ideas out on paper first, then using those ideas to start drawing things out in the Editor (refer to my initial sketch as a guide, if needed).

> **Paper prototyping**
> Paper prototyping is a widely used method in the game design process that helps you test out ideas before committing time to writing code or creating digital art assets. It's also a quick way to validate your gameplay and discover potential problems early on.

While experimenting with your level design, consider that we'll want to guide the player to where we'll be placing collectibles throughout the level in the following sections. Don't be afraid of making any mistakes; making changes in response to playtesting your games is something that you'll have to do regularly – iterating changes to improve gameplay or balance the difficulty.

Before tailoring additional elements of the environment and level design, let's make sure our art assets are looking their best in the game view by adjusting the **Camera** settings.

Camera settings for crisp graphics

The default settings Unity provides in new Scenes are not always ideal for the game assets you'll be working with. Because of this, we'll want to make some adjustments to our sprite sheet image import settings and our scene camera to have them looking their best!

The individual images in our tilemap image are 64 pixels square (note that these sizes could be different, depending on the desired size of the art for the game design dictates). The best way to work with tilemap images is to ensure that the individual sprite size equals the grid size. Since tilemap grid cells are sized to equal one square Unity unit, we'll have to set our pixels per unit to equal one Unity unit.

To set the pixels per unit for the sprites in our *Tower Defense (top-down)* asset, click on the `towerDefense_tilesheet.png` image in the `Assets/Sprites` folder to view the import settings in the **Inspector** window (refer to *Figure 2.3*). Find the **Pixels per Unit** field in the **Sprite Mode** section, and ensure its value is set to `64` (pixels).

We'll now have to address the camera setting so that 64-pixel sprite images are represented on screen at their native resolution. These are the *crisp graphics* we are referring to – the native size. This will require a bit of math and a decision on the preferred resolution for the target platform. Assuming that most players on desktop systems – our target platform for the collection game – have a screen resolution of 1,920 px wide x 1,080 px high, let's use that.

Select the **Main Camera** in the **Scene Hierarchy** window; then, in the **Projection** section, set the orthographic camera projection size (this is a vertical value) by taking the screen resolution height divided by the pixels per unit and dividing the result by 2 – we are dividing by 2 because the size value is half the vertical viewing volume of the camera. Here is what the calculation looks like with the resulting value set in the **Inspector** window – *(1,080 ÷ 64) ÷ 2 = 8.4375.*

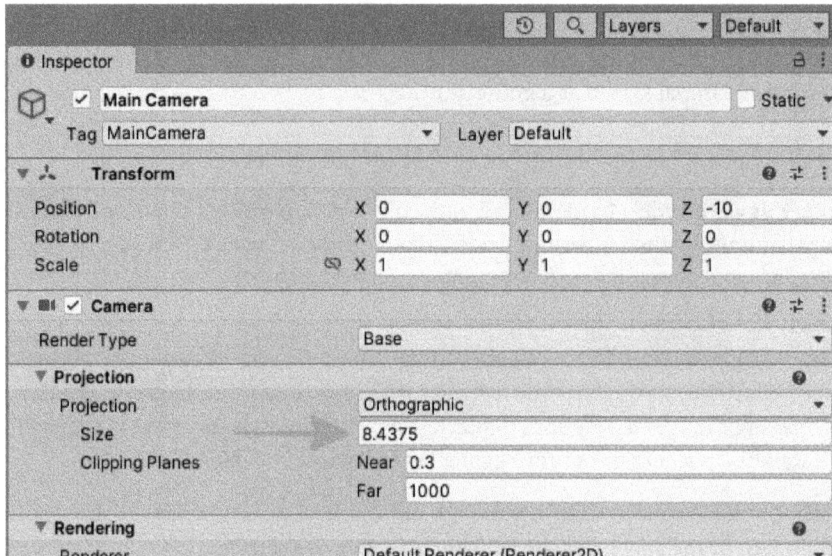

Figure 2.12 – The camera orthographic size

The number of sprites extending the vertical extent is now optimal for crisp graphics! To visualize the result in the Editor, switch to the **Game** view (by clicking on its tab next to **Scene**, just under the main toolbar) and set the **Aspect Ratio** drop-down list value to **Full HD (1920x1080)** – by default; this is set to **Free Aspect**.

In this section, we learned how to draw tiles in the scene View to create the background environment and define the play area for the level. We also learned how to make our art assets look their best! Next, we'll add some 2D lights – leveraging the Universal Render Pipeline's (Universal RP, or URP) 2D features – to enhance the environment design.

Adding 2D lights

The Universal RP 2D renderer allows us to enhance our environment by adding 2D lights! Let's add some 2D lights to the position of some tiles, representing light sources in our environment design. Follow these steps to add some *light sources*:

1. Start by creating a new tilemap for tiles that will draw above the background tilemap, and name it Tilemap - Objects.

2. Set **Sorting Layer** to *Objects*.

3. Create a new **Tilemap Palette** for the **Top-down Tanks Redux** sprite sheet and name it `Environment Objects`.

4. Add the sprites from the `topdowntanks_onlyObjects_default` sprite sheet to create the tiles used for painting.

5. Select *Tilemap - Objects* as **Active Tilemap** in the **Tile Palette** window.

6. Select the star-looking tile and paint a few in the Scene using the **Paintbrush** tool.

> **Important note**
> To focus work on a specific tilemap with the other GameObjects in the scene faded out, use the **Tilemap Focus** mode (the floating overlay in the scene View).

Okay, we have some light sources in the environment; now, let's add some 2D lights to them by following these steps:

1. Create a new empty GameObject in the root of the scene's **Hierarchy** window and name it `Lights`.

2. Right-click on the **Lights** GameObject – to open the **Create** menu – and select **Light | Freeform Light 2D | Circle.**

3. Using the **Move** tool, position it over one of your *star* tiles.

4. You can experiment with what looks good to you. I changed the light's **Blending** section's **Blend Style** value to **Additive** (the default is **Multiply**) to better visualize the example screenshots. For a finished game, I would reduce the global light source intensity and use the **Multiply** blend style on many 2D lights added to the environment to set the right lighting design tone.

> **Additional reading | Unity documentation**
> **Light Blend Styles**: `https://docs.unity3d.com/Packages/com.unity.render-pipelines.universal%4015.0/manual/LightBlendStyles.html`

5. Adjust the **Radius, Inner Spot Angle/Outer Spot Angle, Intensity**, and **Falloff Strength** values to your liking.

6. Repeat as necessary for all of the light sources in your environment design!

You will want to adjust the overall lighting in the scene to use 2D lights to their full effect. Do this by selecting the **Global Light 2D** GameObject in the **Scene Hierarchy** window and adjusting the Light 2D component's **Intensity** value.

The results of adding 2D lights to the scene can be seen in *Figure 2.13* – to the left and right sides of the player character (the ladybug). We'll be using the different types of 2D lights in the coming chapters.

In this section, you learned how to add 2D lights to your environment design to great effect. The following section will tackle a missing requirement to get our level playable by using the **TilemapCollider2D** component.

Making the level playable – Tilemap Collider 2D

We'll need to prepare our level design a bit more before we're able to make it playable – we'll be making the player character controllable in the next chapter by adding a custom script and mapping inputs.

To keep the player from crossing into specified areas in the level, we need to add a special kind of collider to a tilemap. Colliders are a part of Unity's Physics system and provide a way for developers to work with GameObjects, similar to how objects work in the real world.

Colliders prevent things from entering each other, and collisions of objects can all be responded to in code as events – we'll be exploring this more in the upcoming chapters when adding features to our game.

Let's add a new tilemap to our scene to define areas that the player is not allowed to enter. This will be similar to how we created the previous tilemaps, except for having to add a **TilemapCollider2D** component:

1. Create a new tilemap for tiles that will draw above the background tilemap and contain areas the player cannot cross. Name it `Tilemap - Collider`.
2.   Set **Sorting Layer** to *Collider*.
3. Let's use a **Rule Tile** to make drawing complete areas in the level easier. In the **Tile Palette** window, create a new palette and name it `Rules Palette`.
4. Drag in the **Environment Area 1** rule tile we previously created from the `Assets/Tile Palettes/Rule Tiles` folder.
5. Select *Tilemap - Collider* as **Active Tilemap** in the **Tile Palette** window.
6. Select the rule tile and paint out areas in the scene using the **Filled Box** tool.

Now, to make this tilemap interactive, with **Tilemap - Collider** selected in the **Hierarchy** window, go to the bottom of the **Inspector** window and click the **Add Component** button. In the search field at the top of the dialog that opens, type *Tilemap*, and then select the **Tilemap Collider 2D** item to add it – we can use the default values.

Refer to the following figure to get an idea of what we're creating:

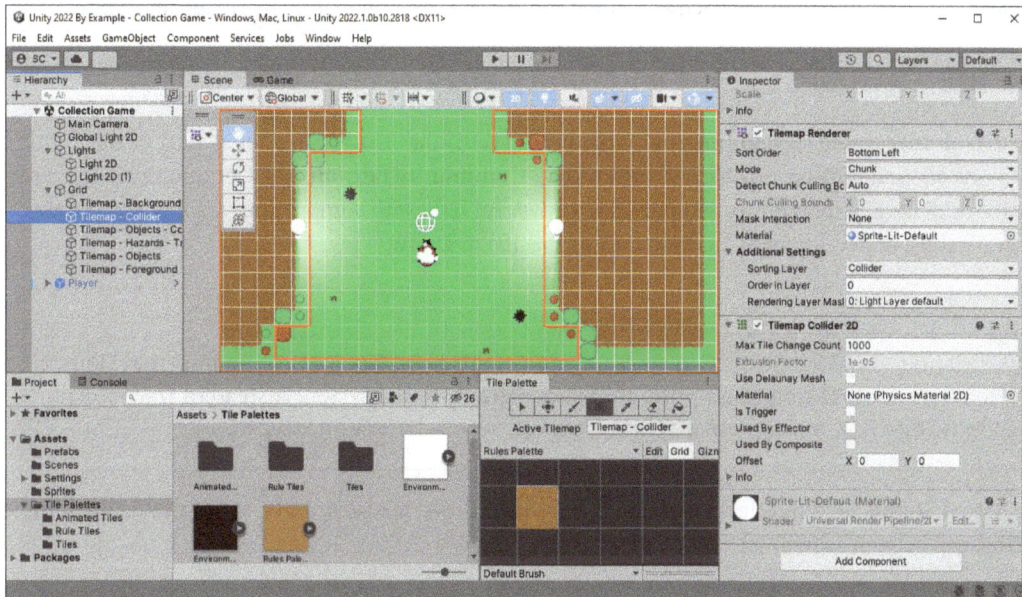

Figure 2.13 – A collider level design example in the scene View

Now that you understand how to add tilemaps, paint tiles in the scene, and sort the tilemap layers to set the proper drawing order, spend some time filling in environmental details. This is your time to experiment with detailing the environment's design and think about how the player will be guided through the level to accomplish the game objective – collecting items.

In this section, you learned some key elements of level design. You understood how to create and work with tilemaps by importing and slicing sprite sheet images. To draw out a game level, you used tiles to paint individual sprites and areas. You completed the design by adding lights and making the graphics crisp. Finally, we made interactable objects through the use of colliders.

We'll start bringing all of these concepts together in the following sections by adding C# scripts to move the player and collect items.

Introduction to creating scripts in C# – IDE, SOLID principles, and design patterns

When making games in Unity, you'll need to create your own functionality specific to the requirements of your game. We'll need to learn how Unity provides programming support for writing scripts using the C# language to accomplish this.

In this section, you'll be introduced to the default code editor, get into the specifics of the C# language, and learn the best-practice approach to writing manageable, maintainable, and extensible code.

The IDE – Visual Studio Community 2022

The default **integrated development environment** (**IDE**) that Unity provides as part of the Unity Editor installation is **Microsoft Visual Studio 2022 Community Edition**. The Community Edition is free for students and individuals and provides a powerful IDE, with comprehensive tools and features for every stage of development, including tools specific to Unity.

Additional reading | Unity Visual Scripting

We will use Visual Studio 2022 to write C# code in this book, but it's worth noting that Unity also offers a visual scripting option that uses a graph-based system instead of traditional code: `https://unity.com/features/unity-visual-scripting`

The Visual Studio 2022 interface, with its standard windows, features, and a script that is open for editing, can be seen as follows:

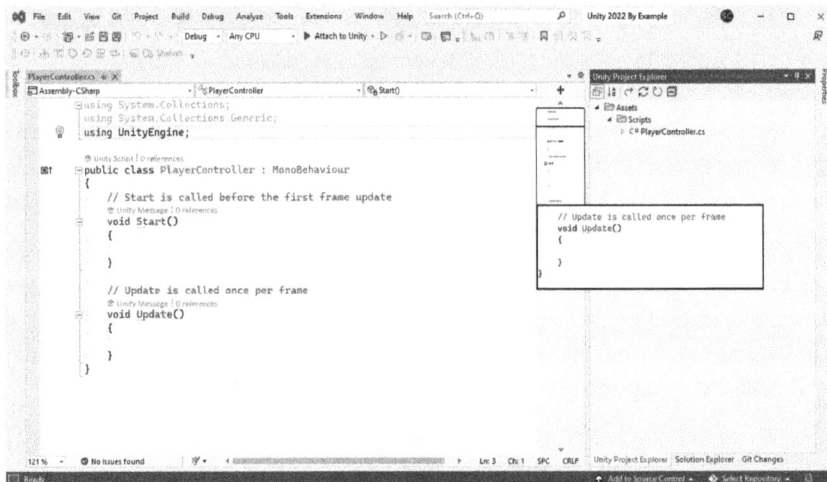

Figure 2.14 – Visual Studio 2022 Community Edition

In addition to an essential script editing capability, let's have a quick look at some of the features that Visual Studio 2022 provides:

- **IntelliSense**: Aids in code completion by providing suggestions for naming, parameter info, and words – saving you keystrokes!

- **IntelliCode**: Enhances your software development with **artificial intelligence** (**AI**) that can automatically complete code – even up to a whole line!

- **CodeLens**: Keeps you focused on your code's structure without leaving the IDE, code references, and contextual information at your fingertips!

- **Live Share**: Provides a real-time collaborative coding session you can share with other members on your project – work together to produce results faster!

- **Integrated debugging**: Allows the developer to control the execution, step through the code, and examine its state interactively while it's running – see what your code is doing while it runs!

> **Additional reading | Visual Studio**
>
> **What's new in Visual Studio 2022**: `https://docs.microsoft.com/en-us/ visualstudio/ide/whats-new-visual-studio-2022?view=vs-2022`

Before opening up the IDE to edit a script, we'll first need to create one! Creating scripts can be performed in a couple of ways. It's good practice to keep your project files organized from the start, so let's first create a new folder named `Scripts` at the root of our project – `Assets/Scripts`.

To create a new script in this folder, do the following:

1. Make sure the `Scripts` folder is the currently selected folder in the **Project** window.

2. Right-click within the **Project** window to open the **Create** menu (or use the + button).

3. Select **C# Script**.

4. Rename the script from its default name of **NewBehaviour** to `PlayerController` (it's currently highlighted to make naming it right away easier).

> **Tip**
>
> Do not use any spaces in your script filenames or start the script's name with a number. Both are naming rule violations for C# classes, so your script will not work. You can use the underscore character in place of spaces. However, this naming convention style is generally frowned upon in the developer community – unless it is the first character of the name for private member variables (which is a common convention in C# and used in this book).
>
> As a reference for a good naming convention standard to use – and stick to (because consistency and good naming improve code readability for everyone) – check out the following guide: `https:// https://www.c-sharpcorner.com/UploadFile/8a67c0/C-Sharp-coding- standards-and-naming-conventions/`

You can also add new scripts (that is, components) to GameObjects directly in the **Inspector** window through the **Add Component** button at the bottom of the window (and when you currently have a GameObject selected). Click the **Add Component** button, and then select **New Script >**.

You'll be prompted to enter the script name, and then click **Create and Add**. You cannot choose the location to save the script with this method – it saves the script in the root of the `Assets` folder – so I prefer the methods previously mentioned to create in a specific folder.

New scripts are generated from the default template Unity provides (you can modify the template to suit your personal needs; see the following *C# Script Templates* callout box). New scripts are also given the `.cs` file extension, denoting them as C# script files (file extensions are not visible in the Unity **Project** window).

We've named it `PlayerController` here, because the first C# code we'll be writing in the next section is to give the ladybug player character we previously created the ability to move around from player input.

> **C# script templates – Unity documentation**
>
> Script templates are what Unity uses to generate the default C# code when new scripts are created, providing commonly used code as a starting point for your scripts. You can modify the templates with your own changes, additions, notes, and so on.
>
> Additional information can be found here: `https://support.unity.com/hc/en-us/articles/210223733-How-to-customize-Unity-script-templates`

Unity has also named the class in the script `PlayerController` as part of the script generation process (a function provided by the script template). Note that the filename and the class name must match for scripts that will need to be added as Components to GameObjects (otherwise, you'll get a **Can't add script component 'ScriptName' because the script class cannot be found.** error if you try). Refer to *Figure 2.14* for an example of a new script's contents.

VS2022 is installed and configured as the default script editor for the C# scripts in your project when the Unity Editor is installed. You can verify this by going to **Edit | Preferences…** and selecting **External Tools** – the **External Script Editor** dropdown will be set to **Microsoft Visual Studio 2022 [version]**.

Now that we have a script created and confirmed our script editor is configured, editing the script is as simple as doing any of the following:

- Double-clicking on scripts in the **Project** Window.
- For components added to GameObjects in the Scene, right-clicking on the component in the **Inspector** window (or using the vertical ellipsis menu) and selecting **Edit Script**.
- Opening from within VS using the **Unity Project Explorer** window.

We have learned how to create a new script and open it for editing in VS, but we still need to understand what the C# code we're looking at means. In the next section, let's dissect the generated C# code for the player controller script.

The C# Language – Object-oriented programming (OOP)

Unity provides coding support through the open source .NET **Common Language Runtime (CLR)** and uses the C# language – these are Microsoft technologies. C# is a managed language, which means that it provides a safe coding environment by – mostly – managing memory and the CLR safely executing the code (that is, not bytecode being executed by the operating system directly).

Code can be compiled to run **just-in-time (JIT)** or **ahead-of-time (AOT)**, depending on which of the two scripting backends that Unity provides is used – Mono or **Intermediate Language to C++ (IL2CPP)**. The primary difference here is in compilation time, but some platform builds require one or the other.

> **Unity is a C++ game engine**
>
> The Unity game engine is written in C++, but C# is used as a "more friendly" coding language. The Unity Object type in C# is linked to a native C++ counterpart object.

Okay, the preceding paragraph was pretty boring. So, now comes the obligatory nerd warning… we're going to talk about some programming concepts, principles, and design patterns in this and the next two sections. While these won't teach you the basics of how to write a section of code that loops or even evaluates variables, I feel they are an important introduction at the onset of a developer's journey.

It's essential to not only learn the basics (such as how to write code that loops and evaluates variables) but also to understand the broader concepts of how code should be structured – for the sake of your future developer self (and possibly for the benefit of your future developer teammates).

While I've attempted to make this section concise and provide relatable real-world analogies, these broader concepts on how code should be structured may be too early an introduction for some of you, so, in that case, don't be afraid – still have a read and do your best to understand, but don't worry if it's difficult.

We'll be implementing all of this as we work through the projects in the book (maybe you'll even recall some of what was stated here or will want to come back and reread).

Okay, with that now out of the way, let's dive into the concepts that will level up your coding skills on your journey as a Unity developer!

The C# language

C# (pronounced as *C sharp*) is a modern, object-oriented, type-safe, managed programming language. **Object-oriented programming** is a computer programming model that organizes software design around objects, which can contain data and code (classes) rather than functions and logic.

C# primarily being a type-safe language means that specific types can only interact through their defined protocols, which ensures each type's consistency. For instance, you cannot write code to interact with a string type the same as if it were a number type.

OOP has four basic concepts for working with objects. We'll review them briefly now and apply these concepts in practice while writing the code for our game in the upcoming chapters:

- **Encapsulation**: Hide the internal data and behavior of an object (class) from other objects and only allow access through public methods; preventing objects from modifying other objects directly reduces the chance of making mistakes external to the object.

- **Abstraction**: An incomplete implementation that hides specific details and only provides required information but is also not associated with any particular instance, since an abstract class is intended only to be a base class of other classes. It provides a template to organize an object hierarchy with required details, such as a spotlight (object) being a type of light (abstract class) that requires a brightness property.

- **Inheritance**: Abstraction is possible due to inheritance because a derived class inherits all properties and methods of the base class when creating a new class from it, which means we can reuse, extend, and modify the behavior of the base class. In the preceding light example, a point light will have the same brightness property as the spotlight if both are inherited from the light abstract base class.

- **Polymorphism**: This is a Greek word, meaning *one name with many forms* or *having many forms*, which we can apply in C# to mean one name with multiple functions. Polymorphism allows a class to have multiple implementations with the same name by overriding it, thanks to inheritance.

  Continuing with the preceding light example, we can implement inherited methods from the base class differently across the spotlight and point light abstractions by having different code for a `ChangeLightRadius()` method – a spotlight has a *cone of light*.

  In contrast, a point light is omnidirectional, so it behaves differently. Polymorphism helps in code reuse since classes, once written, tested, and implemented, can be reused as required and save a lot of time while keeping things more logically structured.

Additional reading | C#

Microsoft C# documentation for object-oriented programming: `https://learn.microsoft.com/en-us/dotnet/csharp/fundamentals/object-oriented/`

The MonoBehaviour class

The `MonoBehaviour` class is the base class that every Unity script assignable to a GameObject derives from – this is Unity's **component-based architecture**. You've seen the word *class* mentioned previously, so for clarity, a class is an object type. You will be creating custom object types that group variables, methods, and events together to give your script data and provide its intended function.

Deriving the class from `MonoBehaviour` – due to inheritance – provides a base set of variables, methods, and events for components added to GameObjects; this is also a requirement – you cannot add a script to a GameObject unless it inherits from MonoBehaviour. We'll be using the functionality the MonoBehaviour base class provides for our custom script components throughout the projects in the book.

> **Important note**
>
> Due to the nature of the MonoBehaviour class requiring it to be an instance of an object in the **Scene Hierarchy**, you cannot use the new keyword to create a new instance of it, as you could for a class that does not derive from anything.

The code for the new PlayerController class we created previously, as generated from the default Unity template, already derives from MonoBehaviour for us – as denoted by : just after the public class declaration line in the following:

```csharp
using UnityEngine;

public class PlayerController : MonoBehaviour
{
    // Start is called before the first frame update
    void Start()
    {
    }

    // Update is called once per frame
    void Update()
    {
    }
}
```

> **// (C#) | Code commenting**
>
> A single-line **comment** is started with two forward slashes (//). The C# compiler will strip out this text and ignore it when running the program.
>
> Don't be afraid to comment on anything and everything in the code to clarify the intent. There are always at least two people looking at your code when programming – you and yourself in 6 months' time (or sometimes on any Monday morning). You *will* forget why you did what you did.
>
> **Self-commenting code/self-documenting code** is something that you may hear when discussing the topic of code commenting, which relies on good class, method, and variable naming. This, of course, is important. Still, my opinion is that you can spend an awful lot of time deliberating what to name things – while interrupting your train of thought – where a simple few words in a comment can instead bring quick clarity to the desired intention.

The `Start()` and `Update()` methods are also provided for us. `MonoBehaviour` provides Message events for initializing any code in `Start()` and running code on each frame update in `Update()`. There are several other Message events that `MonoBehaviour` provides, and we'll be exploring them in the upcoming chapters.

> **namespace (C#)**
>
> `using UnityEngine;` at the top of the preceding `PlayerController` code example is what gives the containing code access to the `MonoBehaviour` base class. **Namespaces** are a way of organizing and providing levels of separation in your code. They are handy for providing containers to control the scope of classes and methods in larger projects. You can, and should, add your own namespaces to your classes.

With the basics of understanding how to create a new C# script, open it in the VS IDE, and identify the essential parts of a `MonoBehaviour`-derived class out of the way, we can now explore some best-practice principles to follow as we write code for our game with SOLID principles.

SOLID principles

SOLID is an acronym adopted to represent the first five principles of object-oriented design by Robert C. Martin and applies to various programming languages, not just C#. They are basic design principles to keep the code you write maintainable and extensible while avoiding refactoring (restructuring existing code) unnecessarily in the future. The resulting code should be easily read and followed by other developers – avoid writing spaghetti code!

> **Spaghetti code**
>
> **Spaghetti code** is a phrase used to describe unstructured and difficult-to-understand-and-maintain source code in your projects. The cause can be attributed to several factors, which usually boil down to not following best practices such as SOLID principles, **don't repeat yourself** (**DRY**), and clean code – code that is formatted correctly and organized in a manner that is easily read or modified by others, or by yourself in 6 months' time!

The SOLID principles are as follows:

- **S – single-responsibility principle(SRP):** Each class or function (a *function* for a language-agnostic term, but I will refer to functions in C# as methods) in your code should only have a single responsibility.

  It means *do one thing* and not everything, including the kitchen sink! Everything within the class should contribute to the one reason it exists. Multiple single-function classes can work together to complete larger, more complex tasks. If something breaks, it's generally easier to know where the bug exists, and it will also be isolated from that class.

- **O – open-closed principle (OCP)**: Classes can be extended but not modified.

  No, this is not a contradiction; how can something be both open and closed simultaneously? Well, it means that a class should be open for extension but closed for modification. Through polymorphism, we can change the behavior of an abstract base class by the inheriting class while leaving the base class unmodified. By designing base classes that *never change*, you prevent undesirable changes that propagate to dependent classes, contributing to code that is not reusable and ultimately harder to work with.

- **L – Liskov substitution principle**: Methods that reference base classes must be able to use derived classes seamlessly.

  Any function that uses a reference to a base class should be able to use a derivative of that class without knowing it. That might sound a bit confusing, so let's use the light example from the earlier OOP concepts – a method that references a light should not care whether that light is a spotlight or a point light. We can substitute a point light for a spotlight, and the code should just work.

  We can accomplish this substitution through the use of an **interface** (a type that is similar to a class, but it only represents a structure of declarations – this is commonly referred to as a **contract**, but you can also think of it as a *blueprint for a class definition*).

  An interface can also be used to implement the **open-closed principle** by having a base class take an interface as an abstract reference for injecting different functionality. The functionality can be extended while the reference in the closed base class remains the same.

- **I – Interface segregation principle**: Several smaller scope interfaces are preferred over a single monolithic interface.

  Classes should not include behaviors they do not use in their single-responsibility role. In Martin's original introduction, he describes it as a *disadvantage of fat interfaces*, where the functions of a large multipurpose interface can instead be broken up into groups of member functions to provide better cohesion. This can prevent classes from using interfaces that they do not require to function.

- **D – Dependency inversion principle**: Use abstractions over direct class references.

  This last principle states that concrete classes should depend upon interfaces – or abstract functions and classes – rather than concrete functions and classes. Abstract classes should also not depend on concrete classes; interfaces should be used here. The advantage here means code requires less work to change because the interface abstractions decouple concrete classes – a change in one also does not break the other. Loosely coupled code is more flexible and easier to test. Yay!

> **Additional reading | SOLID principles**
>
> To read the original quotes and ideas behind the SOLID principles, as introduced by Robert C. Martin, you can refer to this article: `https://learn.microsoft.com/en-us/archive/msdn-magazine/2014/may/csharp-best-practices-dangers-of-violating-solid-principles-in-csharp`
>
> Alternatively, just do some web searching on the subject: `https://www.google.com/search?q=SOLID+programming+principles`

After this brief introduction, you may not completely understand these principles, but don't worry – we'll be using them in practice in the upcoming chapters, so you will learn how they are implemented. With a basic understanding of the SOLID principles, we can now talk about design patterns that implement the ideas the SOLID principles present.

Design patterns

Now that you know what the C# language is and some principles to keep in mind while coding your games, let's see how to apply them in practice.

When writing the code for your game, you can make your own decisions on how to approach the architecture. Within Unity, there aren't any restrictions or specific ways things need to be set up – it can be as messy or as organized as you like.

However, it should be evident that the more disorganized and unstructured your code is, the harder it will be to work with and extend upon in the future. We've already touched on some fundamental principles with OOP and SOLID, but those are just principles to follow. To execute those principles, we use design patterns.

Common design patterns in games development

The following is a list of common design patterns used to solve common problems when writing code – this is not an extensive list by any means, and we're only touching on the reasons for the patterns here:

- **Singleton pattern (a "singleton")**: One of the most popular design patterns in software and games. This pattern creates a single instance of the desired class using a static variable (a single copy of a variable is created and accessible from all objects simply by the declaring class name) – for example, `MyClass.Instance.MyMember` (where `Instance` is the static variable and `MyMember` is any publicly accessible variable or method).

   This pattern gets a bad reputation for misuse, but it can be pretty helpful for small projects – it's quick and easy to implement and use everywhere (hello, game jams!). This reputation is mainly related to more extensive projects where static variables make for code that's harder to troubleshoot and debug – the general sentiment is that when you make something accessible to every part of the code, you're just asking for trouble!

- **Service locator pattern** (a better alternative?): Unity uses this pattern as part of its component structure by using the `GetComponent()` method of its GameObject scripting API. Essentially, the locator will retrieve the reference to the required class or object instance so that it can be consumed in the calling method.

- **Observer pattern (events)**: Another one of the most popular design patterns. This pattern enables a listener to register with and receive notifications from a provider. This is generally performed with a one-to-many relationship, where any number of listeners are notified when the provider object changes state and invokes the notification. This pattern in C# is implemented using events. An event is a special keyword in C# that enforces a proper pattern, where only the declaring class can invoke the notification.

- **State pattern/finite state machine (FSM)**: This pattern allows an object to change its behavior – and/or appearance (think of a GameObject in a Scene) – when its state is changed to one of a finite set of predefined states. A typical use case would be for a `GameManager` class, where you could have states for loading, playing, paused, game over, win, lose, and so on. Events can be triggered when entering and exiting states to extend state-based functionality easily. The object the pattern is implemented on can also delegate state-related behavior to separate classes, such as a `Tick()` method called on every frame update.

- **Command pattern**: This pattern turns a request for executing functions into an object that contains all the required information about the request. The command pattern is popular in C# for when we want to delay or queue a request's execution, or when we would want to keep track of the sequence of operations (such as in a replay system).

  The other benefit of this pattern is that it decouples the invoking class from the object that executes the process – loosely coupled code (one class will not affect another class, reducing dependency) is easier to test and maintain!

- **Manager pattern**: This pattern serves to manage the multiple – usually many – instances of another type of object while providing access to the instances that are being managed. If you have thousands of objects being instantiated in your Scene, with scripts defining an `Update()` method, each one is added to a list of scripts that need to be updated in each frame.

  This pattern can reduce the number of `Update()` methods being called each frame to reduce overhead – these Unity messages are called from native C++ to managed C#, and they have a cost! For example, the manager will be the only object with an `Update()` method that runs on each frame update, while the many instances referenced only have a `Tick()` method that the manager calls.

Additional reading | Programming patterns

Game Programming Patterns (the web version is free): `https://gameprogrammingpatterns.com/contents.html`

Optimization note

If you don't need to use the Unity message events, then remove them from your code – even empty `Start()` and `Update()` methods are still cached and called on every script derived from `MonoBehaviour`, and in the case of `Update()`, it will be called on every frame update!

In this section, you learned about creating scripts and editing them with the VS2022 IDE and received an introduction to the C# language, SOLID principles, and some common design patterns.

We'll dig into specific use cases in the upcoming chapters as we flesh out the features and functionality in our collection game, starting with the next section, where we'll put together a simple player controller for our ladybug character.

Coding a simple player controller with the new Input System

You'll be surprised to find that most of the work to make our player character move has already been done for us by Unity's features. The features that we'll be looking at in this section are as follows:

- **The new Input System**: For receiving keyboard device input from the player.
- **2D physics engine**: For translating input values into movement and providing interaction with objects in the environment.

New Input System

The new **Input System** is a package that provides input device support for controlling objects in your project in a flexible and configurable way. It also replaces the legacy Input Manager. We'll first want to make sure that it's installed to use it:

1. Open the Package Manager from the **File** menu by going to **Windows | Package Manager** and following these steps.

2. Make sure the **Packages** drop-down list is set to **Package: Unity Registry** (just under the window's tab).

3. Find **Input System** in the list of packages (the dialog's left side) and click to select it.

4. Now, click the **Install** button in the window's bottom-right corner.

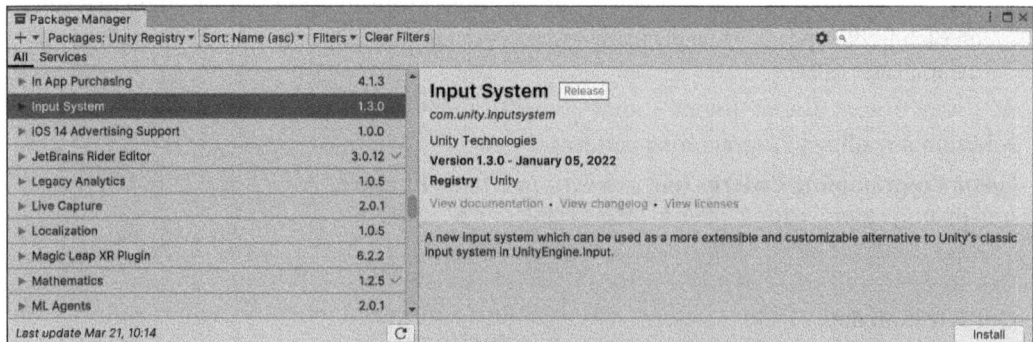

Figure 2.15 – The Input System installation Package Manager

If, once the installation completes, you receive a warning stating **The project is using the new input system package, but the native platform backends for the new input system are not enabled in player settings**, go ahead and click **Yes** to enable the backends. Note that this will restart the Editor – so, if you have unsaved changes in your Scene, click **Save** when prompted.

Additional reading | Unity documentation

Input System: `https://docs.unity3d.com/Packages/com.unity.inputsystem%401.3/manual/`

We're ready to start coding our player movement in the next section with the Input System installed.

Player controller script

Before we jump into coding, we'll first need to understand that there are two ways we can work with the new Input System:

- Receiving input directly from an **Input Device**
- Receiving input through an **Input Action** indirectly

We'll cover both of these approaches in the book, but we'll start with reading directly from an Input Device – the keyboard. Having determined our approach to receiving player input, let's start coding that now, since we've already created our ladybug character's player controller script in a previous section.

Receiving keyboard input

Processing player input through the keyboard directly from the Input Device is straightforward. You simply need to get a reference to the current keyboard device and read the property appropriate for the desired function. We read input in the `Update()` method because it runs every frame.

Let's have a look at this portion of the `Update()` method in the following `PlayerController` code:

```
// Update is called once per frame - process input
void Update()
{
    var keyboard = Keyboard.current;

    // Keyboard connected?
    if (keyboard == null)
        return;     // No - stop running code.

    if (keyboard.spaceKey.IsPressed())
    {
        // Move while holding spacebar key down.
        ...
```

Within the Update() method (between the { } squiggly brackets), we start by assigning a variable (a container for storing data) for the current keyboard device with var keyboard = Keyboard. current; (in C#, every line of code to be executed must end in a semicolon).

> **var (C#)**
>
> The var keyword is an implicitly typed variable that infers the type from the right side of the assignment statement (what comes after the equals sign). This means that we don't need to use an int type for a math expression that returns an integer value – var will infer that it should be of an int type.

Keyboard.current is available to our code from the Input System, simply by including the namespace for InputSystem at the top of our script with the using keyword:

```
using UnityEngine;
using UnityEngine.InputSystem;

public class PlayerController : MonoBehaviour
{
    ...
```

Once the current keyboard device is assigned to our keyboard variable, we can check to make sure a keyboard is connected by testing the variable against a null value, using the == operator (a single = is for assigning a value, whereas a double == is an *is equal?* operation).

Here, we can see that if the keyboard value is equal to null, we stop running the code in this method immediately by using the return keyword:

```
// Keyboard connected?
if (keyboard == null)
    return;     // No - stop running code.
...
```

> **null (C#)**
>
> The null value means there is *no object* assigned.

The last part of the code snippet we'll be looking at here to receive input from the keyboard is checking whether a specified key is currently being held down. We do this by simply stating which key we're interested in – spaceKey – and evaluating the value returned from the IsPressed() method to see whether it is true:

```
if (keyboard.spaceKey.IsPressed())
{
    // Move while holding spacebar key down.
    ...
```

<hr>

if (C#)

An `if` statement tests whether a condition is evaluated to `true` or `false` and executes the directly preceding block or line of code if `true`.

In C#, a variable type that holds a `true` or `false` value is called a `bool` type (Boolean). A `bool` type has a default value of `false` when it's declared without a specific value being assigned.

Moving with 2D physics

We are receiving input from the player and are ready to process it now. Yay!

Technically speaking, following the single responsibility principle, we'd now create a second script to handle the player movement. For now, since our collection game is small, with a single input, it doesn't require a more complex architecture.

When the player is holding down the *spacebar*, let's move them forward while the key is held down and stop them moving when it's released:

```
if (keyboard.spaceKey.IsPressed())
{
    // Move while holding spacebar key down.
}
else if (keyboard.spaceKey.wasReleasedThisFrame)
{
    // The spacebar key was released - stop moving.
}
```

else if (C#)

An `else if` statement tests a new condition only if the previous `if` statement evaluates to a `false` condition and executes the directly proceeding block or line of code if `true`. To evaluate multiple conditions, you can cascade multiple `else if` statements as many times as is required.

Using the `Update()` method is excellent for processing player input. Still, we want to move the player using `2D Physics`, and for that, we need to use `FixedUpdate()` – this is called every physics **Fixed Timestep**, which is at 50 times per second at the default value of `0.02` (this value, generally, doesn't need to be changed).

You might already wonder how we can tell a different Unity message event that is being called *automatically* to execute code from another Unity message event. We'll simply use a `bool` variable as a *flag* to indicate that code should be executed.

Let's create one now:

```
public class PlayerController : MonoBehaviour
{
    private bool _shouldMoveForward;
    ...
```

> **Public and private accessors (C#) | Serialization**
>
> **Accessors** such as **public** and **private** define the scope of a variable for how it can be accessed internally and externally by the class. A public accessor assigned to a variable makes it accessible to other external classes, whereas a private accessor makes it accessible only within the class. If you do not explicitly declare an accessor, the default will be internal (refer to `https://docs.microsoft.com/en-us/dotnet/csharp/programming-guide/classes-and-structs/access-modifiers`).
>
> Another function of declaring a variable public is Unity marking it for serialization. **Serialization** is the process of transforming an object's state into a format that Unity can use – this is required for fields to be available in the **Inspector** window.

We've created the variable as a private field to only be accessible within the `PlayerController` class. We did not assign a value when declaring it so that it will have a starting value of `false` – our flag will not prompt code execution when the program starts.

To execute the code in `FixedUpdate()` to move the player, we'll set it to `true`, with the *spacebar* key being held down and set back to `false` when it is released:

Our player input code will now look like this:

```
        if (keyboard.spaceKey.IsPressed())
        {
            // Move while holding spacebar key down.
            _shouldMoveForward = true;
        }
        else if (keyboard.spaceKey.wasReleasedThisFrame)
        {
            // The spacebar key was released - stop moving.
            _shouldMoveForward = false;
        }
```

And we'll check the value of the `_shouldMoveForward` Boolean flag in `FixedUpdate()` like this:

```
    // FixedUpdate is called every physics fixed timestep
    private void FixedUpdate()
    {
        if (_shouldMoveForward)
        {
```

```
        // Process physics movement.
    }
    else
    {
        // Stop movement.
    }
}
```

> **Tip**
>
> Refer to the Input System documentation for all the possible properties and methods available on a keyboard device. You can also see what is available right within the VS IDE when pressing the . (period/dot) character. IntelliSense will provide a list of all the possible choices for completion (for example, when typing keyboard and then ., the popup list will have both the IsPressed() method and the wasReleasedThisFrame property listed).

Rigidbody2D

Moving objects in the Scene with the ability to interact with other objects is really quite easy when using the physics system – you get a lot of value *out of the box*, since you don't have to program these interactions yourself. For example, moving the player is as simple as setting a velocity value for the direction of movement.

Consider the following addition to our FixedUpdate() method:

```
private void FixedUpdate()
{
    if (_shouldMoveForward)
    {
        // Process physics movement.
        // Up is the direction the object sprite
        // is currently facing.
        Rb.velocity = transform.up * MoveSpeed;
    }
    else
    {
        // Stop movement.
        Rb.velocity = Vector2.zero;
    }
}
```

We've added a line that assigns a value to an Rb object's velocity property, by multiplying the transform.up value with a MoveSpeed variable's value. Rb is a public field we declare, representing the Rigidbody2D component that we'll add to the player GameObject in the **Inspector** window later. Because we declare Rb as public, assignments can also be made in the **Inspector** window.

Let's add the declaration for `Rb` and the `MoveSpeed` variable – with a default value of `10f` (`f` indicates that this is a float value, a numeric value stored in floating-point representation) – to provide some speed to the movement:

```
public class PlayerController : MonoBehaviour
{
    public Rigidbody2D Rb;
    public float MoveSpeed = 10f;
```

> **Tip | VS IntelliSense**
>
> You can type `MoveSpeed` first in the `Rb.velocity` assignment line of code, but VS will complain that it is not declared yet with a squiggly red underline. You can easily add the variable declaration to the code by first clicking on the underlined word, then pressing *Alt/Cmd + Enter*, and selecting the **Generate** field under **MoveSpeed**.

By setting a value to `velocity`, we tell the object to move in a specified direction – that direction is indicated by `transform.up` – and at the speed indicated by the `MoveSpeed` multiplier value. The player sprite will always move in the direction indicated by `transform.up`, no matter its rotation, as seen in the following diagram:

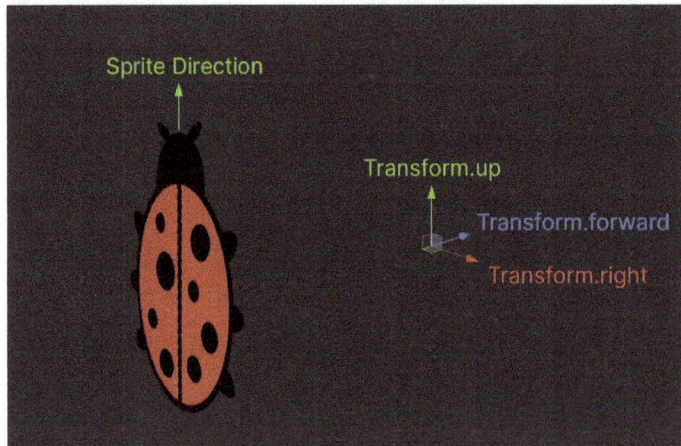

Figure 2.16 – The ladybug Sprite Transform.up direction

> **Transform.up (Unity API)**
>
> This manipulates the position of an object on the *Y* axis (the green axis) in world space. It's similar to `Vector3.up`, but `Transform.up` instead moves the object while considering its rotation.
>
> Other axes available are right (the *X* axis – red) and forward (the *Z* axis – blue).

When the player releases the *spacebar* key, _shouldMoveForward is assigned `false`, which results in Rb.velocity being set to Vector2.zero – velocity is a Vector2 structure, meaning that it can be used to represent 2D positions by X and Y values, as represented by this notation – Vector2(float, float). Vector2.zero is shorthand for Vector2(0, 0).

> **Additional reading | Unity documentation**
> **Vector2**: https://docs.unity3d.com/ScriptReference/Vector2.html

Vector Math, Mathf, and Quaternion

The code, so far, has tackled moving the player character in a forward direction in relation to the direction the sprite is facing. Still, it's currently only facing one direction – up. We'll guide our player around the environment by having it look in the direction of the mouse pointer – steering the forward movement.

We can accomplish this by reading another input device – the mouse – and using some vector math to rotate the player object, according to where the mouse pointer is positioned on the screen.

Have a look at the following LookAtMousePointer() method:

```
private void LookAtMousePointer()
{
    var mouse = Mouse.current;
    if (mouse == null)
        return;

    var mousePos = Camera.main.ScreenToWorldPoint(
        mouse.position.ReadValue());

    var direction = (Vector2)mousePos - Rb.position;
    var angle = Mathf.Atan2(
        direction.y, direction.x) * Mathf.Rad2Deg
            + SpriteRotationOffset;

    // Direct rotation.
    Rb.rotation = angle;
}
```

Let's analyze the code item by item:

1. Just like how we verified we had a valid keyboard, mouse is being assigned and checked for null (stops running the code with return if none is assigned)

2. The `mousePos` variable is assigned by the return value of the `mouse.position.ReadValue()` method, which returns a `Vector2` position, and uses the main camera reference in `Scene` (`Camera.main`) to convert its `Vector2` screen space position to a `Vector3` world point (the 3D coordinate system the Unity engine uses).

3. The direction from the mouse position to the `Rigidbody2D` position on the player is calculated using a simple vector math expression – subtracting one vector from another is a quick way to calculate the direction. Simple!

4. Note that you can only use vector arithmetic operations on the same vector type – here, we have `Vector3` representing the mouse position and `Vector2` representing the `Rigidbody2D` position. To treat the `mousePos` variable as `Vector2` in the expression, we cast the value by prepending (`Vector2`) to it (we only need the x and y values for 2D).

5. Now, we need the rotation angle to apply to the player's `Rigidbody2D`. Unity provides a math library for just such an occasion – `Mathf`. In particular, we're using the `Atan2` method, which returns an angle in radians from a 2D vector.

6. We'll actually want our angle value to be in degrees of rotation – not radians – so again, `Mathf` to the rescue. `Mathf.Rad2Deg` is a conversion constant (equal to *360 / (PI x 2)*) that we'll multiply with the returned radians, and with that, we have our degrees of rotation value. The `SpriteRotationOffset` public variable provides an offset if the sprite's direction is not facing `Transform.right` – the direction the equation results in (we do need a value of -90 degrees here, since our ladybug sprite is pointing up, not right).

7. Finally, we set the resulting angle (in degrees of rotation) to the `Rb.rotation` property!

To make sure our player rotation is being set every frame, add a call to the `LookAtMousePointer()` method at the end of `Update()`:

```
void Update()
{
    ...
    LookAtMousePointer();
}
```

Smoothing rotation

Setting the `Rb.rotation` value for every frame update would be sufficient to make pointing the player in the direction of the mouse pointer work, but we can do better!

We can accomplish a smoother rotation and adjust the rotation speed to alter gameplay using an interpolation method that Unity provides, called **Slerp**. Slerp will smooth the rotation and remove abrupt rotations while providing a better game feel, enhancing the player experience.

> **Slerp (Unity API) | Additional reading | Unity documentation**
>
> **Slerp** is shorthand for **spherically interpolated**. It provides a method that interpolates values between *A* and *B* by an amount of *T*. The difference between this and **linear interpolation** (**Lerp**) is that the *A* and *B* vectors are treated as directions instead of location points.
>
> **Quaternion.Slerp**: https://docs.unity3d.com/ScriptReference/Quaternion.Slerp.html

Replace the direct rotation assignment (`Rb.rotation = angle;`) with the following code snippet:

```
// Interpolated rotation - smoothed.
// Forward (Z-axis) is what we want to rotate on.
var q = Quaternion.AngleAxis(angle,
    Vector3.forward);
Rb.transform.rotation = Quaternion.Slerp(
    Rb.transform.rotation, q, Time.deltaTime
        * LookAtSpeed);
```

Let's break down this code:

1. To interpolate the rotation value, we'll need to use a slightly different approach using `Quaternion` (based on complex numbers that represent rotations that can easily be interpolated and what Unity uses internally to represent all rotations).

2. The first calculation we'll need to get our angle degrees into a quaternion value (`var q`) is using the `AngleAxis()` method to calculate the rotation around an axis. Here, we provide `Vector3.forward` as our rotation axis – referring to *Figure 2.16*, we can see that it will rotate the ladybug sprite on the *XY* plane, which is precisely what we want!

3. Unlike with `Rb.rotation`, which required a float value for the angle degrees, we need to assign a quaternion value now, so we'll have to use the `Rb.transform.rotation` property instead.

4. Finally, we will use `Quaternion.Slerp()` to interpolate the rotation value, producing a smooth rotation from the player character's current rotation to the desired rotation – pointing in the direction of the mouse pointer.

We can change the rotation speed by using the `LookAtSpeed` public variable value. The speed value is multiplied by `Time.deltaTime` to make this a frame rate independent rotation speed (which means it will be consistent across all different performing systems that will run the game).

> **Additional reading | Unity documentation**
>
> **Time.deltaTime**: https://docs.unity3d.com/ScriptReference/Time-deltaTime.html

Our final variable declarations, including the newly added variables for `SpriteRotationOffset` and `LookAtSpeed`, with some default starting values assigned, look like the following:

```
public class PlayerController: MonoBehaviour
{
    public Rigidbody2D Rb;
    public float MoveSpeed = 10f;
    public float SpriteRotationOffset = -90f;
    public float LookAtSpeed = 2f;

    private bool _shouldMoveForward;
    ...
```

> **PlayerController.cs Code**
>
> To view the completed code for the `PlayerCharacter` class, visit the GitHub repo here: `https://github.com/PacktPublishing/Unity-2022-by-Example/tree/main/ch2/Unity%20Project/Assets/Scripts`

This section taught us how to read player input directly using Input Devices, move a `Rigidbody2D` object by manipulating its velocity, and implement smooth rotation using vector math and quaternions.

Summary

This chapter introduced many subjects, including level design concepts, adding 2D artwork to a project, preparing it with Unity's 2D tooling as sprites, and using the prepared artwork to draw a 2D game environment with a **Tilemap**. We also created scripts using the C# language and Visual Studio IDE to add functionality to GameObjects, for reading input and moving the player character with physics.

In the next chapter, we'll add a virtual camera system to follow the ladybug character around the level, add a basic UI, and process conditions for winning and losing the game.

3

Completing the Collection Game

In *Chapter 2*, you were introduced to level design, adding different 2D assets, using assets with Tilemap to create a game environment, and creating scripts using the C# language and the **Visual Studio** (**VS**) IDE to add movement to the player character.

With the player moving around the environment, we'll want a way to follow them visually around the level. This chapter will use Unity's camera system called **Cinemachine** (**CM**) – a powerful camera control feature that makes polished camera movement easy to add and set up.

We'll finish the chapter with an introduction to adding a **user interface** (**UI**). You'll learn how to add text to the screen to track gameplay progress with a timer and score. We'll use Unity's UI system, commonly called **uGUI**, to accomplish this task.

In this chapter, we're going to cover the following main topics:

- Using CM to follow the Player and playtesting
- Game mechanics and how to create with code (components)
- Introduction to uGUI, the timer, counting, and winning

By the end of this chapter, you'll be able to code game mechanics with C# scripts by adding functionality to GameObjects and be comfortable with adding essential UI to your game.

Technical requirements

You can download the complete project on GitHub at `https://github.com/PacktPublishing/Unity-2022-by-Example`.

Using CM to follow the Player and playtesting

CM is expansive in its features – no wonder it's an Emmy award-winning suite of codeless camera tools – but we're only going to focus on one function: the ability to have a camera follow our ladybug around the 2D environment.

First, we'll need to ensure we have the CM package installed by going to **Windows | Package Manager**, selecting **Cinemachine**, and clicking the **Install** button. CM won't help us much without a Player to follow around, so let's get the Player we designed in the previous chapter and import it into our level scene.

Creating a Player Prefab

Unity's **Prefab** system allows you to store configured GameObjects as reusable assets (files) in your project – complete with all its components, assigned values, and any child GameObjects. You can even have Prefabs as children of another Prefab; in this case, we call them **nested Prefabs**.

Additionally, new Prefab assets can be derived from an original Prefab as a **Prefab Variant** (the same base properties but unique variations – when the base Prefab is modified, so are all its derived variants).

One of the powers of Prefabs is the ability to spawn new instances of them into your scene at runtime (think swarms of enemy **non-player characters** (**NPCs**), projectiles, pick-up items, repeating parts of an environment, and so on).

> **Additional reading | Unity documentation**
> **Prefabs:** `https://docs.unity3d.com/2022.3/Documentation/Manual/Prefabs.html`

Follow these steps to create a Prefab of our player character:

1. Open the scene you previously saved with your player character design.

2. Create a `Prefabs` folder at `Assets/Prefabs` and make it the current folder.

3. Click and drag the **Player** GameObject from the **Scene Hierarchy** into the `Assets/Prefabs` folder. In *Figure 3.1*, you can see our freshly created ladybug Player Prefab asset.

4. Now, go back to your game level scene and click and drag the **Player** Prefab from the **Project** window to the **Hierarchy** window (this will place the Player object at position (0, 0, 0) in the scene).

5. You may not be able to see the player character in the scene View because we have not assigned any Sorting Layer for its sprites. Rather than assigning the Sorting Layer to all the sprites individually, we can simply use the **Sorting Group** component as follows:

    I. With the Player selected, in the **Inspector**, click on **Add Component** and type `sorting` in the **Search** field. Select **Sorting Group** in the results.

    II. We've previously established the **Sorting Layers** for the tilemaps. From that, we can determine that **Default** will be suitable for the player character.

    III. Set the **Order in Layer** value to something high such as `1000`, to ensure that the player will always draw as the top-most sprite if we add any additional sprites to this Sorting Layer.

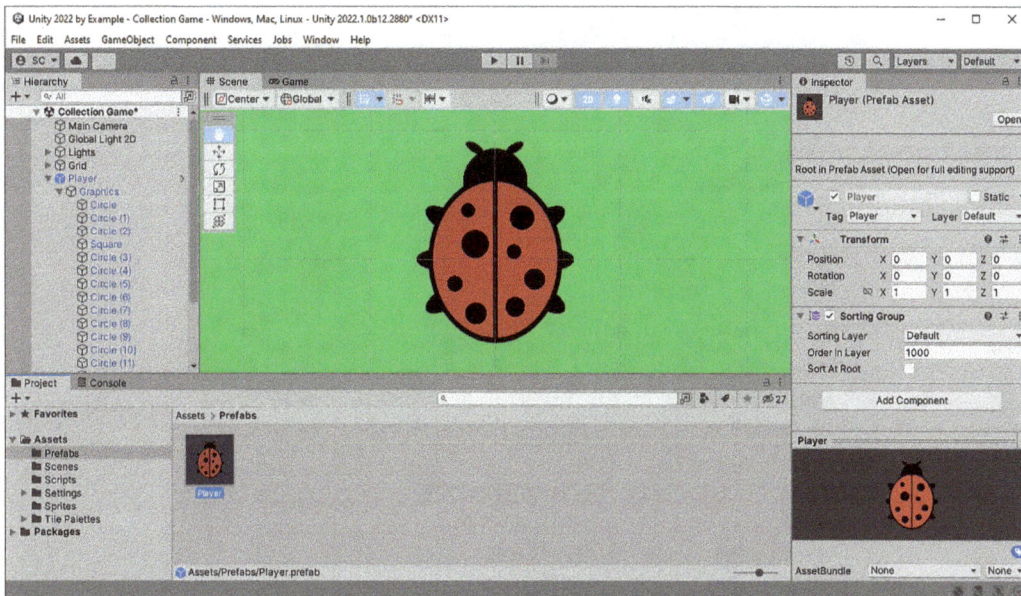

Figure 3.1 – The Ladybug Player Prefab

Tip | Unity documentation

To apply changes to a Prefab in the **Scene Hierarchy** – while the Prefab is not open in **Prefab Mode** (an isolated environment for directly editing the Prefab) – use the **Review**, **Revert**, or **Apply Override** drop-down at the top of the **Inspector** (simply shown as **Overrides**). Alternatively, the quickest way to enter **Prefab Mode** is by double-clicking on a Prefab in the Project window.

Editing a Prefab in Prefab Mode | Editing in Isolation: `https://docs.unity3d.com/2022.3/Documentation/Manual/EditingInPrefabMode.html`

Creating a 2D follow camera

With CM, it's trivial to create a follow camera for a 2D environment. Now that we have our player in the scene, we can add a CM camera that will follow the player around as it moves with the following steps:

1. Create a new CM **virtual camera (vcam)** from the **Hierarchy** window's **Create** menu or the main **File** menu by going to **GameObject | Cinemachine | 2D Camera** – for the first vcam added to the scene, this also adds a **Cinemachine Brain** component to **Main Camera** (CM controls **Main Camera** through one or many vcams).

2. With the new vcam selected, drag the **Player** object from the **Hierarchy** to the **Follow** field in the **CinemachineVirtualCamera** component (to assign its reference), as seen in the following screenshot:

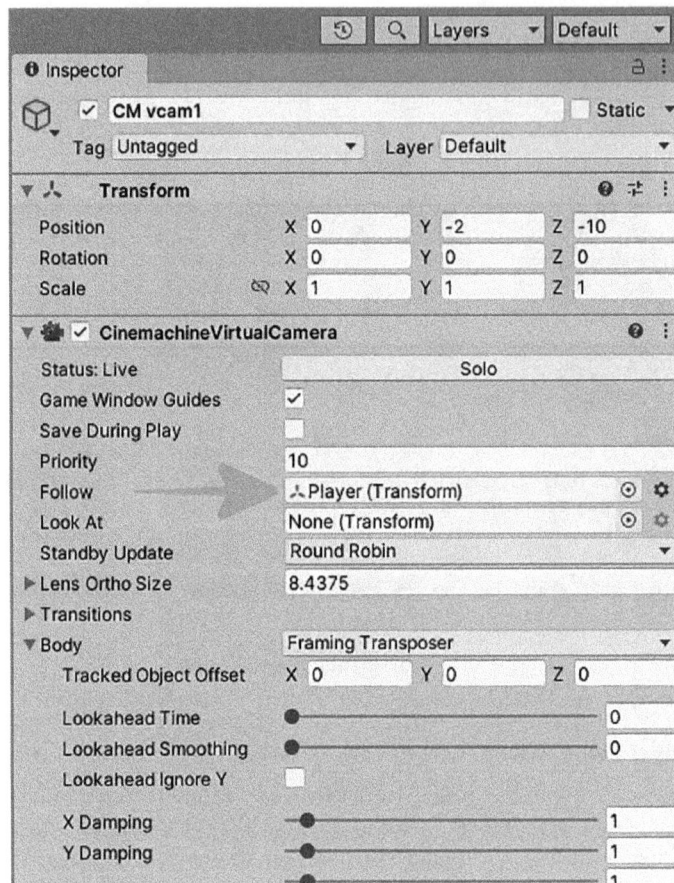

Figure 3.2 – The CM Follow Player Inspector assignment

3. Within the **Body** section, expand the arrow to the left of **Body** if necessary; you can adjust XY damping, screen position, dead zones, and so on. Play around with these settings and fine-tune your camera to follow your preferences.

Additional reading | Unity documentation

Cinemachine: https://docs.unity3d.com/Packages/com.unity.cinemachine%402.3/manual/index.html

Cinemachine feature: https://unity.com/unity/features/editor/art-and-design/cinemachine

Cinemachine for 2D Tips and Tricks: https://blog.unity.com/technology/cinemachine-for-2d-tips-and-tricks

We can now explore the entire environment knowing that the camera will follow us around. Yay!

Playtesting the level

The last step to make our level playable is to add the necessary components (that is, **PlayerController** and physics), adjust the component values, and assign the component references.

Let's do this now with the following steps:

1. Open the **Player** Prefab in **Prefab Mode** by selecting it in the **Hierarchy** or selecting the **Prefab** asset in the **Project** window, and then click **Open** at the top of the **Inspector**.

2. Add a **CapsuleCollider2D** component by clicking **Add Component**, searching for `collider`, and selecting it in the results – this will provide physics interactions with other colliders in the environment via Unity's built-in 2D physics engine (Box2D).

3. Adjust the collider to fit the graphics for your player using both the **Offset** and **Size** fields, then use the **Edit Collider** button to move it manually into position in the scene View.

4. Add a **Rigidbody 2D** component by clicking **Add Component**, searching for `rigidbody`, and selecting it in the results – this will provide the physics properties to our player character, such as how it's affected by gravity, its mass, drag, position/rotation constraints, and the type of **Collision Detection** to employ:

    I.   Enable **Auto Mass** by clicking the checkbox as this will automatically set the mass to the 2D collider's mass.

    II. Set the values for **Linear Drag** (`100`) and **Angular Drag** (`1`) to refine the player movement (adjust to your preferences).

    III. Most importantly, disable **Gravity** by setting its value to `0` – in a top-down environment, everything is already *sitting on the ground*, so we don't assign any gravity (if we did, the object would fall toward the bottom of the screen since gravity is based on the y axis; this works well for a 2D side-on game view).

5. Add the **PlayerController** script by using the **Add Component** button in the **Inspector** or dragging the script from the **Project** window onto **Player** in the **Hierarchy** window or, with the **Player** object already selected, into the **Inspector**.

6. Assign the **Rigidbody 2D** component to the **Rb** field by clicking on the component's **Rigidbody 2D** title and dragging it to the **Rb** field (where it currently says **None (Rigidbody 2D)**) – this creates a direct reference to the component that our `PlayerController` script uses.

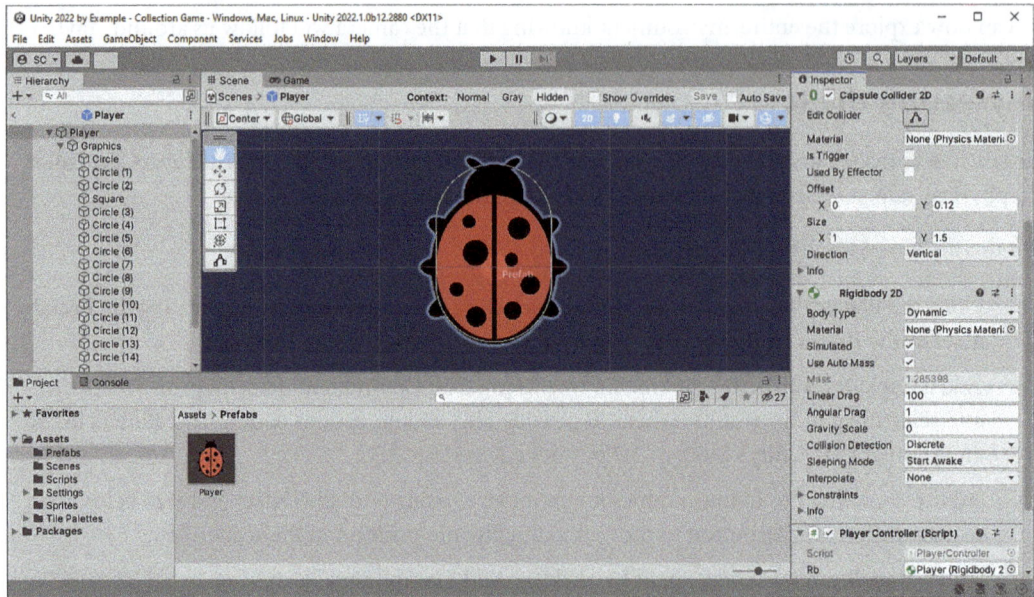

Figure 3.3 – The Player Prefab components configuration

That's it! We can now press the **Play** button on the main toolbar to playtest our level design.

The ladybug will rotate to face the mouse pointer's position and move in the direction it's facing while the spacebar is being held down. Tiles on the **Tilemap** with **TilemapCollider2D** assigned will prevent the player from moving into those areas.

Now is an excellent time to adjust the values on the **Rigidbody 2D** fields for mass and drag and test changing the values for the move, and look at rotation speeds on `PlayerController`. Here, you can see the default values we started with:

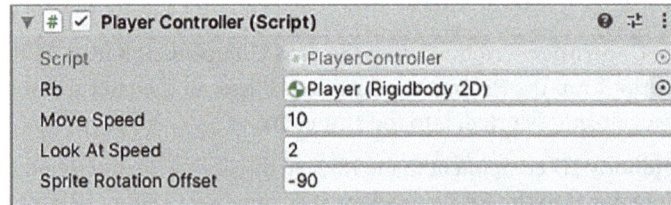

Figure 3.4 – The PlayerController Inspector field values

Making these adjustments affects the game feel and the overall player experience, so spend time fine-tuning these values until it just *feels right* to you as the game designer. Have fun!

> **Tip**
> When you find values that you like and want to quickly go back and forth between them during testing (without having to write them on a scratch pad somewhere), you can use **Presets** (the *slider icon* on the right side of the **Component** title bar): `https://docs.unity3d.com/2022.3/Documentation/Manual/Presets.html`

In this section, you learned how to read input directly from an input device to move the player character with a custom C# script while using physics, you explored the timesaving and powerful features adding a CM vcam provides, and you understood that playtesting is about refining values to get the right game feel for a good player experience.

> **Bonus activity**
> Make the ladybug player character move while also holding down the left mouse button.

In the next section, we'll implement some interactions for the player by coding collecting pickups.

Game mechanics and how to create with code (components)

This section will explore the concept of game mechanics and what that means for the gameplay in the *Outer World* collection game we're making. We'll define precisely the mechanic we'll be using in a brief expansion on the **Game Design Document** (**GDD**) and finish with writing the code for the mechanic by adding a new Prefab with a custom component that uses a 2D physics interaction event.

What is a game mechanic?

A **game mechanic** can loosely be defined as the rules that dictate the gameplay and how the player is supposed to engage with the mechanics sufficiently to provide a joyful and entertaining experience. Examples of some popular types of gameplay mechanics include collecting, moving, shooting, and building things.

Some game mechanic rules that dictate gameplay are followed by the game and not the player, such as the game only unlocking a new level when the player completes the current level.

> **Buff & Nerf | game balance**
> Changing elements of mechanics to a player's advantage is known as the **Buff** and to their disadvantage is the **Nerf**, keeping the game balanced: `https://www.inverse.com/gaming/nerf-buff-meaning-video-games-coined-series`

Let's explore adding our game's primary mechanic now.

Adding to our GDD

The game mechanic we're going to add to our GDD and implement is the collection of pickup items in the environment. We can describe it as follows:

Name of Game	Outer World
What is the core game mechanic for the collection game?	The player will find and collect "water diamonds" throughout the environment by touching them until all are collected, or the countdown timer expires.

Table 3.1 – The GDD game mechanic addition

Additional reading | Game mechanics

Will Wright (*The Sims* creator), *5 Tips for Writing Game Mechanics*: `https://www.masterclass.com/articles/will-wrights-tips-for-writing-game-mechanics#will-wrights-5-tips-for-writing-game-mechanics`

Now that we have our core game mechanic defined, let's put the code together to make it work!

Bonus activity

By exploring the idea of adding additional physics-based objects to our level, design a game mechanic that involves the player pushing around tile blocks that we'll call *toolboxes*. The toolboxes could inconvenience the player and slow them down in pursuit of collecting the *water diamonds* or provide additional gameplay by requiring them to be moved into specific locations.

Collecting pickups

A lot of the basics for working within Unity and creating component-based scripts have been covered by this point, so we're going to be moving a bit faster in putting things together – while not spending as much time explaining the small details.

The primary objective of our core mechanic is collecting objects. We'll accomplish this by taking advantage of the interaction events the 2D physics engine provides. Here we go!

Creating a collectible item Prefab

First, we need a sprite for our collected object – the *water diamond*. You can either create your sprite art within Unity (as we did with the ladybug character) or import one (the suggested size is a 64 px square). Create a new Prefab for the collection object with the following steps:

1. Add your *water diamond* sprite to the scene.

2. Parent the **Graphics** object's sprite to an empty GameObject (container).

3. On the parent GameObject, add a primitive collider that best fits the shape of your *water diamond* sprite – I chose **CapsuleCollider2D**, as seen in *Figure 3.5*.

4. Enable the **IsTrigger** field on the **CapsuleCollider2D** component since we don't want any collision interactions to occur between the diamond and the Player; we just want to receive a message for the interaction event.

Additional reading | Unity documentation

The simple takeaway here is that when `IsTrigger` is enabled, no physical collisions between objects will occur – objects won't move in response to forces applied.

`Collider.isTrigger:https://docs.unity3d.com/2022.3/Documentation/ScriptReference/Collider-isTrigger.html`

---

5. Let's not forget to ensure the sprite is drawn at the correct depth in our `Sprite Renderer`'s **Sorting Layers**. Let's use the **Objects** layer and set the **Order in Layer** value to something high such as `100`.

6. If you created your sprite within Unity using the Sprite Creator tool's shapes, select the **Graphics** parent object and add a **Sorting Group** component to assign the Sorting Layer.

7. Drag the root object the sprite was parented to into the `Assets/Prefabs` folder in the **Project** window.

Tip | Unity documentation

Objects in the **Hierarchy** window turn blue and have the right caret (arrow) to their right when the GameObject is made into a Prefab. Clicking the caret will take you into **Prefab Mode** in the context of the scene (as shown in *Figure 3.5*).

Editing a Prefab in Prefab Mode | Editing in context: `https://docs.unity3d.com/2022.3/Documentation/Manual/EditingInPrefabMode.html`

You can verify your collectible **Water Diamond** Prefab setup with the following screenshot:

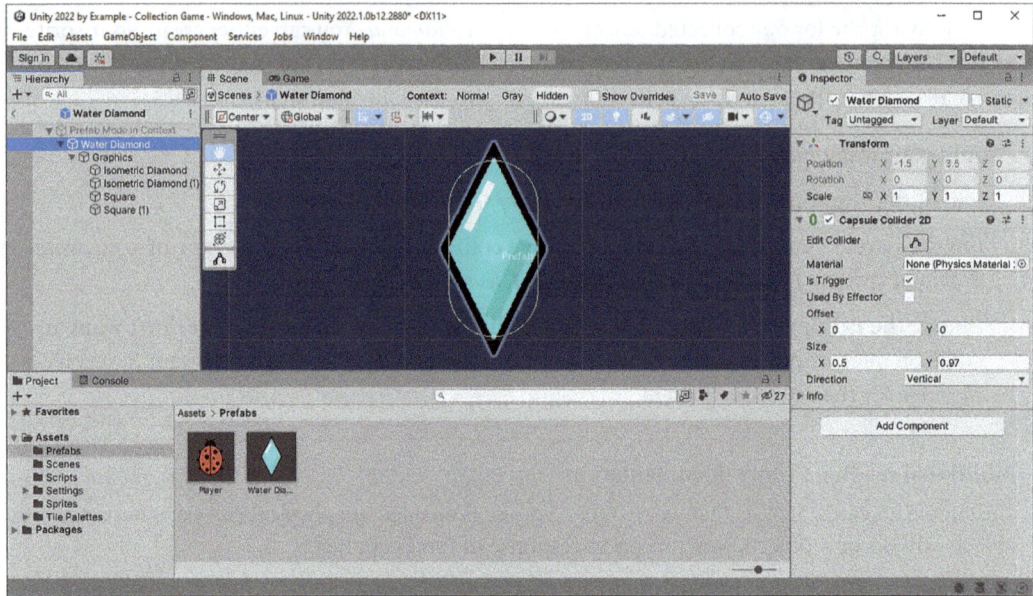

Figure 3.5 – Water Diamond Prefab with collider

> **Additional reading | Unity documentation**
>
> **Collider 2D**: https://docs.unity3d.com/2022.3/Documentation/ScriptReference/CapsuleCollider2D.html
>
> **Optimizing Physics Performance**: https://docs.unity3d.com/2022.3/Documentation/Manual/iphone-Optimizing-Physics.html
>
> Note that the performance order of 2D colliders, from the fastest to slowest, is: **Circle**, **Capsule**, **Box**, **Composite**, **Polygon**, and then **Edge**.

Okay, we have our Prefab for the collectible item's sprite; now, let's code the interaction.

Creating the CollectItem component

Create a new script in the `Scripts` folder and name it `CollectItem`. Replace the generated code with the following template:

```
using UnityEngine;
public class CollectItem : MonoBehaviour
{
    void Start()
    {
```

```
        Debug.Log($"{gameObject.name}'s Start
            called", gameObject);
    }

    private void OnTriggerEnter2D(Collider2D collision)
    {
        Debug.Log("Collision message event triggered!");
        Destroy(gameObject);
    }

    private void OnDestroy()
    {
        Debug.Log("Destroyed");
    }
}
```

Let's have a look at the code line by line:

1. The `Debug.Log()` line in `Start()` will just output a message to our **Console Window** (aka **Console**) with the name of the GameObject the `Start()` method was called on. We can use this to verify that all of the collectible item objects we'll add to the scene are *registering* themselves. We'll be using this to increment the count for total items needing to be collected in the level (the 10 in *0 of 10*) when we add our UI in the next section.

$ string interpolation | C#

The dollar sign ($) in `Debug.Log($"{gameObject.name}'s Start called");` is for identifying an interpolated string. The text within { } (open/close squiggly brackets) in the string literal (represented by two double quotes (" ")) is a string result of the inserted code (usually a variable or expression).

Refer to the C# language reference for additional reading: `https://docs.microsoft.com/en-us/dotnet/csharp/language-reference/tokens/interpolated`

Because we've added a second parameter to the `Debug.Log()` call (`gameObject`), when the **Console** log item is clicked on, the corresponding object in the **Hierarchy** will be highlighted. The `gameObject` (camelCase) keyword refers to the current GameObject (PascalCase) the component is running on (also provided by the `MonoBehaviour` base class).

Optimization note | Unity documentation

Note that you'll want to comment out `Debug.Log()` statements for performance reasons once you've finished testing and they are no longer needed.

Debug: `https://docs.unity3d.com/2022.3/Documentation/Manual/class-Debug.html`

---

2. `OnTriggerEnter2D(Collider2D collision)` is where the magic happens. This Unity message is called when a collision occurs between two physics objects where one of the objects has a collider `IsTrigger` value enabled. We receive a reference to the other object's collider with the collision parameter value (we'll be using `collision` to detect whether the other object is the Player in the next section).

3. `OnDestroy()` is another Unity message that is called when the GameObject in the scene is destroyed – by using the `Destroy()` method, as you can see in the `OnTriggerEnter2D()` method previously, when the Player collides with the collectible item, it will process the collection of the item in `OnTriggerEnter2D()` and then remove the item from the scene by calling `Destroy(gameObject)` (so items cannot be collected again).

4. Save the `CollectItem` script and add it as a component to your **Water Diamond** collectible item Prefab (do that in **Prefab Mode** or the **Hierarchy**; but, either way, don't forget to save or apply your overrides). Note that any component being added that is intended to work with `Collider` components must be on the same GameObject as the collider – you can see in the **Inspector**, in *Figure 3.6*, where the `CollectItem` script is just below the `CapsuleCollider2D` component.

5. Have a look at the following screenshot displaying the `Debug.Log()` output in the **Console** while playtesting and running the ladybug character into the *water diamond* sprites:

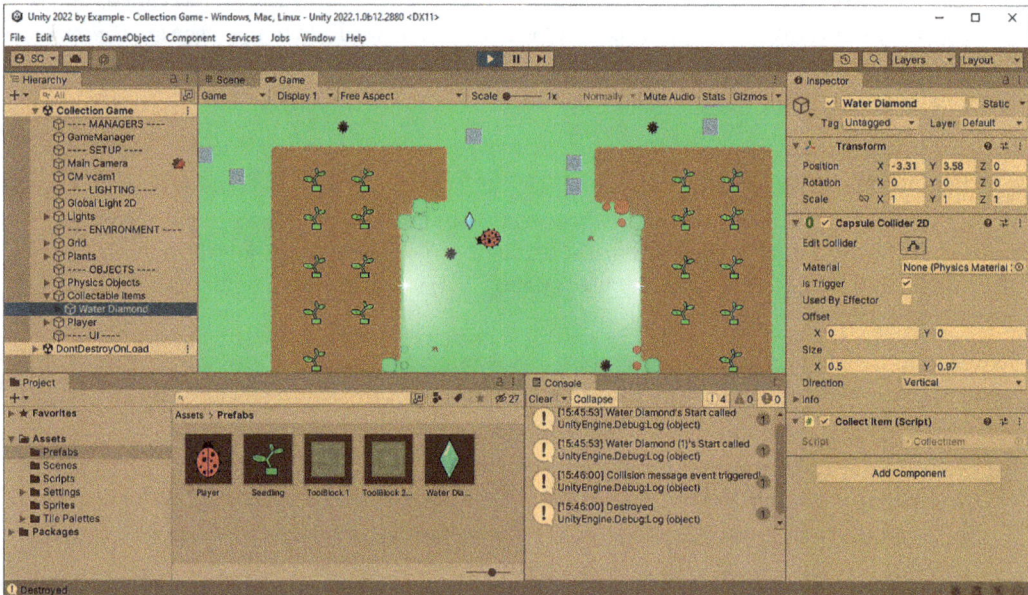

Figure 3.6 – Playtesting the collectible items

> **Important note**
>
> Two additional things worth mentioning in the previous screenshot are as follows.
>
> The **Editor** is tinted orange because I've assigned a color to show when entering **Play Mode** –
> set colors in **Edit** | **Preferences** | **Colors** then **General** | **Playmode tint**. This tint *should* remind
> you that you are in **Play Mode**, and most changes you make to **Component** values will not be
> saved when stopping (changes in file-based assets will be saved, however).
>
> I've added some empty GameObjects to the **Scene Hierarchy** to organize objects into logical
> groupings. For any object added for purely organizational purposes (that is, no child assets
> are being used in the game), you can assign it an `EditorOnly` tag so that it is not included
> in the final game build (saving some resources in the process).

You can go ahead and duplicate the **Water Diamond** Prefab by selecting one in the **Hierarchy** and
pressing *Ctrl/Cmd + D* several times and repositioning it throughout the level. Playtest to get a feel
for how well you can move around the level to reach each of the items!

> **Tip**
>
> You can organize all of the **Water Diamond** collectible items in the scene in an empty root
> GameObject – just make sure the root object is at (0, 0, 0) before dragging in the items.
>
> You can quickly reset the `Transform` component of a GameObject in the **Inspector** by
> clicking the vertical ellipsis button (to the right of the **Help** and **Presets** buttons in the title
> bar) and choosing **Reset**.

With that, our game is shaping into something fun! But what about adding to the challenge? We'll use
hazards to slow the player when they touch one… look out!

Hitting hazards

We've previously discussed hazards in the game, and level design, so now let's look at implementing
them before coming to a win/lose condition in the final section of this chapter.

First, let's define our nerf mechanic by adding it to our GDD document:

Name of game	Outer World
What is a nerf mechanic for the player in the collection game?	The player's speed will be decreased when they touch a "toxic puddle" in the environment – increasing the challenge of collecting all "water diamonds" before the timer expires.

Table 3.2 – The GDD nerf mechanic addition

Okay, we need to add *toxic puddles* to the level then! We could do these the same way we created the **Water Diamond** Prefabs and populate the level in that fashion, but let's take a different approach here and revisit our tilemaps.

Add toxic puddles in challenging locations throughout the level by following these steps:

1. Create a new **Tilemap** in the **Hierarchy** (as a child of **Grid**) named `Tilemap - Hazards - Trigger`.

2. Open the **Tile Palette** (**Window** | **2D** | **Tile Palette**) and make sure the active tilemap is the tilemap we just created in the previous step.

3. Select **Environment Objects** from the available **Palettes** dropdown menu list.

4. Select the *black splotch/puddle* sprite from the available tiles and use the **Paintbrush** tool (shortcut key *B*) to scatter these tiles in strategic spots throughout your level (my suggestion would be to place them near water diamonds!).

5. Since we want to make the toxic puddles interactable (via the physics engine), add a `TilemapCollider2D` component to the **Tilemap** GameObject in the **Hierarchy**.

6. Compared to the previous collision objects' **Tilemap**, the difference is that now we're going to enable the `TileMapCollider2D.IsTrigger` value so that we can use the `OnTriggerEnter2D()` event to respond to the collision.

Here is what the setup looks like with the new hazards tilemap, toxic puddles painted in the level (selected in the Tile Palette), and `TileMapCollider2D`'s **Is Trigger** box ticked:

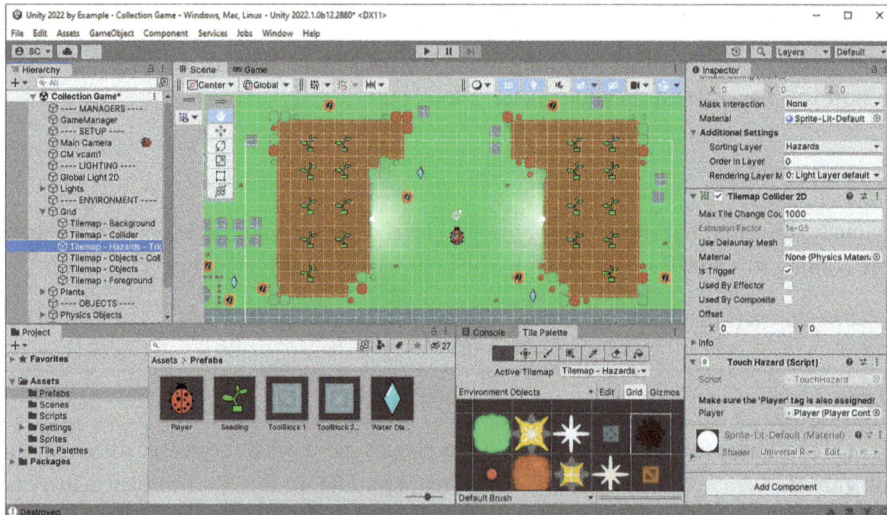

Figure 3.7 – The Hazards Tilemap trigger collider setup

In *Figure 3.7*, you can see that I've already added a `TouchHazard` component for the Player interaction with the hazard tilemap's tiles.

Let's have a look at the `TouchHazard` script:

```
using UnityEngine;
public class TouchHazard : MonoBehaviour
{
    [Header("Make sure the 'Player' tag is also
        assigned!")]
    public PlayerController Player;

    private void OnTriggerEnter2D(Collider2D collision)
    {
        if (collision.CompareTag(Player.tag))
        {
            Debug.Log("You touched a toxic puddle!");
            Player.SlowPlayerSpeed();
        }
    }
}
```

Now, let's break it down point by point:

1. The first new line you'll notice here is we've added some *header information*. [Header] is an attribute that we can decorate the public field with to show some information to the developer – or designer - viewable in the **Inspector** along with a field. You can see this message in the **Inspector** in *Figure 3.7*.

> **Additional reading | Unity documentation**
>
> **Attribute**: `https://docs.unity3d.com/2022.3/Documentation/Manual/Attributes.html`

2. The first public variable we're declaring here – `Player` – is for the `PlayerController` reference. Instead of using a GameObject reference here – requiring an additional step to access a component added to the GameObject – we're using the object type of the component to access its public members directly.

3. We can see here again that we're using the `OnTriggerEnter2D` message called when the 2D collider on the tilemap reports a collision.

4. Now that we have a collision occurring, we will do a quick test with an `if` statement to see *who* collided with the hazard tiles by using the `collision.CompareTag()` method.

> **Additional reading | Unity documentation**
>
> **CompareTag**: `https://docs.unity3d.com/2022.3/Documentation/ScriptReference/GameObject.CompareTag.html`

We're specifying which tag we want to detect by passing in the player's tag assignment with `Player.tag`. So, let's make sure the **Player** Prefab has the **Player** tag assigned to it now!

5. With the **Player** object selected in the **Hierarchy**, go to the top of the **Inspector** and, using the **Tag** drop-down (*A*), select the **Player** tag from the list and then apply the change to the **Player** Prefab by using the **Overrides** drop-down and clicking **Apply All** (*B*), as seen in the following figure:

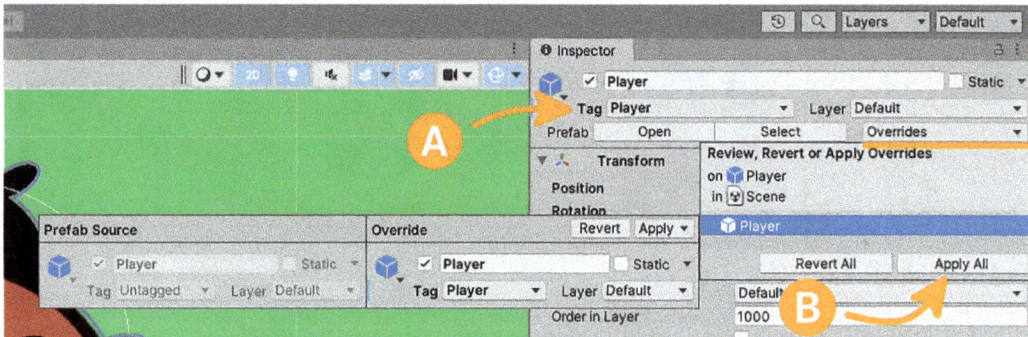

Figure 3.8 – Set Player tag in Inspector and apply override

Additional reading | Unity documentation

Tags: `https://docs.unity3d.com/2022.3/Documentation/Manual/Tags.html`

Tags and layers: `https://docs.unity3d.com/2022.3/Documentation/Manual/class-TagManager.html`

---

6. If the result of our `if` statement returns `true` (the tags matched), then the **Player** object has touched a toxic puddle, so the player's movement will be nerfed – slowed down for some time. The more the player touches them, the more challenging it will be to collect all the water diamonds before the timer expires!

    I.   We'll first use the `Debug.Log()` statement to notify us in the **Console** that the collision on the Player has occurred.

    II. The next line calls `SlowPlayerSpeed()` on the `Player` object reference. Accessing public members of an object is performed with **dot notation** – the name of the object, followed by a dot (.), followed by the name of the member, as in `Player.SlowPlayerSpeed()`.

    III. Let's have a look at the *additions* required for adding the `SlowPlayerSpeed()` public method to our `PlayerController` class:

```
public class PlayerController : MonoBehaviour
{
```

```
    ...
    [Header("Hit Hazard Speed")]
    public float SlowedSpeed = 2f;
    public float SlowedTime = 5f;    // Seconds

    private float _speedStartValue;

    private void Start()
    {
        _speedStartValue = MoveSpeed;
    }
    ...
    public void SlowPlayerSpeed()
    {
        MoveSpeed = SlowedSpeed;
        Invoke(nameof(RestoreSpeed), SlowedTime);
    }

    private void RestoreSpeed()
    {
        MoveSpeed = _speedStartValue;
    }
}
```

IV.   We can see a few new variables are required to support slowing the player character's speed for a short duration. The public variables, assignable in the **Inspector**, are as follows:

i.   SlowedSpeed: A float variable to specify the new slower move speed of the player, with a default value of 2f.

ii.   SlowedTime: A float variable to specify the duration (in seconds) to keep the speed at the SlowedSpeed value, with a default value of 5f (5 seconds).

V.   The Start() method is used to cache (retain) the original MoveSpeed value of the player.

VI.   The SlowPlayerSpeed() method has been added – called from the TouchHazard class – to set the current MoveSpeed of the player to the SlowedSpeed value to slow down movement.

VII.   Following the MoveSpeed assignment, the Invoke() method (MonoBehaviour) is being used with a delay time value (the second parameter) to wait to call the new RestoreSpeed() method until the time specified by the SlowedTime variable has passed.

VIII.   We're also using the C# nameof() expression with the RestoreSpeed() method here, so we aren't using a string literal for the method name.

> **Additional reading | Unity documentation**
>
> **Invoke**: `https://docs.unity3d.com/ScriptReference/MonoBehaviour.Invoke.html`

IX.   Finally, we've added the `RestoreSpeed()` method, which simply resets the `MoveSpeed` variable back to its original value – resetting player movement back to normal.

With that, we can now enable the hazards interaction by referring to *Figure 3.7*:

1.   Add the **TouchHazard** component to the **Tilemap - Hazards - Trigger** object.

2.   Drag the **Player** object from the **Hierarchy** to the **Player** field in the **TouchHazard** component.

3.   Verify the **SlowedSpeed** and **SlowedTime** values in the **PlayerController** component on the **Player**.

You'll now want to playtest your level and adjust the speed and duration of the hazard effect until it suits you (I've chosen a **SlowedSpeed** value of 2 and a **SlowedTime** value of 5, as my initial values); move the hazards and diamonds around the level to make it more challenging or fun. Again, just have fun with this part!

In this section, we understood what a game mechanic is and how to define it in the GDD, and we learned how to implement a collection mechanic and hazard that affects gameplay at our level. Next, we'll be applying the finishing touches to the collection game by adding a UI, countdown timer, and win condition.

Introduction to uGUI, the timer, counting, and winning

The player needs to know what's going on in the game, and the way we inform the player of this critical information is with a UI. Unity provides easy-to-use tooling for creating a UI, and it's commonly referred to as **uGUI** (or **Unity Graphical User Interface**) – pronounced *You-gooey*. Formally, it's just **Unity UI** (searching *ugui* in the documentation returns *No results found*).

Unity UI is a GameObject-based UI system, so you'll already be familiar with adding UI components to display and interact with UI controls in different ways.

> **Additional reading | Unity documentation**
>
> **Unity UI**: `https://docs.unity3d.com/Packages/com.unity.ugui%401.0/manual/index.html`
>
> *UI How Tos*: `https://docs.unity3d.com/Packages/com.unity.ugui%401.0/manual/UIHowTos.html?q=unity%20ui`

The three key pieces of information the player needs to see in the collection game are as follows:

- The number of items collected versus the total number to collect
- The time remaining to collect all the items
- The winning state (win/lose) when the game ends

The first UI object we'll discuss adding is the Canvas because it's what's responsible for drawing our UI.

Canvas

The **Canvas** is a defined area where all the UI widgets are contained – think of it as a piece of paper where everything to be drawn will be contained on the page. As mentioned previously, the Unity UI is GameObject based, so the Canvas is a GameObject with a Canvas component. All UI widgets added to the scene must be children of a Canvas.

Adding a UI Canvas to our scene is as simple as going to **UI | Canvas** in the **Create** menu. The first time a Canvas is added to the scene, it also adds an **EventSystem**. We'll have to select the **EventSystem** and update it to use the new Input System. Click **Replace with InputSystemUIInputModule**, as seen in *Figure 3.9*:

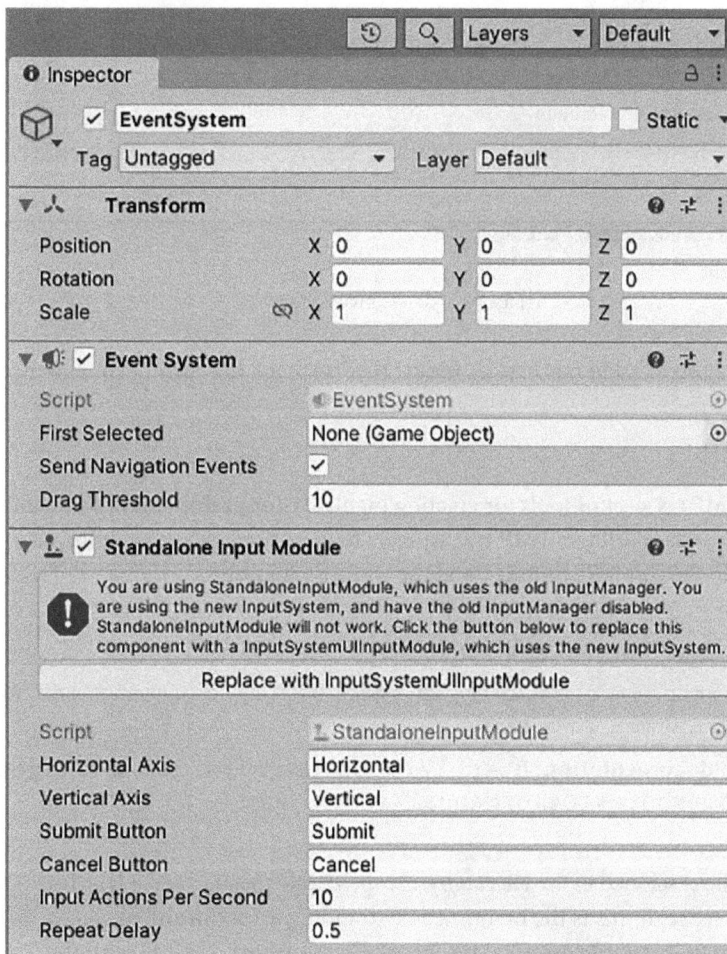

Figure 3.9 – Update EventSystem for new Input System

With the **Canvas** added and the Input System now sorted, let's set our Canvas to scale the UI to the screen using the **Reference Resolution** setting. As before, we're standardizing on a player screen of Full HD (1920x1080), so enter the values in the fields accordingly. We also want to balance the resolution matching mode (**Match Width Or Height**) with a **Match** value of 0.5, as shown in the following figure:

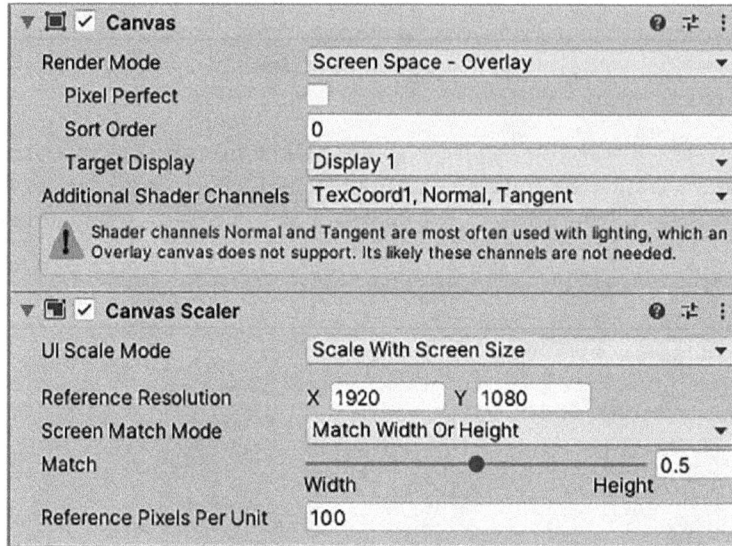

Figure 3.10 – Canvas Scaler

The Canvas is all set, so let's add our first UI widget for displaying the current count of items collected!

TextMesh Pro

TextMesh Pro (**TMP**) is a set of tools for creating highly customizable and high-quality 2D and 3D UI text in your projects. We'll use TMP text widgets to display all gameplay information, including the items collected, the countdown timer, and the win/lose message.

> **Additional reading | Unity documentation**
>
> **TextMeshPro**: https://docs.unity3d.com/2022.3/Documentation/Manual/com.unity.textmeshpro.html
>
> **TextMesh Pro documentation**: https://docs.unity3d.com/Packages/com.unity.textmeshpro%404.0/manual/index.html

With the Canvas object selected in the **Hierarchy**, use the **Create** menu to add a **UI | Text – TextMeshPro** text widget to the scene. If this is the first time you're adding a **TextMeshPro** widget to your project, you'll be prompted to install the essential resources. Click **Import TMP Essentials**:

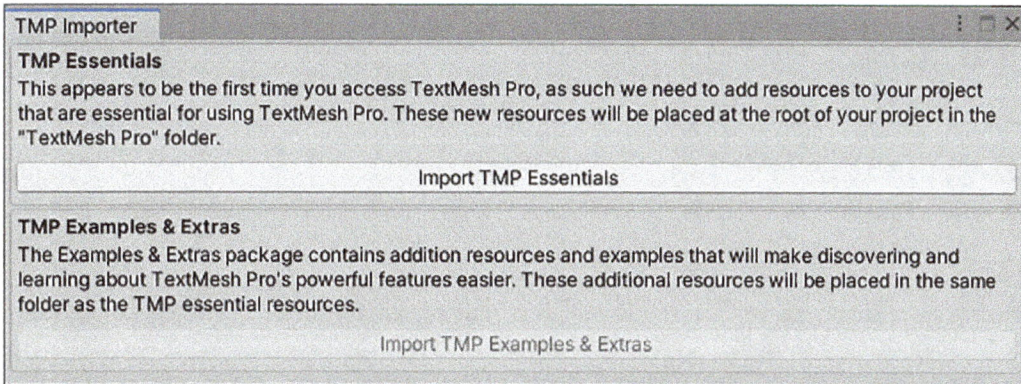

Figure 3.11 – The TMP Importer dialog

Use the following steps to make the collected item text shown in the bottom-left corner of the screen:

1. Rename the newly added TMP text object `Collected Count Text`.

2. To anchor the text to the bottom-left corner of the screen, use **Anchor Presets** (*A*), hold both the *Shift* (**Also set pivot**) and *Alt* (**Also set position**) keys down (*B*), then click the bottom-left anchor (*C*):

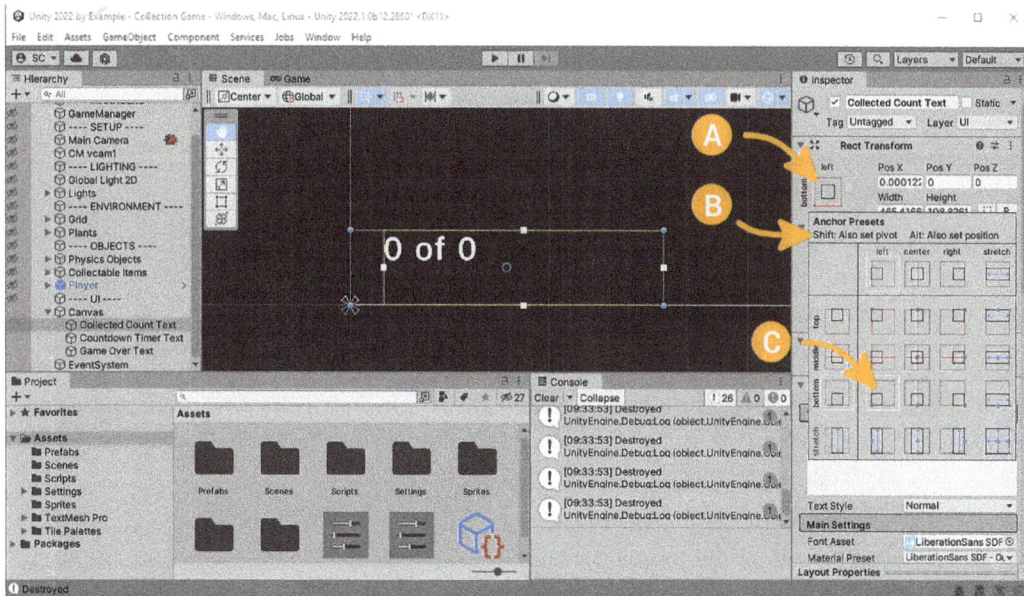

Figure 3.12 – TextMeshPro rect anchor setting

3. Use the following settings to provide default text to display and style the text (refer to *Figure 3.12* as a reference):

I.   **Default text**: 0 of 0

II.   **Font Style: Bold**

III. **Font Size: 48**

IV.   **Extra Settings | Margin | Left: 50**

With that, we're ready to have the text updated every time an item is collected. Let's see how to tie it all together in the next section.

Updating the pickup count

The problem we're going to solve now is *how is UI text updated when an item is collected?* We can refer to design patterns for a solution! We previously covered both the Singleton and Observer patterns, and either one will suit our needs here.

To provide an example in practice, we will use both, starting with the Singleton.

The pickup count will be displayed as **2 of 10** – meaning 2 items of a total of 10 have been collected. So, we'll first need the total count of collectible items in the scene. The GameManager, will be responsible for holding the data (variables) for the collectibles, so this will become a Singleton instance.

The Start () method for each collectible item in the scene will be called when the game first runs, so we'll use this to add each collectible item to the total items count.

Let's first create the GameManager script and make it a Singleton instance by adding the following lines to the variable declaration section (near the top of the class):

```
// Singleton instance.
public static GameManager Instance
    { get; private set; }
private void Awake() => Instance = this;
```

The GameManager class itself is declared a public static property – we want to directly access the class from another type using the Instance variable. We've let get remain public, while declaring set as private so external classes will only be able to read.

In the Awake () method, we're setting the Instance variable to this, the current class instance. We're using an expression body here (indicated by =>) instead of the usual { } squiggly brackets (block body) to keep the code concise, since it's only a single expression that's needed.

> **Property (C#)**
>
> A C# property is a class member that uses explicit accessors for the get and set methods to read or write data values.

Now, we can add the variable for storing the total number of collectible items, the public method for incrementing the total, and the call to the incrementing method from the CollectItem class by following these steps:

1. Add a private _totalCollectibleItems integer variable to the variable declaration section. Its accessor is private because it's encapsulated only to change its value within the declaring class (with OOP encapsulation, it's considered bad practice to modify variable values outside the class):

```
private int _totalCollectibleItems;
```

2. Add a public AddCollectibleItem() method with no return type (void), as an expression body, to increment the _totalCollectibleItems current value. The ++ unary arithmetic operator increments the operand by 1. The public accessor means that we can call this method outside the declaring class, which we will do next from the CollectItem class via the Singleton instance:

```
public void AddCollectibleItem() =>
    _totalCollectibleItems++;
```

3. Now, in the CollectItem class, add the following to the bottom of the Start() method for calling the GameManager Singleton instance to add the collectible item to the total items count:

```
void Start()
{
    ...
    GameManager.Instance.AddCollectibleItem();
}
```

With that, we now know how many **Water Diamond** collectible items are in the scene for the player to collect. While adding them to our level design, we don't need to know beforehand how many were added and manually specified somewhere; we're always guaranteed to have the total count since Start() will be called on every object instance in the scene. Easy-peasy!

The second part of the UI display for the collectible items requires the number of items currently collected. We'll do that now with the second design pattern we're covering in practice: Observer.

Incrementing collected item count

We've already seen how to access the GameManager Singleton instance via a static variable. We're going to do something similar with the CollectItem class to know when any item has been collected.

Typically, every instance of an object in the scene is independent of one another – each with its independent members. We want GameManager to register and listen (that is, observe) for any one of the items being collected, so we're going to accomplish this by invoking a static event – a static declaration will make it common across all class instances, so we won't need a reference to any individual item's instance.

Let's go ahead and add our static event to the CollectItem class, then register the listener from GameManager to take action when the event is invoked:

1. Add the following public static variable declaration to CollectItem:

```
public static event UnityAction OnItemCollected;
```

The event keyword is added to the declaration to enforce invocation only from within the declaring class – we don't want external classes to be able to trigger the event!

We're using UnityAction here as a method delegate, so we can use it to refer to calling multiple methods through it or even pass it as a method argument. It's also common practice to preface naming event delegates and methods with the word *On* (as in OnItemCollected).

> **Additional reading | Unity documentation**
>
> **UnityAction**: https://docs.unity3d.com/2022.3/Documentation/ScriptReference/Events.UnityAction.html

2. Add the following line to the top of the OnTriggerEnter2D() method to invoke or *trigger* the event (also known as **event invocation**) when the item is collected:

```
OnItemCollected?.Invoke();
```

The question mark and dot (?.) appended to OnItemCollected is a null-conditional operator – it will only perform the operation if the operand does not evaluate to null. In this case, it means if there are no registered listeners, we don't want to throw a NullReferenceException error when calling Invoke (executing the event).

> **NullReferenceException (C#)**
>
> A **NullReferenceException** (NRE) is an error that can occur when trying to access a member of a null object (not assigned).

3. Moving into the GameManager class now, we add a private variable to track the number of items collected by the player:

```
private int _collectedItemCount;
```

4. The class encapsulates this variable – because its accessor is set to `private` – so, let's add an `ItemCollected()` method that increments the count:

    ```
    private void ItemCollected() =>
        _collectedItemCount++;
    ```

    We're simply incrementing the count by 1 every time an item is collected by using the ++ arithmetic operator after the variable (postfix).

5. What is the `ItemCollected()` method called now? Well, that's where registering with the `CollectItem.OnItemCollected` event comes in. We'll use the `OnEnable()` and `OnDisable()` Unity message events – these are called when the behavior is first enabled and active, then becomes disabled, respectively:

    ```
    private void OnEnable()
    {
        CollectItem.OnItemCollected += ItemCollected;
    }

    private void OnDisable()
    {
        CollectItem.OnItemCollected -= ItemCollected;
    }
    ```

The += operator specifies adding an event handler method when subscribing to events. We're registering the `ItemCollected` method to be called when the `CollectItem.ItemCollected` event is invoked (that is, registering a listener).

> **Important note**
>
> To avoid resource leaks, unsubscribing from events should always be performed!

The -= operator specifies removing an event handler method to stop subscribing to events. We're unregistering the `ItemCollected` method so that it will no longer be called when the `CollectItem.ItemCollected` event is invoked. Suppose we don't unregister the event handler, and the object holding the handler method is destroyed. In that case, it will throw an **NRE** error (i.e., you can't call a method on an object that is `null`) – and we quite certainly won't be happy with that!

> **Tip**
>
> Don't forget, if the handler method isn't already created, you can have VS generate the method for you by clicking on the `ItemCollected` text, then using *Alt/Cmd + Enter* and selecting **Generate method ItemCollected**.

Now that we have our total and collected item values calculated, let's update the UI display.

Updating the UI

The text component displaying the collected item progress has already been added to the scene as **Collected Count Text**. Now, we need a way to reference and update the text displayed every time a **Water Diamond** is collected by the player.

We'll create a new `UIManager` class to keep UI functionality separate from `GameManager` – following the single responsibility principle.

Create a new script called `UIManager` and add the following code:

```
using UnityEngine;
using TMPro;
public class UIManager : MonoBehaviour
{
    public TextMeshProUGUI CollectedItemText;

    void Start()
    {
        CollectedItemText.text = string.Empty;
    }

    public void UpdateCollectedItemsText(
        int currentItemCount, int totalCollectableItems) =>
            CollectedItemText.text =
                $"{currentItemCount} of
                    {totalCollectableItems}";
}
```

Here's what we'll do when writing the new `UIManager` script.

1. The first thing we'll do is add a `using` statement for `TextMeshPro` (`TMPro`) so that we can reference the TMP text component's properties.

2. We'll then add a public variable for the **Collected Count Text** TMP object to be assigned to in the **Inspector** (for referencing).

3. When the game starts playing, we reset the `CollectedCountText` text to not display anything by using `string.Empty` (a zero-length string equivalent to `" "`) – this will clear out any temporary display text we entered at design time for adjusting position and display values.

4. Lastly, the `UpdateCollectedItemsText()` method is added that sets the `CollectedCountText` text. It has two method parameters, one for the current item count and the other for the total collectible items in the scene. We're again using string interpolation to combine the variables' values with *of* for display.

Let's return to the Unity Editor to complete the `UIManager` by adding it to the **GameManager** object as a new component. Then, drag in the reference for the **Collected Item** TMP text object to assign it to the `CollectedCountText` field, as seen in the **Inspector** in the following figure:

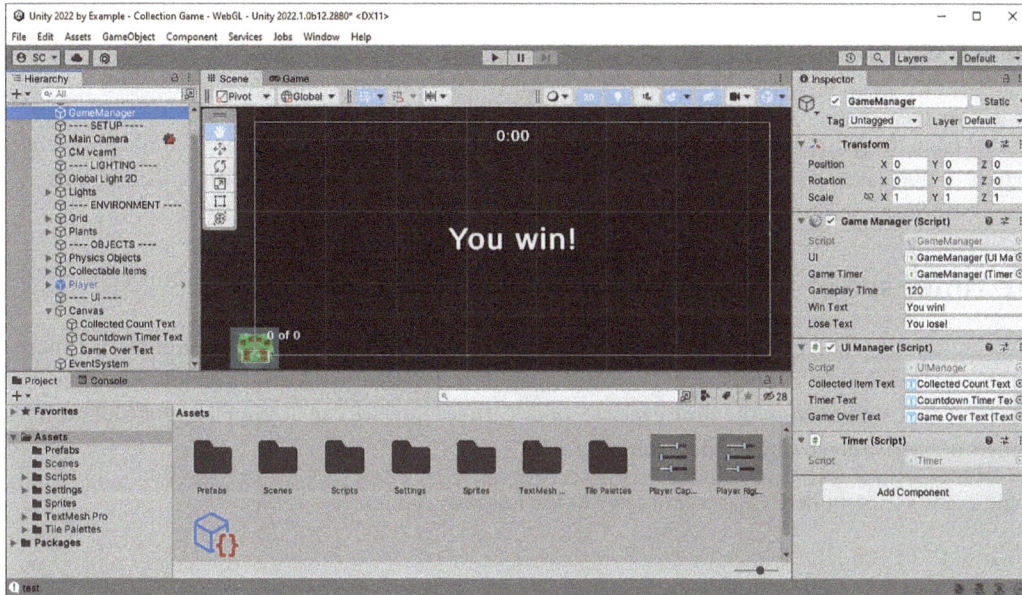

Figure 3.13 – The Manager Inspector assignments

The last thing we have to do to get our text displayed when collecting an object is to wire up the `UI.UpdateCollectedItemsText()` call in `GameManager.ItemCollected()`:

1. In `GameManger`, add a public variable to hold the reference to the `UIManager` component:

    ```
    public UIManager UI;
    ```

2. Change the `ItemCollected()` method to now call the `UI.UpdateCollectedItemsText()` method passing in an incremented _ `collectedItemCount` and the `_totalCollectibleItems` values:

    ```
    private void ItemCollected() =>
        UI.UpdateCollectedItemsText(
            ++_collectedItemCount,
              _totalCollectibleItems);
    ```

The `++` operator in front of `_collectedItemCount` means that we will pre-increment its operand (prefix) and then read the value. We want the already incremented value to be passed to the method.

-- (C#)

Note that the - - unary arithmetic operator is also available in C#, and instead of incrementing the operand by 1, it decrements it. The same rules apply to a pre- or post-decrementer assignment.

With that, the only remaining task is to drag UIManager to the UI field on the GameManager component to assign its reference (refer to *Figure 3.13*). Save your files and playtest!

We now have the game keeping track of the number of collected water diamonds and displaying progress to the player. It's not quite a game yet, so we'll add a simple countdown timer that we'll later tie in to win/lose conditions in the next section.

The Timer script

Referring to our GDD in *Table 3.1*, we know that the game will be lost if the countdown timer reaches 0 before the player collects all the water diamonds. For that, we'll need a simple timer script that subtracts 1 second from the time every second. Let's see how we can accomplish that and notify listeners of both time-changing and expiring events.

Create a new script named Timer and refer to the following code:

```
using UnityEngine;
using UnityEngine.Events;
public class Timer : MonoBehaviour
{
    public event UnityAction<int> OnTimeUpdate;
    public event UnityAction OnTimeExpired;

    private int _timeSeconds, _timeCurrent;

    public void StartTimer(int time)
    {
        _timeSeconds = time;
        _timeCurrent = 0;
        InvokeRepeating(nameof(UpdateTimer), 1f, 1f);
    }

    public void StopTimer() => CancelInvoke();
    private void UpdateTimer()
    {
        _timeCurrent++;
        OnTimeUpdate?.Invoke(_timeCurrent);

        if (_timeCurrent >= _timeSeconds)
```

```
        {
            StopTimer();
            OnTimeExpired?.Invoke();
        }
    }
}
```

Let's have a look at the code and review what's new:

1. The `OnTimeUpdate` event has a parameter added: `UnityAction<int>`. This allows passing an integer value to the listener when the event is invoked (you can add up to four types as parameters with `UnityAction`). We use the passed-in `int` value (seconds) for the displayed time remaining in the UI.

2. We've declared two `int` type variables in a single line, `_timeSeconds` and `_timeCurrent`, by separating the variable names with a comma.

3. `StartTimer()` is a public method that GameManager will call to start the timer countdown and begin the gameplay.

4. `InvokeRepeating()` is similar to `Invoke()` except, you guessed it, with a repeating time specified. We use this to update the timer value every second.

> **Additional reading | Unity documentation**
>
> **InvokeRepeating:** `https://docs.unity3d.com/2022.3/Documentation/ScriptReference/MonoBehaviour.InvokeRepeating.html`

`InvokeRepeating()` takes a string representation for the method's name as a parameter, just as `Invoke()` does, and any time you're typing a string literal into the code, it should raise a flag – hardcoded strings or *magic strings* are pure evil! So, instead of typing a string, we'll use the `nameof` expression to return the method name as a string. Sweet!

> **nameof (C#)**
>
> The `nameof` expression obtains a string constant from a variable, type, or member and is evaluated at compile time. This can be used instead of a string literal. It ensures no spelling errors and allows VS CodeLens to produce references.

5. The `StopTimer()` method calls `CancelInvoke()`, which will stop the timer counting down (by halting the `InvokeRepeating` process).

6. The `UpdateTimer()` method is responsible for incrementing the timer and invoking the `OnTimeUpdate` event, passing the `_timeCurrent` value.

7. Finally, we test for a timer expired condition using the >= comparison operator: is _timeCurrent greater than or equal to _timeSeconds? We use >= instead of == for working with an edge case where the time incremented may skip a second, and we'd miss an equal condition. Also, the timer will be stopped, and the OnTimeExpired event will be triggered.

Whew! Not a long script, but several vital parts make it all function to work well with our other components. Speaking of, we can now integrate the Timer code with GameManager and UIManager to start the running of the timer and update the time remaining text in the UI.

Additions to GameManager

With the Timer script all ready, let's connect the Timer script in the GameManager class to start running the timer and handling events for time updates and countdown expiration.

The first thing we'll tackle is starting the timer countdown when the game begins. The timer countdown starts by calling the Timer.StartTimer() public method through a direct reference assigned in the **Inspector**, which is initiated from the Start() method of the GameManager class.

Follow these steps to see how it's set up:

1. Add a public variable, GameTimer, to reference the Timer component added to the **GameManager** GameObject in the **Hierarchy**:

    ```
    public Timer GameTimer;
    ```

2. Add a public int variable, GameplayTime, to specify the amount of time in seconds the game will count down from – this is the total time the player will have to collect all the water diamonds. The best way to know what value to set is to playtest the game!

    ```
    public int GameplayTime;
    ```

3. The timer is started from the Start() method when the game is run. We call Invoke() on the StartTimer() method with a 2-second delay – giving the players time to orient themselves a bit before gameplay starts:

    ```
    private void Start() =>
        Invoke(nameof(StartTimer), 2f);
    ```

4. Add the StartTimer() method, which performs the actual call to the Timer's StartTimer() method, passing in the amount of time to count down from:

    ```
    private void StartTimer() =>
        GameTimer.StartTimer(GameplayTime);
    ```

5. Lastly, back in the **Editor**, add the Timer component to the **GameManager** GameObject in the **Hierarchy** and assign the reference to it in the **GameTimer** field (refer to *Figure 3.13*).

The preceding steps have integrated `Timer` with our `GameManager` and start the time counting down when the game begins. But we still have to handle the Timer's events, so let's first add the listener for the time being updated:

1. In the `OnEnable()` method, add the listener for the `OnTimeUpdate` event with the handler method specified:

    ```
    private void OnEnable()
    {
        ...
        GameTimer.OnTimeUpdate += TimeUpdated;
    }
    ```

2. Remember, when we subscribe to an event, we should also unsubscribe from it to prevent memory leaks, so add the following code to `OnDisable()`:

    ```
    private void OnDisable()
    {
        ...
        GameTimer.OnTimeUpdate -= TimeUpdated;
    }
    ```

3. Create the `TimeUpdated()` method (*Alt/Cmd + Enter*), where we'll execute the code for handling when the timer is updated with a new time:

    ```
    private void TimeUpdated(int seconds) =>
        UI.UpdateTimerText(seconds, GameplayTime);
    ```

Same as before, we're calling a public method on our UI reference for updating the countdown timer text while adding the values for the seconds passed in from `Timer` and the total gameplay time.

> **Note on code architecture**
>
> We could have added a listener to the `Timer.OnTimeUpdate` event right within `UIManager` to update the countdown timer text, but I decided to only have one reference to the `Timer` in the project and handle all timer events in `GameManager`. This will keep things cleaner, and it is more straightforward to troubleshoot any issues since the timer-related code is all within one class.

We'll add the `UpdateTimerText()` method in the next section, where we'll update `UIManager`.

Additions to UIManager

Let's get right to adding the additional variables and methods to support updating the timer text and the game over messages we'll finish in the next section:

1. Open up the `UIManager` script and make the following revisions:

    ```
    public TextMeshProUGUI TimerText;
    public TextMeshProUGUI GameOverText;

    void Start()
    {
        ...
        TimerText.text = string.Empty;
        GameOverText.text = string.Empty;
    }

    public void UpdateTimerText(int currentSeconds,
        int totalSeconds)
    {
        var ts = System.TimeSpan.FromSeconds(
            totalSeconds - currentSeconds);
        TimerText.text = $"{ts.Minutes}:{ts.Seconds:00}";
    }

    public void SetGameOverText(string text) =>
        GameOverText.text = text;
    ```

2. Add the two `TextMeshProUGUI` public fields for the text displayed in the UI for the timer and game over text.

3. In `Start()`, we'll clear out any placeholder text used during design time, so no text will be displayed until it's updated.

4. Add the `UpdateTimerText()` method with the current and total seconds parameters to calculate the time remaining in the countdown to display.

    Note that we're using `System.TimeSpan` to make things easy. We can then use the `ts` variable when displaying the text, formatting it to show minutes and seconds, where seconds will always be displayed with two digits (`00` being specified).

5. Finally, add the `SetGameOverText()` method for, you guessed it, updating the `GameOverText` text for displaying the game over message.

To finish up the revisions, we'll go back into the **Editor** and create two new **Text - TextMeshPro** UI widgets as children of **Canvas** in the **Scene Hierarchy** (refer to *Figure 3.13*):

1. To display the timer in the UI, name it *Countdown Timer Text* and anchor it to the top center of the screen. Size the text to something easily visible to the player.
2. To display the game over text in the UI, name it *Game Over Text* and anchor it to the center of the screen. Give the text a large size, covering most of the screen, so there is no mistaking when the game is over and the player has either won or lost.

When you've finished the UI design for these text widgets, drag them to the **TimerText** and **GameOverText** fields on the UIManager component.

A game just isn't a game unless you can win or lose playing it! So, let's finish up the collection game in the next section.

Winning the game

Everything we require to determine a win-and-lose condition is already in place. Let's start with the win condition: they win if the player collects all the water diamonds before the countdown timer expires.

This is a simple equality conditional statement: does the accumulated item count equal the total collectible items count? If we check this condition when an item is collected, and it equates to `true`, we know we've won the game.

In GameManager, revise the ItemCollected() method to include this check, and we'll call a Win() method if true.

You'll first have to refactor the expression body to a block body to include additional lines of code in the method (hint: VS can refactor this for you and add the squiggly brackets). It will now look as follows:

```
private void ItemCollected()
{
    UI.UpdateCollectedItemsText(
        ++_collectedItemCount, _totalCollectibleItems);

    if (_collectedItemCount == _totalCollectibleItems)
        Win();
}
```

That was easy! Now, let's state the losing condition: they lose if the countdown timer runs out before the player collects all the water diamonds. This requires just a bit more work, but we're just going to use what we've already implemented, specifically, the Timer.OnTimeExpired event.

In `GameManager`, again, let's add a listener to `OnEnable()` for the time expiring and assign a handler method:

```
private void OnEnable()
{
    ...
    GameTimer.OnTimeExpired += TimeExpired;
}
```

Again, unsubscribe:

```
private void OnDisable()
{
    ...
    GameTimer.OnTimeExpired -= TimeExpired;
}
```

Create the `TimeExpired()` method as an expression body where we'll just call `Lose()` – the time has expired; if the player had already collected all the items – and won – this method would never be reached:

```
private void TimeExpired() => Lose();
```

Finally, as the very last steps for completing the collection game, referring to the additions to the following `GameManager` script, we'll add both the `WinText` and `LoseText` public fields that hold the messages to be displayed when the game ends (as seen in *Figure 3.13*):

```
public string WinText = "You win!";
public string LoseText = "You lose!";
```

If you don't assign some default text values to the variable declarations here, then don't forget to assign them in the **Inspector**!

We'll also add the `Win()` and `Lose()` methods:

```
private void Win()
{
    GameTimer.StopTimer();
    UI.SetGameOverText(WinText);
    Time.timeScale = 0f;
}

private void Lose()
{
    UI.SetGameOverText(LoseText);
    Time.timeScale = 0f;
}
```

A quick breakdown of these two new methods is as follows:

1. The `Win()` method starts off by stopping the timer by calling `GameTimer.StopTimer()`. This will cancel invoking the update timer method since our game is over now.

2. Both of these methods call `UI.SetGameOverText()` but with the corresponding argument to display the correct message, `WinText` and `LoseText` respectively.

3. They each set the `timeScale` variable of the running game `Time` to 0, which has the effect of pausing the game (with some exceptions; see the following callout block on `Time.timeScale` for more information).

   Note that setting `timeScale` to zero will technically have the same result as calling `GameTimer.StopTimer()` here, since the only option at this point, is to restart the game. Still, I like to be explicit with some statements so that the intent is clearly visible in the code (so long as it's not something that would otherwise negatively impact things, of course, heh).

Additional reading | Unity documentation

`Time.timeScale`: `https://docs.unity3d.com/2022.3/Documentation/ScriptReference/Time-timeScale.html`

Since we broke things up into smaller chunks of code quite a bit in this section, you may want to refer to the completed `GameManager` script, which can be found in the GitHub repo here: `https://github.com/PacktPublishing/Unity-2022-by-Example/tree/main/ch3`

With that, our game is complete. Yay! However, one more crucial step is needed before we can share our game with the world… and that is making sure others find the game challenging and fun to play.

Playtesting

Now is the time to playtest and change the parameters of your level design, change the paths where the player can go, move objects and hazards, move the collectibles, and tweak the gameplay time to make the game challenging and fun for players.

Have other people play your game – watch them play, if you can, to get an idea of how they play your game and get their feedback. Implement and iterate on changes in response to playtesting to improve your game!

Downloading the completed game code

Remember, the example code for this book can be downloaded from the GitHub repo here: `https://github.com/PacktPublishing/Unity-2022-by-Example`

Summary

This chapter was an introduction to the power of the CM camera system for quickly adding a Player follow camera, implementing game mechanics for collecting items that engage the player, displaying key game progression elements to the player with Unity UI, and how to finish and balance our collection game with both win and lose conditions.

In the next chapter, we'll introduce the next project in the book by defining the GDD for the *Outer World* 2D adventure game, learning how to import original artwork to use with additional Unity level building tooling, and exploring more aspects of level design and game polish.

Part 3:
2D Game Design Continued

In this part, you will be introduced to a different perspective for a 2D game – the side-on view – for an adventure game in which the player is challenged to navigate an interactive environment while trying to find pieces of a "puzzle key" that will allow you to gain access to a habitat station. You will gain the knowledge to set up a player character with different animation states driven by the new Input System. You'll also learn to define gameplay mechanics relating to the environment and level design, as well as code player interactions.

Shooting, included as a player verb specified in the GDD, is implemented with the new Unity Pool API, which provides an efficient and native code approach to reducing **garbage collection** (**GC**). Introducing a data-driven and extensible approach to game architecture using ScriptableObjects will give you the knowledge to create flexible and customizable game systems.

This coding-heavy section will have you implementing the systems for the game's core loop, including a spawning system, quest system, and event system. The event system will tie code together loosely, following good programming practices (SOLID principles). You will complete a quest mission and trigger the final events to produce a win condition.

This part includes the following chapters:

- *Chapter 4, Creating a 2D Adventure Game*
- *Chapter 5, Continuing the Adventure Game*
- *Chapter 6, Introduction to Object Pooling in Unity 2022*
- *Chapter 7, Polishing the Player's Actions and Enemy Behavior*
- *Chapter 8, Extending the Adventure Game*
- *Chapter 9, Completing the Adventure Game*

4

Creating a 2D Adventure Game

In *Chapter 3*, you were introduced to Cinemachine for creating a quick but powerful camera-follow system. We learned how to implement a game mechanic for collecting items and displaying game progression with Unity UI, and how to approach game balance while considering win-and-lose conditions.

By completing the collection game, the previous chapters provided foundational knowledge that we'll continue to build upon in this chapter as we start to make a 2D adventure game. We'll explore importing artwork and creating assets to build out a 2D side-on environment and level design using additional Unity 2D tooling, namely **Sprite Shape**.

We'll finish the chapter by adding dynamic moving platforms and triggers that will provide secondary actions to create a more engaging player experience, and some optimization techniques to keep the game performant and polished for increased immersion.

In this chapter, we're going to cover the following main topics.

- Extending the **Game Design Document (GDD)** – Introducing the 2D adventure game
- Importing assets to use with Sprite Shape – A different kind of 2D environment builder
- Level and environment design – Guiding the player
- Moving platforms and triggers – Creating a dynamic interactable environment
- Adding polish to our environment to immerse the player and optimizing

By the end of this chapter, you'll have another example of a GDD to use for your projects and be able to import and use original artwork to create an environment that guides the player while designing a level using Sprite Shape. You'll also be able to create an interactive and dynamic moving environment that is optimized and visually polished.

Technical requirements

To follow along in this chapter with the same artwork created for the project in the book, download the assets from the GitHub link provided in this section. To follow along with your own artwork, you'll need to create similar artwork using Adobe Photoshop, or a graphics program that can export layered Photoshop PSD/PSB files (for example, Gimp, MediBang Paint, or Krita).

Additionally, to follow along with the player input section, you'll want a compatible game controller for your system (although this is optional since keyboard input will also be provided).

You can download the complete project on GitHub at `https://github.com/PacktPublishing/Unity-2022-by-Example`.

Extending the GDD – Introducing the 2D adventure game

We're continuing with the next game concept now – while continuing with the overall theme for the projects in the book – so let's go ahead and update the GDD and extend upon it where needed.

Let's first update the following overview sections that we've previously covered; this will also serve as your introduction to the game:

Name of game	Outer World
Describe the gameplay, the core loop, and progression.	Find your way to the habitat station while searching for parts of the key required to gain access to the entryway and neutralize infected robots that attempt to halt your mission.
What is the core game mechanic for the adventure game?	The player will repeatedly engage in battle with infected robots that impede progress to the habitat station.
What is the secondary game mechanic for the adventure game?	The player will search the environment for hidden parts of a key. The pieces will need to be combined correctly as input to gain access to the entryway of the habitat station.
What systems need to be implemented to support the game mechanics?	The player movement, equipping a weapon with ammo reloading and shooting capabilities, a pickup with inventory, puzzle solver, health, and damage

Table 4.1 – The GDD for the adventure game

We've added a new section to the end of the preceding table that takes the game mechanic details a step further into the *development requirements* realm. With some systems defined, we can start thinking about the code architecture and what *problems need solving* – this is where we can apply some design patterns.

Now, instead of a nerf, this time, let's add a buff for the player:

What is a buff mechanic for the player in the adventure game?	The player will be able to collect energy shards (*water diamonds*) scattered throughout the environment, which, when a certain quantity has been collected, will give a power-up state to all of the weapons (increasing damage dealt).

Table 4.2 – Adding a buff to the GDD

Our player character and the enemies in the game will be much further developed compared to the collection game. Let's add sections in the GDD for these characters' bios and details about the player's challenge structure:

Main character: **Describe the main character of the game and how they drive the story. Who is this player character?**	**Type**: Kryk'zylx race of humanoids. **Backstory**: The people of Kryk'zylx have outgrown their home planet and are searching the galaxy for suitable planets to colonize. Scouts are sent to establish habitat stations on planet surfaces with the potential to sustain life. **Goals**: Establish and maintain a habitat station with an automated crew of construction and maintenance robots. **Skills**: Power suit jumping and charging. **Weaknesses**: Atmosphere not breathable.

Kryk'zylx scouts must wear power suits to survive the planet's hostile atmosphere outside of the habitat station. Here is an example of a power suit helmet:

Figure 4.1 – Power suit helmet

And now, we'll continue by adding the sections for the details of the player's challenge structure:

What is the main character's challenge structure?	Navigate platforms, make their way past infected robots, and solve the critical puzzle.
Enemy A: **Describe the first enemy in the game and how they drive the story. Who is this enemy?**	**Type**: Construction Robot, Biped **Backstory**: Robot deployed on pre-colonization missions for habitat station construction and maintenance. **Goals**: Construction and maintenance. **Skills**: High mobility, including onrough terrain. **Weaknesses**: Long charging.
Enemy B: **Describe the second enemy in the game and how they drive the story. Who is this enemy?**	**Type**: Maintenance Robot, Wheeled **Backstory**: Robot deployed on pre-colonization missions for habitat maintenance and support. **Goals**: Maintenance and personnel support. **Skills**: Quick charging. **Weaknesses**: Limited mobility.

Table 4.3 – Adding character and enemy bios

We will take the document further and introduce a section that gives the environment some attention. The description here will help maintain the game's visual direction while the environment and level are being designed and the art assets are created:

Describe the environment in which the game takes place. What does it look like, who inhabits it, and what are the points of interest?	The game takes place on the surface of a prospective planet to colonize, even though this particular planet does not have a breathable atmosphere. The planet comprises areas of purple-red rock and thick vegetation (which moves in such a way as to suggest it may have the capacity to think).
Describe the game level(s).	The game level is a combination of static and moving platforms with obstacles needing to be overcome or avoided by the player as they make their way to the habitat station.

Table 4.4 – Environment and level

Let's add a section on the input controls. Previously, we used the keyboard and mouse for the collection game, but this time, we'll also be adding support for input from a game controller:

Define the input/control methods actions.	**Keyboard**: *W*, *A*, *S*, and *D* keys to move; the space bar to jump; the mouse to aim, the left mouse button to shoot the primary weapon, and the right mouse button hold/release to launch the secondary weapon while aiming with the mouse (the left mouse button to cancel); and the *E* key to interact with things. **Game Controller**: The left stick/D pad to move; *X* to jump; the right stick to aim and the right trigger or *Y* to shoot; the right shoulder hold/release to launch the secondary weapon while aiming with the right stick (the right trigger to cancel); and button *A* to interact.

Table 4.5 – Input/control methods actions

Finally, let's define how all the pieces interact to make a complete game experience for the player:

How do all the pieces interact?	The player interacts with the environment through exploration, discovering what is required to reach and enter the habitat station using parts of a puzzle key found throughout the level while fending off robots that have become infected by the strange plants that cover the planet's surface.

Table 4.6 – Putting all the pieces together

> **Full GDD for the 2D adventure game**
>
> To view the full GDD document for the 2D adventure game, visit the project GitHub repo here: `https://github.com/PacktPublishing/Unity-2022-by-Example/tree/main/ch4/GDD`

Refer to the GDD any time you are unsure about what comes next. So long as you're following what you've written, the next steps should come as a natural progression you can iterate upon. However, don't feel like you're locked into what you've written as your first draft of the GDD – as a living document, let ideas organically change as you work through them, and new ideas come to light (just as I have done throughout writing this book!).

In this section, you learned how to extend upon our GDD to include additional details for the main characters in the game and describe the game world and how everything works together to create an immersive experience for the player. We'll continue importing original artwork to start building the game level with Sprite Shape in the next section.

Importing assets to use with Sprite Shape – A different kind of 2D environment builder

The artwork that we'll be using for the 2D adventure game is an original artwork and the result of a collaboration with the artist, Nica Monami. The art assets have all been created specifically for this project in the book. Nica has lent her talent for creating fantastic painterly-looking art to create a unique environment for the game, and I am very excited to be working with these assets.

Figure 4.2 – Original game artwork

Adventure game 2D art assets

To follow along in this chapter, download the art assets from the project GitHub here: `https://github.com/PacktPublishing/Unity-2022-by-Example/tree/main/ch4/Art-Assets`

Nica Monami has permitted the use of the provided game art for learning purposes only; commercial use is strictly prohibited. Nica's portfolio can be viewed on ArtStation at `https://www.artstation.com/dnanica213`

In addition to Sprite Shape – the 2D feature we'll be using to build out most of the adventure game's level – we'll cover most of the tooling provided by the **2D Animation package**. We'll first cover importing and performing any required prep for the artwork to get started.

Importing and preparing the artwork

Let's start by creating a new Unity project for the adventure game, again using the 2D **Universal Render Pipeline** (**URP**) Core template. We'll continue exploring different gameplay styles by making the 2D adventure game a side-on orthographic game view (similar in style to many other Mario Bros or Metroid-inspired platformer games).

Once the project opens, import the artwork under a new `Assets/Sprites` folder.

> **Dutiful organizational reminder**
>
> Maintaining an adequately named folder structure can help keep things sorted and easy to find and work with later.

Viewing all of the imported Sprite assets – in the event you did lose track of something or work with many assets at once – is easily accomplished with the **Search by Type** option in the project window's **Search** tool, as seen in the following figure:

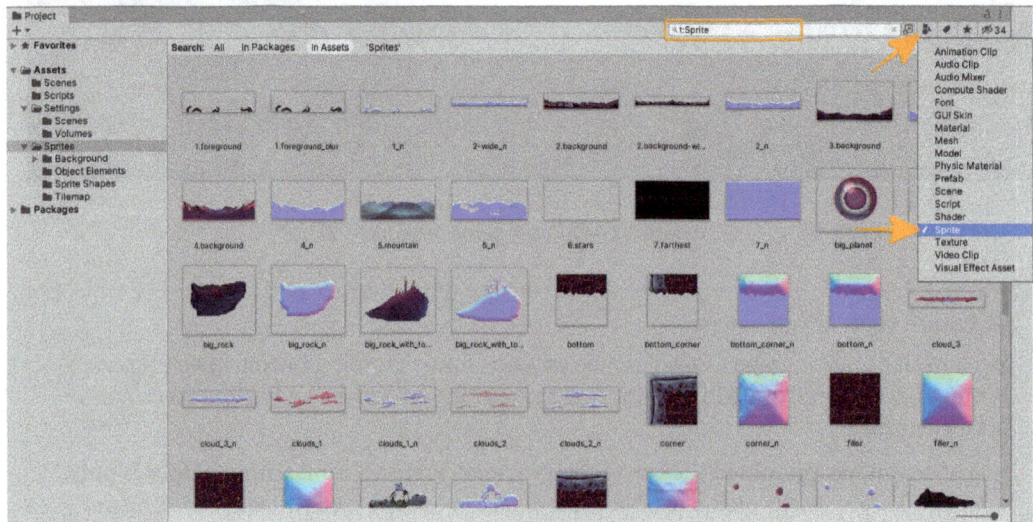

Figure 4.3 – Filtering on Sprite in the project window

The image assets imported will serve different purposes and use additional Sprite tooling. Here is a quick overview of how we'll be working with the imported assets, by folder, throughout the remainder of this chapter:

- `Assets/Sprites/Sprite Shapes`: Images in this folder will be used with Unity's 2D **Sprite Shape** tool. Sprite Shape is a spline-based tool that provides the ability to make open-ended paths or enclosed shapes that can be used as parts of the level that the player character can walk on, adding background elements or quickly decorating the environment.

> **Additional reading | Unity documentation**
>
> *2D Sprite Shape:* `https://docs.unity3d.com/Packages/com.unity.2d.spriteshape%409.0/manual/index.html`

- `Assets/Sprites/Sprite Skins`: Images in this folder will be used with Unity's 2D **Sprite Skin** tool. We'll use the **Skinning Editor** feature within the **Sprite Editor** window to create geometry for which we'll add weighted bones so that we can animate the sprite within the scene View using the `SpriteSkin` component.

> **Additional reading | Unity documentation**
>
> *Sprite Skin*: `https://docs.unity3d.com/Packages/com.unity.2d.animation%409.0/manual/SpriteSkin.html`

- `Assets/Sprites/Tilemap`: Images in this folder will be used to create a **Tilemap**; we are already familiar with this 2D feature from building the collection game in *Chapter 1*.

- `Assets/Sprites/Background`: Images in this folder will be used to create our game environment's layered background. We won't use any particular Unity feature this time, but we will use a script to add parallax movement.

> **Background parallax**
>
> Parallax is a technique applied to background images that moves images further away from the camera across the frame more slowly than foreground images, creating a sense of depth in 2D scenes. This technique was popularized in 2D video games starting in the early 1980s (although it was limited in the number of background planes due to hardware limitations of the time – thankfully, we're not stuck with the limitations of those that came before us because we have no such limitations today).

- `Assets/Sprites/Background/Clouds`: Images in this folder will be used to create clouds that continuously scroll and loop across the sky in the background of the game environment, furthering the sense of depth and immersion. We'll use Unity's **Spline** package and the corresponding **Spline Animate** component to make this quick and easy work (similar to the moving platforms mentioned earlier).

- `Assets/Sprites/Object Elements`: Images in this folder will be used to detail the environment and provide some possible special areas of interest. There is nothing special here, just the Sprite Renderer and 2D colliders to facilitate interactions with the player character.

With the fundamental review of the imported assets by folder finished, let's do the prep work for the sprites that require it before usage.

Preparing the artwork for Sprite Shape

The Sprite Shape feature creates dynamic and *flexible* shapes that can either be open paths or closed and filled areas. The sprite that is assigned for the Sprite Shape's spline path is tiled and deformed along the defined outline. Because of the tiling, we need to take some special steps to prepare the sprite used here by adding a border at the locations on the left and right edges where it will tile seamlessly.

Follow these steps while referring to *Figure 4.4* to ensure that the sprite will be compatible and work well with Sprite Shape:

1. Select the sprite in the project folder.

2. In **Import Settings (Inspector)**, click the **Sprite Editor** button.

3. Click on the image and drag in both the left and right sides by using the green boxes or typing in values into the **L** and **R** fields of **Border** in the **Sprite** dialog – for the tileable position in the sprite. Click **Apply** when finished.

4. Set the **Mesh Type** to **Full Rect** (this is required for use with Sprite Shape and will produce a warning on the Sprite Shape asset if not set accordingly).

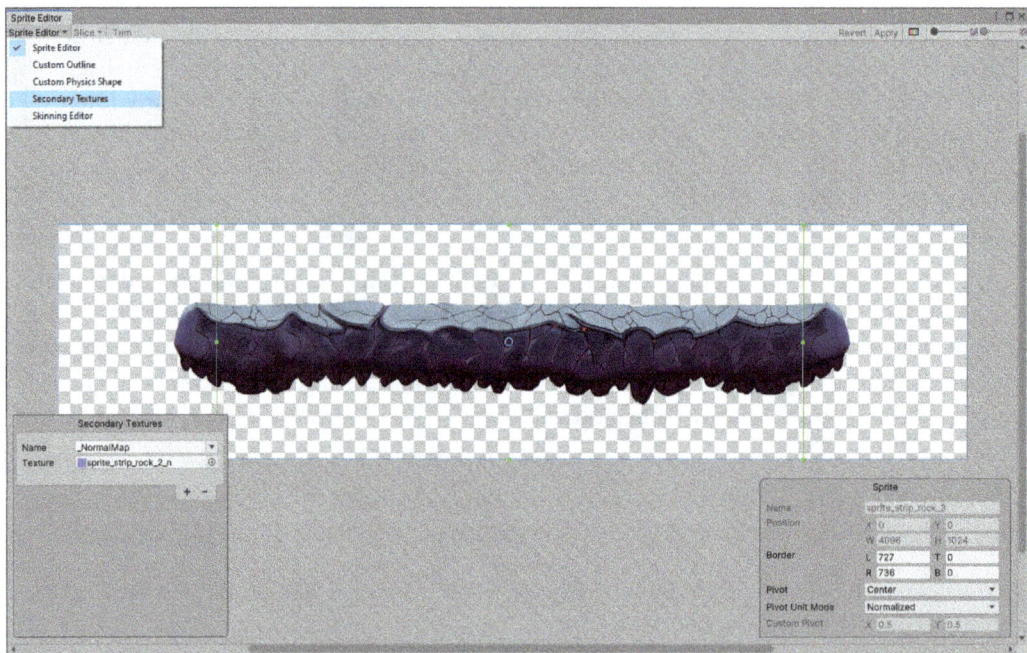

Figure 4.4 – Sprite Editor for Sprite Shape art

> **Package Manager samples**
>
> Most packages in the Unity Package Manager have additional content as additional installations on their respective Package Manager pages. In Sprite Shape's case, samples and extras that demonstrate many of its capabilities can be imported. I encourage you to check out the package samples for additional learning.

Preparing artwork for Sprite Skin

The Sprite Skin feature is commonly used to create rigged limbed characters and is imported from a layered Photoshop PSB/PSD file (using the **PSD Importer** package). We'll use this 2D animation package feature in *Chapter 5*, when we create the adventure game's player character. Still, for now, in this chapter, we're just going to add some movement to some environmental elements.

We'll be animating individual PNG images imported into the `Sprites/Sprite Skins` folder. Aside from keeping these as separate images (for the sake of simplicity), there is nothing extra we need to do with the import settings.

Preparing artwork for Tilemap

We used Tilemap extensively in *Chapter 2*; you should be a pro with it by now! Let's quickly recap the image preparation process by creating a new Tile Palette for the `Tilemap-01` sprite sheet image imported to `Sprites/Tilemap`, as seen in the following figure:

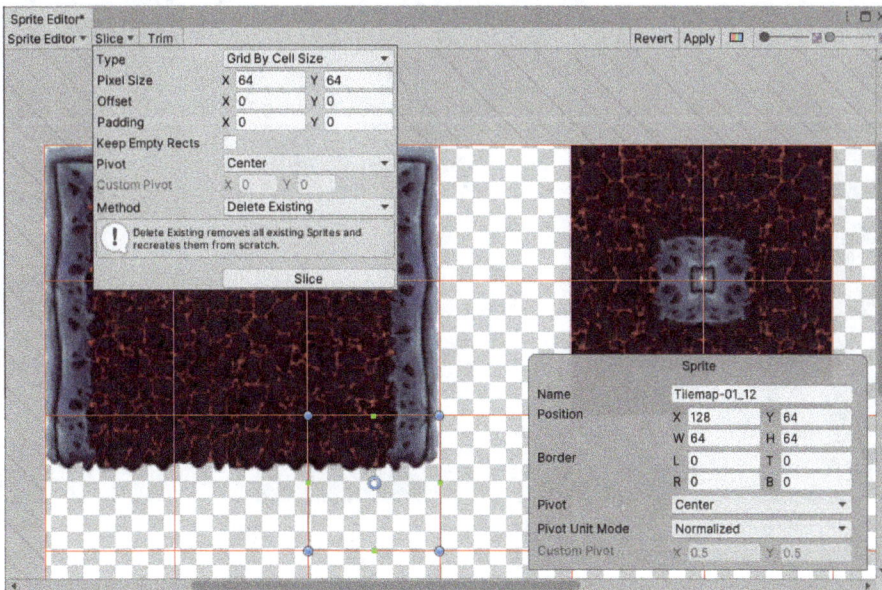

Figure 4.5 – Slicing the Sprite Sheet for Tilemap tiles

Use the following steps to prepare the artwork for use with Tilemap:

1. Select the sprite sheet image in the project folder.

2. In **Import Settings (Inspector)**, set **Sprite Mode** to **Multiple**.

3. Apply the changes and click the **Sprite Editor** button.

4. Select the **Slice** dropdown menu and then select **Grid By Cell Size** for **Type** of slicing.

5. The supplied Tilemap image uses a 64x64 pixel size grid (same as the collection game), so verify the correct size and click **Slice**.

6. Click **Apply**, or close **Sprite Editor** and save, and that's it for the Tilemap prep work!

Normal maps

A **Normal map** is a special type of image where encoding in the RGB channels represents the direction a pixel faces. It's easy to add fake volume and details to achieve a simulated 3D effect using a sprite's pixels and 2D lights with URP.

You'll need to use special third-party software to create normal maps outside of Unity, as the following figure shows for a rock that we'll use in the environment:

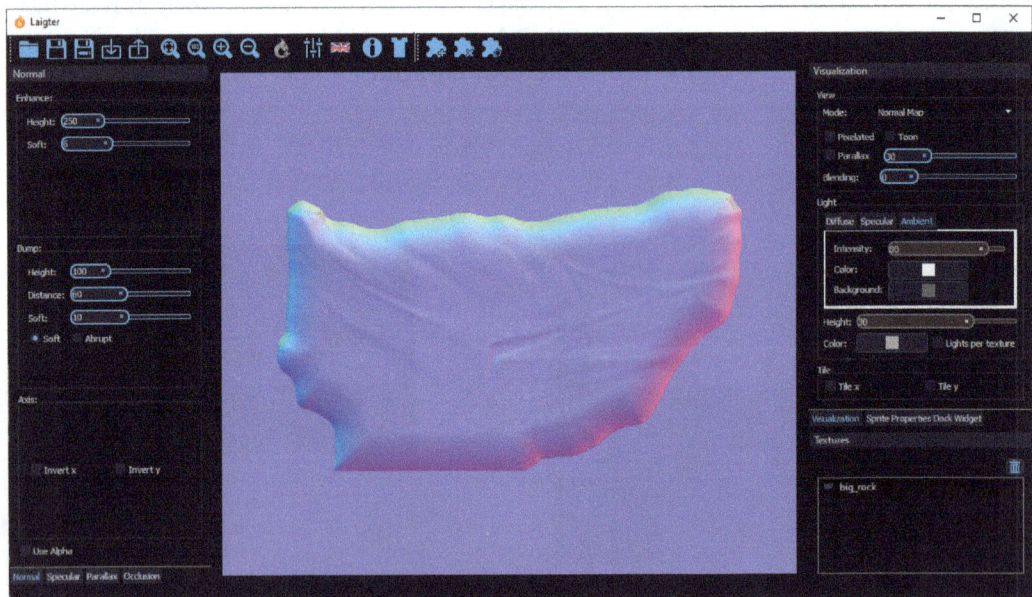

Figure 4.6 – Normal map image of a rock in Laigter

> **Normal maps generation software**
>
> The program used to create all of the 2D sprite normal maps for the 2D adventure game's artwork is Laigter (available for free at `https://azagaya.itch.io/laigter`). Other programs that can export 2D normal maps include Photoshop, SpriteIlluminator, Sprite Dlight, and Sprite Lamp.

Note that when importing 2D sprite normal maps, **Texture Type** for the image import settings will be **Sprite (2D and UI,)** and not **Normal Map** – because sprites that will be affected by 2D lights need to have the normal map texture assigned in the Sprite Editor (2D lights will also need to have normal maps enabled for affecting the sprites).

All of the images provided in the example 2D adventure game project have accompanying normal maps. For your own images, go ahead and generate the normal maps and then assign them to the sprites by following these steps:

1. Select the sprite in the project folder.
2. Click the **Sprite Editor** button in **Import Settings (Inspector)**.
3. From the **Sprite Editor** dropdown menu, select **Secondary Texture**.
4. In the **Secondary Texture** dialog that opens, select **_NormalMap** from the **Name** dropdown menu.
5. Assign the sprite's normal map image by dragging it into the **Texture** field from the project window.

Refer to the preceding *Figure 4.4* for an example of the **Secondary Texture** dialog.

With that, we've successfully prepared the project's imported art assets to be used with the 2D tooling used in the coming sections. In the next section, we'll take another look at level design before putting the art to use in creating different types of platforms.

Level and environment design – Guiding the player

Let's take a quick break from the technical aspects of the project and talk about some game design concepts again before moving on to the features we'll use to build the game level.

Having the GDD is great, but it doesn't provide any concrete visuals for communicating the theme and style of the game. Art can quickly elicit emotional responses and build excitement that is hard to compare to the written words of a GDD. For example, have a look at this concept art created for the *Outer World* game:

Figure 4.7 – Original "Outer World" environment concept art

You should immediately have a sense of what it would be like to be a player in this environment! Games are often marketed early on with an artist's rendering for visualizing concepts to gain interest and excitement for a project. They are also used internally to inspire the production team to build the product.

With excitement buzzing, let's introduce a new game design principle to guide players and get them to explore this unique environment.

Signposting

We've discussed guiding the player in *Chapter 2*, by introducing shapes to nudge the player in the desired direction. The game design principle we'll present here is focal points in the environment, which is called **signposting**. Signposting helps the player know what they should do or tells them their destination.

We want the player to have a goal in mind as they are playing – the player should hardly ever experience the feeling of being lost in the game (only when they are spending a long time not progressing, point them in the right direction to avoid player frustration or worse, leaving the game). This type of signposting is also referred to as a **journey**.

The journey we're setting our player on in the adventure game can be seen in the following figure, where a habitat station is visible in the distance in the background:

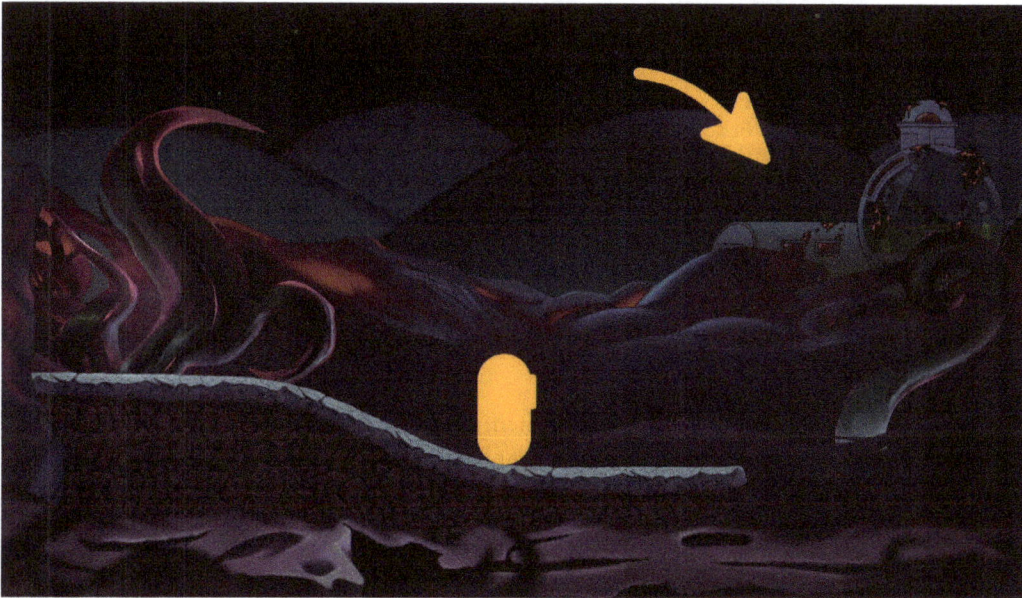

Figure 4.8 – Signposting the habitat station for the player

The player has a goal now. Always consider designing the level with a purpose… what do we want the player to try to do, what should the player accomplish, or where should the player get to?

Keep these things in mind when designing your level using the tools we'll introduce in the following sections, starting with adding the platforms that the player will walk on.

Creating platforms

You can't have a platformer game without platforms! Unity's 2D tooling provides different tools for building platforms for 2D games. We've already used Tilemap to create a rigid and grid-based level design for the collection game. While we'll still use Tilemap for sections of the adventure game level, we'll start by creating platforms without the same constraints using Sprite Shape.

Creating a closed Sprite Shape profile

We can use two types of shapes when creating with Sprite Shape: open and closed shapes. An open shape provides a sprite outline of the spline path, whereas a closed shape includes a fill texture for creating an enclosed shape where both sides and a bottom sprite can also be defined.

Let's start the level design with a closed Sprite Shape platform where the player will spawn. To do so, we'll start by creating the Sprite Shape profile asset in the project before adding the platform to the scene by following these steps:

1. In the project window, within the `Assets/Sprites/Sprite Shapes` folder, select **Create** | **2D** | **Sprite Shape Profile** and name it *Platform Closed 1*.

2. Assign the sprite for the top edge of the platform by, first, clicking on the blue border of the circle that defines **Angle Ranges**.

3. Adjust the **Start** and **End** angles by either clicking and dragging on the small handles from the bottom of the circle or typing values into the fields provided just below the circle. We'll use a **Start** value of 45 and an **End** value of -45.

4. With the region for the top edge now defined, assign the `sprite_strip_rock` sprite by dragging it from the project window and into the **Sprites** list field (replacing the `SpriteShapeEdge` placeholder).

5. Do the same for both the left and right sides of the Sprite Shape, using 45 and 135 for the angles, respectively, and `sprite_side_rock` as the sprite, as seen in the following figure:

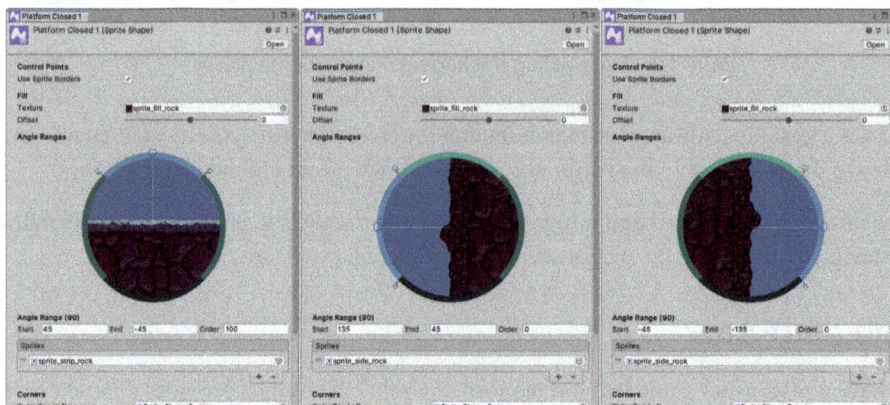

Figure 4.9 – Closed Sprite Shape asset properties (top, left, and right)

6. The last step for creating a closed Sprite Shape is to assign the fill texture, so assign the `sprite_fill_rock` sprite to the **Fill** | **Texture** field (replacing **Sprite Shape** fill placeholder).

    If the size of the sprites assigned is not to your liking, you can revisit the sprite import settings and change the **Pixels Per Unit** value to something that looks better (this can be adjusted at any time). The images being used for Sprite Shape are high resolution, so you have the flexibility to change the size without a loss in display quality.

7. Similarly, we can adjust the scale of the fill texture, but we'll do that on the `SpriteShapeController` component when added to the scene – use the **Fill** | **Pixels Per Unit** field to adjust the fill texture scale.

8. With the profile asset created, we can now make the first platform for our player by clicking **GameObject | 2D Object | Sprite Shape | Closed Shape**.

9. Assign the Sprite Shape asset **Platform Closed 1** – that we just created – to the **Profile** field of the `SpriteShapeController` component. Adjust the spline shape by manipulating and adding or removing knots, as seen in the following figure:

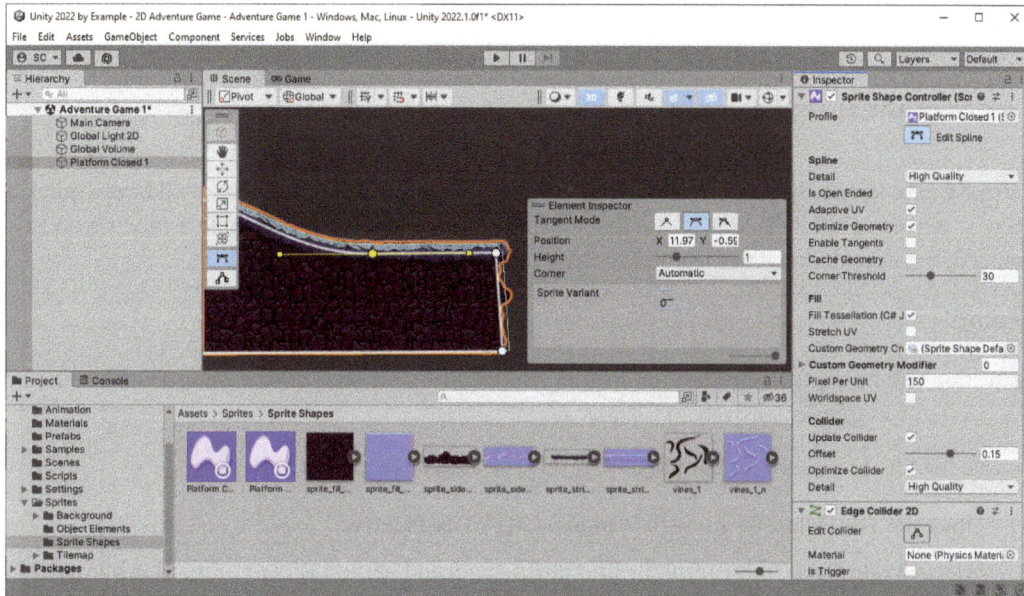

Figure 4.10 – Closed Sprite Shape platform in the scene

The last step required to make this closed Sprite Shape a walkable platform for the player character is to add a collider. For Sprite Shape platforms, we use the **Edge Collider 2D**. Add it to the platform object by using the **Add Component** button at the bottom of the **Inspector** window and searching `edge`.

10. Once added, select **Sprite Shape Controller | Collider | Updated Collider** to enable it. Then, adjust the **Offset** value to line up with the top edge sprite – this can also be adjusted later when the player character is added to the scene.

> **Optimization note**
>
> By default, the Sprite Shape API allows you to change the spline nodes at runtime. If you don't require runtime changes, you can bake or cache the geometry of the spline for a performance boost. So, if you don't need to modify the spline at runtime, select **Sprite Shape Controller**, enable **Edit Spline**, and click *Cache Geometry* to bake the mesh. Otherwise, install the Burst package (version 1.3 or later) from the Package Manager to improve performance when modifying the spline at runtime.

Another thing to note is that Sprite Shape's Renderer works like any other Sprite Renderer component and utilizes **Sorting Layers**. We'll be using **Sorting Layers** throughout the project to properly layer the different artwork that will make up the level's platforms and overall environment. We'll keep all the level's platforms on the default layer, so we have nothing to change.

Bonus activity

Sprite Shape can also be used to distribute a variety of sprites along an updateable spline path. Explore Sprite Shape's features to discover how you can decorate the platforms with the *vine* and *toxins* assets provided.

Now that you've learned how to create flexible shaped platforms using Sprite Shape, we'll have a quick review of setting up a Tile Palette to use for drawing platforms with Tilemap before moving on to creating dynamic and interactive features in the level.

Tilemap

This section will be a quick refresher for creating platforms in our level design using Tilemap since we've covered this topic extensively in *Chapter 2*.

To create the tiles to be used for painting using the Tile Palette, follow these steps:

1. Open the Tile Palette by going to **Window | 2D | Tile Palette**.

2. Select **Create New Palette** from the dropdown menu and name it `Environment` while using the default properties for a Rectangle Grid.

3. Click the **Create** button and save it to `Assets/Sprites/Tilemap/Tile Palettes`.

4. From the `Asset/Sprites/Tilemap` folder, click and drag the `Tilemap-01` image into the **Tile Palette** window and save the tiles to the `Assets/Sprites/Tilemap/Tiles` folder when prompted.

Now that the tiles we'll use for the Tilemap portion of our level are created, let's go ahead and make a platform to the right of the Sprite Shape platform using these steps:

1. In the **Hierarchy** window, create a new Tilemap by using the **Create** or **GameObject** menu and selecting **2D Object | Tilemap | Rectangular**.

2. We'll use the same **Default** Sorting Layer on **Tilemap Renderer** as we did for the Sprite Shape Renderer.

3. Draw the platform using the brush and flood fill tools while using the correct tiles for the top, sides, bottom, and corners, as shown in the following figure:

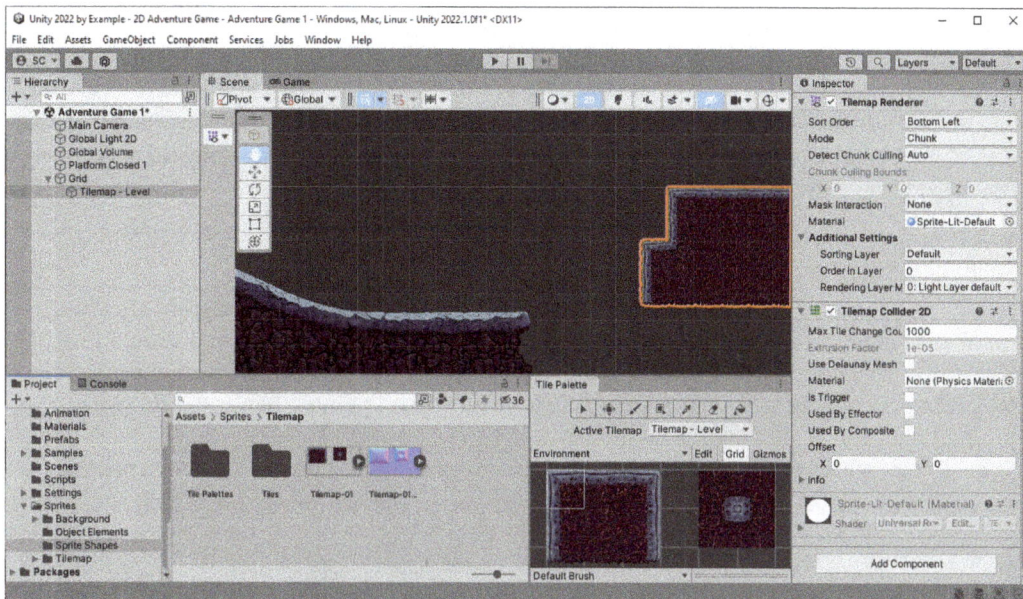

Figure 4.11 – Tilemap platform added to the scene

4. As a last step, add a **Tilemap Collider 2D** component to the Tilemap object, so it's walkable by the player character.

We now have two different styles of platforms we can use to create visually distinct areas in the level. In the next section, we'll take platforms to the next level by introducing movement and triggering interactions in the environment with a C# script.

Moving platforms and triggers – Creating a dynamic interactable environment

Adding moving platforms to our game's level adds to the visual interest and provides additional challenges for the player. Unity again provides tooling that makes creating moving platforms a simple and straightforward task right in the scene View and without needing to write any code.

We previously used Sprite Shape to make a closed platform, but Sprite Shape also allows the creation of open shapes, and that will be perfect here for making a small platform that we can move.

Moving a Sprite Shape platform with Splines

Introduced for the Unity 2022 tech stream, we have a new 2D package called **Splines**. We'll be using Splines to create the paths our platforms will move along – this is where the `Spline` component is used. And the `Sprite Animate` component is used for moving a platform along the spline path and doing so without using any code. Nice!

Splines

Smooth shapes can be created with splines by having a generated line pass through any arbitrary number of control points. Different approaches for interpolating the line and providing adjustments between hard edges and rounded corners are available, such as Catmull-Rom (helpful in calculating a curve that passes through all control points), Bézier curves (which provide handles to adjust the tangents of the line in relation to points), and B-Splines (similar to Catmull-Rom splines but the generated line does not necessarily pass through the control points).

The Unity 2022 `Splines` package specifically supports Linear, Catmull-Rom, and Bézier types.

To use Splines in our project, we'll need to verify that the package is installed. Open **Package Manager** (**Window | Package Manager**), select **Unity Registry** in the **Packages** dropdown menu, then either search for `splines` or find **Splines** in the **Packages** list to select and install.

The first step of creating a moving platform is creating the platform.

Creating an open Sprite Shape profile

We already had an introduction to Sprite Shape when we created the closed platform in the *Creating a closed Sprite Shape profile* section. Creating an open Sprite Shape is even easier!

Use the following few steps to create an open Sprite Shape profile asset:

1. In the project window, within the `Assets/Sprites/Sprite Shapes` folder, select **Create | 2D | Sprite Shape Profile** and name it *Platform Open 1*.

2. Assign the `sprite_strip_rock` sprite by dragging it from the project window and into the **Sprites** list field (replacing the `SpriteShapeEdge` placeholder).

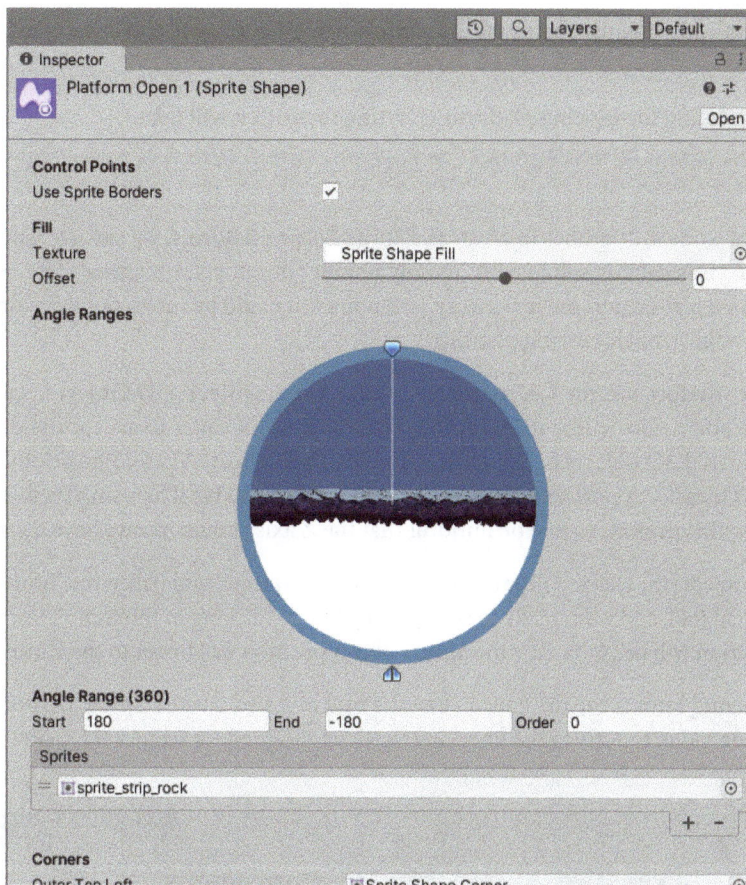

Figure 4.12 – Open Sprite Shape asset properties

3. With the profile asset created, we can now add our first open Sprite Shape moving platform to the scene by clicking **GameObject | 2D Object | Sprite Shape | Open Shape**.

4. Assign the **Platform Open 1** profile asset to the `SpriteShapeController` component's **Profile** field, then use the **Shape** editing tool in the scene Toolbar overlay to make a small straight platform (as seen in *Figure 4.13*). Hint: Use only two knots with **Tangent Mode** set to **Linear**.

    As before, the last step required to make this open Sprite Shape a walkable platform for the player character is to add a collider – we also use the **Edge Collider 2D** for open Sprite Shapes. Add it to the platform object by using the **Add Component** button at the bottom of the **Inspector** window and searching `edge`.

5. Once added, ensure the **Sprite Shape Controller | Collider | Updated Collider** field is enabled, then adjust the **Offset** value to line up with the top edge sprite (this can be adjusted at any time).

Lastly, let's keep our moving platforms on the same Sorting Layer as all the other platforms in the level: **Default**.

The next step in creating the moving platform is setting the path it will take.

Spline path

Much like how we work with splines to create the Sprite Shape platforms, we can use Unity's **Splines** feature tooling to create generic spline paths for any number of gameplay reasons. Here, we're simply building a path for a platform to move between two points (it could be more, but we're only creating a simple vertical or horizontal moving platform).

In the **Hierarchy** window, use the **Create** menu or click **GameObject | 3D Object | Spline | Draw Spline Tool…** to add a new Spline to the scene (rename it or hit *Enter* to accept the default). Yes, Splines are considered 3D objects because the spline path's knots are Vector3 positions (relative to the GameObject's transform position, so they are in **Local Space**). We'll be using them solely in 2D (*X* and *Y* axes) for this project, so just be mindful that the *Z* axis should always have a zero value.

With the **Spline** object still selected in **Hierarchy**, the **Draw Splines** tool (from the **Toolbar** overlay) becomes the current tool, and we're ready to start placing knots. Clicking in the scene View will add the first knot (its position will be (0, 0, 0) – the **Spline** object position will be set to the clicked position).

Click to add a second knot some distance below the first one and use the **Sprite Inspector** overlay to set its *X* position value to zero so the knots are vertically aligned, as seen in the following figure:

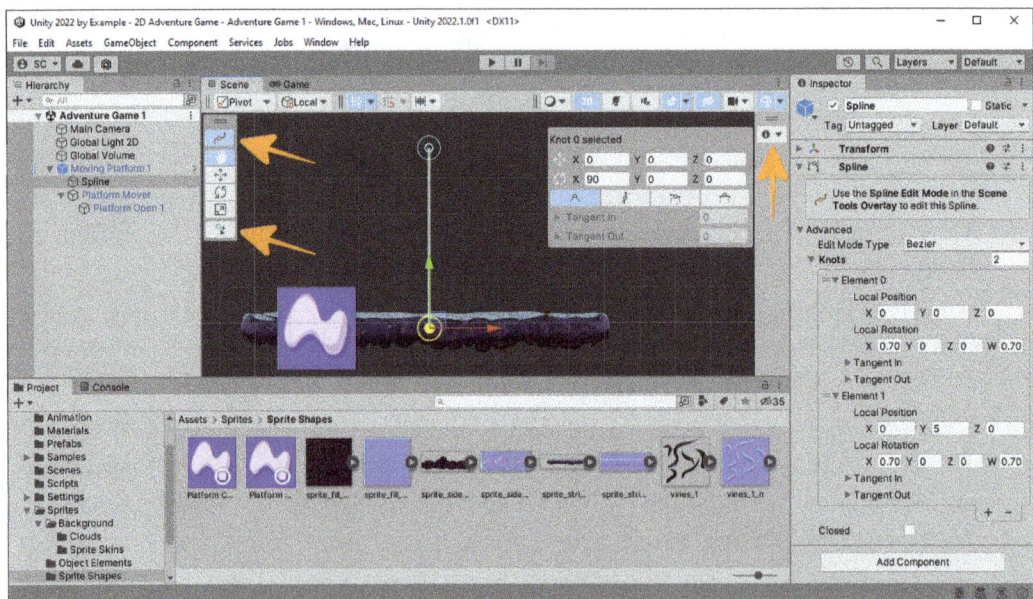

Figure 4.13 – Working with Bezier Splines for an open Sprite Shape platform

Note that the **Spline Inspector** overlay may be docked to the side of the scene View when the Spline tool is selected and as shown in *Figure 4.13*. You can click on the icon in the overlay toolbar or drag it into the scene View window to work with the selected Spline knot's properties.

If you need to edit the position, or the knot's **Tangent Mode**, use the **Spline Transform** tool, and for adding knots to the spline, use the **Draw Splines** tool (both are available in the **Toolbar** overlay when a Spline object is selected in **Hierarchy**).

Additional reading | Unity documentation | New in Unity 2022

Splines: `https://docs.unity3d.com/Packages/com.unity.splines%401.0/manual/getting-started-with-splines.html`

Now it's time to get things moving!

Spline Animate

The `Splines` package provides additional components covering some common use cases for spline paths. We will use one of them here – the **Spline Animate** component.

A bit of housekeeping is in order to make our moving platform not only easy to animate and work with but also so that we can create a prefab that will make it quick and easy work to add additional moving platforms throughout the level.

Right now, we have an open Sprite Shape platform object and a Spline path object at the root of the hierarchy. To make this a reusable prefab, where we have the most control over the position of the spline path and platform, we'll want to end up with this object hierarchy:

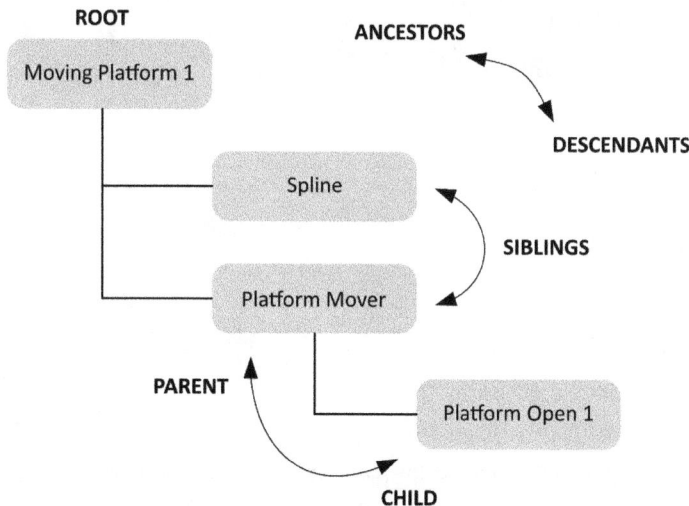

Figure 4.14 – Hierarchy relationships

Use the **Create Empty** or **Create Empty Parent** options available in the **Hierarchy** window's **Create** menu to accomplish this – empty GameObjects are a great way to organize or provide an additional way to manipulate objects!

Just be sure the Spline and platform object's transform positions are at (0, 0, 0) since we should be using the root object (Moving Platform 1) transform position to place the platform in the level.

Part of the reason for creating the object hierarchy this way was to add the object we'll move with the **Sprite Animate** component – the **Platform Mover** object. Add the **Spite Animate** component now by selecting **Platform Mover** and clicking **Add Component** in the **Inspector** window. Assign the **Spline** object to the **Spline** field, select **Spline Object** in the **Align To** field, then set **Loop Mode** to **Ping Pong** (so the platform moves back and forth between the spline knots).

Adjust **Duration** and **Easing** to your liking using the following figure as a reference.

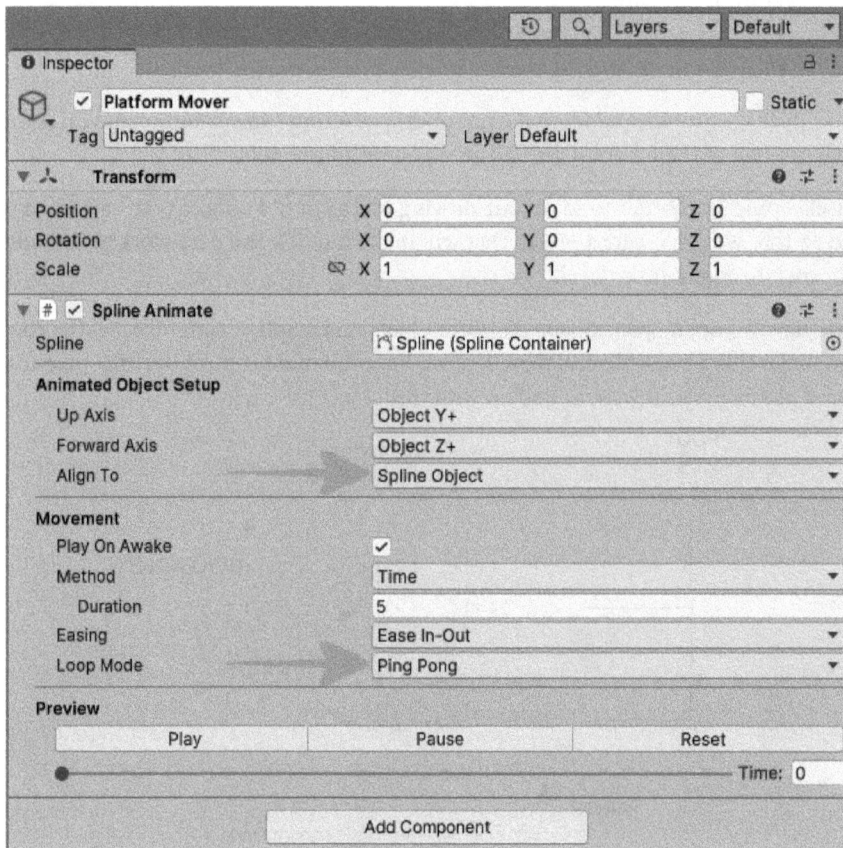

Figure 4.15 – Moving the platform with Spline Animate

And that's it! I said it would be easy to get our platform moving… it just required a bit of setup. You can preview the movement of the platform right in the scene View without having to enter **Play Mode**. Use the **Play**, **Pause**, and **Reset** controls to adjust the duration, type of easing, position of the spline knots, and offset position of **Platform Open 1** until you're satisfied with the positioning and movement.

Finish up by dragging the `Moving Platform 1` object to the `Assets/Prefabs` folder in the project window to create the prefab asset.

Triggering actions in the level

Games often have interactions that are triggered in the environment when the player moves into a specific area or touches something– think rocks falling, doors opening, alerting the enemy, turning lights on/off, transitioning to a cutscene, or whatever your game requires. This is pretty trivial to add if we leverage the Unity physics engine similar to how we previously performed collecting items in the collection game.

The process is as follows:

1. We first add a sprite for the object to our game level.
2. Create a collider volume around the object to define a trigger area – using the least expensive 2D collider type, the better (e.g., `CircleCollider2D`).
3. Enable **IsTrigger** on the collider so that it does not physically interact with any other objects in the scene.
4. Add a component that will invoke an event when the Unity physics message, `OnTriggerEnter2D()`, is called when the player enters the collider.

We're also going to take a designer-friendly approach and make the action assignable in the **Inspector** window so we won't have to create a custom script for every type of triggered event. This will make the triggered event component reusable anywhere in the game we want the player to trigger an interaction. Yay!

UnityEvent

The event delegate type we'll use here is **UnityEvent** – we've previously used **UnityAction**. The difference is that **UnityEvent** is serialized and becomes available in the **Inspector** to assign public methods.

Additional reading | Unity documentation

UnityEvent: `https://docs.unity3d.com/2022.3/Documentation/ScriptReference/Events.UnityEvent.html`

In Unity 2022, the **UnityEvent** list in the **Inspector** is now reorderable!

Let's see how this works by creating a new script named `TriggeredEvent`, with the following code:

```
[RequireComponent(typeof(Collider2D))]
public class TriggeredEvent : MonoBehaviour
{
    [Tooltip("Requires the player character
        to have the 'Player' tag assigned.")]
    public bool IsTriggeredByPlayer = true;
    public UnityEvent OnTriggered;

    private void OnTriggerEnter2D(Collider2D collision)
    {
      if (IsTriggeredByPlayer &&
          !collision.CompareTag(Tags.Player))
              return;

      OnTriggered?.Invoke();
    }
}
```

Let's break this down, especially since there are several new items worth noting:

1. The class declaration is decorated with the `RequireComponent` attribute. This means that when the `TriggeredEvent` script is added as a component to a GameObject, it will require that the specified type exists as a sibling component on the GameObject. If possible, the required component will be added.

2. In our case, we only specified the base `Collider2D` type that all of the 2D colliders inherit from – this allows us to add any type of collider as a trigger volume for the triggered event. If we try to add the script to a GameObject that does not already have a 2D collider added, we'll receive this error:

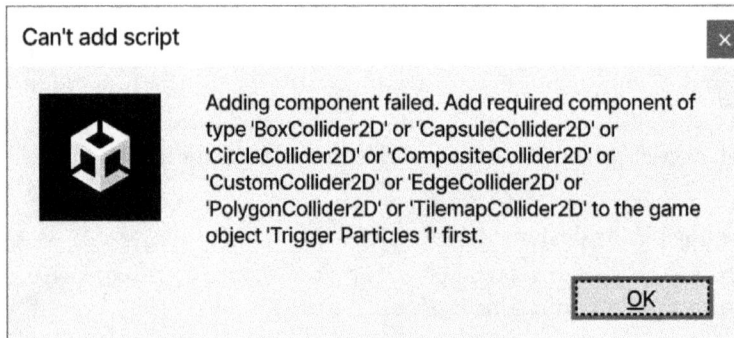

Can't add script

Adding component failed. Add required component of type 'BoxCollider2D' or 'CapsuleCollider2D' or 'CircleCollider2D' or 'CompositeCollider2D' or 'CustomCollider2D' or 'EdgeCollider2D' or 'PolygonCollider2D' or 'TilemapCollider2D' to the game object 'Trigger Particles 1' first.

OK

Figure 4.16 – Can't add script for the required component

The solution is to add a 2D collider (appropriate for the player interaction scenario) to the GameObject before adding the `TriggeredEvent` script. For this example, we've added a `CircleCollider2D`, placed and sized it in front of a peculiar rock formation, and triggered a particle system to start playing (as seen in *Figure 4.16*).

3. Next is the `IsTriggeredByPlayer` Boolean that we'll set in the **Inspector**. A few lines below it will be used to determine whether only the player can trigger the interaction. This field is optional but a quick and easy addition that could provide some exciting behavior if, say, enemies or other objects enter the collider (think outside the box here for all the different things that could be triggered in the environment).

4. As just discussed, the `UnityEvent` declaration is what allows us, as the developer and designer, to add the specific interaction – one or many – in the **Inspector** that is invoked when triggered (as seen in the **Inspector** under the darker gray **On Triggered** () section in *Figure 4.16*).

5. The `if` block determines whether we'll continue running the triggered code or not if we require the event only to be triggered by the player. So if `IsTriggeredByPlayer` is set to `true`, the conditional AND operator (`&&`) will evaluate whether the object collided is tagged `Player`. And if not, it will stop executing this method's code (by using the `return` statement).

&& (C#)

The conditional AND (`&&`) operator means both statements must logically equate to `true` for executing the code in the block. Note that this has "short-circuiting" logic, meaning that if the first expression evaluates to `false`, then the second expression will not be evaluated.

The other thing to note here is that `Tags.Player` will give you a compiler error because the `Tags` type has not been defined yet (identified by a red squiggly underline). Let's fix that now. We'll use the built-in refactoring tooling in the IDE by either right-clicking on the word **Tags** (or clicking anywhere on **Tags** and pressing *Alt + Enter*, or *Ctrl + .*, depending on your IDE).

In the dialog that opens, select **Refactoring...** | **Generate type 'Tags'** | **Generate class 'Tags' in a new file**.

Optimization note

For performance reasons, you should always use `CompareTag()` over evaluating the `.tag` property of a GameObject with the `==` operator.

With the new file created, add a constant for the `Player` tag string like the following:

```
internal class Tags
{
    // Ensure all tags are spelled correctly!
    public const string Player = "Player";
}
```

Note that this class does not inherit from `MonoBehaviour` because we won't be adding it to a GameObject in the scene. It will be used solely when we need to specify an object's tag anywhere in our code – minimizing the use of string literals throughout the code, issues related to simple spelling mistakes that are easy to overlook, and providing CodeLens references for everywhere it's being used. Anytime you need to reference a tag for any other object in the game, just add it here.

Const (C#)

A variable declared as a constant is an immutable value that does not change for the program's life (that is, known at compile time). A `const` (`string`, `int`, `float`, etc.) declared with a public accessor is available to other classes without an instance reference – similar to how a static variable would be used.

---

6. The last line invokes the event(s) assigned in the **Inspector** with `OnTriggered?.Invoke();` – note that the `?.` (null-conditional) operator is optional here, but I always include it for invoked member types for code consistency.

Now, with the setup shown in the following figure, when the player enters the collider, particles will spawn from the rock!

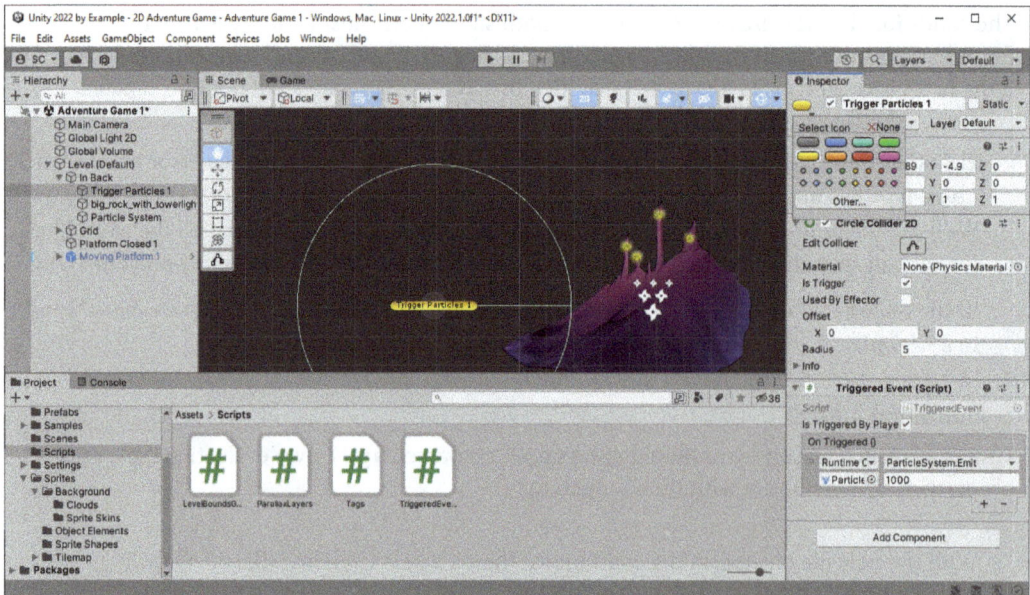

Figure 4.17 – TriggerEvent collider with scene View icon

With this kind of reusable component for triggering any type of event, it is now trivial to add interactions throughout the game level by both developers and designers working on the project. Easy-peasy!

In this section, we've now finished up with all the 2D feature tooling used to create all the walkable platforms, including those moving, as well as interactivity in the level. In the next section, we'll look at finalizing the game's environment, adding polish, and optimizing the draw calls for better performance.

Adding polish to our environment to immerse the player and optimizing

With the tools and techniques introduced in the previous sections, you should now explore building out your level design. Closed Sprite Shapes for the primary platforms, open Sprite Shapes for platforms that are both static and moving, and Tilemap platforms mixed in for a particular area of the level – all used to give the player an experience and challenge on their journey to the habitat station we provide a signpost for in the background.

Platforms on a blank screen are rather dull. Let's add some polish to the level by adding a background and some movement.

Polishing the environment

With your game level now well on its way to being defined (we'll, of course, be making plenty of adjustments once we add the player character and can start play-testing the level), let's have a look at finishing and polishing the environment, starting with adding the background and foreground elements to more fully flesh out the game design – providing immersion and setting the tone for the player experience.

Background parallax effect

As mentioned previously, we'll be applying a relatively old 2D game design technique called parallax to layered images in the background to create depth and a sense of immersion in the environment. We'll accomplish this with a custom C# script since Unity does not provide a 2D feature specifically geared toward tackling this technique.

Let's start by creating a new C# script in the Assets/Scripts folder and name it ParallaxLayers. This script will be a reusable component that we can use to add parallax movement to any number of layered images. In the 2D adventure game, we'll use this for the background and a few foreground elements.

Let's have a look at the following code:

```
public class ParallaxLayers : MonoBehaviour
{
    public List<Layer> Layers;
    [System.Serializable]
    public class Layer     // Nested class.
    {
```

```
        public Renderer Image;
        [Tooltip("How far away is the image from the
            camera?"), Range(0, 10000)]
        public int Zdepth;
    }

    private Camera _camera;
    private Vector3 _cameraLastScreenPosition;

    private void Awake()
    {
      _camera = Camera.main;
      _cameraLastScreenPosition =
          _camera.transform.position;
    }

    private void LateUpdate()
    {
      if (_camera.transform.position.x ==
          _cameraLastScreenPosition.x)
              return;

      foreach (var item in Layers)
      {
        float parallaxSpeed = 1 - Mathf.Clamp01(
            Mathf.Abs(_camera.transform.position.z
                / item.Zdepth));
        float difference = _camera.transform.position.x
            - _cameraLastScreenPosition.x;
        item.Image.transform.Translate(
            difference * parallaxSpeed * Vector3.right);
      }
      _cameraLastScreenPosition =
          _camera.transform.position;
    }
}
```

Let's break down the code item by item:

1. The first public variable declared is a list of the images we'll use for the parallax layers that will
    be moved at different speeds based on their depth (this is how the parallax effect is achieved;
    objects in the distance appear to move slower than nearby objects). We use C# List<T>
    specifying Layer as the type, where Layer is a custom class [object] we declare to hold the

reference for a specific image and the depth value assigned to it for calculating its movement speed relative to the camera movement.

> **List (C#)**
>
> `List<T>` is a collection of strongly typed objects that can be added, accessed by index, or returned by methods it provides for searching. Methods are also provided for sorting and manipulating objects in different ways.
>
> Additional reading: `https://docs.microsoft.com/en-us/dotnet/api/system.collections.generic.list-1`

2.   Next is the `Layer` nested class declaration with public fields for `Image` with a type of `Renderer` (all sprites, meshes, lines, trails, and particles are drawn with a component derived from the `Renderer` class) and `ZDepth`, which is an `int` value for the image's distance away from the camera.

3.   The `ZDepth` variable declaration is also decorated with the `Tooltip` and `Range` attributes (these can be comma delimited in a single `[]` statement). The `Tooltip` attribute displays the specified message text when the mouse hovers over the field label in the **Inspector**. The `Range` attribute will limit the values valid for the field between the range specified.

> **Additional reading | Unity documentation**
>
> *Attributes*: `https://docs.unity3d.com/2022.3/Documentation/Manual/Attributes.html`

4.   The `Layer` class declaration itself is decorated with the `System.Serializable` attribute. A class declaration is not available in the **Inspector** unless we mark it for serialization – even though the `Layers` accessor is public, that alone is not enough to serialize the class.

5.   The private `_camera` member is used to cache the main camera reference because we're getting values from it multiple times in our calculation, and we want this to be performant.

6.   `LateUpdate()` is executed after `Update()` and after the internal animation update is processed. After everything in the game loop has been updated, we can use `LateUpdate()` to affect those updates further. In this case, we want to move the background layers only after the camera movement.

> **Additional reading | Unity documentation**
>
> *LateUpdate*: `https://docs.unity3d.com/2022.3/Documentation/ScriptReference/MonoBehaviour.LateUpdate.html`
>
> **Order of execution for event functions**: `https://docs.unity3d.com/2022.3/Documentation/Manual/ExecutionOrder.html`

7.   Did the camera move? A quick evaluation on the `_cameraLastScreenPosition` Boolean value will tell – if not, don't execute further (using the `return` statement). Don't run code that you don't have to!

8.   This is where the magic happens... using the `foreach` iterator, we'll go through each of the `Layer` items added in the **Inspector** and move it the calculated distance relative to how far the camera moved:

```
foreach (var item in Layers)
{
```

-   `parallaxSpeed`: we start by getting a ratio for the depth of the image from the camera position (order of operations for math), returning the absolute value (a positive number), clamping the value to a range of 0 to 1, and finally subtracting the clamped value from 1 to get a percentage of the camera's movement to move the image:

```
float parallaxSpeed = 1 - Mathf.Clamp01(
    Mathf.Abs(_camera.transform.position.z /
        item.ZDepth));
```

-   `difference`: How far the camera moved from its current position compared to its last position:

```
float difference = _camera.transform.position.x
    - _cameraLastScreenPosition.x;
```

-   `transform.Translate()`: Move the image horizontally on the *X* axis (due to `Vector3.right`) by the distance the camera moved multiplied by the speed factor (effectively, speed values closer to 1 move relatively the same distance as the camera):

```
item.Image.transform.Translate(
    difference * parallaxSpeed * Vector3.right);
```

---

foreach (C#)

The `foreach` statement iterates over a collection of items and executes its code block for each. Each list element is of the type specified when the list is declared.

Additional reading on C# iteration statements: `https://docs.microsoft.com/en-us/dotnet/csharp/language-reference/statements/iteration-statements`

---

9.   Finish off by updating the variable that holds the camera's last position after moving.

> **int versus float divide by zero**
>
> You might be thinking, *Hey, wait a second, Scott, with the* Range *attribute, you're allowing a zero value for distance, and to calculate the parallax speed, we're dividing by the distance. So, if the distance is zero, then won't that throw a divide-by-zero exception error?* What an astute observation to make that point; thank you for asking! You would be correct if the camera's `transform.position.z` value is an integer type. It would absolutely throw an exception, but it is a float value, so dividing by zero will never throw an exception because, in C#, floating-point types are based on the IEEE 754 standard, which allows for numbers representing infinity and **Not a Number (NaN)**.
>
> **Additional reading**: `https://docs.microsoft.com/en-us/dotnet/api/system.dividebyzeroexception?view=net-6.0`

Now that we have completed our parallax effect script, let's add the background layers to the scene and assign the field values.

The following **Inspector** screenshot illustrates a set of five background images that will display a layered parallax effect, with the distant images having a ZDepth value of 10,000 – meaning they'll appear to be stationary – and images closer to the camera having values less than 1,000 – meaning they'll move more closely to the camera movement:

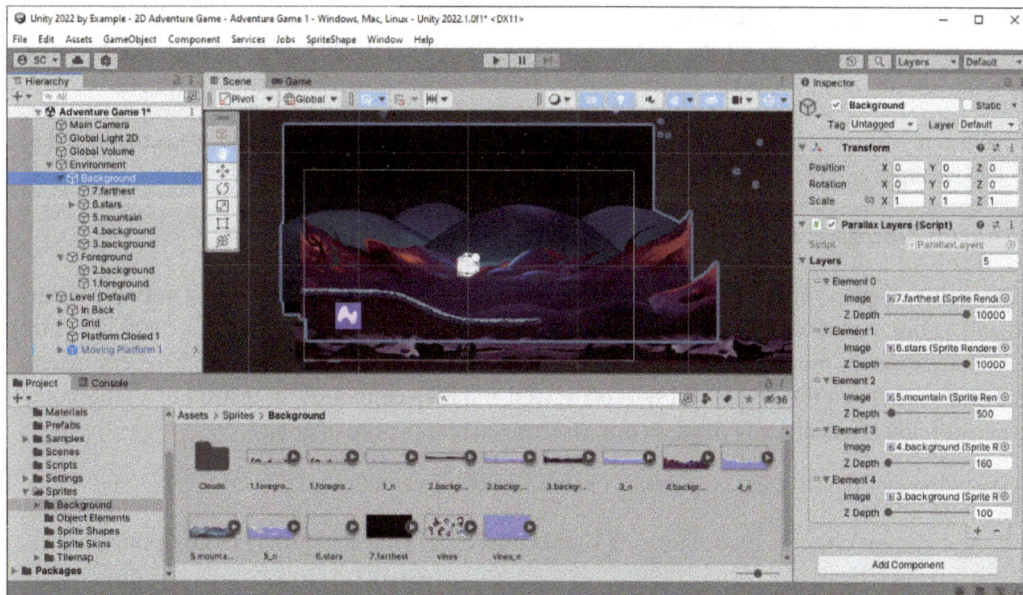

Figure 4.18 – Parallax background layers assigned to the Inspector

The preceding figure also shows an object hierarchy we can use to organize the images for the background layers (and foreground images using the same technique). Drag the background images from `Assets/Sprites/Background` into the scene and parent the objects in **Hierarchy** to a **Background** empty GameObject.

With the objects in the scene, you can position them and use the Sorting Layer and Order in Layer (Sprite Renderer) to set their display. Be sure to add a new **Background** Sorting Layer and position it at **Layer 0** so that our platforms' **Default** layer will draw in front of the background images.

This is more art than technical so play around with the position and scale, background image order, and so on, until things look good!

Now, add the **ParallaxLayers** component to the **Background** object (drag the script from the project window or use the **Add Component** button) and the individual background sprite images to the **Layers** list (use the + button to add an item, then assign to the **Image** field by dragging the sprite image from **Hierarchy**), and set the `ZDepth` value for each image – assign higher values for the images further in the distance.

You can test the parallax effect by entering **Play Mode** and scrubbing the Main Camera's transform position **X** value (when you hover the mouse cursor over the **X** label, you'll see a cursor with left/right arrows; clicking and dragging the mouse left/right will change the value). You'll notice that when you reach a specific value, the right edge of the background image will be exposed – you'll be making adjustments to the background image position and size, as well as the `ZDepth` value so that it works within the bounds of your level design.

Let's continue adding polish to the environment with some clouds animated to move across the sky. This will be super easy by again leveraging the **Spline Animate** component.

Animated clouds

Earlier in this chapter, you learned how to easy it is to create a moving platform using a simple spline path and a component to animate its position. We'll use the same approach to create two layers of different-style clouds moving across the sky at different speeds. The only difference here, compared to the platform, is that the clouds will start off-screen and loop continuously, instead of a ping-pong movement (clouds usually only move in one direction across the sky, after all).

Use the following steps to create two layers of clouds moving across the sky:

1. Add a Spline object to the scene using **GameObject** | **3D Object** | **Spline** | **Draw Spline Tool…**:

    I.   Add a knot beyond the extent of the level on both the left and right sides, high enough that they are in the *sky* area of the background environment. Remember, these spline knot positions can be adjusted later should the extent of the level need to be revised.

II.   Note that the knot rotation should be correct when clicking within the scene with the **Draw Splines** tool in **2D view** mode, but if you don't see the circular outline of the knot, then the rotation may not be correct. If necessary, click the knot, and in the **Spline Inspector** toolbar overlay, reset the rotation value to (90, 0, 0).

2.   Add cloud sprites to the scene and parent along with the Spline to a new empty GameObject named Clouds 1 (you can add as many cloud layers as desired, just increment the count, as shown in *Figure 4.18*). We have two cloud layers, and multiple cloud sprites added for the Clouds 2 object.

3.   Add a **Spline Animate** component to the cloud image GameObject and set the following properties:

I.   Assign the Spline object from **Hierarchy** to the **Spline** field.

II.   Set **Align To** as **Spline Object**.

III. Set the **Duration** field to a value that sets the clouds moving slowly across the sky – make sure to set this value to a different time, so multiple cloud layers are offset from one another!

IV.   Set **Loop Mode** as **Loop Continuous** because we want the clouds to restart the animation over again on the right side once they reach the end knot on the left side.

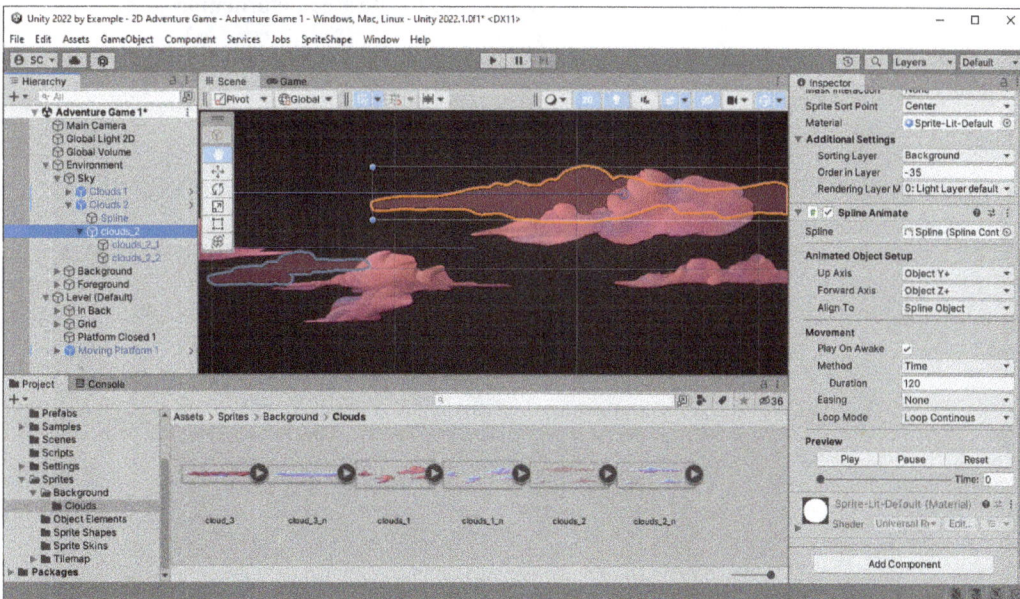

Figure 4.19 – Animated background clouds setup

This is a quick and easy way to bring more life to the environment. The same is true for the next 2D tooling feature we'll introduce to add more polish.

Animated environment art with sprite skin

Let's now set up some of the plant entity's vines to sway and move in different ways in the background, giving the player a sense of presence by the unnatural forces at work here.

We've already imported and prepared artwork to use with **Sprite Skin**, so let's select one of the vine sprites in the Assets/Sprites/Sprite Skins folder and open **Sprite Editor**.

Use the following steps to create a mesh geometry that is deformable by a set of weighted bones:

1. From the **Sprite Editor** dropdown menu, select **Skinning Editor** and enable the **Visibility** tool (top-right of the window next to **Revert and Apply**).

2. Under the **Bones** section (left side of the window), select the **Create Bone** button.

3. The next step is essential and not apparent in the **Skinning Editor** workflow – double-click on the sprite!

    You should now have a cursor with a red dot at the tip. Start creating bones by dragging the bones to the desired length, click to add a new bone, and hit the *Esc* key when finished. Refer to *Figure 4.19* for an example of how to approach the bones layout (keep in mind that it is more performant to use fewer bones while still allowing for the desired deformation).

4. When satisfied with the bones, select **Auto Geometry** under the **Geometry** section and click the **Generate For Selected** button:

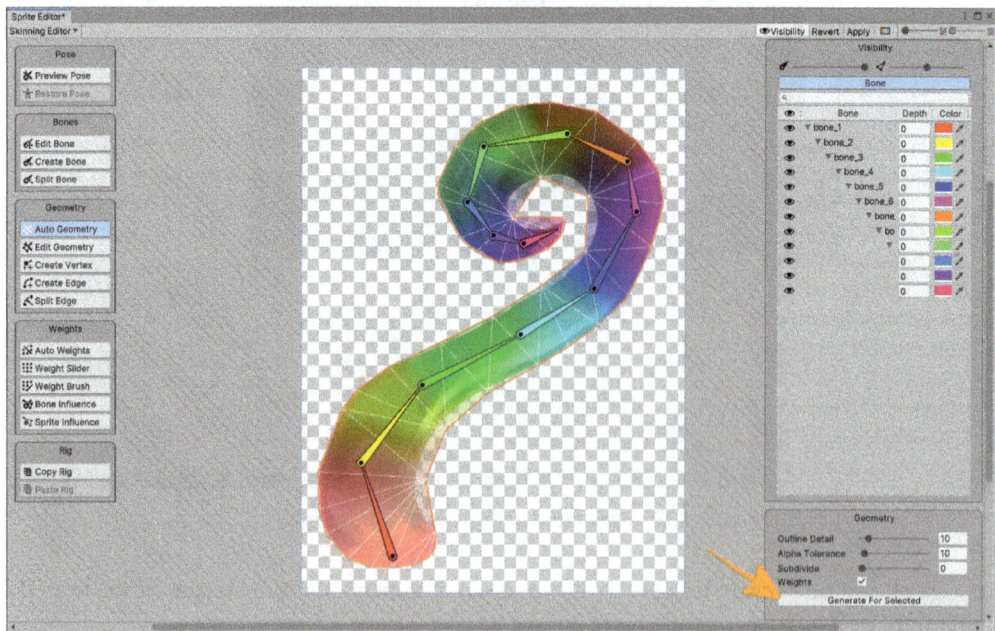

Figure 4.20 – Animating a vine with Sprite Skinning

5. Lastly, select **Auto Weights** under the **Weights** section and click the **Generate** button (bottom-right corner of the window).

   At this point, click the **Apply** button, and we have created our first Sprite Skin. Yay, ready for animation! You can test the deformation in **Skinning Editor** by clicking the **Preview Pose** button under the **Pose** section.

Now that our vine is *skinned*, it's ready for animating!

Animating a Sprite Skin

Let's now add the vine sprite to the game environment as a background element and prepare it to be animated:

1. Drag the vine sprite into the **Scene Hierarchy** as a child of one of your background images (so that it will move along with the background parallax), position, and set the Sorting Layer and Order in Layer accordingly.

2. With the sprite selected, in the **Inspector**, add a **Sprite Skin** component by using the **Add Component** button and searching for `skin`.

3. In the **Sprite Skin** component, click the **Create Bones** button.

Now we need to add animation to the vine. While the vine is selected, open the **Animation** window using **Window | Animation | Animation** and then click the **Create** button. This will create an animation asset file (`.anim`) that we'll name `Vine 1 Idle` and save to the `Assets/Animation` folder. Put your animator hat on now… because it's time to animate!

Using *Figure 4.20* as a reference, by clicking the red record button (the **Animation** window), a keyframe will be recorded in the **Animation** timeline any time you rotate or position a bone (using the same transform tools we manipulate any object with). Scrub the timeline and repeat the process to get the desired movement for the vine.

Again, this is more art than technical and takes some trial and error. With practice, this becomes more intuitive and quicker to achieve good results. Use the playback controls and adjust accordingly – you've got this!

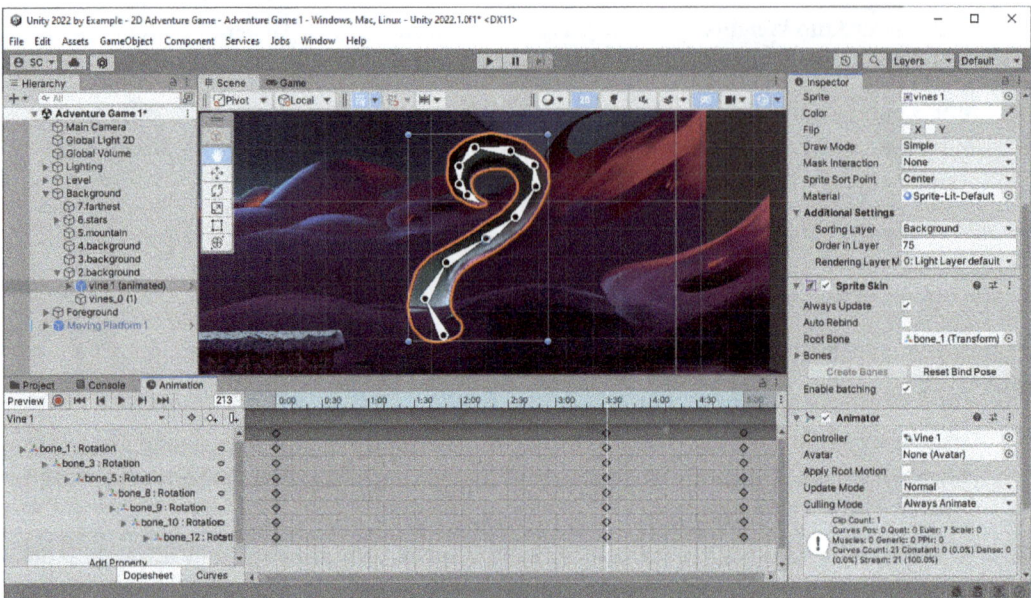

Figure 4.21 – Animating the vine with keyframed bone rotation

---

Additional reading | Unity documentation

Animation: `https://docs.unity3d.com/2022.3/Documentation/Manual/AnimationSection.html`

From the static pictures in this book, you can't get a proper sense of how creepy these vines look when animated! Enter **Play Mode** to experience this yourself in your level design, or make sure to check out the completed project code or play the game online from the GitHub repo.

Let's look at how to keep the vine animations performant in our game now.

Sprite Skin performance

As the last step for increasing animation performance with Sprite Skin, install the **Burst** and **Collections** packages to enable **deformation batching**:

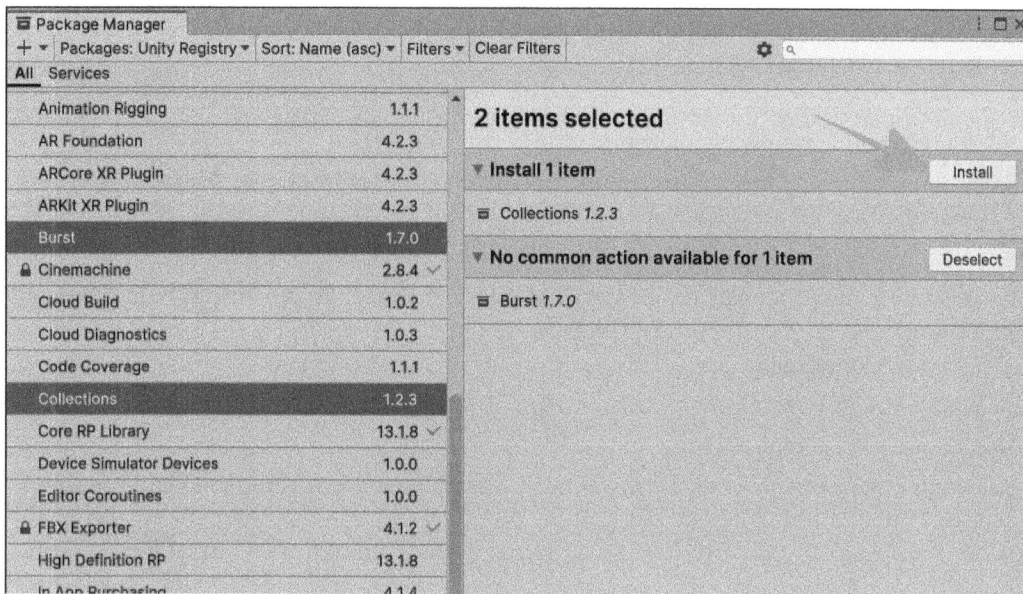

Figure 4.22 – Installing Burst and Collections for optimization

Continuing on the performance optimization topic, before we move on to importing and setting up the player character and enemies for the game, let's look at optimizing the sprite draw calls to help keep our framerate from dropping to unacceptable levels.

Optimizing draw calls

The most impactful performance optimization we can make is addressing the draw calls. This was discussed briefly in a callout in *Chapter 2* in the *Importing Sprites* section when we created a **Sprite Atlas**. Not surprising, we're going to do the same thing here now.

Sprite Atlas

Use the **Create** menu in the project window's Assets/Sprites folder, then choose **Create** | **2D** | **Sprite Atlas** and name it Sprite Atlas. Note that we won't be adding any of the 2D sprite normal maps as those will be handled internally with the Sprite Atlas in the same way they are in **Sprite Editor**.

Assign the individual sprite images from the Assets/Sprites folders to the **Objects for Packing** list (click and drag images or folders to the list). If you want to assign folders to the list, move the normal maps to a sibling folder (such as Assets/Sprites/Background Normal Maps).

When there are too many images to fit into a single texture, Sprite Atlas will create additional textures and indicate that with a dropdown indicator labeled **#0** in the title bar of the **Preview** window. In some cases (for example, larger games), it may be desirable to use multiple Sprite Atlases where you will determine which one to bind at runtime.

Note for compatibility reasons, while using **Sprite Shape,** we'll have to make sure to disable both **Allow Rotation** and **Tight Packing.**

Additional reading | Unity documentation

Sprite Atlas: `https://docs.unity3d.com/2022.3/Documentation/Manual/class-SpriteAtlas.html`

In this section, we looked at polishing the visuals of the *Outer World* 2D adventure game by adding a parallax background effect and animated elements. We then looked at keeping our game performant by reducing sprite draw calls.

Downloading the completed game code

Remember, the example code for this book can be downloaded from the GitHub repo here: `https://github.com/PacktPublishing/Unity-2022-by-Example`

Summary

This chapter first covered the game, level, and environment design by introducing new elements for the GDD to cover the larger scope of the adventure game, importing and prepping artwork for use with Unity's 2D features, and introducing a new level design principle to guide the player to a goal.

We continued with the imported artwork by creating static and moving platforms to challenge players on their journey. Then, we immersed the player in the game world by setting up a parallax background and animated environmental elements.

Finally, we optimized the draw calls for the sprites that define all the elements of the level and environment with a Sprite Atlas.

In the next chapter, we'll set up the player character with an imported rig using the PSD Importer, which will allow a quick setup of the character for animating. We'll also learn how to add a player weapon that shoots projectiles in an optimized way by implementing another feature new for Unity 2022, and start introducing the enemy characters the player will be fending off on their journey to the habitat station.

5
Continuing the Adventure Game

In *Chapter 4*, we covered the larger scope of the adventure game by extending the **Game Design Document** (**GDD**) for the game and explored the level and environment design while also considering new principles to guide the player. We also imported and prepped artwork for use with additional 2D tooling that will bring the game to life.

We also challenged the player by exploring and adding moving platforms and interactive elements in the level design, immersing the player with a parallax effect on a layered background, and optimizing the sprite draw calls to keep things performant.

With the game level and environment established, we can now move on to creating our player character using the 2D Animation package.

In this chapter, we're going to cover the following main topics:

- Setting up the player character with PSD Importer
- Using an Input Action Map
- Moving the player with a player controller script
- Animating the character with Mecanim

By the end of this chapter, you'll be able to set up a 2D sprite-based character rigged for animation and driven by player input. You'll also be able to assign and transition between different animations required for the player's current state.

Technical requirements

To follow along in this chapter and use the same artwork that was created for the project in this book, download the adventure game 2D art assets from the following GitHub link: `https://github.com/PacktPublishing/Unity-2022-by-Example/tree/main/ch5/Art-Assets`.

To follow along with your own artwork, you'll need to create similar artwork using Adobe Photoshop or use a graphics program that can export layered PSD/PSB files (e.g., Gimp, MediBang Paint, Krita, and Affinity Photo).

You can download the complete project from GitHub at `https://github.com/PacktPublishing/Unity-2022-by-Example`.

Setting up the player character with PSD Importer

Creating our player character for the 2D adventure game will be a multi-step process. In this section, we'll cover all the steps required to make an animated 2D character controlled by the player.

We'll start with the import settings for the artwork and setting up the character bones that will allow us to animate. You'll repeat these actions many times throughout a project while importing artwork and assets. **PSD Importer** is an asset importer that can work with multilayered PSB/PSD files to create a Prefab of sprites based on the source layers.

The import options allow Unity to generate both a sprite sheet and a character rig that arranges the sprites according to their original position and layer order, significantly simplifying the creation of a sprite-based animated character.

Let's go ahead and import the artwork for the player character. Here, we will set up the source file so that it uses PSD Importer to create the actor (the Prefab created based on the multilayered Photoshop file is called an **actor**):

1. Create a new folder in the `Assets/Sprites/Character` directory.
2. Import `PlayerCharacter1.psd` into the newly created folder.
3. Select the imported file and, in the **Inspector** window, change **Importer** to **UnityEditor.U2D.PSD.PSDImporter** in the dropdown list.

PSD Importer provides two new options that appear as tabs once it's selected as the importer:

- **Settings**: This is where you will set the properties for how the file will be imported. The fields under **Settings** are similar to the default texture importer, with the addition of the **Layer Import** (when **Texture Type** is set to **Multiple**) and **Character Rig** sections
- **Layer Management**: This is where you can customize which layers from the Photoshop file are imported

With the default import settings, we are already in good shape to continue with the player character setup since we'll be using all the layers. The importer will keep the layer positions and sorting order from Photoshop so that our character sprites are arranged correctly for our actor.

We won't have to recreate the player character in Unity from the individual sprites that make up the character – arms, torso, legs, head, and so on. So, we are ready for the next step – rigging the character by adding bones.

Additional reading | Unity documentation

Preparing and importing artwork: `https://docs.unity3d.com/Packages/com.unity.2d.animation@9.0/manual/PreparingArtwork.html`.

PSD Importer Inspector properties: `https://docs.unity3d.com/Packages/com.unity.2d.psdimporter@8.0/manual/PSD-importer-properties.html`.

Rigging the actor

With the actor (the player character PSD file) still selected in the **Project** window, click the **Open Sprite Editor** button in the **Inspector** window. The default view is the sliced sprite shape representing our Photoshop layers.

Don't worry – as I said, we won't have to work with our character in a dismembered way, as shown in *Figure 5.1*:

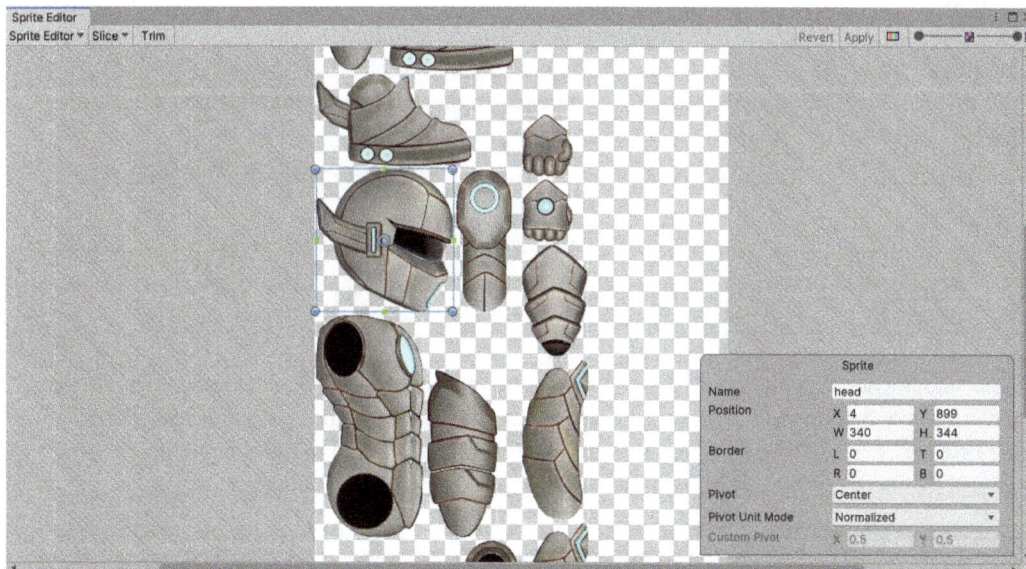

Figure 5.1 – Actor sprite sheet

If the artwork is shown with the individual sprites representing the Photoshop layers, then we need to switch to **Sprite Skinning Editor** (part of the 2D Animation package), where we'll continue setting up the animation rig for the actor.

From the **Sprite Editor** dropdown, select **Skinning Editor** – the sprites should now represent the character, as seen in the original Photoshop source file. Sweet!

The entire workflow of setting up the actor for animations looks like this:

1. **Create a rig**: A skeleton made up of individual bones.

2. **Generate mesh geometry**: This will influence the sprite's position, rotation, and, optionally, deformation.

3. **Adjust bone influence**: Change the attributes of a sprite to specify which bone or bones affect it.

4. **Adjust weights**: How much a bone, or bones, influence a given sprite (the entire sprite, or just part(s) of it).

5. **Preview pose**: Test, test, and test again regarding the rig and sprite bone, geometry, and weight assignments until everything is looking and working correctly when bones are moved or rotated; this is a cyclical process.

6. **Inverse kinematics (IK)**: You can apply IK to a chain of bones to automatically calculate positions and rotations that make it much easier to animate the actor's limbs.

Additional reading | Inverse kinematics (IK)

IK is a technique that's used in computer animation and robotics to control the position and orientation of limbs. It calculates the joint angles required to place the end effector in a given position and orientation – endpoints such as hands and feet. It's useful for creating natural and realistic human motion, especially for complex poses.

You can read more at `https://docs.unity3d.com/Manual/InverseKinematics.html`.

---

7. **Animate**: Create animations for the actor's different gameplay states, such as idle, walking, jumping, and attacking.

Game development is a multidisciplinary craft that combines both technical and artistic skills. Creating a character rig with individual bones to affect individual sprites combines both skill sets. Thankfully, the rigging process is non-destructive, so we'll be able to go back and make changes at any time to adjust and fix anything that doesn't seem to be working well or look correct.

The general creation process we'll follow when creating the rig is to start with the pelvic bone – this bone will represent the *true center* of the character. Moving or rotating this bone will affect all child bones in the rig, moving the entire actor.

This first bone we'll create can be seen in *Figure 5.2* as the small red bone pointing upwards in the pelvic area of the character. In the bone hierarchy (as shown in the **Visibility** window), this bone, bone_1, is the root bone.

Let's create our bones! Perform the following steps to develop the whole character rig:

1. We should already be in **Skinning Editor**; if not, select it from the **Sprite Editor** dropdown.

2. To view the hierarchy of the bones being created (or to manage the visibility of the sprites you're working with), toggle on the **Visibility** panel (the top right of the window next to **Revert** and **Apply**).

3. Now, under the **Bones** section (the left-hand side of the window), select the **Create Bone** button – your cursor should now have a red dot at the tip.

4. Start at the character's pelvis and click to start creating the root bone. Click again slightly above the first click point to make the small red root bone – refer to *Figure 5.2* for the general size and orientation.

5. Continue by clicking again to create a bone that will represent the lower portion of the torso (yellow bone) and again to make a bone for the upper part of the torso (green bone); this will allow us to bend the character's torso (for example, arching forward or backward).

6. Moving up the spine, create two additional bones – a small bone for the neck and another for the head (refer to *Figure 5.2*).

7. Right-click or press the *Esc* key to stop creating bones.

Our actor now has a spine (if you'll forgive the pun). Creating the limbs will follow a similar process, but the critical difference in creating bones for limbs is to first select the bone that will parent the limb bones. Follow these steps:

1. With **Create Bone** still selected, click on the upper torso bone (green). You should now see a semi-transparent bone coming from the bone that was just clicked (this indicates parenting; you can also see this in the **Visibility** window's **Bone Hierarchy** view).

2. We'll start by creating a limb for the character's left arm (the arm behind the torso), so click where the left shoulder joint should be to begin making bones.

3. Note that to make placing bones easier for sprites behind others, you can toggle the visibility off for the sprites in front by using the **Visibility** panel.

4. Keep clicking to create bones for the upper arm, lower arm, and hand (refer to *Figure 5.2*).

5. Right-click or press the *Esc* key to stop creating bones.

6. Continue creating the other limb bones for the right arm and both legs by selecting the parent bone again – for the legs, parent to the pelvic bone (the red root bone we started creating the rig from).

Your actor's entire skeleton rig should look similar to the rig shown in *Figure 5.2*:

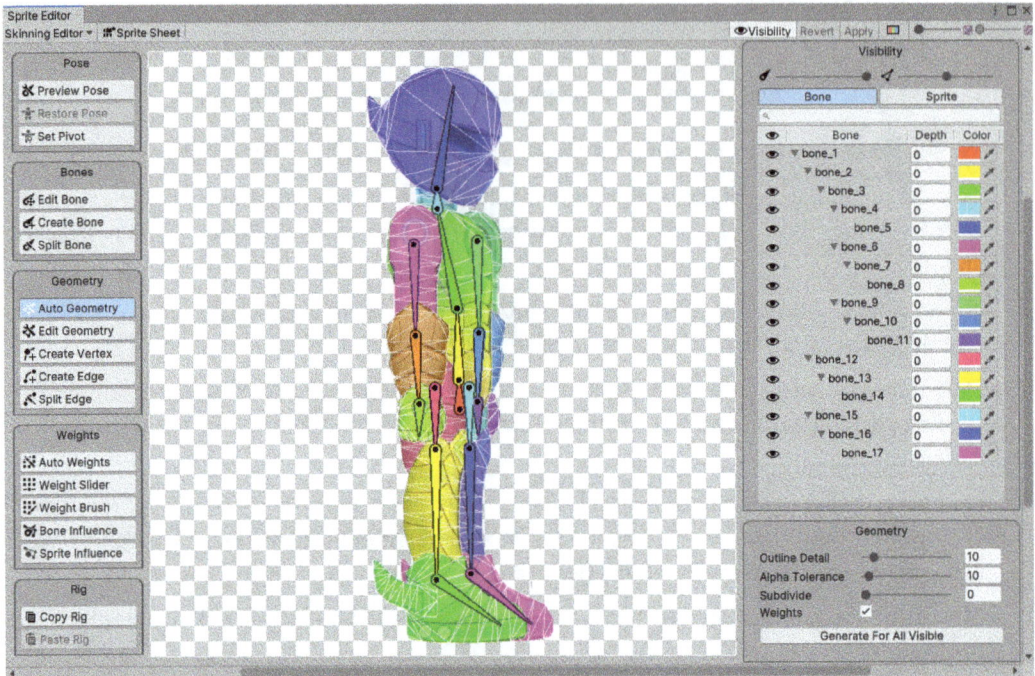

Figure 5.2 – The actor rig and Auto Geometry with Weights

If you want to have your bones named more intelligibly – making it easier for you to know which bones you are working with – you can easily rename them while using the **Edit Bone** tool. Click on a bone to select it and then use the **Name** field in the **Bone** dialogue (on the right-hand side of the window and under **Visibility**) to specify a new name (for example, bone_14 to `foot_right`).

Creating the rig is just the first step in making a fully animatable character. Let's take the next step by assigning the sprite's mesh geometry to the rig's bones – as you continue to work with the rig, it may help to know that the sprite mesh geometry a bone influences will be the same color as the bone.

Generating the sprite mesh geometry

Perform the following steps to create the sprite mesh geometry and apply default bone weights that will influence the sprites that we'll start with:

1. First, double-click in the checkered area to unselect all sprites (noting that this action is not apparent).

2. Select the **Auto Geometry** button under the **Geometry** section (the left-hand side of the window).

3. In the **Geometry** panel at the window's bottom-right section, ensure **Weights** is enabled; this will automatically generate bone weights along with the mesh geometry, saving a step.

4. Finish up by clicking the **Generate For All Visible** button.

> **Additional reading | Unity documentation**
>
> 2D Animation | Editor tools and shortcuts: `https://docs.unity3d.com/Packages/com.unity.2d.animation@9.0/manual/SkinEdToolsShortcuts.html`.

After clicking the **Generate For All Visible** button for the geometry, all your sprites will have taken on coloring from the associated bones. You'll notice that it does not look like the solid colors seen in *Figure 5.2* since the generated weights were distributed across adjacent bones; instead, the coloring will be a gradient of the adjoining bone colors.

This is due to both the overlapping sprites and Unity wanting to blend the sprite deformation across the bones, which is desired in some cases. We'll be addressing this shortly in the *Editing sprite bone weights* section for the torso and neck sprites. For the remainder of the sprites, we only want one bone to influence a sprite (represented by the solid color) for this particular character.

We'll fix this by adjusting the bone influence for each sprite.

Adjusting the bone influence

The process we're going to follow is to remove – or set – bones in the **Bone Influences** property of the sprites. This will ensure that our character's sprites are not deformed or distorted while positioned or rotated by only assigning a single bone.

For a different style character, we may very well want a single sprite to be deformed by multiple bones (as we will do shortly with the torso and neck), but that is not what we want for the style character we're building here.

Perform the following steps to assign the correct bone influence for all of the sprites (again, except for the torso and neck):

1. Select the **Bone Influence** button under the **Weights** section (the left-hand side of the window).

2. Double-click on a sprite. For example, in *Figure 5.3*, we'll double-click on the *wing* for the right foot.

    The panel in the bottom right of the window will now show the selected sprite and the bones influencing it.

3. Since we only want one bone to influence the sprite, we'll select `bone_13` (the yellow bone) in the list and then click the minus (-) button on the small tab below the list to remove it.

The sprite will turn the same color as the remaining assigned bone, which is green in this case, as seen in *Figure 5.3*:

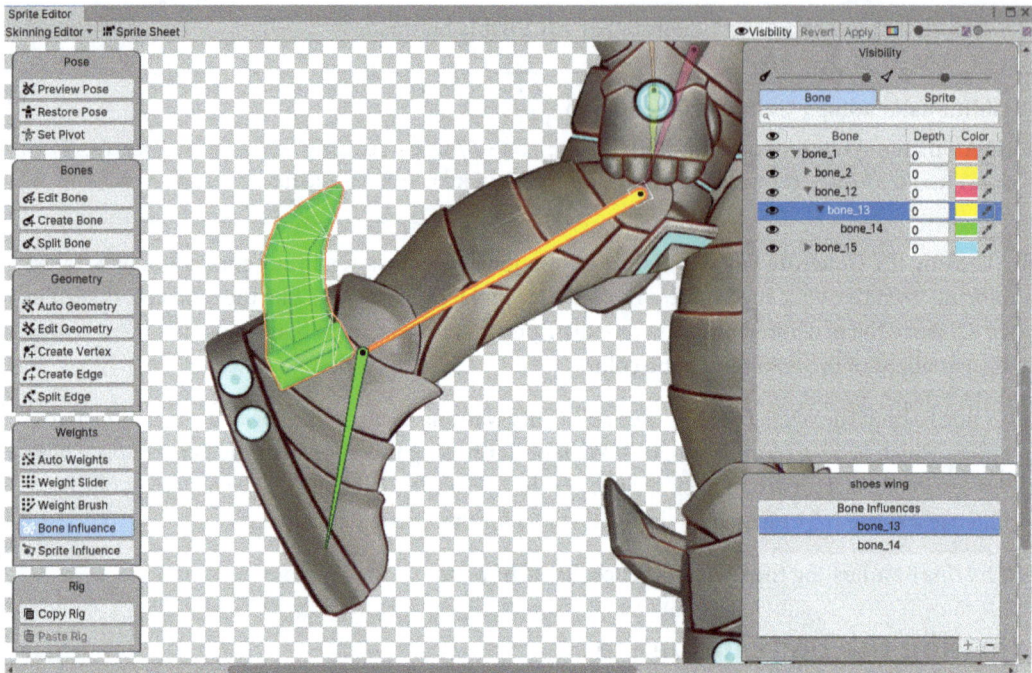

Figure 5.3 – Bone Influence editing

Note that if there is more than one bone to remove, you can select multiple bones by holding the *Ctrl/ Cmd* key down while clicking and then clicking the minus (-) button.

This workflow was for assigning a single bone to a sprite. Next, we'll learn how to use multiple bones to influence the deformation of a sprite.

Editing sprite bone weights

We'll use two weight tools to assign the bones and adjust the weight influence each bone has over the sprite: **Weight Slider** and **Weight Brush**.

We'll start with **Weight Slider** to adjust the overall influence of the bones and then perform any fine-tuning with **Weight Brush**.

The **Auto Weights** generation may have already performed an excellent job but let's examine the results and make some adjustments, starting with the neck sprite:

1.   Double-click on the neck sprite (note that you can use the mouse scroll wheel to zoom in on a sprite to work with it).

2. **Auto Weights** assigned three bones with influence over the neck sprite, so this sprite should have a color gradient representing the color of the three bones: the (green) upper torso, (cyan) neck, and (blue) head bone (see *Figure 5.4*).

3. Select the **Weight Slider** button under the **Weights** section (the left-hand side of the window).

4. In the **Weight Slider** panel in the window's lower-right corner, you can now adjust the amount of influence a selected bone has over the sprite; drag the **Amount** slider left or right to increase or decrease the influence.

5. Preview the effect of weight changes to see how the mesh deforms by rotating the bones:

   You can rotate and move bones by simply clicking and dragging on them (the mouse cursor will show a rotate or move icon, depending on where you are hovering on the bone). Adjust the **Weight Slider** amount until you get overall pleasing results – this operation is a lot more artistic than technical:

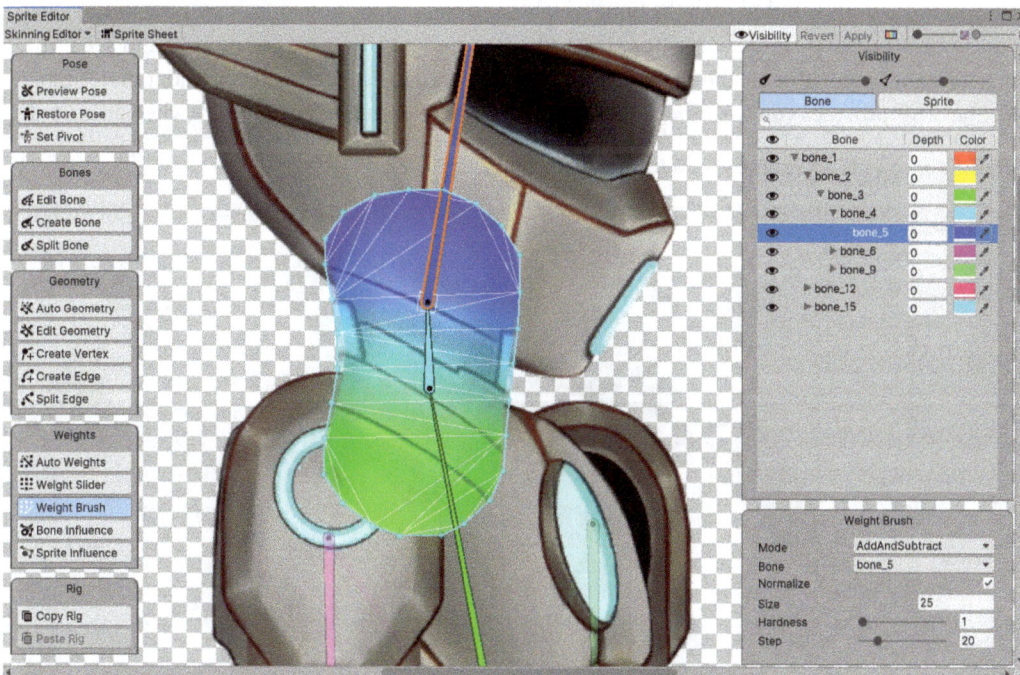

Figure 5.4 – Editing weights with Weight Brush

You may find it challenging to get good results using **Weight Slider** alone. In this case, continue to fine-tune the bone influence for any problem areas using **Weight Brush**.

6. With the sprite still selected, select the **Weight Brush** button under the **Weights** section (the left-hand side of the window).

7. Click on the bone you want to paint weights for.

8.   You can now adjust the **Weight Brush** properties in the **Weight Brush** panel in the window's lower-right corner or start painting weights. The sprite mesh geometry will update in real time as you paint.

9.   Continue to rotate the bones by hovering your mouse over the bone to show the rotate icon. Then, click and drag to rotate the bone to test the weight painting and adjust it until you've achieved the desired results.

Note that you can undo bone rotations using *Ctrl/Cmd + Z*. If you've made many bone rotations and aren't sure how much influence is being applied, then you can use the **Reset Pose** button under the **Pose** section (the top-left corner of the window) to start over.

Continue to repeat this process for the actor's torso. Once you're finished with the torso, this is probably a great time to think about saving your work. Use the **Apply** button in the toolbar on the right-hand side.

When you're relatively satisfied with the bone weights and want to see how things are progressing, you can do some pose tests! Doing some pose tests allows you to test the range of motion and see if the sprites are associated with only the correct bones that influence the sprite weights without undesirable distortion. This provides you with the first indication of how the character may look when animated.

Figure 5.5 represents an example of a test pose for the actor:

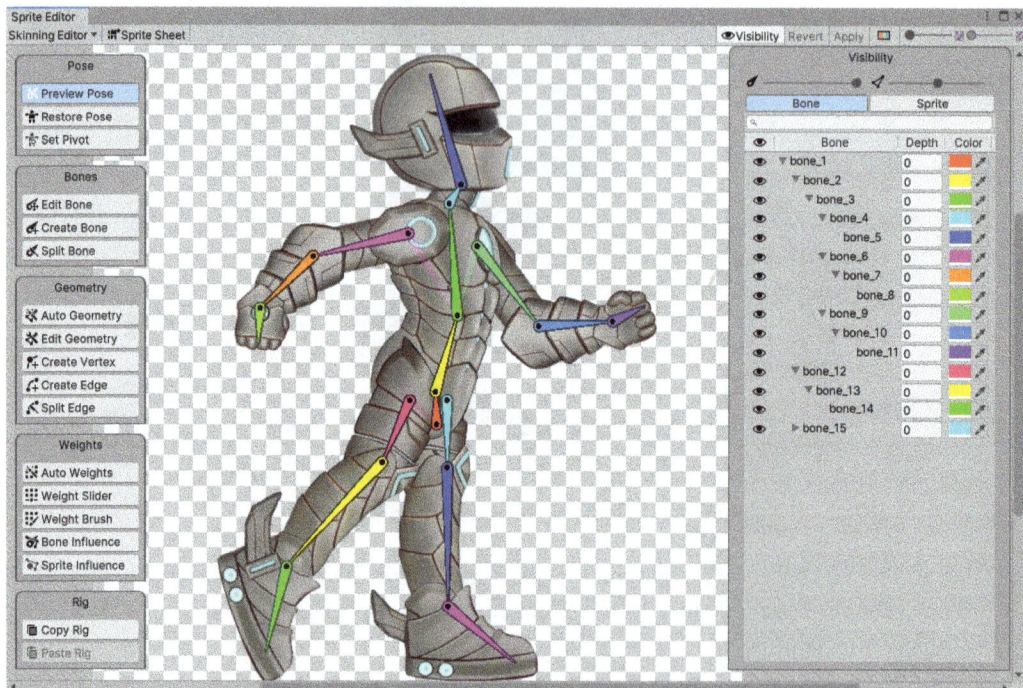

Figure 5.5 – Pose test for the actor

Being able to pose our character is great, but for the limbs, in particular, it can be time-consuming and challenging to get good results by rotating all the bones individually. Thankfully, there is a better way – using **IK**.

Setting up inverse kinematics (IK)

We've now finished working with **Skinning Editor** and will continue working with the actor in the scene to add IK. So, let's create a Prefab from the actor for our player character:

1. Drag the `PlayerCharacter1` asset from the **Project** window into the **Hierarchy** window or **Scene** view (optionally, you can create a new empty scene to work in).

2. Parent the actor to a new empty GameObject and name it `Player`.

3. Then, drag the GameObject into the `Assets/Prefabs` folder to create the Prefab.

Since we'll be using an IK setup to position our actor for animation, we'll want to change the actor's rig to make posing the limbs – in particular, the legs – much better. We could have done this when we first created the actor's rig, but it's easier to show the reason for this change while implementing IK, and it shows that we can go back and make changes to the bone structure at any time and make corrections where necessary.

Next, we'll be adding a new root bone to the rig using the same **Create Bone** workflow we used to create the actor's complete skeleton rig:

1. Start a new bone at the feet of the actor and name it `root_bone`.

2. Change the color to gray (just as an indication to us that no sprites are being influenced by it).

3. Now, click and drag the existing rig (`bone_1`) to parent it to the new root bone, as shown in *Figure 5.6*:

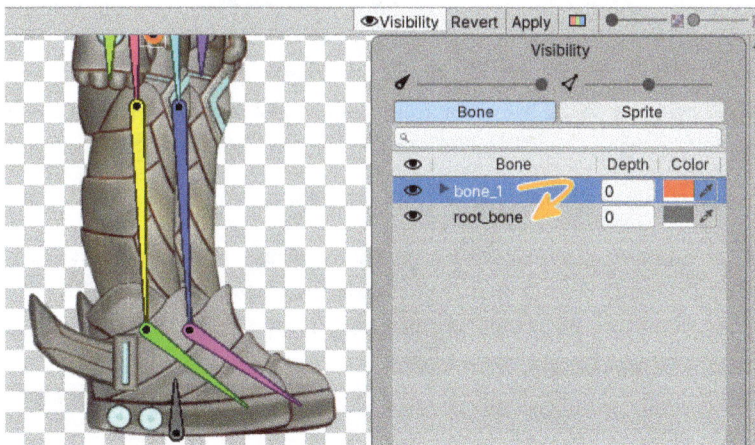

Figure 5.6 – Adding root_bone for IK and the parent existing rig

Be patient with me – this will all make sense in a bit when we add **IK Limb Solvers** for the legs.

Adding IK Solvers

We will make IK-constrained limbs for the arms and legs, starting with the right arm (the arm in front of the actor).

First, let's add GameObjects that act as our IK Effectors (the transform the IK Solver solves for) and work in tandem with the limb solver's Target:

1.   Select the forearm bone of the arm in front of the character's torso – bone_7 in my case.

2.   Right-click on it in the **Inspector** window and click **Create Empty** to add a child GameObject – this will make the new object a sibling of the hand bone.

3.   Rename it IK Effector and then position it at the tip of the forearm bone.

4.   Using the red arrow of the position tool gizmo, drag it down, extending it past the hand just a bit. Positioning the effector outside of the sprites will make them more visible and easier to click on to manipulate (this step is purely personal preference; you may also leave it at the tip of the forearm bone).

5.   Repeat these steps for the other arm (the arm behind the actor) and both legs.

6.   With our effectors in place, add the IK Manager 2D component to the root bone and start adding **IK Solvers** for the limbs by clicking the plus (+) button in the tab and then selecting **Limb**.

We're making this selection because it's meant for posing joints specifically for arms and legs (also known as a **two-bone Solver**).

> **Additional reading | Unity documentation**
> IK Solvers: `https://docs.unity3d.com/Packages/com.unity.2d.animation@9.0/manual/2DIK.html#ik-solvers`.

Adding an **IK Solver** property to the list will automatically add a new GameObject to the bottom of the actor object hierarchy named **New LimbSolver2D** with a LimbSolver2D component added.

Perform the following steps to complete the IK limb setup for the front arm:

1.   Rename New LimbSovler2D to Front Arm LimbSolver2D and keep it selected so that the LimbSolver2D component is visible in the **Inspector** window.

2.   Click and drag the IK Effector object for the front forearm to the **Effector** field of the LimbSolver2D component.

3. Once this **Effector** has been assigned, we can click the **Create Target** button – which visually sets a color to the dot at the base of the bone for all bones in the IK chain. This will also create a `_Target` child GameObject for the limb solver that we can now manipulate the position of in the **Scene** view to pose the limb.

4. Click and drag the circle icon that has now appeared on the **Target** transform in the **Scene** view to test out the IK limb.

5. Note that the default value for **Flip** is disabled, which may work just fine for how the target was created, but if the limb is bending backward as you drag the target around, then enable **Flip** to solve this issue.

Figure 5.7 illustrates the results of these steps:

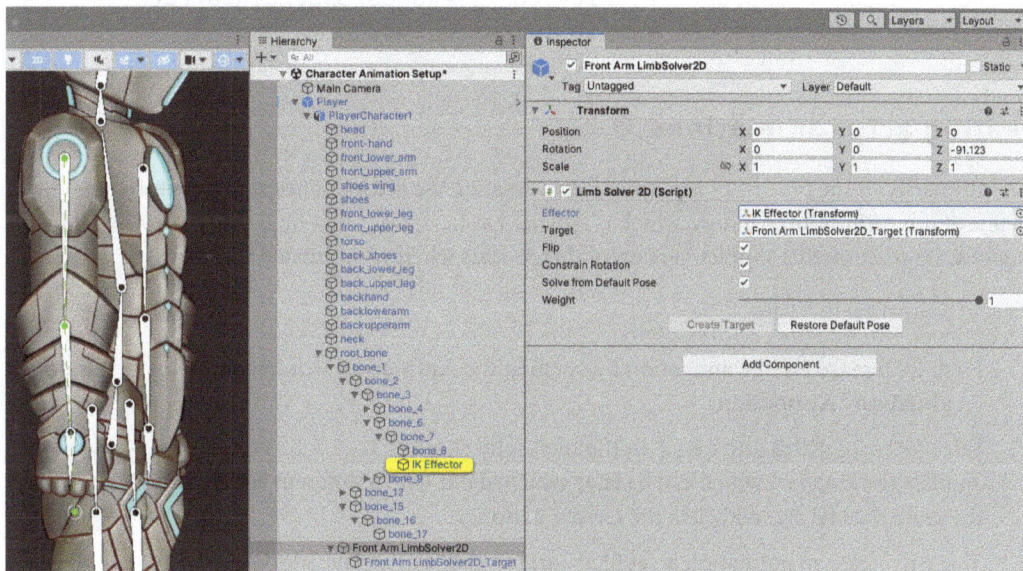

Figure 5.7 – IK Effector and limb solver setup

The `Front Arm LimbSolver2D_Target` object's transform position is what is keyframed during animation.

> **Important note**
>
> Once the IK Solvers have been set up, we can no longer scale the player character with the **Pixels Per Unit** value in the **PSD Importer** settings – this is due to the IK targets being based on the transform local positions, which won't update when the sprite's PPU is scaled. Instead, change the transform scale on the parented `PlayerCharacter1` object.

You can now follow these steps any time you need to create an IK limb on a skinned actor rig. Go ahead and make IK limbs for the back arm and both legs. With that, we're ready to animate!

> **Reminder**
> Don't forget to apply changes to the `Player` Prefab!

> **Additional resources | 2D characters**
> Another option for getting a head start on creating your rigged and animated 2D characters is to work with existing example characters from either the Dagon Crashers or Lost Crypt sample projects that Unity provides on the Asset Store:
>
> Dragon Crashers: `https://assetstore.unity.com/packages/essentials/tutorial-projects/dragon-crashers-2d-sample-project-190721`.
>
> Lost Crypt: `https://assetstore.unity.com/packages/essentials/tutorial-projects/lost-crypt-2d-sample-project-158673`.

Creating actor animations

Actor animation is the process of repositioning or rotating the actor's bones at different times on the timeline of the **Animation** window, which is recorded as keyframes that hold the transform data. In *Chapter 4*, we performed a simpler version of this process when we animated the vines.

Let's start animating the actor:

1. First, we'll make an idle animation by opening the **Animation** window by going to **Window | Animation | Animation**.

2. Select `PlayerCharacter1` in the hierarchy (not the root `Player` object; we want to animate the *graphics*, which can be later switched out with a different actor that may not have the same skeletal rig) and click the **Create** button.

3. This will create an animation asset file (`.anim`) that we'll name `Player Idle` and save to the `Assets/Animation` folder.

Now, put your animator hat on again, because it's time to animate!

The first thing to note while animating with IK limbs is why we added the new root bone in the *Setting Up inverse kinematics (IK)* section. Select the **Move** tool in the **Scene** view and click on the actor. With the bones now visible, select the pelvic bone (`bone_1`, our *original* root bone).

You'll notice that a square outline has appeared at the larger end of the bone (the base) and that hovering the mouse pointer within this square will show a move icon. Hovering over the part of the bone outside of the square will show a rotate icon. Click and hold the left-mouse button within the square and slowly drag it down.

The whole actor will start to move down except for the feet – they will remain firmly grounded in place. This would not be possible without IK and the new root bone we added. Yay!

One more thing to note with the IK limbs before letting you lose to create your idle animation – you can position the limbs in two ways:

- By clicking and dragging the IK circular targets in the **Scene** view

- By selecting the ... `LimbSolver2D_Target` objects in the **Hierarchy** window and using the **Move** tool in the **Scene** view (or entering **Transform** values in the **Inspector** window)

Do not try to keyframe individual bones in the IK chains, the IK Effector objects, or the objects with the `LimbSolver2D` components! Create your idle animation by following these steps:

1. Click the red record button (the **Animation** window; **Keyframe Recording** mode) to start recording keyframes (any time a bone is rotated or repositioned, a key will be created).

2. Pose your actor into a nice starting position for a simple idling in place.

3. Scrub the timeline ahead about 1.5 seconds and make a second pose by lowering the torso bending into the knees, bending the arms, and tilting down the head slightly – something like that.

4. Now, to animate between these two poses, click and drag a rectangle around the starting frame's keys – press *Ctrl/Cmd + C* to copy those keys.

5. Scrub the timeline to 3 seconds and press *Ctrl/Cmd + V* to paste the keys.

6. Click the red record button again to stop recording, then use the **Play** button to check the results.

You know the drill – animation is more art than technical, so fiddle with the poses and timing until you get good results. *Figure 5.8* illustrates a completed actor idle animation:

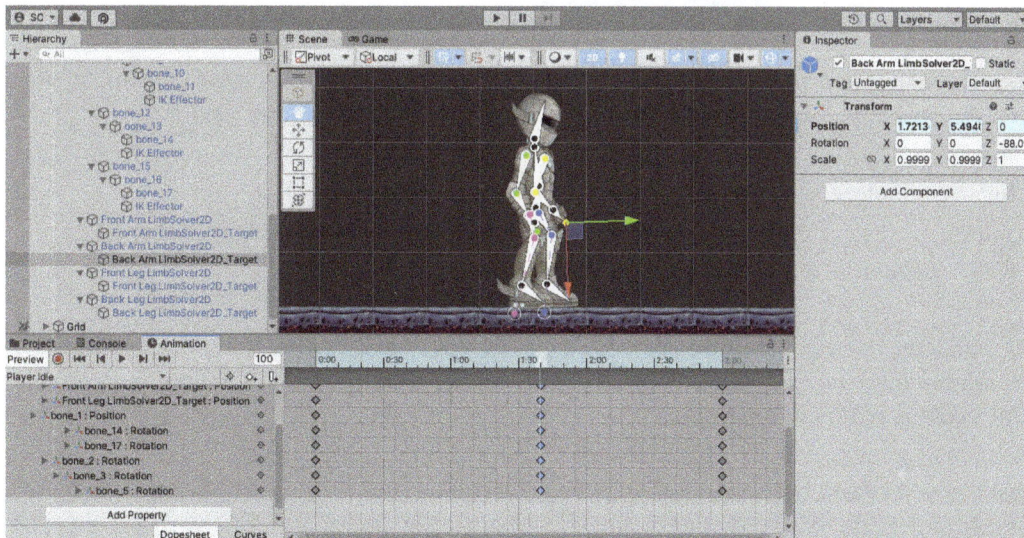

Figure 5.8 – Actor keyframe idle animation

Note that in *Figure 5.8*, I've changed the color of the IK Solvers (the `IKManager2D` component on the `root_bone` object) to give the limb bones their own color – another way to visually organize the different components that make up the actor and make it a bit easier to work with.

With our base idle animation added to the character, let's look at how to drive animation and apply movement to our character from player input.

Using an Input Action Map

We'll continue to use the new **Input System** for the adventure game player input. So, let's first make sure we have the `Input System` package installed:

1. Open **Package Manager** by going to **Window | Package Manager**.
2. The default **Packages** location is **In Project**, so if you see **Input System** in the list under the **Package – Unity** section, then we're all set!
3. If you can't find **Input System**, then switch the **Packages** dropdown to **Unity Registry**, select **Input System** from the list, and click **Install** (the bottom-right corner of the window). See *Chapter 2*, section *New Input System* for a refresher.

For the collection game, we received input directly from **Input Device**; this time, we will use the **Input Action** approach. For that, we're going to need an **Action Map** asset.

Thankfully, creating a default action map for the player character is quite trivial – with most of the device inputs we'll need for the keyboard and a game controller already made!

Follow these steps to accomplish this:

1. The magic happens via the `Player Input` component. So, add it to the root object of the `Player` Prefab by using the **Add Component** button in the **Inspector** window.

> **Important note**
> You can work with the `Player` Prefab by adding it to a scene or working on it directly by double-clicking the Prefab in the **Project** window.

2. Once the `Player Input` component has been added, click the **Create Actions…** button, as shown in the following screenshot. This will create a default set of **Input Action Maps**, **Input Actions**, and **Input Bindings**:

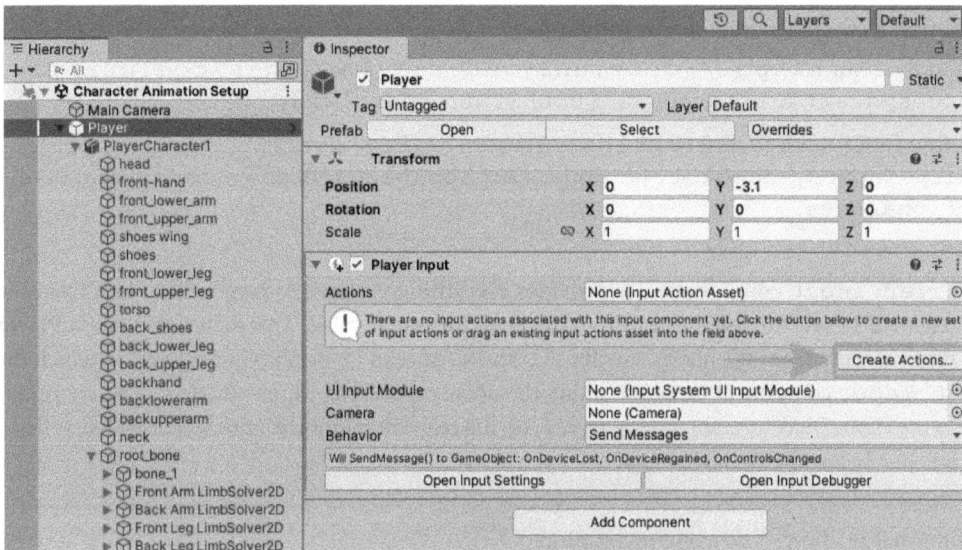

Figure 5.9 – Player Input | Create Actions

3. You will be prompted to save the new `.inputactions` asset. Select the `Assets/Settings` folder and use `Player Input` as the name for it. This asset will then be connected to the `Player Input` component and bring up the **Input Actions editor**.

Here, we can see all of the actions pre-populated for the player action map, as shown in *Figure 5.10* – everything we need plus more!

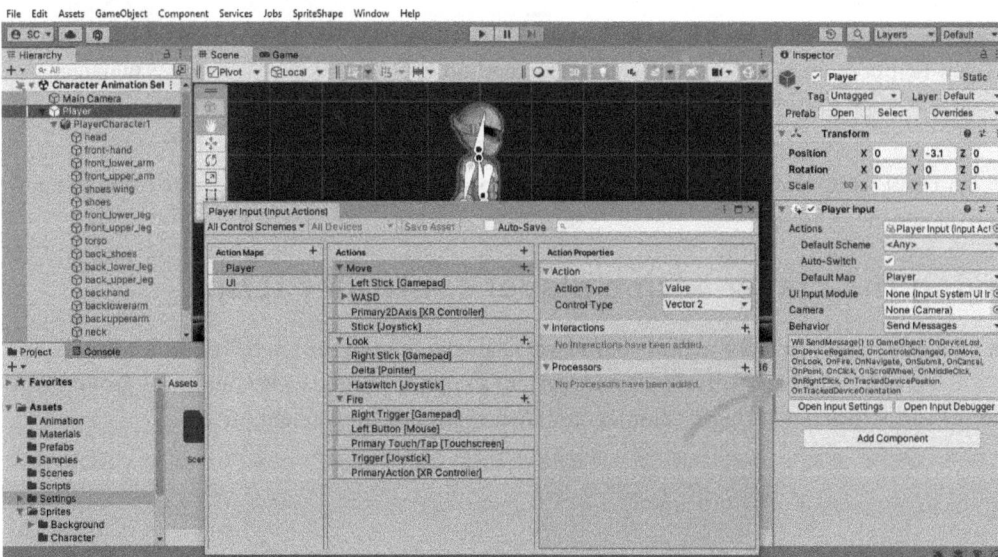

Figure 5.10 – Player Input Action Maps (Input Actions)

> **Additional reading | Unity documentation**
>
> Input System: `https://docs.unity3d.com/Packages/com.unity.inputsystem%401.4/manual/QuickStartGuide.html`.
>
> Editing Input Actions Assets: `https://docs.unity3d.com/Packages/com.unity.inputsystem%401.4/manual/ActionAssets.html#editing-input-actionassets`.

The `Player Input` component also provides everything we need to respond to input from the player. *Figure 5.10* sets the **Behavior** field to the **Send Messages** value. This means that when **Player Input** receives input from the player, it will use `SendMessage()` on the GameObject to which the `Player Input` component belongs, calling the name on every component (`MonoBehaviour`) added to the GameObject. As shown in *Figure 5.10*, the relevant method names will be called as listed in the box just below the **Behavior** field.

> **Additional reading | Unity documentation**
>
> Notification behaviors: `https://docs.unity3d.com/Packages/com.unity.inputsystem%401.4/manual/Components.html#notification-behaviors`.

For simplicity's sake, we'll stick with the **Send Messages** behavior. In cases where input events could occur quite rapidly, **Invoke Unity Events** would be preferred over `SendMessage()`, which is slower due to having to use **Reflection**.

> **Reflection | C#**
>
> `SendMessage()` heavily relies on **Reflection** to find the method to call at runtime. **Reflection** is slow (supposedly 3x slower than calling the method directly) because it requires managed code to read (search) its metadata to find assemblies.

With our input set up and we know how to respond to the inputs we've defined, we're ready to write our player controller code!

Moving the player with a player controller script

We'll need more than just input to move our player character around the level. We'll have to configure the `Player` object to work with the **Physics** system so that it interacts with our level's ground and platforms and interactivity trigger volumes, and then applies the movement force.

Configure the `Player` Prefab's root GameObject with the following components while referring to *Figure 5.11*:

1. Add a `Rigidbody2D` component (by pressing the **Add Component** button) with these values:

    I.   **Use Auto Mass** enabled: Physics objects react more believably in the physics simulation if they have proper mass. This setting will estimate the mass based on the density and area of the accompanying collider.

    II. **Linear Drag** = 1: We'll add some drag to the player's movement to provide additional constraints on the move to help the player feel more grounded and not very floaty.

    III. **Constraints | Freeze Rotation Z** enabled: We want to prevent the player from spinning around and have them always stand vertically (perpendicular to the ground plane).

2. Add `CapsuleCollider2D` (by pressing the **Add Component** button). Use the **Edit bounding volume** button and modify the collider's shape until it encompasses the actor.

    This represents the player's hitbox, which may need to be adjusted over time as you playtest:

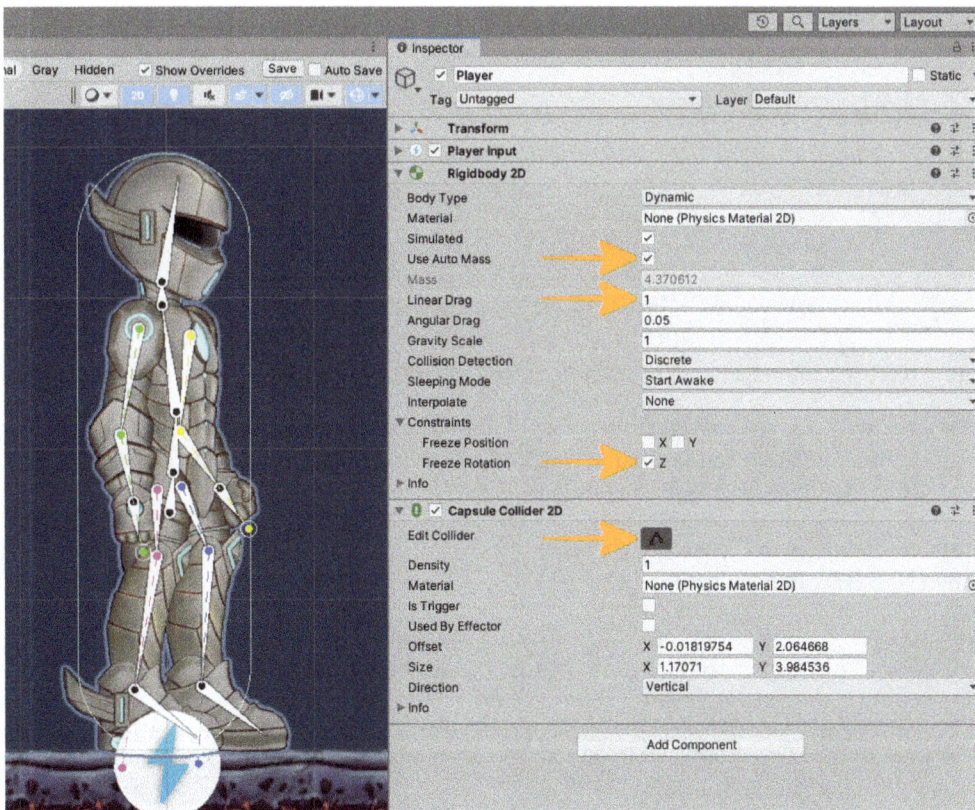

Figure 5.11 – Player physics setup

With the required **Physics** components now added and configured for the player, let's have a look at taking player input and applying the movement force through code since the `Player Input` component is only responsible for reading the device input and invoking the assigned action.

Processing Player Input

While we could use the device input directly, as we did in the *Player controller script* section of *Chapter 2*, it's essential to understand the value of using **Input Action Map**. Here are a few examples:

- Multiple device inputs can be simultaneously configured without the need to change the input handling code

- A different action map can be switched to, depending on the current input actions required

- Key rebinding can be implemented without changing the input handling code

Armed with this knowledge, you can determine the best approach for different use cases (are you prototyping? Are you engineering a flexible solution for a commercial title?).

Previously, we continuously polled the `isPressed` key state directly on the input device in the `Update()` loop to control the player movement, like so:

```
void Update()
{
    if (keyboard.aKey.isPressed
        || keyboard.leftArrowKey.isPressed)
            _moveHorizontal = -1.0f;

    ...
}
```

Since we're receiving our input handling from `SendMessage()` now, we'll need to make some changes.

`SendMessage()` is called anytime there is input. Input is not continuously polled! It behaves like an event whenever an input action occurs, which means that for the *Move* action, `OnMove()` will be called when a key is pressed and when it's released.

The **Inspector** window in *Figure 5.10* shows all the `SendMessage()` method names that were invoked for the defined input actions.

`SendMessage()` will also pass an `InputValue` parameter representing the **Action Type** and **Control Type** – `Vector2` in the case of *Move*, which holds the horizontal and vertical input values.

We'll use the horizontal (X-axis) value to determine if a **move left/right** button was pressed or released since we want to move our player while the button is being held down (or while holding a game controller's stick in a certain direction).

Our move method (OnMove()) will be called with a Vector2 X-axis value for a horizontal input of -1 and 1 for left and right, respectively, when a button is pressed, and a value of 0 when the button is released.

Now that we have all this information, we can begin to code our input handler!

Creating the PlayerController script

Let's start by creating a new C# script named PlayerController in the Assets/Scripts folder.

We can now add our input handler method for the *move* action – OnMove() – and assign the InputValue parameter's Vector2 value to a private member variable named _movementInput. We'll use this later when we calculate the movement to be applied to the player's **Rigidbody2D** velocity.

Here's what our initial code looks like:

```
public class PlayerController : MonoBehaviour
{
    private Vector2 _movementInput;

    void OnMove(InputValue value)
    {
        var move = value.Get<Vector2>();
        _movementInput = (move.x != 0f)
            ? new Vector2(move.x, 0f) : Vector2.zero;
    }
}
```

We read the input values of the input using the Get() method of the value parameter variable. We're only interested in the horizontal movement here and will set the _movementInput value depending on the input state: pressed or released.

So, if move.x equals 0, then the button key was released (think of it as IsPressed == false). Then, using a ternary operator (?:), we'll assign Vector2.zero to make sure that any/all input is ignored.

When move.x is not equal to zero, we have a button key press input value that represents the direction (-1: *left*, 1: *right*) and assign it to _movementInput with a new Vector2 of the direction and no vertical value: new Vector2(move.x, 0f).

With this approach, we can still respond to vertical input later, such as one for a jetpack! Now that we have a direction, we can apply force (in the form of velocity) to move the player character in that direction.

Let's add the following code to handle this:

```
private Rigidbody2D _rb;

void Awake() => _rb = GetComponent<Rigidbody2D>();
void FixedUpdate() => UpdateVelocity();

private void UpdateVelocity()
{
    var velocity = _rb.velocity;
    velocity += Time.fixedDeltaTime * _movementInput;
    _rb.velocity = velocity;
}
```

Let's break this code down item by item:

- The `_rb` variable: This private member variable will hold a reference to the `Rigidbody2D` component that we previously added, and we'll be setting the velocity here to move the player. We once made this a `public` field that could be assigned in the **Inspector** window, but we'll keep it private this time to encapsulate it.

- The `Awake()` method (expression body): Since our `Rigidbody2D` variable is private, we're only using this Unity message event to get the reference to the `Rigidbody2D` component on this object at runtime using `GetComponent()`.

> **Additional reading | Unity documentation**
> `GameObject.GetComponent`: `https://docs.unity3d.com/2022.3/`
> `Documentation/ScriptReference/GameObject.GetComponent.html`.

- The `FixedUpdate()` method (expression body): This Unity message event is called at every **Physics Fixed Timestep** and should always be used when applying physics. Here, we're simply calling the `UpdateVelocity()` method.

- The `UpdateVelocity()` method: This method will calculate and apply the velocity vector to the Player's `Rigidbody2D` component.

  - `var velocity`: This holds the current value of the `Rigidbody2D` component's velocity that we'll modify and assign back.

  - The `velocity` calculation: This adds the movement input vector multiplied by `fixedDeltaTime` to the velocity variable. Multiplying by `fixedDeltaTime` ensures frame rate independence.

- The `_rb.velocity` assignment: Here, we simply assign the calculated velocity back to the `Rigidbody2D` component, which moves the player. Remember that `FixedUpdate` should always be used for executing physics-related code, especially for applying force to a `Rigidbody` component!

Author's note

I've read discussions on whether it's necessary to apply `deltaTime` in `FixedUpdate` because it runs at a consistent framerate. Anyway, I err on the side of knowing that the `FixedUpdate` interval is always relative to the in-game time. Also, the Unity-provided example code always includes `deltaTime`, and being consistent in including it will ensure we achieve frame rate independence across all devices.

Additional reading | Unity documentation

`Time.fixedDeltaTime`: `https://docs.unity3d.com/2022.3/Documentation/ScriptReference/Time-fixedDeltaTime.html`.

Time and Frame Rate Management: `https://docs.unity3d.com/2022.3/Documentation/Manual/TimeFrameManagement.html`.

Multiplying by zero always gives zero, so `Vector2.zero` with `_movementInput` will assign a velocity value of zero, stopping the player's movement.

We've applied velocity to the player, but don't get too excited. This will likely have no net effect on moving the player because the velocity value is too small! Let's fix that by applying an acceleration multiplier to the velocity.

Add the following member variables and make the changes that are shown here to the `UpdateVelocity()` method:

```
[Header("Movement")]
[SerializeField] private float _acceleration = 0.0f;
[SerializeField] private float _speedMax = 0.0f;

private void UpdateVelocity()
{
    var velocity = _rb.velocity;
    velocity += _acceleration * Time.fixedDeltaTime
        * _movementInput;
    velocity.x = Mathf.Clamp(velocity.x,
        -_speedMax, _speedMax);
    _rb.velocity = velocity;
}
```

Let's break down these changes:

- We've added two new variables, `_acceleration` and `_speedMax`, which we've set to private but made available to assign in the **Inspector** window by introducing a new attribute. By using the `[SerializeField]` attribute to decorate a private accessor field, we're telling Unity to serialize it and make it available for assignment in the **Inspector** window without making it public. This allows us to follow good OOP practices such as **encapsulation** so that other classes won't be able to modify the field's values:

    - `_acceleration`: This applies our overall velocity amount (force) for how quickly we'll reach the maximum speed value.

    - `_speedMax`: The maximum speed of the player character's movement.

- In the `UpdateVelocity()` method, we've made the following changes:

    - We've modified the line that adds to the velocity vector by multiplying by the `_acceleration` value.

    - We've ensured that `velocity.x` will be clamped to the maximum speed value by using the `Mathf.Clamp()` method.

Additional reading | Unity documentation

`SerializeField`: `https://docs.unity3d.com/2022.3/Documentation/ScriptReference/SerializeField.html`.

You can now add the `PlayerController` script to the root of the `Player` Prefab, add your player to a scene with some ground (don't forget to add a collider), and playtest. You can move the player with the *A* and *D* keys, left and right arrow keys, or a game controller's left stick. You can adjust values for acceleration, max speed, and the **Rigidbody2D** properties for mass and linear drag.

PlayerController.cs code

To view the completed code for the `PlayerController` class, visit the following GitHub repository: `https://github.com/PacktPublishing/Unity-2022-by-Example/tree/main/ch5/Unity%20Project/Assets/Scripts`.

Playtesting means tweaking these movement variables repeatedly until the player control *feels right* to you. I've currently settled on values of 30 for acceleration and 8 for max speed. As you build out the game, you'll find yourself going back and tweaking these values many, many more times – game feel is crucial to player satisfaction.

Physics materials

While playtesting, you may notice that the player may slide for a bit when you release the key (or controller stick). This can be corrected by freezing the player in place when the key is released, but I believe a better way is to change the properties of the ground the player is currently standing on. By adjusting the friction of the ground, we can provide different types of ground, including ice, where the player would slide when input is released.

We're going to add a **physics material** to the ground to provide *higher default friction*. That way, we can have areas with less friction later to change up the gameplay, and so on.

Follow these steps to create a new physics material and apply it to the ground objects in your level:

1. In the **Project** window, in the `Assets/Settings` folder, create a new folder and name it `Physics Materials`.

2. Right-click within the new folder, select **Create | 2D | Physics Material 2D**, and name it `Default Friction`.

3. Select the newly created physics material and, in the **Inspector** window, assign a **Friction** value of `10` (this, like all other gameplay variables, is subject to change with playtesting).

4. Now, assign the physics material to your ground objects by dragging it into the **Material** field of the `Collider` component, as seen in the following figure:

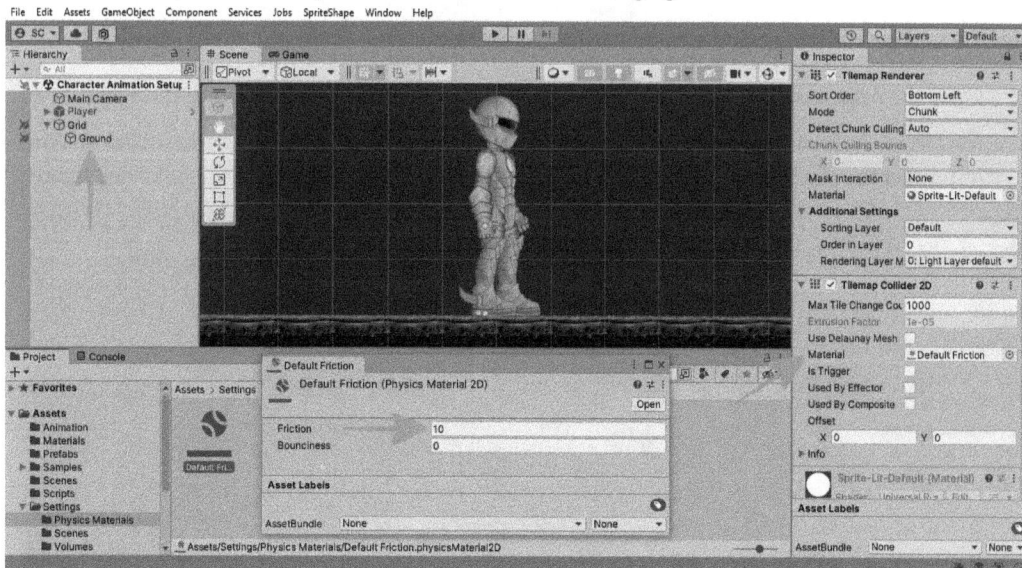

Figure 5.12 – Assigning a physics material to the ground collider

Continue to playtest and adjust these values.

We'll finish up the initial player controller next by adding a run animation and having the player always face the direction of movement.

Animating the character with Mecanim

In the *Creating actor animations* section, we added an **Animator** component to the `PlayerCharacter1` object when we created the idle animation for the actor. The `Animator` component allows us to assign animation to a GameObject in the **Scene** view – it's the interface responsible for controlling Unity's **Mecanim** animation system (on the actor or any other GameObject you'd like to animate).

An **Animator Controller** asset was also created in the same folder the idle animation was saved to and automatically referenced in the **Controller** field of the `Animator` component. This controller asset defines what animations to use and when and how to transition and blend them.

> **Additional reading | Unity documentation**
> Animator: `https://docs.unity3d.com/2022.3/Documentation/Manual/ class-Animator.html`.

To continue setup in the `Animator` component, we'll need a second animation to transition to when our player character moves or, let's say, runs.

Follow these steps to create an additional animation for the actor:

1. Open the **Animation** window, then select the `PlayerCharacter1` object in the **Scene** view.

2. Now, in the **Animation** window, click the **Animation Clips** list dropdown (the top left of the window just under the playback controls), then click **Create New Clip…**, as seen in the following figure:

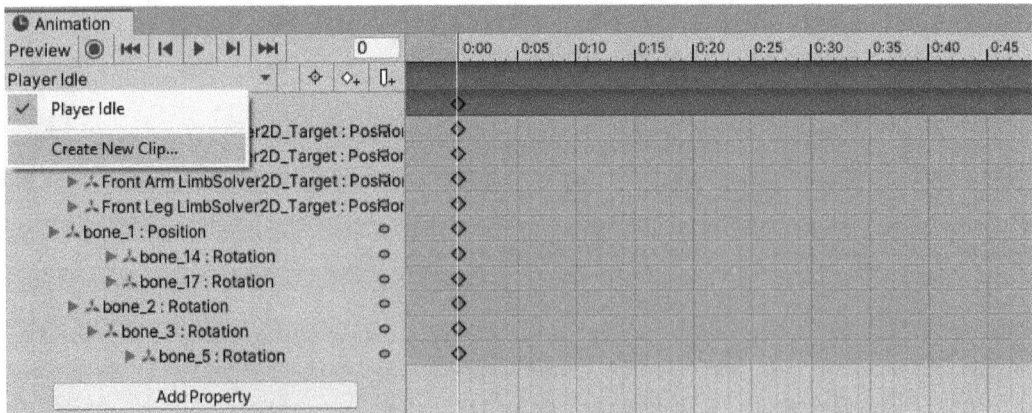

Figure 5.13 – Create New Clip…

3.   Save the new clip as `Player Run` in the `Assets/Animation` folder.

4.   Animate a run cycle. You've got this!

Seriously though, animation is a skill you'll need to grow into. While I can create animations, I still need lots of practice in this area to get better at it (I'll likely employ a skilled character animator for the final actor animations in the finished project files for this book, and these will be available for you to use and learn from as well).

Additional resources | 2D character animation

Another option for getting a head start on character animations is to work with existing example characters from either the Dragon Crashers or Lost Crypt sample projects that Unity provides on the Asset Store.

Dragon Crashers: `https://assetstore.unity.com/packages/essentials/tutorial-projects/dragon-crashers-2d-sample-project-190721`.

Lost Crypt: `https://assetstore.unity.com/packages/essentials/tutorial-projects/lost-crypt-2d-sample-project-158673`.

We can now move on to wiring up the transition for the *idle to run* animation.

Transitioning animation states

To open the **Animator** window, you can either double-click on the **Controller** field of the `Animator` component on the `PlayerCharacter1` object or double-click on the `PlayerCharacter1` asset in the `Assets/Animation` folder in the **Project** window.

With the **Animator** window open (you may want to dock it so that you can still see all the Editor windows clearly), you should see the default states (**Any State**, **Entry**, **Exit**) as well as states for the animation clips currently on the actor (**Player Idle**, **Player Run**):

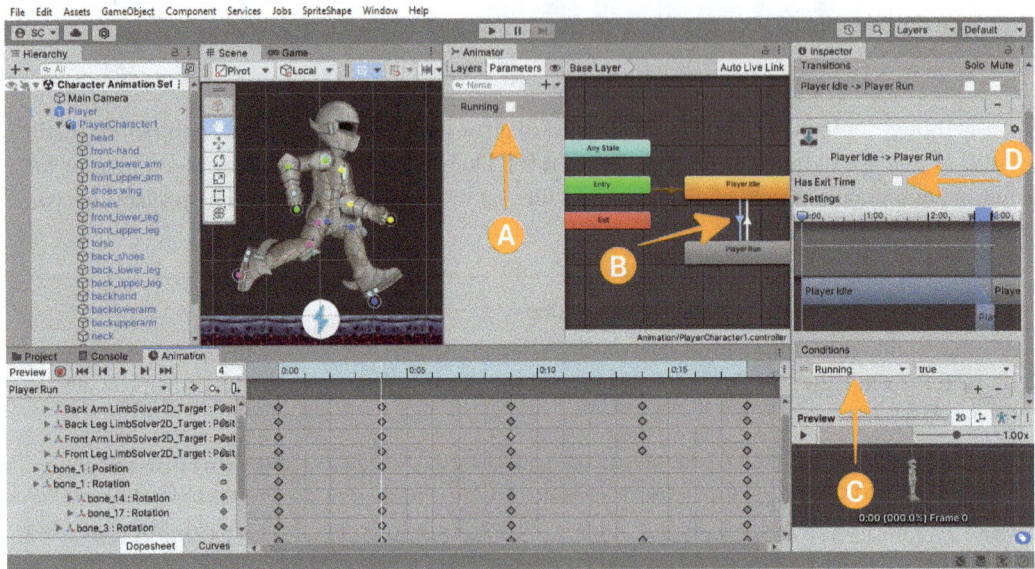

Figure 5.14 – Actor animation transition with Mecanim

Note that you can navigate the **Animator** window using the mouse scroll wheel to zoom in/out and use the *Alt* + left-mouse button/*Option* + left-mouse button shortcut to pan the view. Click and drag on any State Node to reposition it in an orderly fashion.

> **Additional reading | Unity documentation**
> The **Animator** window: `https://docs.unity3d.com/2022.3/Documentation/Manual/AnimatorWindow.html`.

Follow these steps to create a **Transition** from the **Player Idle** to **Player Run** states:

1. First, create a Boolean parameter for whether we are running or not by executing the following steps:

    I.   First, select the **Parameters** tab (*A* in *Figure 5.14*), then click the plus (+) dropdown and select **bool**.

    II. A new parameter will be added and highlighted for renaming. Name it `Running` – we want a Boolean parameter to hold the state for if the player is running (`true`) or not running (`false`).

2. Right-click on the **Player Idle** node and select **Make Transition** from the popup menu.

3. With the **Transition** arrow attached to the mouse pointer, click on the **Player Run** node.

4. Now, click on the newly created transition line (*B* in *Figure 5.14*) to view its properties in the **Inspector** window.

5. Within the **Conditions** list, click the plus (+) button to add a new condition – **Running** should be selected by default since it is our only parameter. Make sure the value is `true`; meaning, transition to the assigned node when the **Running** bool value becomes `true`.

6. Go back to *step 2* and repeat the instructions to transition back to **Player Idle**, but when the **Running** condition is `false`.

7. The last step is to disable **Has Exit Time** (*D* in *Figure 5.14*) for both transitions. We want the state to exit (the animation to stop) as soon as the condition is met and not wait for the animation to play completely.

Since we haven't wired up the animation states to the player input yet, you can test manually to make sure the transition is working well by entering **Play Mode** and, while viewing both the **Game view** and **Animator** windows, toggling the **Running** parameter on/off. The player should change from an *idle* to *run* animation and back every time you toggle the parameter. Yay!

Now, let's make this transition based on the player's input.

Changing the animation state with code

The first thing we'll need is a variable to hold the reference to the `Animator` component (assignable in the **Inspector** window). Then, when we receive player input for moving the character left or right, we will set the `Running` parameter to `true`, and when the player stops moving the character, we will set the `Running` parameter to `false`. We'll make quick work of this and accomplish it with only one line of code!

Add the following code to the `PlayerController` script:

```
[Header("Actor")]
[SerializeField] private Animator _animator;

private void UpdateVelocity()
{
    ...
    // Update animator.
    _animator.SetBool("Running", _movementInput.x != 0f);
}
```

The `_animator` variable declaration is our reference to the `Animator` component in the `Player` object. We marked it as `private` so that no other class can modify it but decorated it with the `[SerializeField]` attribute so that it's serialized and assignable in the **Inspector** window.

In the `UpdateVelocity()` method, we added a call to `_animator.SetBool()` and passed in the `Running` string to identify the bool parameter we'd like to set. We'll pass in the evaluation result of the `_movementInput.x != 0f` expression for the bool value.

What this means is that if our player is receiving movement input (value is not zero), then we are moving (`Running` equals `true`); otherwise, (that is, if `_movementInput.x == 0f`), we are standing still (`Running` equals `false`).

Once you've added and saved this code, drag the child `PlayerCharacter1` object to the `PlayerController` component's **Animator** field using the **Hierarchy** and **Inspector** windows in the **Editor** view.

Enter **Play Mode** and move the player left and right to test that the animation is transitioning from idle to run and back. You'll likely notice that even though the animation is transitioning, the actor is facing the wrong direction while moving to the left!

We'll address this next to finish up the player movement.

Flipping the player character

While the **Animator** handles changing animation states for us quite nicely, it does not handle flipping the direction the actor is facing based on player input. It's pretty simple to add logic to our `PlayerController` script to ensure the player is always facing the direction of movement.

Open the `PlayerController` script and, at the end of the `OnMove()` method, add a call to a new method named `UpdateDirection`. Create the `UpdateDirection()` method with the following code:

```
void OnMove(InputValue value)
{
    …

    UpdateDirection();
}
private void UpdateDirection()
{
    if (_movementInput.x != 0f)
    {
        transform.localScale = Vector3.one;
        if (_movementInput.x < 0f)
            transform.localScale = new
                Vector3(-1f, 1f, 1f);
    }
}
```

The simple trick we're employing to flip the direction the player character is facing is to set the `localScale` X value to -1 for the player object's **Transform** if the movement input value is less than zero (that is, input indicating moving to the left).

The first `if` statement in `UpdateDirection()` checks if we have any input for moving the player. Remember, a value of zero means that the player released the direction key (or game controller stick).

If the movement's input horizontal value is not zero, we first set a default scale for facing right (an X value of 1). If it turns out that the movement input is for facing left (an X value of -1), then we set `localScale` to `Vector3` with an X-axis value of -1. Easy-peasy.

Bonus activity

Based on the techniques learned in this chapter for processing player input, moving the player via its `Rigidbody2D` component and animating the player using the **Animator** all within the `PlayerController` script, add the ability for the player character to jump. Might as well jump, go ahead and jump!

I didn't necessarily leave you hanging here. If you're looking for a lifeline to solve this problem, you can view instructions for completing this jumping bonus activity by visiting the following GitHub link: `https://github.com/PacktPublishing/Unity-2022-by-Example/tree/main/ch5/Unity%20Project/Assets/Scripts/Jumping`

In this section, you learned how to create animations and apply them to the player, as well as how to transition and change the animation state with code, all while using Mecanim. We finished off by learning how to flip the facing direction of the player character.

Summary

This chapter walked us through the complete setup of an animated 2D player character, including importing art and setting up an animatable rig via PSD Importer, setting up IK Solvers, and creating and applying animations to the player using Mecanim.

We continued by adding movement ability from player input by using an Input Action Map asset with the new Input System, coding a simple player controller script, processing input, and changing animations based on the current player action, also using Mecanim.

In the next chapter, we'll add a weapon for the player so that they can shoot projectiles efficiently.

Introduction to Object Pooling in Unity 2022

In *Chapter 5*, we imported and prepped artwork for use with additional 2D animation tooling, which is bringing the game to life. We also processed player input using an input action map – instead of reading device input directly – with the new Input System, and we made a `PlayerController` script to move the player.

We dove deeper into Mecanim as we learned how to transition between animations and drive animation state changes from code.

In this chapter, you will be introduced to object pooling while we use this optimization pattern for the player's shooting mechanic, and we'll accomplish that using Unity's object pooling API. The object pooling software design will be based on a pooled player shooting model UML diagram.

In this chapter, we're going to cover the following main topics:

- The object pooling pattern
- A pooled player shooting model

By the end of this chapter, you'll be able to create an optimized shooting mechanic for a ranged weapon.

Technical requirements

To follow along in this chapter with the same artwork created for the project in the book, download the assets from the following GitHub link: `https://github.com/PacktPublishing/Unity-2022-by-Example/tree/main/ch6`.

To follow along with your own artwork, you'll need to create similar artwork using Adobe Photoshop, or a graphics program that can export layered Photoshop PSD/PSB files (e.g., Gimp, MediBang Paint, Krita, and Affinity Photo).

You can download the complete project on GitHub at `https://github.com/PacktPublishing/Unity-2022-by-Example`.

The object pooling pattern

The **object pooling** design pattern is a type of **creational** or **abstract factory** design pattern that uses a stack to hold a collection of initialized object instances. It is excellent for use in situations when you will have either a large number of objects that need to be spawned or objects that will be created and destroyed rapidly.

Since we will be shooting projectile objects from a weapon – which can be performed at a high rate by the player – this is a great place to apply object pooling because repeatedly instantiating and destroying objects comes with a high cost. In this case, object pooling provides a way to optimize CPU, memory, and **garbage collection (GC)**.

Rather than creating a new projectile object directly every time the player needs to shoot, we'll instead reuse an already instantiated projectile object by requesting it from the objects in the pool. As such, the Object Pool provides methods for requesting (getting) and returning (releasing) objects. So, for example, for a pool of 10 projectile objects, we'll get one at a time from the pool when the player shoots and return the projectile that was shot when it expires (e.g., hit something).

If you recall from our **Game Design Document (GDD)** in *Chapter 4*, specifically *Table 4.1*, we defined shooting capabilities, so we'll use object pooling to implement this mechanic in a performant and optimized way, using Unity's new **object pooling API**.

The Unity object pooling API

Unity has added a new namespace to the engine – `UnityEngine.Pool` – that includes several new classes to implement the object pooling pattern. For our requirement of a weapon to shoot bullets, we'll use the `ObjectPool<T0>` class.

> **Additional reading | Unity documentation**
> `ObjectPool<T0>: https://docs.unity3d.com/2022.3/Documentation/ScriptReference/Pool.ObjectPool_1.html`

The following is a list of required actions we'll need when working with the object pool:

- `Creating` (instantiating): Making a new object instance in the pool available.
- `Getting` (requesting): Retrieving an available object instance from the pool (or creating and returning a new one if more are needed).

- Releasing (returning): Putting an active object instance back into the pool for reuse when it's finished with.

- Destroying (removing): Removing an instantiated object from the pool completely if it grows over its size limit.

Thankfully, or by design, the ObjectPool<T0> class provides everything we need, such as creating a pool and taking and returning items to the pool. Now, let's create a new object pool for our projectiles.

Creating a new object pool

Let's have a look at the following code that creates a new object pool of BulletPrefab projectiles (of type ProjectileBase; there'll be more on this shortly in the *Creating the pooled player shooting model* section):

```
private void Start()
{
    _poolProjectiles = new ObjectPool<ProjectileBase>(
        CreatePooledItem, OnGetFromPool,
        OnReturnToPool, OnDestroyPoolItem,
        collectionCheck: false,
        defaultCapacity: 10,
        maxSize: 25);

    ProjectileBase CreatePooledItem() =>
        Instantiate(_weapon1.BulletPrefab);

    void OnGetFromPool(ProjectileBase projectile) =>
        projectile.gameObject.SetActive(true);

    void OnReturnToPool (ProjectileBase projectile) =>
        projectile.gameObject.SetActive(false);

    void OnDestroyPoolItem(ProjectileBase projectile) =>
        Destroy(projectile.gameObject);
}
```

In the preceding code, we can see a method (a local function in this case) declared for each of the required ObjectPool parameters, corresponding to the actions we listed previously.

A local function (C#)

The new `ObjectPool` creation code uses local functions instead of a common approach, using lambdas (anonymous delegates), so that we, for one, avoid unnecessary memory allocations. We create a local function by declaring a method inside the body of an already existing method; this also limits the scope of a local function to only being able to be called from within the method, which promotes encapsulation over using private member methods (we don't need these methods outside the scope of setting up the object pool, and they only need to be called once).

A delegate has to be created when using a lambda, which is an unnecessary allocation if a local function is used. Allocations to capture local variables are also avoided, as local functions are really just functions; no delegates are necessary. In addition, calling a local function is also cheaper, and performance can be increased even further if in-lined by the compiler (eliminating call-linkage overhead).

Also, local functions just look better! They provide better code readability and verbose parameter names – a lambda anonymous delegate would obscure each parameter type! (Can you tell I'm just a bit biased here?)

Here's some additional reading on the subject: `https://docs.microsoft.com/en-us/dotnet/csharp/programming-guide/classes-and-structs/local-functions`.

Here are some interesting design notes from when local functions were added to C# 7: `https://github.com/dotnet/roslyn/issues/3911`.

I've declared the following local functions within the `Start()` method:

- `CreatePooledItem`: This will instantiate an object of type `ProjectileBase` when a new item is needed. This is a bullet Prefab we have a reference to on the player's weapon.

- `OnGetFromPool`: We'll use `_poolProjectiles.Get()` to return a `ProjectileBase` object instance, while this method calls `gameObject.SetActive(true)` to enable the object for use.

- `OnReturnToPool`: Calling `_poolProjectiles.Release(projectile)` will execute `projectile.gameObject.SetActive(false)` on the object instance passed in, making sure it's inactive (disabled) while sitting in the pool waiting to be retrieved.

- `OnDestroyPoolItem`: Calling `Destroy(projectile.gameObject)` when an item is removed from the pool means the object will no longer exist in the scene.

To clarify some of the preceding actions, *instantiate* means an object is created and exists in the Scene. When the instantiated object's active state is `SetActive(true)`, it is visible in the Scene, and code is executed.

Setting the GameObject as `SetActive(false)` will ensure that it doesn't display in the Scene, and for each component, the `Update()` method will no longer be called.

Additional parameters affecting the object pool

Beyond the preceding action methods, we have three additional parameters that affect how the pool functions. They are as follows:

- `collectionCheck`: We can save some CPU cycles if we set this parameter to `false`, as it won't check whether an object was returned to the pool already (be cautious with this value, since it will throw errors if you try to release an item already in the pool).

- `defaultCapacity`: You should set this value to the number of projectiles we'll generally need to have available on screen simultaneously (you can best determine this number by playtesting the rate of fire).

- `maxSize`: This value will prevent the pool from growing too large and getting out of hand. Any instances above this number will be destroyed instead of being returned to the pool (exceeding the maximum size often will trigger unwanted garbage collection, and resizing is an expensive operation – more CPU cycles – so you'll want to fine-tune this value by playtesting too).

Let's put our new Object Pool to good use now by implementing the shooting mechanic, using a pool of bullet projectiles.

A pooled player shooting model

We will use an **OOP (object-oriented programming)** design approach to the player shooting setup, allowing for easy future extensibility of new types of weapons and projectiles.

Creating the pooled player shooting model

Let's consider the following class diagram:

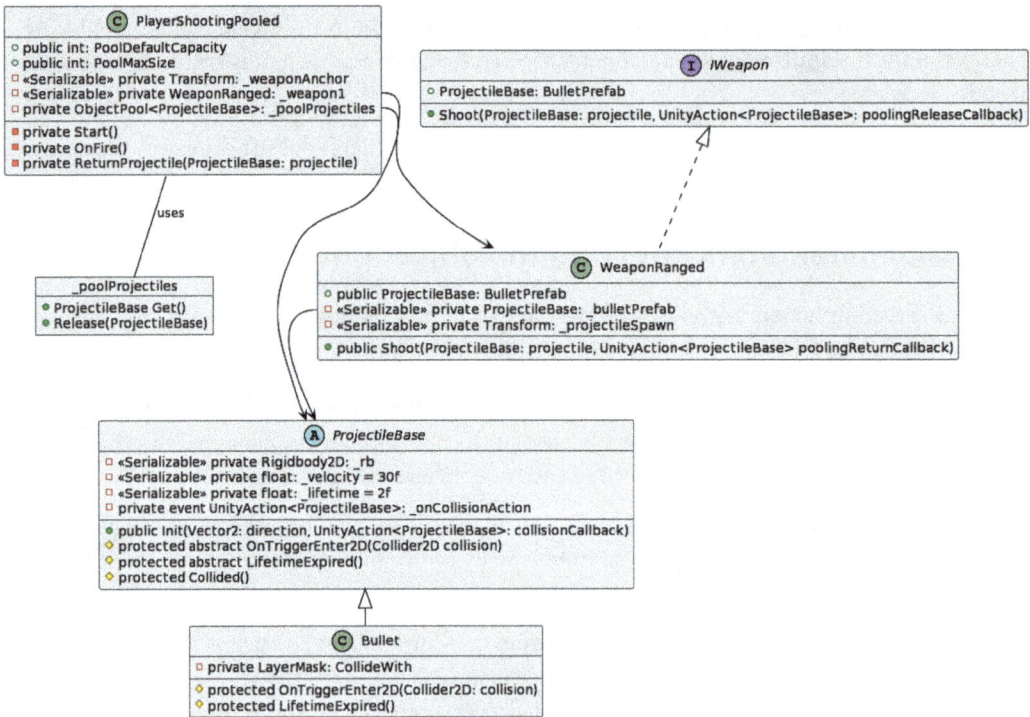

Figure 6.1 – A pooled player shooting UML class diagram

Figure 6.1 presents a UML diagram. **UML** stands for **Unified Modeling Language**, and it's a standardized approach to specifying and visualizing the relationships of artifacts within software projects. There are several types of UML diagrams, each serving a specific purpose. The class diagram we're using displays the static structure of a system, including classes, attributes, methods, and their relationships. It's one of the most widely used diagrams in software architecture.

> **Additional reading | UML diagram**
>
> UML: `https://www.uml.org/what-is-uml.htm`
>
> PlantUML Language Reference Guide: `https://plantuml.com/guide`
>
> PlantText UML Editor: `https://www.planttext.com/`

Okay, we've modeled the system for the pooled player shooting part of our game's code, but what does it mean? Referring to the diagram in *Figure 6.1*, let's break down the structure with these points:

- The `PlayerShootingPooled` class (*C*) – responds to `SendMessage()` of player input's `OnFire()` event to shoot the weapon's projectile:

    - This uses the `_poolProjectiles` object, which represents the stack of instantiated `ProjectileBase` objects (that is, `Bullet`)

    - It gets the `Bullet` (*C*) object Prefab (derived from the `ProjectileBase` (*A*) class type) for use in the `_poolProjectiles` (the `ObjectPool<ProjectileBase>` type) stack

    - It has a reference to the current ranged weapon (the `WeaponRanged` (*C*) class type) equipped to the player

- The `WeaponRanged` class (*C*) – represents any number of ranged weapon types we can equip the player with:

    - It implements the `IWeapon` (*I*) interface, meaning that we must declare the same members (like a contract). Therefore, any classes that implement the interface will have the same members available (this allows us to swap out object types without having to change code that consumes the interface members).

- The `Bullet` class (*C*) is a component added to our bullet Prefab and assigned in the Inspector to the `WeaponRanged`-serialized `_bulletPrefab` field. The bullet Prefab is retrieved via the public `BulletPrefab` property (encapsulating the private variable) for consumption in the `PlayerShootingPooled` class:

    - This extends the inherited abstract class, `ProjectileBase` (making `Bullet` a subclass); we cannot instantiate a class-declared `abstract` and must instead use the derived class. However, members declared in the `base` class are available in the derived class.

> **Important note**
>
> When we say *serializable*, we're indicating that we'll be able to make an assignment in the Inspector – a private field in most cases (in a C# script, the private field is decorated with the `[SerializeField]` attribute).

After reading through the preceding details and reviewing the class diagram, you may have started visualizing what our code should look like … … …

Well, you can stop daydreaming about it now because the code follows.

Create a new C# script in Assets/Scripts, named PlayerShootingPooled.cs:

```csharp
using UnityEngine.Pool;
public class PlayerShootingPooled : MonoBehaviour
{
    [SerializeField] private WeaponRanged _weapon1;
    public int PoolDefaultCapacity = 10;
    public int PoolMaxSize = 25;
    private ObjectPool<ProjectileBase> _poolProjectiles;

    private void Start() {}      // new ObjectPool<>();

    private void OnFire() =>
        _weapon1.Shoot(_poolProjectiles.Get(),
            ReturnProjectile);

    private void ReturnProjectile(
        ProjectileBase projectile) =>
            _poolProjectiles.Release(projectile);
}
```

We've declared the _weapon1 variable to assign the reference to the WeaponRanged Prefab object attached to the player's hand in the Editor (at design time). We also declared public variables PoolDefaultCapacity and PoolMaxSize, with default values of 10 and 25, respectively, for the default and maximum sizes of our private ObjectPool, declared as _poolProjectiles.

We'll then use the Start() code from the previous *The Unity object pooling API* section (excluded from the preceding code) and declare an OnFire() method, to be called via PlayerInput SendMessage() when the player presses the *fire* button. Within OnFire(), we will provide a Bullet instance, returned from calling _poolProjectiles.Get(), to shoot with.

Lastly, we will declare the ReturnProjectile() method, since it will be invoked by the _weapon1. Shoot() callback when the bullet is finished... doing what bullets do.

> **A note on code architecture**
>
> When we create the pool item, we could pass a reference to _poolProjectiles and call Release() on it directly, but if we make this an event, we can provide ReturnProjectile() as a callback. Plus, we have the option of adding any additional callbacks. I don't have any immediate plans for it. Still, it's nice to consider the options to create a flexible approach, without having to refactor the code later to introduce the ability and possibly break functioning/tested code in the process.

Now, we will create a new C# script in Assets/Scripts named WeaponRanged.cs:

```csharp
using UnityEngine.Events;
public class WeaponRanged : MonoBehaviour, IWeapon
{
    [SerializeField] private ProjectileBase _bulletPrefab;
    public ProjectileBase BulletPrefab => _bulletPrefab;

    [SerializeField] private Transform _projectileSpawn;

    public void Shoot(ProjectileBase projectile,
        UnityAction<ProjectileBase> poolingReturnCallback)
    {
        projectile.transform.position =
            _projectileSpawn.position;

        projectile.Init(_projectileSpawn.right
            * transform.root.localScale.x,
                poolingReturnCallback);
    }
}
```

The WeaponRanged script is where we'll assign a reference to our bullet Prefab in the Inspector, using the serialized private field, _bulletPrefab – a weapon that shoots will require something to shoot, after all.

We've kept _bulletPrefab encapsulated and then allowed only read (getter) access to the reference via the BulletPrefab public property. So, *encapsulation* here means we don't want any other classes to have access to set a bullet reference. The weapon will manage its own projectiles (although we could still add functionality later to assign a new bullet Prefab via a public setter method, such as WeaponRanged.SetBulletPrefab(GameObject) or similar).

The Transform variable, _projectileSpawn, provides a location where we'll spawn the bullet Prefab that is shot from the weapon – we will set this up in the WeaponRanged 1 Prefab later.

Lastly, the Shoot() method sets the position of the projectile that the pool provided to the projectile spawn position, and then it calls the Init() method on it (presumably to fire it off in the correct direction by applying some force to it).

We will also provide a reference to poolingReturnCallback so that the bullet Prefab can be released back to the pool when it collides with another object, or its lifespan expires.

Now, we will create a new C# script in `Assets/Scripts/Interfaces` named `IWeapon.cs`:

```
using UnityEngine.Events;
internal interface IWeapon
{
    ProjectileBase BulletPrefab { get; }
    void Shoot(ProjectileBase projectile,
        UnityAction<ProjectileBase>
            poolingReleaseCallback);
}
```

The `WeaponRanged` class implements the `IWeapon` interface to satisfy the contract, which means `WeaponRanged` must implement the `BulletPrefab` property and the `Shoot()` method declared in the `IWeapon` interface. Note that interface members in C# are public by default!

The weapon and object pooling portions of the class diagram in *Figure 6.1* are now satisfied. Let's finish up the class diagram implementation with the projectile that will be shot from the weapon.

Create a new C# script in `Assets/Scripts` named `Bullet.cs`, as follows:

```
public class Bullet : ProjectileBase
{
    [SerializeField] private LayerMask CollideWith;

    protected override void
        OnTriggerEnter2D(Collider2D collision)
    {
        if ((CollideWith
            & (1 << collision.gameObject.layer)) != 0)
                base.Collided();
    }

    protected override void LifetimeExpired()
        => base.Collided();
}
```

The `Bullet` class extends the `ProjectileBase` class, meaning that it will inherit all the members and/or be required to **override** members. You can declare unique properties for a derived class that differentiate it from other derived classes (a tenet of **inheritance** in OOP design).

We're overriding the `OnTriggerEnter2D()` method – we're required to, since it's declared as `abstract` in the inherited `ProjectileBase` class – to perform the specific Bullet collision action.

Note that we're also calling the `Collided()` method in the inherited class by using the `base` keyword. `Collided()` is declared as `virtual`, which means we can redefine it in a derived class while also using it for the same basic/default functionality.

This bit of code – ((CollideWith & (1 << collision.gameObject.layer)) != 0) – in the OnTriggerEnter2D() method evaluates whether the GameObject the Bullet collided with is included in the layers selected in CollideWith LayerMask. For example, we'll select *Environment*, *Wall*, *Ground*, *Enemy*, and so on, but not *Player* for Bullet to collide with.

Additional reading | Unity documentation

LayerMask: https://docs.unity3d.com/2022.3/Documentation/ScriptReference/LayerMask.html

Bitwise and shift operators: https://learn.microsoft.com/en-us/dotnet/csharp/language-reference/operators/bitwise-and-shift-operators#left-shift-operator-.

We've seen how Bullet extends the ProjectileBase class (via OOP inheritance), so let's have a look at the ProjectileBase class now.

Create a new C# script in Assets/Scripts named ProjectileBase.cs:

```csharp
using UnityEngine.Events;
public abstract class ProjectileBase : MonoBehaviour
{
    [SerializeField] private Rigidbody2D _rb;
    [SerializeField] private float _velocity = 30f;
    [SerializeField] private float _lifetime = 2f;

    private event
        UnityAction<ProjectileBase> _onCollisionAction;

    public virtual void Init(Vector2 direction,
        UnityAction<ProjectileBase> collisionCallback)
    {
        _onCollisionAction = collisionCallback;
        _rb.velocity = direction * _velocity;
        Invoke(nameof(LifetimeExpired), _lifetime);
    }

    protected abstract void
        OnTriggerEnter2D(Collider2D collision);

    protected abstract void LifetimeExpired();

    protected virtual void Collided()
    {
        CancelInvoke();
```

```
            _onCollisionAction?.Invoke(this);
    }
}
```

Here, we can see our bullet projectile's default properties and behavior. By declaring `ProjectileBase` as an abstract base class, it cannot be instantiated (made a new instance of) directly, so we must declare a new class that extends or inherits from it.

Note that any derived classes can still be referenced by the base class it extends (in OOP, this is the **polymorphism** principle). The derived class must override any members declared as abstract but can optionally override members declared as virtual (while also being able to call base virtual methods for default behavior).

Let's break down the code's relevant items:

- Variable declarations for `_rb`, `_velocity`, and `_lifetime` provide a reference to the `RigidBody2D` component of the projectile Prefab, as well as configurable values for the rate of velocity and lifetime of the object

- When `Init()` is called, the passed-in callback action is assigned to `_onCollisionAction` to be invoked later, and then `_rg.velocity` is set for `direction` and rate of `_velocity`, firing it off in the direction the weapon is pointing

- Then, we have two abstract methods that must be overridden in the derived class(es) (for example, `Bullet`):

  - `OnTriggerEnter2D()`: This Unity message event is called when another object collides with the object.

  - `LifetimeExpired()`: In `Init()`, we invoke this method with a delay so that the projectile object only exists active in the Scene for a certain amount of time (releasing it back to the pool to be reused). The `_lifetime` value should be adjusted during playtesting so that the weapon's range works well in-game.

- Finally, we have the `Collided()` method, which first cancels calling the `LifetimeExpired()` method at the `_lifetime` value (for example, as we don't want to call `LifetimeExpired()` if `Collided()` was already called by a collision event) and then invokes the `_onCollisionAction` callback (releasing the object back to the pool).

> **Pooled player shooting model code**
>
> The complete aforementioned pooled Player shooting code can be downloaded from the GitHub repo here: `https://github.com/PacktPublishing/Unity-2022-by-Example/tree/main/ch6/Unity%20Project/Assets/Scripts`.

Following the design model from the UML class diagram in *Figure 6.1*, we have now finished writing the code portion and are ready to move on to implementing it with the player.

Implementing the pooled shooting model

Now, let's see how we implement the pooled player shooting model by creating the required Prefabs – a weapon and a projectile – for integration with the player. The Prefabs we'll need, along with the steps to create them, are as follows:

1. `Bullet 1`: The first projectile Prefab asset we'll shoot from a ranged weapon:

    I.   Import the `bullet1` artwork from the project files to the `Assets/Sprites` folder and set the PPU to `1280` (to set our larger bullet sprite to a good in-game size, which will be possibly adjusted later in relation to the size of the weapon firing the bullet). Also, set **Max Size** to `64` for optimization, since it is a relatively small and fast-moving sprite.

    II. Create an empty GameObject in the **Hierarchy**.

    III. Drag the bullet sprite from the **Project** window onto the new empty object to make it a child object (graphics should be parented to a base object), and then rename the parent `Bullet 1`. Remember that you can easily parent a GameObject in the **Hierarchy** by right-clicking on it and selecting **Create Empty Parent**.

    IV. Add `Rigidbody2D` and `CircleCollider2D` components to the parent object to enable **Physics** (i.e., for collision detection).

    i.   For the `Rigidbody2D` component, set **Mass**, **Angular Drag**, and **Gravity Scale** to 0, and enable **Constraints | Freeze Rotation Z** (we don't want the bullet to rotate; for a 2D game, this is the Z axis).

    ii. For the `CircleCollider2D` component, set **Radius** to a value that surrounds the core part of the bullet sprite (for example, `0.13`), as shown in *Figure 6.2*:

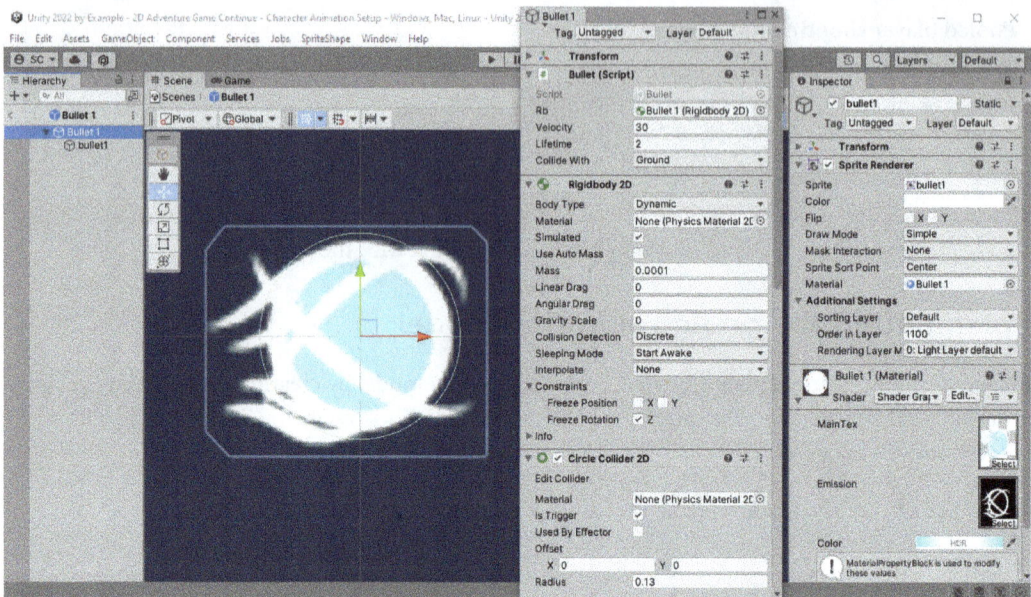

Figure 6.2 – the bullet Prefab setup

V. Now, add the `Bullet` script to the parent object, and assign the `Rigidbody2D` field by clicking and dragging the `Rigidbody2D` section title to the field.

VI. Set **Velocity** and **Lifetime** to some initial starting values, and then assign the layers that the bullet should collide with, using the **CollideWith** field (note that you may need to add a new layer before you can assign it).

VII. Lastly, drag the parent `Bullet 1` object from the **Hierarchy** to the **Project** window to create a Prefab asset in the `Assets/Prefabs` folder. Our bullet Prefab is now ready to be used with the weapon Prefab we'll create next.

2. `WeaponRanged 1`: A ranged weapon that the player will hold and shoot a bullet projectile from:

I. Import the `gun1` weapon artwork, and in the Sprite Editor, set a custom pivot (see *A* in *Figure 6.3*) on the handle so that when it is attached (or spawned) to the player, it is in the correct position and can rotate on a pivot point (as you might expect it to behave).

Pivot | Unity documentation

When working with a GameObject, the pivot serves as the reference point to position, rotate, and scale it. In Unity, when using the Transform tools, you can switch between the pivot or center of a GameObject in the Tool Settings overlay.

Position GameObjects | Gizmo handle position toggles: `https://docs.unity3d.com/Manual/PositioningGameObjects.html`

II.   Drag the gun sprite into the **Hierarchy** and parent to an empty GameObject (making the gun sprite a child object of the empty object), and then name the parent WeaponRanged 1.

III.   Now, add a new empty GameObject, named ProjectileSpawnPoint, as a sibling to the gun sprite object that we'll use as the **Transform** position to spawn the projectile. Position this GameObject at the front of the gun's muzzle (see B in *Figure 6.3*):

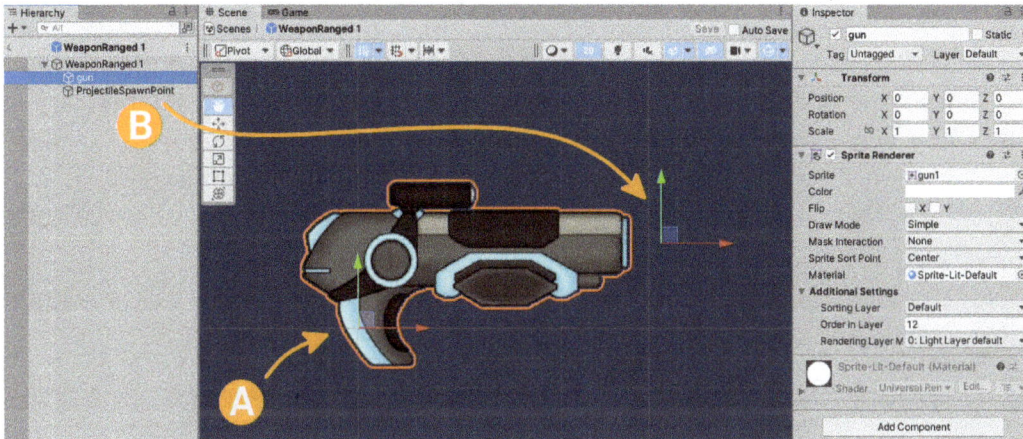

Figure 6.3 – The weapon Prefab sprite and projectile spawn point setup

IV.   Now, add WeaponRanged as a component to the parent GameObject.

V.   The only two assignments we need to make on the WeaponRanged component are the **Bullet Prefab** and **Projectile Spawn** fields:

i.   Assign the bullet Prefab asset – the projectile the weapon will shoot – by dragging Bullet 1 from the **Project** window to the **Bullet Prefab** field.

ii.   Assign the ProjectileSpawnPoint object to the **Projectile Spawn** field by dragging it from the **Hierarchy**.

iii.   Lastly, drag the parent WeaponRanged 1 object from the **Hierarchy** to the **Project** window, creating a Prefab asset in the Assets/Prefabs folder.

Our pooled player shooting model is ready to be used with the player character. Yay!

Adding pooled shooting to the player character

We'll dig right into adding the weapon to our player. You can either make sure the player is in the current Scene or double-click on the Player Prefab in the **Project** window to open it up in **Prefab** Mode.

We'll use the player's (actor's) bones to ensure the weapon follows the character's hand while animating. Perform the following steps to anchor the weapon to the player's hand:

1. Find the hand bone under the `root_bone` hierarchy. In our case, it's `bone_8` as shown in *Figure 6.4*.

2. Add an empty GameObject as a child of `bone_8`, and name it `Weapon_Attachment`; this will serve as the attachment point for the weapon. Making it separate from the bone provides additional options for positioning/rotating. Also, naming it with an `_Attachment` suffix means we can easily search for any/all objects in the **Hierarchy** that serve as attachment points.

3. You can now go ahead and drag the `WeaponRanged 1` Prefab from the **Project** window, dropping it directly on the `Weapon_Attachment` object (this becomes a nested Prefab, which means we can configure its properties at any time, independent of the `Player` Prefab).

The following screenshot shows our player setup with the weapon in hand. In the following *Figure 6.4*, you can see that I have also temporarily dragged in a `Bullet 1` Prefab to check its scale in relation to the weapon (seen in context with the player character):

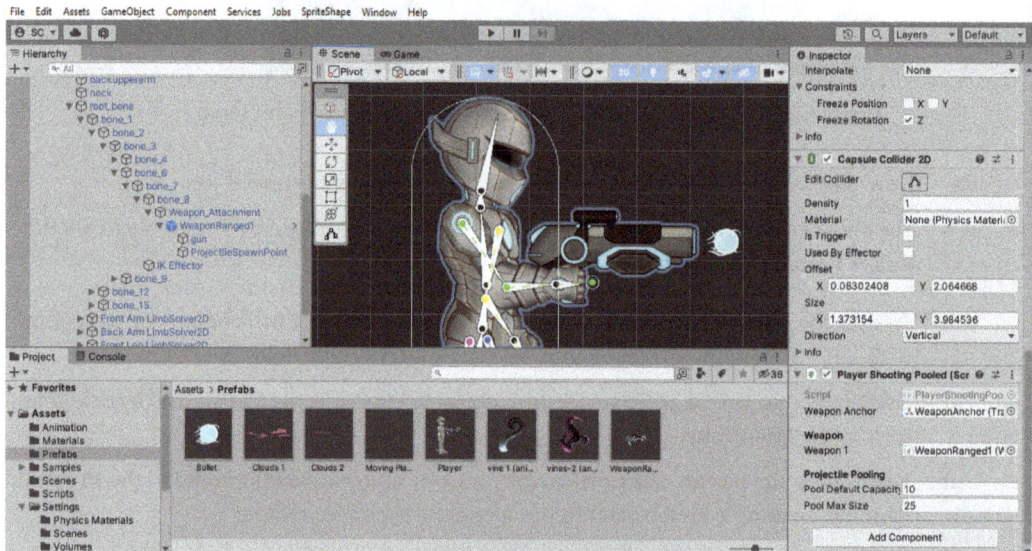

Figure 6.4 – The weapon Prefab added to the player

As a final step, we need to add the `PlayerShootingPooled` script as a component to the root object of our `Player` Prefab. Then, we will drag the `WeaponRanged 1` object from the **Hierarchy** to the **Weapon 1** field on the component (as shown in *Figure 6.4*).

Here's what our Prefab components – based on the classes in our UML diagram – look like in the Inspector with all their respective field assignments:

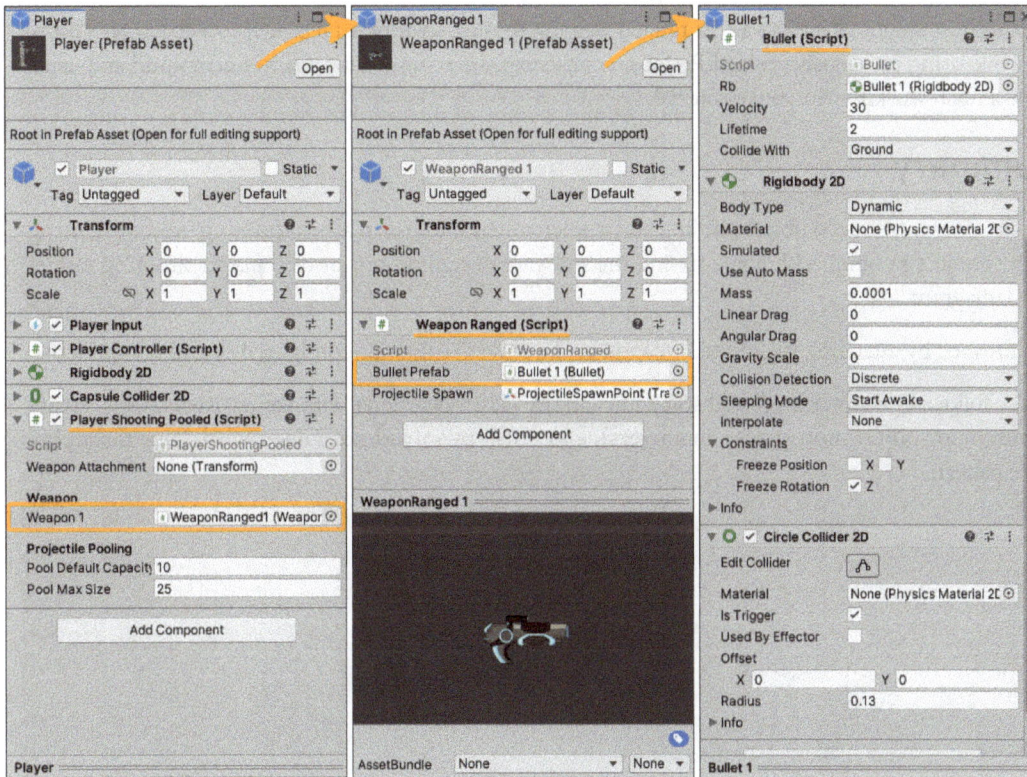

Figure 6.5 – Prefab configuration for the pooled player shooting setup

> **New to Unity 2022**
>
> This may not be new specifically to the 2022 tech stream, but I believe it's a workflow improvement worth mentioning. You can simultaneously open up a focused **Inspector** window for different assets or GameObjects, without constantly changing your selection and using the Inspector. First, select the object, and then either right-click and choose **Properties…** (at the bottom) or press *Alt/Cmd + P*.

Let's go ahead and playtest the results of our efforts!

Referring to *Figure 6.4*, enter **Play** mode, and then disable the **Animator** component on the `PlayerCharacter1` object to pose the actor's arm with the weapon attachment in a shooting position – using the IK `LimbSolver2D` target makes this easy-peasy.

We'll give the player a proper shooting animation that points the weapon in a direction in *Chapter 8* so that we can target those pesky robot enemies (not that it's their fault!).

This section taught us how to add a ranged weapon to the player by attaching it to a character's limb, all while using previously created Prefabs. I can't stress how important it is to understand and use a good Prefab workflow in your projects!

Summary

This chapter introduced object pooling and implemented it for the player's shooting mechanic, using Unity's object pooling API, while basing the software design on our pooled player shooting model UML diagram.

We finished by attaching the ranged weapon to the player, using our configured Prefab components.

In the following chapter, we'll add some juice to the player character with some visual effects, create some enemy NPCs (non-player characters), and finish by introducing enemy behavior through a state pattern.

7

Polishing the Player's Actions and Enemy Behavior

In *Chapter 6*, we added a ranged weapon shooting mechanic in an optimized way using Unity's new **object pooling API**.

In this chapter, we will continue to add some much-needed polish to the player character with visual effects! We'll also create some enemy **non-player character** (**NPC**) variants and finish up the chapter by introducing enemy behavior through a state pattern.

In this chapter, we're going to cover the following main topics:

- Polishing with **Shader Graph** and **Trail Renderer**
- Enemy Prefabs and variants – Configuring with **Scriptable Objects** (**SOs**)
- Implementing basic enemy behavior using a **finite state machine** (**FSM**)

By the end of this chapter, you'll be able to quickly add some visual effect polish to the game's sprites using a custom shader and make several enemy variants configured with varying properties and simple state-based behavior.

Technical requirements

Download the assets from the following GitHub link to follow along in this chapter with the same artwork created for the project in the book.

To follow along with your own artwork, you'll need to create similar artwork using Adobe Photoshop or a graphics program that can export layered Photoshop PSD/PSB files (e.g., Gimp, MediBang Paint, Krita, and Affinity Photo).

You can download the complete project on GitHub at `https://github.com/PacktPublishing/Unity-2022-by-Example`.

Polishing with Shader Graph and Trail Renderer

To really sell the game experience to the player, we can employ some easy-to-execute visual polish. Thankfully, Unity provides visual effect features through some built-in components and includes relevant features as part of its render pipeline.

Here, we'll look at adding a few simple effects to the player, bullet projectile, and the overall visual look.

Enabling post-processing

To take advantage of any of the effects we'll be implementing in this section, we'll first want to enable **post-processing** as it applies to the **Universal RP** (**URP**) (reminder: this is the render pipeline we are working with).

Specifically, the effects we'll be making will use **high dynamic range** (**HDR**) color values that will work with a post-processing bloom effect to make our 2D lights and parts of our sprite assets glow.

> **HDR color | Unity documentation**
>
> HDR color values offer a wider range of luminosity than standard colors, resulting in more accurate depictions of color and brightness, more vibrant colors, improved support for bloom and glow effects, and reduced banding.
>
> High dynamic range: `https://docs.unity3d.com/Manual/HDR.html`

For a look at what's to come in the following sections, as seen in *Figure 7.1*, we'll apply **Vignette** to the screen and add **Bloom** to make the white parts of the bullet sprite glow (exaggerated for clarity).

Note that we can't make the bullet glow with **Bloom** alone, so we'll cover how to accomplish this, plus **Vignette**, in the following sections:

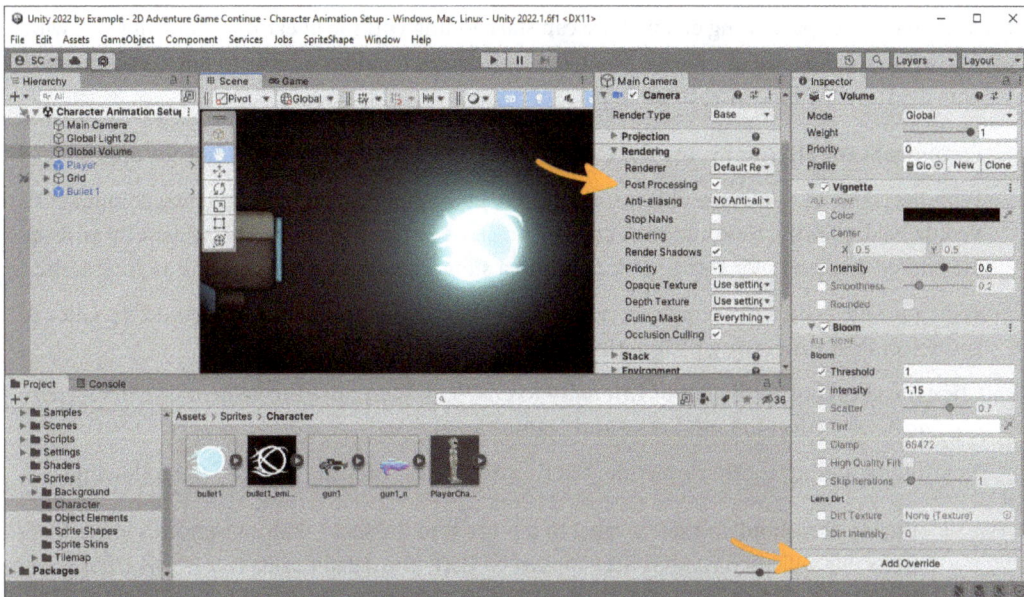

Figure 7.1 – Post-processing effects applied to the scene

A word of caution

Some post-processing effects are performance-heavy and are not suitable for all distribution platforms (mobile in particular), so pay attention to your **frames per second (FPS)** stats when adding new effects!

To enable post-processing in our game, follow these steps:

1. For the camera in the current scene, select **Main Camera** in the **Hierarchy**.

2. Under the **Rendering** section (refer to *Figure 7.1*), enable **Post Processing**.

3. For **Render Pipeline Asset**, go to the `Assets/Settings` folder and select the `UniversalRP` asset.

4. Under the **Quality** section, enable **HDR**.

Now that we have post-processing enabled, we can start adding **volume overrides** that will work with the effects we'll be adding to our objects. To do that, we'll need to add a volume to our scene and add the volume overrides by following these steps:

1. From the **Scene Hierarchy**, use the **Create** menu and select **Volume | Global Volume**.

2. With the `Global Volume` object still selected, in the Inspector, on the `Volume` component's **Profile** field, click the **New** button to create a `Volume Profile` asset. This will create an asset file named `Global Volume Profile` in a subfolder with the same name as your scene.

3. With a profile assigned, we now see a new **Add Override** button (refer to *Figure 7.1*). Click it and select **Post Processing | Vignette**.

4. Click the checkbox to the left of **Intensity** to enable it, then raise the value to see the *screen-edge-darkening* effect in both **Scene** and **Game** views.

5. Now, repeat adding another override, but this time select **Bloom**. Then, do the following

    I.   Enable **Threshold** and set its value to `1` – pixels in the scene with a brightness lower than this value will not have the effect applied by URP. The default value is `0.9` (which is great in most cases), but I've decided to give almost any object the potential to contribute glow to the overall look, so I raised it slightly.

    II. Enable **Intensity** – you'll have to play around with this value to get the look you're after. I've set it initially to `1.15` to make things glow pretty good as an initial visual check, but I will be dialing this down later, I'm sure.

Additional reading | Unity documentation

Post-processing and full-screen effects: `https://docs.unity3d.com/2022.3/Documentation/Manual/PostProcessingOverview.html`

Post-processing in the Universal Render Pipeline: `https://docs.unity3d.com/Packages/com.unity.render-pipelines.universal%4015.0/manual/integration-with-post-processing.html`

Now that things look more cinematic and have the potential to emit some cool glow effects, let's finally set up our bullet Prefab sprite!

Applying glow to the bullet with Shader Graph

You previously saw how we added a secondary texture to a sprite when we added sprite Normal maps in *Chapter 4* (for giving a pseudo-3D effect to sprites).

We'll now use a **secondary texture** that will allow us to specify areas of a sprite that we want to make glow – or, more precisely, *emit light*, by setting a higher brightness value (that is, a pixel brightness value greater than 1, which will be possible by using an HDR color value).

In *Figure 7.2*, you can see the artwork for the bullet as well as the bullet's **emission map** – a black-and-white image where the white areas define parts that will be emissive:

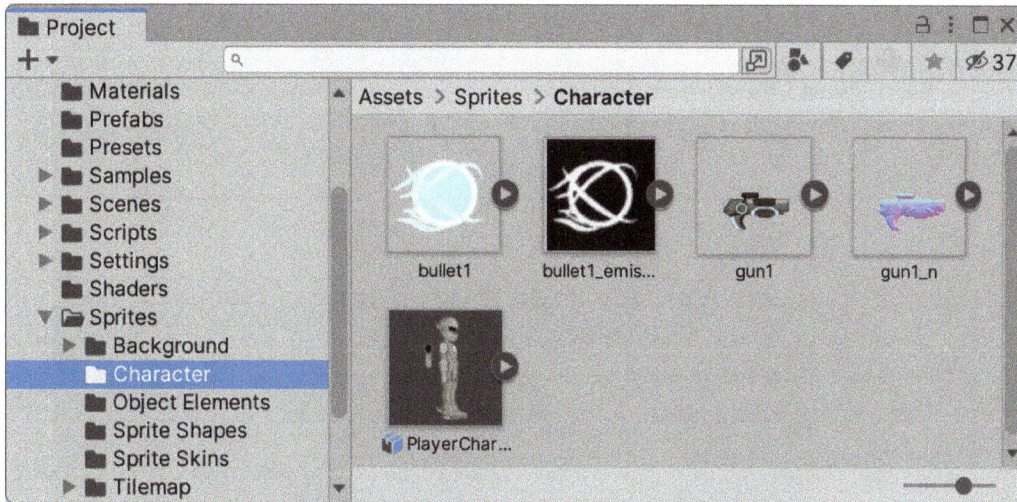

Figure 7.2 – Emission map for bullet sprite

As you can see from the screenshot, we've created an emission map named `bullet1_emission` that represents only the swirly lines of the bullet sprite. Going back to the `bullet 1` sprite, select it so that we can add the emission map as a secondary texture by following these steps:

1. In the sprite's **Import Options** section in the Inspector, click the **Sprite Editor** button.

2. Select **Secondary Textures** from the **Sprite Editor** drop-down menu (top-left corner of the window).

3. In the **Secondary Texture** dialog, click the plus (+) button to add a new texture.

4. Enter `_Emission` in the **Name** field.

5. Drag the **bullet1_emission** sprite from the **Project** window into the **Texture** field.

6. Click the **Apply** button to save the changes (or simply close the window and choose **Save**).

As a refresher, you can refer to *Chapter 4, Figure 4.2*, for an example of the **Secondary Texture** dialog.

With our sprite ready to go, we can move on to creating a new shader for applying our effect.

Creating a new Shader Graph 2D material

Creating custom shaders used to be complicated by hand-coding them in a special shader language, but **Shader Graph** allows custom shaders to be authored visually in real time using a node-based system, making the process more accessible to artists and developers alike.

We'll use **Shader Graph** to create and connect nodes visually to build out our emission shader. Note that this shader can be the basis for any number of materials we can use to make different sprites with emission maps glow!

> **Additional reading | Unity documentation**
>
> Getting started with Shader Graph: `https://docs.unity3d.com/Packages/com.unity.shadergraph%4014.0/manual/Getting-Started.html`

Let's go ahead and create this glow shader now with the following steps:

1. First, create a new `root` folder in the **Project** window as `Assets/Shaders`.

2. Create a shader graph in the new folder by using the **Create** menu, going to **Shader Graph | URP | Sprite Unlit Shader Graph**, and naming it `SpriteEmission_Unlit`.

3. Now, open the **Shader Editor** by clicking the **Open Shader Editor** button in the Inspector or double-clicking on the asset.

The custom sprite shader we're going to create is actually pretty simple and will only require a few nodes, as seen in the completed shader in the following screenshot:

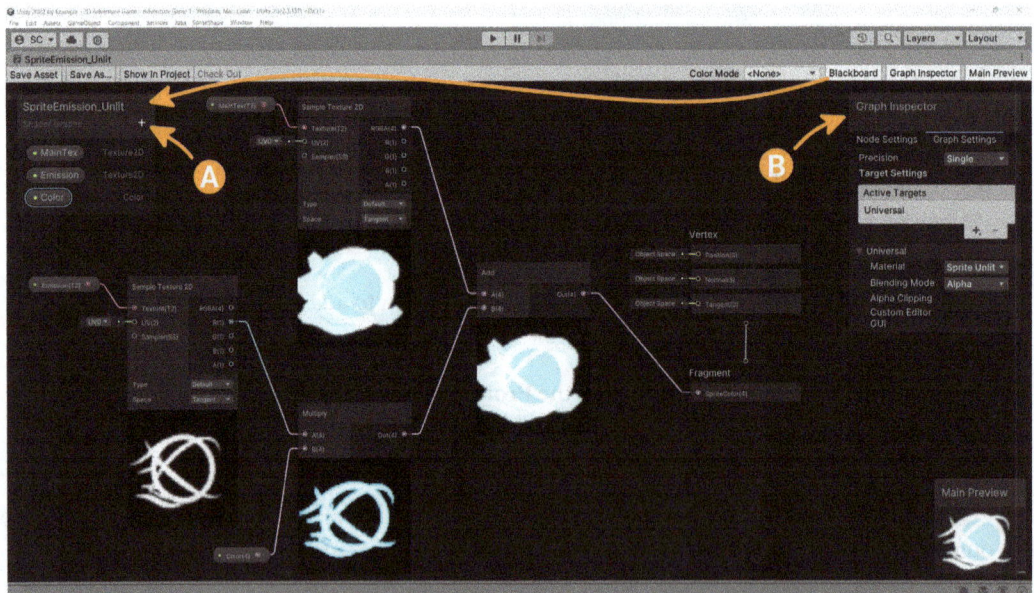

Figure 7.3 – Shader Graph sprite emission shader

Let's construct this shader by following these steps:

First, starting in the **Blackboard** (*A* in *Figure 7.3*) using the plus (+) button, add the following properties:

- **Texture2D**: **Name** = `MainTex` and **Reference** = `_MainTex` (these fields are assigned in the **Graph Inspector** indicated by *B* in *Figure 7.3*), then drag in the bullet sprite as the **Default** texture

- Note that it is essential that **Reference** is spelled correctly since this is what the shader uses internally for the texture reference

- **Texture2D**: **Name** = `Emission` and **Reference** = `_Emission` (note this is what we previously used as the name for the emission map secondary texture in the **Sprite Editor**), then drag in the **Bullet Emission Map** sprite as the **Default** texture

- **Color**: **Name** = `Color` and **Reference** = `_Color`

These properties will be exposed in the Inspector so that we can assign values. **Color** is the only property we'll need to adjust since `_MainTex` and `_Emission` will be obtained from the sprite.

Next, create and connect the nodes that make up this simple shader:

1. Right-click anywhere and select **Create Node** from the pop-up menu (or just press the spacebar).

2. Start typing to search for the desired node to add. In our case, start typing `sample texture 2d` and select it from the list under **Input | Texture**. Let's add two of these texture nodes – we'll need one for **MainTex** and one for **Emission** (refer to *Figure 7.3* as a reference).

3. Let's complete the **MainTex** path first. Create an **Add** node next and then make the following connections:

   MainTex(T2) → **[SampleTexture2D] Texture(T2) | RGBA(4)** → **[Add] A(4) | Out(4)** → **[Fragment] SpriteColor(4)**

> **Important note**
>
> A node is indicated as **[Node]**, connecting lines are shown by → (clicking and dragging on the little circles on the nodes), and input/output is displayed by | (input on the left side, output on the right side).

4. Complete the **Emission** path next by using the second **SampleTexture2D** node we previously added. Create a **Multiply** node and make the following connections:

   Emission(T2) → **[SampleTexture2D] Texture(T2) | R(1)** → **[Multiply] A(4) | Out(4)** → **[Add] A(4)**

5. Finally, we'll finish up the shader by connecting the **Color** property:

   Color(4) → **[Multiply] B(4) | Out(4)** → **[Add] B(4)**

> **Downloading the completed shader graph**
>
> For the completed sprite emission shader, visit the project GitHub repo here: `https://github.com/PacktPublishing/Unity-2022-by-Example/tree/main/ch5/Unity%20Project/Assets/Shaders`

Before we can apply our fancy new emission shader to our bullet Prefab, we'll have to create a new **material** based on this shader – that's how it works. Rendering in Unity (and most digital content creation software) is performed with materials, shaders, and textures, each contributing their part to what the end user sees on screen.

> **Additional reading | Unity documentation**
>
> Graphics: `https://docs.unity3d.com/2022.3/Documentation/Manual/Graphics.html`

Create a new material based on this new shader graph we just created by right-clicking on the shader asset named `SpriteEmission_Unlit` in the `Assets/Shaders` folder of the **Project** window, then follow these steps:

1. Go to **Create | Material** and give it the name `Bullet 1`.

2. Then, create a new folder at `Assets/Materials` and move the material into that folder.

3. Assign the material to the bullet sprite by first opening the `Bullet 1` Prefab (either in the scene or in **Prefab Mode**).

4. With the bullet sprite visible in the scene view, drag the `Bullet 1` material from the **Project** window and onto the sprite. Unity provides a visual indicator for what the new material would look like if applied to the sprite before you commit to releasing the mouse button (a pretty neat trick; this also works the same for assigning materials to 3D objects).

5. Now, this is where the magic happens. Select the `bullet1` sprite in the **Hierarchy** to show its Inspector, expand the **Material** section at the bottom, and then click on the color picker (note that you should exit **Prefab Mode** because post-processing is not visible there). Then, do the following:

   I. Set the **Color** field to an appropriate glow color and then kick up the **Intensity** value for the desired amount of glow (as seen in *Figure 7.4*, but remember that the post-processing **Bloom** effect's **Intensity** value is also at work here).

This is the final result of our efforts:

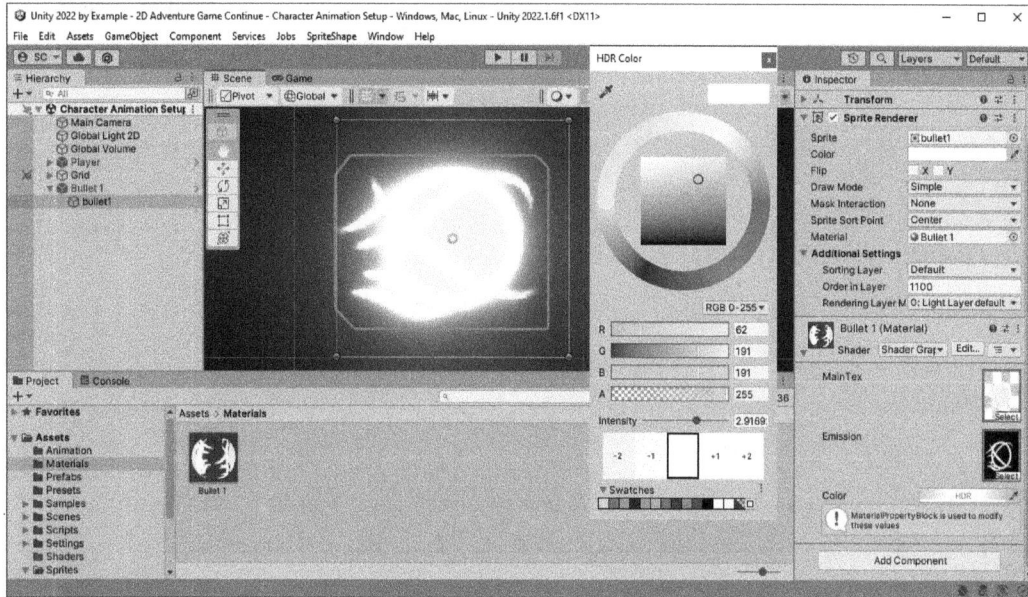

Figure 7.4 – Bullet material HDR color setting

We are looking good – yay!

> **Bonus activity**
> Add a material based on the **SpriteEmission_Unlit** shader for both the player character and weapon!

As you can see from our simple shader, it doesn't take much effort to pump up the visuals of your games in Unity! For another quick win on the game's visuals, let's add a subtle light effect to the player character to make it pop out in the environment.

Adding a 2D light to the player

This is a very quick and easy effect to add but with a huge payoff. All we're going to do here is add a **2D light** as a child of **Player**. With the light parented to **Player** – and within the **Prefab Hierarchy** – it will be *attached* to the object.

Use these steps to add the light:

1. Right-click on the root object of the Player Prefab in the scene.

> **Working with Prefabs**
>
> A reminder for working with Prefabs: if you add the **Light** object in **Prefab Mode** (by double-clicking on the Prefab in the **Project** window), you won't be able to visualize changes to the light settings. You can, however, enter the **Prefab Isolation** mode in the scene by clicking the chevron icon (>) in the **Hierarchy** window and still be able to visualize changes. Lastly, you can modify the Prefab directly in the scene but just remember to apply **Overrides** to save changes to the Prefab.

2. From the menu, select **Light | Spot Light 2D**.

3. Adjust the values to your liking. Referring to *Figure 7.5*, these are the settings I used:

    *   Positioned the light on the chest of the character

    *   **Intensity** = 0.6

    *   **Radius** = 1.5 (*Inner*), 7 (*Outer*)

    *   **Target Sorting Layers** = **Background**, **Default** (includes **Player** but not **Foreground**)

    *   **Volumetric** = 0.01 (just add a touch here for a small contribution to the environment lighting)

    *   **Normal Maps** = **Fast** (*Quality*), 3 (*Distance*)

The before and after can be seen in *Figure 7.5*. Notice how the player pops out of the environment on the right, whereas things look rather flat on the left:

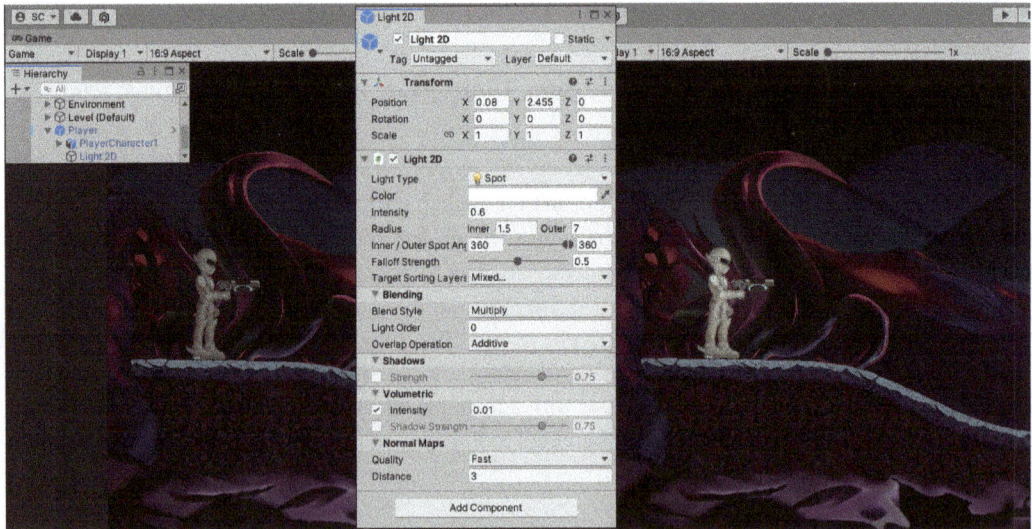

Figure 7.5 – Making the player pop out with a 2D light

That was easy! Let's gain another quick win with an additional visual effect that is simple to add!

Polishing is easy with Trail Renderer

The `Trail Renderer` component creates, well, a trail following a moving object. It's a great way to add more impactful movement to things, and it can be as subtle or over the top as you wish by adjusting only a few settings.

We'll get right to it.

Adding a trail to Prefabs

Follow these steps to add a trail to our bullet:

1. Open the **Bullet 1** Prefab for editing.
2. Open the **Create** menu and add **Effects | Trail** (parented to the root as a sibling of the **bullet1** sprite).
3. Adjust the values to give a nice trailing effect! Here are the values I used as a starting point:

    - **Width** = (0.0, 0.2), (0.5, 0.0) – using this curve will ensure that the trail doesn't extend to be too long
    - **Time** = 0.2
    - **Color** = white with an alpha gradient going from 35 to 0
    - **Materials** = **Sprite-Unlit-Default** (changed from the default shader so that the trail is not affected by scene lighting; found in To be consistent with the asset paths, the slashes should be forward, as in: `Package/Universal RP/Runtime/Materials`)
    - **Lighting** = **Off** (cast shadows)

These settings, and the results, can be seen in *Figure 7.6*:

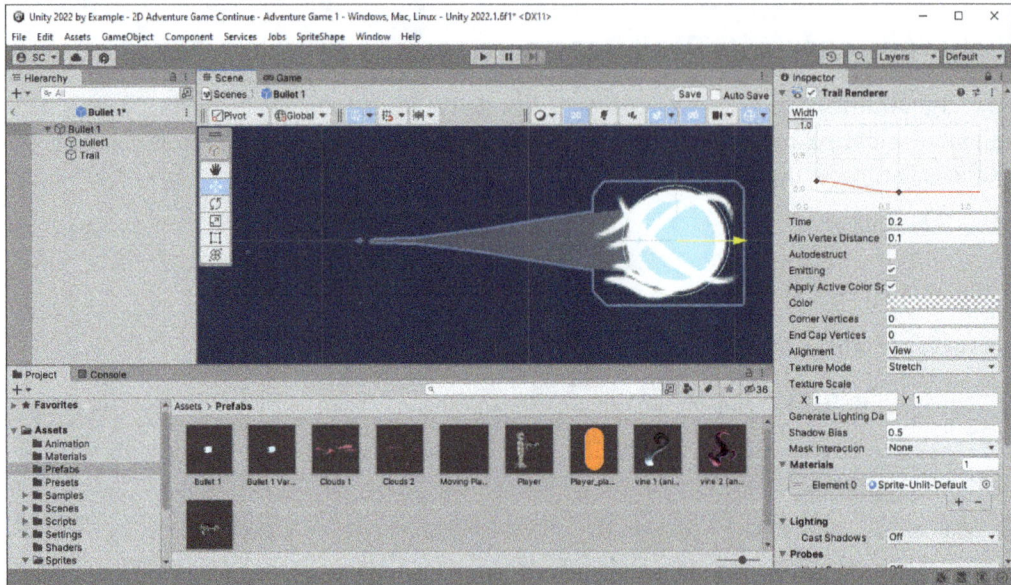

Figure 7.6 – Visualizing Trail Renderer component on bullet Prefab

Don't forget that you'll also have to set the **Order in Layer** option to a value that works with the `Bullet 1` Prefab's sprite renderer, the ground, and foreground/background sorting layers.

Clearing trails for pooled Prefabs

The `Trail Renderer` component creates a trail of polygons behind the bullet GameObject (that's how the trail is rendered), which will still be there even if the GameObject is deactivated in the scene.

That's a problem for us because that's exactly what we're doing with object pooling: deactivating the GameObject projectile when it's returned to the pool.

Fortunately, the `Trail Renderer` component provides a `Clear()` method for clearing the trail. All we need to do is call this method when we initialize the projectile, and voilà! Problem solved.

Add the following code to the `ProjectileBase` class:

```
public virtual void Init(Vector2 direction,
    UnityAction<ProjectileBase> collisionCallback)
{
    // If there is a Trail Renderer component on
    // this GameObject then reset it.
    if (TryGetComponent<TrailRenderer>(out var tr))
        tr.Clear();
```

```
    ...
}
```

Here, we're simply using `TryGetComponent<TrailRenderer>()` to see if a `TrailRenderer` component has been added to the GameObject and only returning a reference to it, with an `out` parameter, if it exists. If the component does not exist, then an allocation will not be made, unlike with `GetComponent()`.

We use an `if` statement to evaluate the `bool` return value from `TryGetComponent()`, so only if we have a `Trail Renderer` component on the projectile will the `Clear()` method be called (essentially, resetting it).

> **Additional reading | Unity documentation**
> `Component.TryGetComponent: https://docs.unity3d.com/2022.3/`
> `Documentation/ScriptReference/Component.TryGetComponent.html`

In this section, we learned how to enable post-processing and add effect overrides such as **Vignette** and **Bloom**, making our bullet glow and our player pop with light. We finished up with an easy-peasy trail effect on our bullet as well!

Next, we'll add a configurable enemy character and variants of the same.

Enemy Prefabs and variants – Configuring with SOs

Instead of a GameObject that has to live in a scene, we can create a file-based asset that can be referenced from any GameObject, including Prefabs, anywhere in the game, called an SO.

This being a single asset reference means no additional allocations are needed, and the same values are used no matter how many objects in the scene are spawned that reference it. Pretty cool!

An SO, as a small and efficient data container, also allows the separation of data from the code that consumes it. Data can be updated from backend cloud systems without recompiling the code or building a new distribution of the entire game.

Being able to respond to data changes for games in production – where an issue affecting players may need to be solved quickly – is an excellent application of SOs.

Others have used SOs for middleware components and even fully decoupled event systems that are designer-friendly because they allow configuration at design time in the Editor (that is, a developer is not required to wire up new events between objects, and so on).

Unity has also built complete game architecture based on SOs in its *Open Projects* development program. And there would be something amiss if I didn't mention Ryan Hipple's (Schell Games) now infamous *Game Architecture with Scriptable Objects* talk at Unite Austin 2017 (link provided in the

Additional material – Unity documentation callout box), where he describes how SOs can be used for building more extensible systems and data patterns.

> **Additional material | Unity documentation**
>
> ScriptableObject: `https://docs.unity3d.com/2022.3/Documentation/Manual/class-ScriptableObject.html`
>
> Open Projects: `https://unity.com/open-projects`
>
> Unite Austin 2017 - Game Architecture with Scriptable Objects: `https://youtu.be/raQ3iHhE_Kk`

With that bit of introduction out of the way, we'll go on now to create our first SO and use it to configure the traits of our enemies.

Creating an enemy Prefab with configurations

An SO is created in a similar way to a `MonoBehaviour` script with some exceptions:

- The SO must inherit from `ScriptableObject` instead of `MonoBehaviour`.

- It cannot be attached to a GameObject (as a component). Instead, it is saved as a file asset and referenced by components as a field exposed in the Inspector.

- It does not receive all of the same Unity message events that a `MonoBehaviour` script does (missing, most notably, are `Start()`, `Update()`, and `FixedUpdate()`).

- It can create new custom assets based on the `ScriptableObject` class by using `CreateAssetMenu`. In contrast, `MonoBehaviour` can only be configured in the scene and saved as a Prefab (it can then be edited in the **Prefab Mode**).

The *script template* for a new enemy configuration data SO asset would look like this:

```
using UnityEngine;
[CreateAssetMenu(fileName = "New EnemyConfigData",
    menuName ="ScriptableObjects/EnemyConfigData")]
public class EnemyConfigData : ScriptableObject
{
    public float Speed, AttackRange,
        FireRange, FireCooldown;
    public bool CanJump;
    public float JumpForce;
}
```

Here, we can see the `[CreateAssetMenu()]` attribute that will create a new menu entry in the Editor to facilitate making new file assets based on this `EnemyConfigData` SO.

Going to the **Project** window and selecting **Create | ScriptableObjects | EnemyConfigData**, we can create multiple enemy configuration assets:

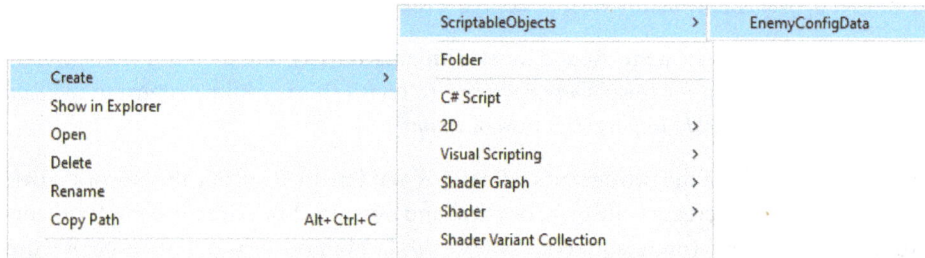

Figure 7.7 – Create ScriptableObjects asset menu

The member fields declared for `Speed`, `AttackRange`, `FireRange`, and so on provide configurable data for different types of enemies. There aren't any currently defined in the previous code, but you can also create methods (for encapsulating fields, returning calculations, helper methods, and ticks as examples).

Create a new script named `EnemyConfigData` in a new `Assets/Scripts/Data` folder using the preceding `ScriptableObject` template.

Regarding the adventure game, and referring to our **Game Design Document** (**GDD**) in *Chapter 4, Table 4.3*, we'll introduce two kinds of enemy characters (including the dull and uninteresting *maintenance robot*).

Go ahead and create two enemy configuration data assets named `Enemy A Config` and `Enemy B Config` in a new `Assets/Data` folder with some default values assigned to give each robot unique characteristics, as seen in the following screenshot:

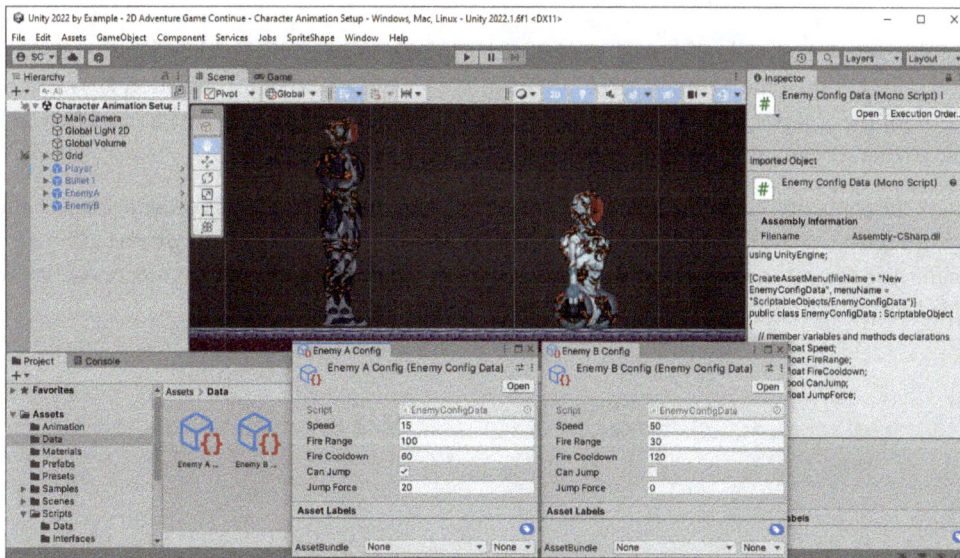

Figure 7.8 – Multiple assets for enemy configurations

As you can also conclude from the preceding screenshot, the evil alien plant entity has had its way with the maintenance robots, which are now under its control! No – not that exactly, but I've already imported and set up the enemy actors we'll be applying the configuration data to.

The process to import, rig, create a Prefab, add **inverse kinematics(IK)**, and then add animation is the same workflow we performed for the player character. To revisit this workflow, return to the *Setting up the player character with PSD Importer* section of *Chapter 5*.

Go ahead and create Prefabs of the two enemies – **Enemy A** and **Enemy B** – using the artwork provided in the GitHub project repo; create your own, or cheat and download the already completed enemy Prefabs from the project repo (preferably, don't choose this last option, as you'll need the practice).

> **Adventure game 2D art assets**
>
> To follow along in this chapter, download the art assets from the project GitHub repo here: `https://github.com/PacktPublishing/Unity-2022-by-Example/tree/main/ch5/Art-Assets`

Whew! Now that we have enemy Prefabs, we can add a component that will utilize the enemy configuration data. Create a new script named `EnemyController` in the `Assets/Scripts` folder – note that we're back to creating a `MonoBehaviour` script now:

```
Public class EnemyController : MonoBehaviour
{
    [SerializeField]
    private EnemyConfigData _config;
}
```

By now, this simple script should make a lot of sense to you. We added a field declaration named `_config` and of the type `EnemyConfigData`. We gave it an explicit protection keyword of `private` for the accessor so that no other script can reach it but added the `[SerializeField]` attribute so that we can make an assignment in the Inspector.

To achieve the configuration in the following screenshot, drag the `EnemyConfigData` script to the `Enemy B` Prefab (on the root object), then drag the `Enemy B Config ScriptableObject` asset in the `Assets/Data` folder to the **Config** field of the `EnemyController` component. We've just added variable configuration data to our enemy! Repeat for `Enemy A`:

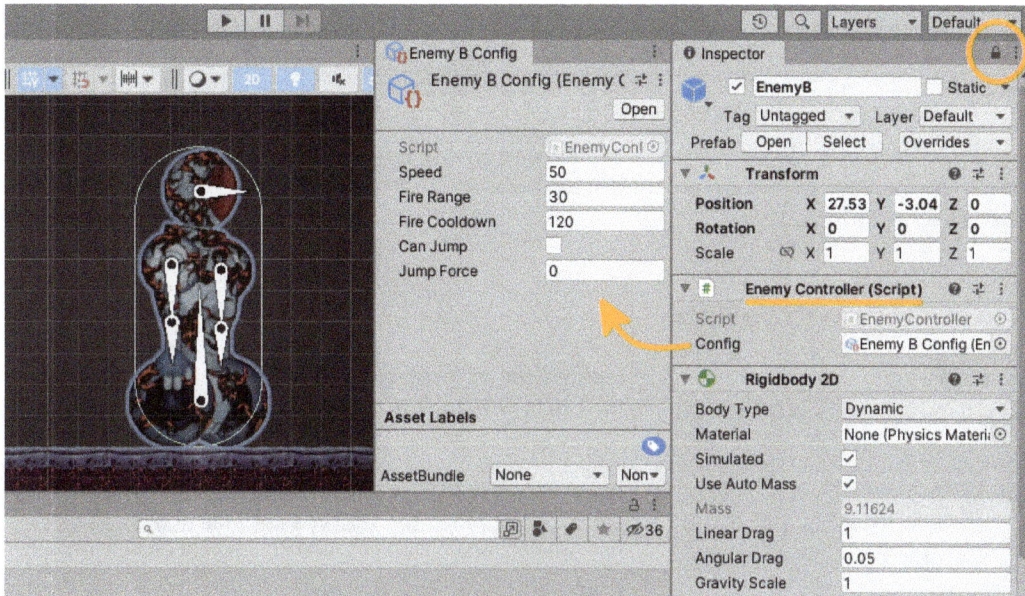

Figure 7.9 – Enemy B configuration

To quickly access the `ScriptableObject` asset for editing the values, double-click on the `ScriptableObject` asset reference assigned to the **Config** field. Alternatively, you can open a **Focused Inspector** by right-clicking on the asset in the **Project** window and selecting **Properties**.

This way, you can view and edit the data without the Inspector changing to other selected objects/assets as you work (an example of this can be seen in *Figure 7.8* with both of the enemy configurations open).

Opening a **Focused Inspector** is also an alternative to locking the Inspector to a single object, which you can do by clicking the little lock icon at the top right of the **Inspector** window (as shown in *Figure 7.9*).

Now that we have our enemy characters, let's see how to easily add some variation by extending upon these base Prefabs with Prefab Variants!

Creating an enemy variant for alternate enemy types

When Unity finally added native support for **Nested Prefabs** (circa Unity 2018.3), it also included a great new feature called **Prefab Variants**, which are extremely useful for having a unique set of variations that are all based on the same basic Prefab properties.

A base Prefab will have all the basic behaviors required by the object, and then several variations can be created that override properties to make changes to the behavior or appearance of the object.

In our case, using an enemy Prefab as an example, assigning a unique set of configuration values through an `EnemyConfigData` asset overrides the base Prefab. This may also include changes in color, artwork, or components.

We accomplish varying the configuration data by creating additional `EnemyConfigData` assets for each of the different enemy characteristics we're looking to have in the game.

> **Additional reading | Unity documentation**
>
> *Prefab Variants*: `https://docs.unity3d.com/2022.3/Documentation/Manual/PrefabVariants.html`

Creating a Prefab Variant

Let's create a variation of the `Enemy B` Prefab with an increased difficulty level for the player by making it move faster, shoot further, and have a quicker rate of fire by shortening the cooldown period.

To do that, let's take these steps:

1. First, we'll duplicate the `Enemy B Config` `EnemyConfigData` asset file in the `Assets/Data` folder:

    I.   Click on the `Enemy B Config` asset in the **Project** window to select it, press *Ctrl/Cmd + D* to make a duplicate, and then rename it `Enemy B Config Difficult`.

    II. On the new asset, adjust the values for **Speed**, **Fire Range**, and **Fire Cooldown** (I used `70`, `60`, and `60`, respectively).

2. Now, change to the `Assets/Prefabs` folder.

    I.   Right-click on the `Enemy B` Prefab, then select **Create | Prefab Variant** and rename it `Enemy B Difficult`.

    II. With the new variant selected and the **Inspector** window showing, go back to the `Assets/Data` folder and drag the `Enemy B Config Difficult` asset to the **Config** field of the `EnemyController` component.

We now have an `Enemy B` Prefab with different SO configuration data assigned to it, but the remainder of the enemy object is exactly the same since we only overrode the field assignment.

Figure 7.10 shows what our new enemy Prefab variant looks like in the Editor with the difficult enemy configuration SO data assigned: *A* is the difficult enemy Prefab, *B* shows that this Prefab has a `Base Prefab` it is a variant of, and *C* shows the SO **Config** field assignment override:

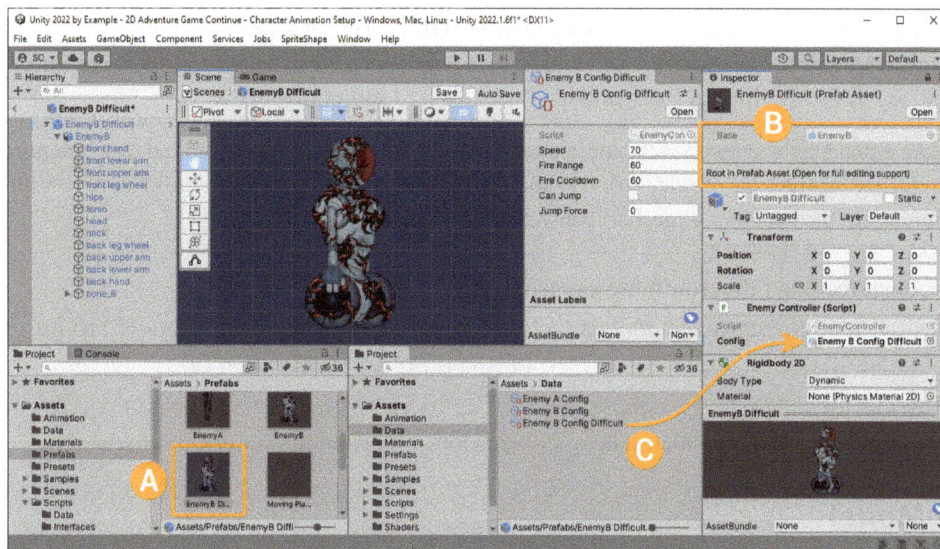

Figure 7.10 – Difficult Enemy B Prefab Variant setup

We can repeat this process for creating different types of enemies or any other type of Prefab that we'll use in the game. Think about maybe having the maintenance robots start the game without any plant infestation and then gradually build up the infestation as the game progresses. We can use Prefab Variants for the different stages by simply overriding the art assets. Prefab Variants can make progressive changes to art easy to achieve!

By using the SO to override the configuration data, we have separated the data from the Prefab asset (including various components, art assets, sound, effects, and so on). The SO data is a small object that can be updated outside of having to edit the Prefab, which makes things more accessible to designers and non-programmers.

Also, if only the data needs to change, then it's a tiny update that can be pushed to games in production without having to push the entire Prefab asset.

From this section, having a set of enemies with unique variations is great, but it would be even better if they had some behavior based on their configuration data!

The next section will look at adding behavior to the enemy robots.

Implementing basic enemy behavior using an FSM

In *Chapter 2*, we briefly introduced the State Pattern, so we'll now look at how to implement this design pattern for keeping the state of our enemy characters. Expressly, using an FSM, we can declare the fixed set of states (that is, finite) our enemy can be in at any given time – and the FSM is only going to do precisely these things.

The first implementation of our FSM will not adhere to SOLID principles very well, but it will hopefully be a simple enough introduction that will make sense practically. We can also use it as an example to point out any flaws with the approach and later refactor it to something better.

I should note that maybe we won't refactor it later... sometimes a simple approach is all that is required, and refactoring for the sake of refactoring is just wasted effort where that time could be better spent on tightening up the core game mechanic instead, for example.

> **Additional reading | Programming patterns**
>
> Finite state machine explained: `https://www.freecodecamp.org/news/finite-state-machines/`

Let's start by having a look at which states – or behaviors – we want for our enemy characters.

State Model

Referring back to our GDD in *Chapter 4*, *Table 4.1*, where we loosely defined the behaviors of our enemy during engagement with the player, we can derive the following minimum states being required: **Idle**, **Patrol**, **Attack**, **Dead**.

Now, we can take those states and design a UML state diagram for our enemy behaviors. Conditions for determining when to change between the states are also defined:

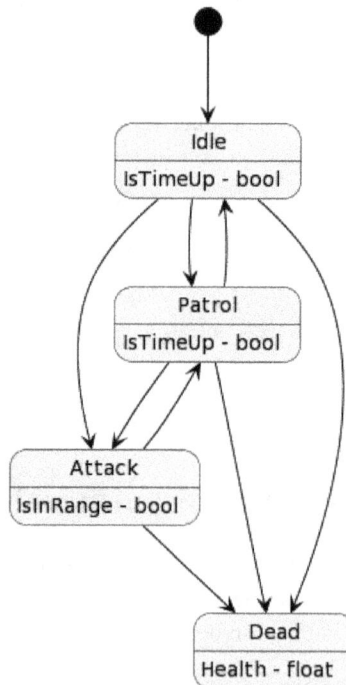

Figure 7.11 – Enemy behavior State Model

Referring to the diagram and using the conditions to determine when to transition from one state to another, we can observe the following:

- The default starting state is **Idle**
- While in the **Idle** state, we will transition to **Patrol** when a timer expires
- While in the **Patrol** state, we will transition back to **Idle** when a timer expires (for example, **Idle** --> **Patrol** --> **Idle** --> **Patrol**)
- While in the **Idle** or **Patrol** state, we will transition to **Attack** when the player is in range
- While in the **Attack** state, we will go back to **Patrol** when the player moves out of range
- We will transition from any state to **Dead** when health is zero

Now that we know which states we need and which conditions change state, we can move on to coding up this thing!

A simple FSM pattern

We'll make our first FSM using enums for defining our finite states and code the state-changing logic within the EnemyController class. This code is similar to what you'll likely run across in projects for beginners because it's easy to understand and simple to work with (and, as mentioned previously, may be all that is required).

But it does have some limitations and drawbacks. The first one is that the state machine is integral to the EnemyController class, violating the **Single Responsibility Principle** (**SRP**).

Let's have a look:

```
public class EnemyController : MonoBehaviour
{
    ...
    public enum State { Idle, Patrol, Attack, Dead }
    private State _currentState;

    private void Start() => ChangeState(State.Idle);

    void Update()
    {
        switch (_currentState)
        {
            case State.Idle:
                // UNDONE: Do stuff --> change state?
                break;
```

```
        case State.Patrol:
            // UNDONE: Do stuff --> change state?
            break;

        // And so on.
    }
}

public void ChangeState(State state) =>
    _currentState = state;
}
```

As we can see, we have defined the FSM states from our State Model in the `public enum State { }` declaration line. Our FSM can only be in one of these defined states. Then, we'll use the `_currentState` variable to keep track of our current state.

Let's skip down to the `ChangeState()` method, where we can see we'll call it by passing in a `State` to set it as the current state of the FSM (that is, transition to a different state). In `Start()`, you can see how we call `ChangeState(State.Idle)` to set our initial (default) state for **Enemy** to **Idle**.

Lastly, in the `Update()` method (called every frame update), we have a `switch` statement that has an implementation for each state declared in the `State` enum. When `_currentState` is equal to one of the defined enum states, we *do stuff* – you can relate the `switch` statement to a block of `if`, `else if`, `else if` statements (but it's undoubtedly more readable for when we don't need to test expressions for ranges of values or conditions).

---

switch (C#)

A `switch` statement is a *selection control* statement that tests an expression (similar to an `if` statement) and executes the matching block of code defined by the cases (and as terminated by the `break` statement). A default case can be defined if there is no match for the expression.

I don't necessarily have a problem with this oversimplified approach to a State Pattern, but the `switch` statement quickly gets ugly. With many states, it can start getting hard to manage, as we'll see when we start adding conditions and behaviors to it, but at least it doesn't rely on an arbitrary number of Boolean variables to attempt and maintain some form of state (and without having to wrangle two of those variables becoming true simultaneously – yikes!).

One problem with the enum-based approach that breaks the *O* in SOLID (**open-closed principle, or OCP**) is that we cannot change the conditions (or possibly behavior) of a single state without modifying the `EnemyController` class (that is, closed for modification). Preferably, we'd like only to change the affected state's code and not touch anything else!

Code that functions and has nothing to do with the state should not have to be retested (for example, acceptance and regression testing) or, in a team environment, even be code reviewed while committing the modification for the state to the **version control system (VCS)**.

At the very least, this enum-based approach provides a structure that makes the code more readable than without it, and it's simple to include the State Pattern without first having to write boilerplate code for the state machine.

You may be able to imagine how we could extend this State Pattern by encapsulating all of a particular state's behaviors (and data) into a single class. If so, great! If not, no worries, since we'll tackle refactoring this FSM later on in *Chapter 13*!

Okay – with the caveats out of the way, let's change some states!

Changing state behaviors

Now that we have our finite set of states defined and coded in our simple FSM pattern, you may wonder how we add in the conditions for changing state. Wonder no more – it's pretty simple!

We'll start by adding the required fields and assigning default values to evaluate our conditions for the first two states.

Idle and patrol behaviors

Referring to the **State Model** in *Figure 7.11*, let's have the enemy patrol a path between two positions (we'll define in the level) after remaining idle (stationary) for a specific time. Then, after patrolling for a particular time, we'll have our enemy return to idling, and repeat the process indefinitely while the enemy is not attacking the player or, well, is dead.

For this, let's first add two variables to our `EnemyConfigData` SO that will specify our times for how long the enemy will idle and patrol:

```
public class EnemyConfigData : ScriptableObject
{
    ...

    [Header("Behavior Properties")]
    public float TimeIdle = 5f;
    public float TimePatrol = 15f;
}
```

Don't forget that you can override these default time values in the Editor by selecting the `Enemy B Config` asset in the `Assets/Data` folder and changing the values.

Now, back in the `EnemyController` class, we'll need a way to track when we changed into a state to calculate the time elapsed:

```
private float _timeStateStart;

public void ChangeState(State state)
{
    _currentState = state;
    _timeStateStart = Time.time;
}
```

We've added the `_timeStateStart` float variable that we set to the current game time in the `ChangeState()` method (which has been changed to a body block from an expression body). Calling the `ChangeState()` method provides a way for us to do things *on the state entered* rather than just setting the `_currentState` variable directly to the new state we're transitioning to (which doesn't afford us the option).

With the required fields and assigned values now set up, we can proceed to evaluate the conditions for changing to/from the **Idle** and **Patrol** states:

```
...
void Update()
{
    switch (_currentState)
    {
        case State.Idle:
            // UNDONE: Do stuff.

            // Change state?
            if (Time.time - _timeStateStart
                >= _config.TimeIdle)
                    ChangeState(State.Patrol);
            break;

        case State.Patrol:
            // UNDONE: Do stuff.

            if (Time.time - _timeStateStart
                >= _config.TimePatrol)
                    ChangeState(State.Idle);
            break;
...
```

As you can see in the preceding code, we have a new `if` block that evaluates the current time, `Time.time`, minus the time we transitioned into the current state, `_timeStateStart`. If the difference is greater than or equal to our configured time to remain idle, `_config.TimeIdle`, then call the `ChangeState()` method to transition to the **Patrol** state. Easy-peasy!

Similarly, we will evaluate transitioning from **Patrol** to the **Idle** state, this time using the `_configTimePatrol` value (that is, how long the enemy should patrol for). And now, you may be wondering: how do we make the enemy character actually patrol?

Implementing behavior

For the **Patrol** state, what we actually mean, considering we have a 2D game, is moving between two set points in the level. This code could again be written directly within the `EnemyController` class, but instead of doing that again, and getting too far away from Single Responsibility, let's at least abstract the behaviors into their own classes.

We'll therefore define behaviors via an interface so that we can swap out the behavior code, should we need or want to change it without modifying the class that implements it.

So, create a new C# script named `IBehaviorPatrolWaypoints` in the `Assets/Scripts/Interfaces` folder and add the following interface declaration for a *patrol waypoints behavior*:

```
public interface IBehaviorPatrolWaypoints
{
    Transform WaypointPatrolLeft { get; }
    Transform WaypointPatrolRight { get; }

    void Init(Rigidbody2D rb, Vector2 direction,
        float acceleration, float speedMax);
    void TickPhysics();
}
```

Here, we can see that we've declared two points in the level that will create the path the enemy will patrol between as `WaypointPatrolLeft` and `WaypointPatrolRight`. We'll place an empty GameObject at each position in the level and assign their references to these fields in the Inspector.

To help visualize the concept of a patrol path better, referring to the following diagram, the dots (blue) represent the waypoints (that is, the empty GameObjects), and the dashed line (orange) represents the patrol path created from the waypoints:

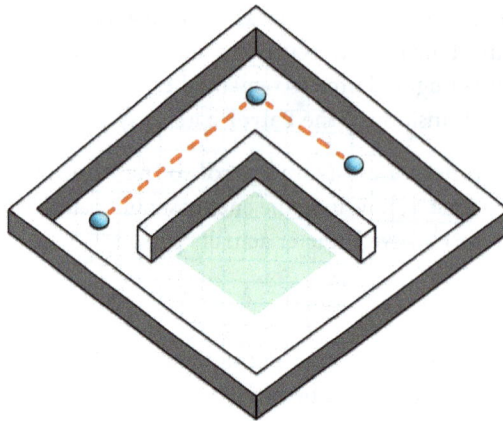

Figure 7.12 – Waypoints and patrol path

(Spoiler… players should hide in the square area (green) just south of the wall to avoid patrolling agents' gaze. Look out!)

Back to the code, we then have an Init() method that will be called from the implementing class that will pass in (or, you might say, *inject*) the required dependencies the behavior needs to function as parameters.

Last is the declaration for the TickPhysics() method that will be called by the implementing class's FixedUpdate() method to perform the actual function of moving the enemy between the waypoints (using physics).

Back in the EnemyController class now, let's create a variable of type IBehaviorPatrolWaypoints. We'll obtain the reference to an instance of it – which should exist as a sibling component on the EnemyController object – by using TryGetComponent():

```
// Implemented behaviors.
private IBehaviorPatrolWaypoints _behaviorPatrol;
private void Awake()
{
    ...
    // Get behaviors and initialize.
    if (TryGetComponent<IBehaviorPatrolWaypoints>(
        out _behaviorPatrol))
    {
        _behaviorPatrol.Init(_rb, _movementDirection,
            _config.Acceleration, _config.SpeedMax);
    }
}
```

This is how we can swap out behaviors with other behaviors of the same type without modifying the implementing class. By using Unity's GetComponent() architecture – which allows a form of **composition**, we can create different components (inheriting from MonoBehaviour) that implement IBehaviorPatrolWaypoints and obtain the component instance via the interface.

Composition (OOP)

Composition is usually referred to as a *has-a-relationship* and is achieved by a class using instance variable(s) that refer to another object (of a class or another class). The term *combining* is also used to describe composition in **object-oriented programming (OOP)** because it deals with bringing multiple objects together to achieve an outcome.

Since IBehaviorPatrolWaypoints is simply a behavior that moves the object it is added to between two points, and it doesn't explicitly have to be used on the enemy, it can be used on any object we want to have this behavior.

Okay – for the last part of implementing the behavior, we'll need to have the TickPhysics() method tied into the FixedUpdate() method of the EnemyController class so that it can perform its behavioral function. And we do it like this:

```
private void FixedUpdate()
{
    if (_currentState == State.Patrol)
        _behaviorPatrol?.TickPhysics();
    else
        _rb.velocity = Vector2.zero;
}
```

What we have here again, same as in Update(), is having to check the current state we are in before calling the tick method for the current behavior to execute (that is, more code smell since the state check is occurring in multiple places in the class).

Code smell

A **code smell** refers to source code that is a *quick-to-spot indicator* for deeper or interesting problems that may exist within the code base. These are not bugs nor errors but violations of fundamental principles that decrease the quality of the code base. By nature, they are unquantifiable and subjective to the developer's experience.

If _currentState is Patrol, then call _behaviorPatrol?.TickPhysics(). We used the null conditional operator (?.) on the behavior variable (an interface) in case the enemy object was not implemented with an IBehaviorPatrolWaypoint component. You may prefer, however, not to use ?. and let it throw an error while playtesting in the Editor to validate the object configuration. It's up to you how you prefer to work (this would commonly be referred to as *developer style*).

> **Null conditional operator (?.) and Unity objects**
>
> Note that null propagation is incompatible with Unity objects because Unity overrides the null comparison operator (to properly return null for objects that were destroyed but not yet garbage collected). Your IDE should provide a warning when attempting to use the ?. operator on Unity objects (if not, then get a new IDE). The correct approach would be to simply use a null comparison. Example:
>
> NO
>
> ```
>     _player?.Jump();
> ```
>
> YES
>
> ```
>     if (_player != null)
>         _player.Jump();
> ```

The final step for fully implementing the behavior is to create a class that implements `IBehaviorPatrolWaypoints` and contains the movement logic. I'll present the completed code later in *Chapter 13*, when we refactor and extend upon the enemy behavior logic, but for now, I'm going to challenge you to create the script yourself!

To give you a hint, take a look at how we added movement to the player (remember – use physics; don't manipulate the transform position directly!) and start with this class declaration for a new component script named `PatrolWaypoints`:

```
public class PatrolWaypoints
    : MonoBehaviour, IBehaviorPatrolWaypoints
{
    // Do move between waypoints stuff.
}
```

Or, you can cheat. I mean, take a peek now at the final code in the GitHub project files for the book here: `https://github.com/PacktPublishing/Unity-2022-by-Example/tree/main/ch7/Unity%20Project/Assets/Scripts/Behaviors`.

We'll implement every other behavior in the same fashion, adding on a new state conditional check within each method that requires processing for the current state, as well as doing the following:

- Declaring an instance variable and using `GetComponent()` to obtain the reference
- Initializing the behavior instance and providing required dependencies via its `Init()` method
- Adding a call to its `TickPhysics()` method to `FixedUpdate()` to perform its function

> **Optimization note**
>
> With our base class – `EnemyController`, in this case – implementing the behaviors, it means none of our behaviors needs to have its own `FixedUpdate()` Unity message event being registered and called; so, having a `tick` method being called instead is a bit more efficient as it reduces the interop overhead (C# code being called from the C++ engine code) – and it could certainly add up if we have many enemies!

As you can see, however, this is, in fact, getting ugly with the enum-based State Pattern. For now, let's continue with setting up the condition for transitioning to the **Attack** state.

Behavior for attacking the player

We'll need some additional references for the **Attack** behavior – specifically, the player. After all, you can't attack what you can't see.

In *Chapter 4*, we assigned the `Player` tag to our `Player` object in the Inspector to determine if it was the player that interacted with a collision event. Well, we're going to use the tag again now, but differently.

We'll use the tag as a parameter for the `FindWithTag()` method to obtain a reference to our `Player` object in the scene.

> **Additional reading | Unity documentation**
>
> `FindWithTag:` `https://docs.unity3d.com/2022.3/Documentation/`
> `ScriptReference/GameObject.FindWithTag.html`

Instead of assigning the reference to a field exposed in the Inspector, we need to get the `Player` reference dynamically because our enemy Prefabs may be spawned into the scene at runtime. Hence, it's impossible to make that assignment (scene references cannot be made in file-based assets such as Prefabs and SOs).

Still within `EnemyController`, once we have the reference to `Player`, let's check the distance to the player and change the state to **Attack** if the player is within our arbitrary range value:

```
private GameObject _player;

private void Awake() =>
    _player = GameObject.FindWithTag(Tags.Player);

private bool IsPlayerInRange(float rangeAttack)
{
    var distance =
        Vector2.Distance(transform.position,
```

```
                  _player.transform.position);
        return distance <= rangeAttack;
}
```

We first declare a `_player` variable to hold the GameObject reference to the `Player` object. Then, in `Awake()`, we assign the player found in the scene by using `FindWithTag()` and passing in our previously declared string constant, `Tags.Player`.

> **Optimization note**
>
> `FindWithTag()` is a slow command, but we'll do it in `Awake()` to get the reference to Player before gameplay stats (i.e., caching the reference). Generally, you wouldn't want to do this during gameplay because it's slow, and definitely not in `Update()` since that is called every frame!

Next, we add the `IsPlayerInRange()` method for calculating the distance to the `Player` object using advanced trigonometric geometry mathematics! Nah – you can already see in the previous code that we simply use the `Vector2.Distance()` method and pass in the current position of both `Enemy` and `Player` to get a float value of the distance between them. Easy-peasy!

Previously, we used the `return` keyword to stop the code in a method from executing further. Here, we're doing the same thing (sort of; in this case, it's the last statement), but because the method signature for `IsPlayerInRange()` is declared as `bool` and not `void`, we need to return a bool value, the result of evaluating the distance returned from `Vector2.Distance()` being less than or equal to (`<=`) the specified `rangeAttack` value.

> **Question**
>
> Is the `EnemyController` class the best place to declare the `IsPlayerInRange()` method? Does this follow the SRP? If we need to change the logic for calculating the player distance, will it negatively affect other code?
>
> While you ponder that, I'll leave it right where it is in `EnemyController` (for now?).

And, finally, with our dependencies now in place, let's wire up the condition for transitioning into and out of the **Attack** state:

```
void Update()
{
    switch (_currentState)
    {
        case State.Idle:
            // UNDONE: Do stuff.

            // Change state?
            if (IsPlayerInRange(_config.AttackRange))
                ChangeState(State.Attack);
            else if (Time.time - _timeStateStart
```

```
              >= _config.TimeIdle)
                  ChangeState(State.Patrol);
          break;
      ...

      case State.Attack:
          // UNDONE: Do stuff.

          if (!IsPlayerInRange(_config.AttackRange))
              ChangeState(State.Patrol);
          break;
      ...
```

If we refer back to the **State Model** diagram in *Figure 7.11* for which state should transition to another, we can see that **Idle** and **Patrol** should check for an `is [player] in range` condition and transition to the **Attack** state if so.

We'll do this by adding to our `Change state?` if block, so we add `IsPlayerInRange()` and pass in the `AttackRange` value configured for this enemy object. If within range, boom: `ChangeState(State.Attack)`!

Conversely, in the **Attack** state, we want to transition back to **Patrol** should the player move outside the attack range value. We prepend an exclamation mark (`!`) to the `IsPlayerInRange()` call to reverse the result (the logical `not` operator returns `false` if the result is `true`) – which changes the evaluation from *is player in range* to *is player not in range*.

Not shown in the preceding code (for brevity), but the **Idle** and **Patrol** states will have the same `IsPlayerInRange()` condition check (just as our State Model says it should).

Now let's see how we handle our final state from our State Model: the **Dead** state (that is, in fact, quite final).

Dead state

The enemy would now be able to attack the player while in range – we'll look at implementing the exact attack behavior in the following chapter, *Chapter 8*, when the player attacks and immobilizes the infected robot enemy; however, we can, for now, provide a **Dead** state to handle that.

When we change to this state, we'll simply destroy the **Enemy** GameObject like so:

```
void Update()
{
    switch (_currentState)
    {
        ...
        case State.Dead:
            Destroy(gameObject);
            break;
    }
}
```

Using `Destroy()` like this is fine since we can still add *on-death* effects to the enemy using its `OnDestroy()` Unity message event.

> **Additional reading | Unity documentation**
> `MonoBehaviour.OnDestroy: https://docs.unity3d.com/2022.3/`
> `Documentation/ScriptReference/MonoBehaviour.OnDestroy.html`

The condition, however, needs to be checked outside the `switch` statement because we want to check for some health value becoming zero all the time – no matter the current state. We can do this by adding an `if` statement after the `switch` statement – the `switch` section is only for processing the current state!

```
void Update()
{
    switch (_currentState)
    …

    // Any state.
    if (_health <= 0)
    {
        ChangeState(State.Dead);
    }
}
```

Don't worry too much about the `_health` variable at the moment; we're going to implement health and damage systems in *Chapter 8*. As you can see in the preceding code, we're simply checking if this enemy's health equals or dips below zero and changing to the **Dead** state if so.

This section introduced you to the State Pattern, a State Model UML diagram, and setting up a simple FSM for managing different states based on the State Model for our enemy characters.

Summary

In this chapter, we added some polish to the shooting and player character by introducing URP post-processing, **Shader Graph**, 2D lights, and the **Trail Renderer** effect. Phew! Having these features out of the box with Unity allows us to add visual quality to our games with little effort.

We continued by adding some configurable enemy characters to the game by creating two enemy Prefabs and assigning unique configuration variables to each via the ScriptableObject assets. The enemy objects were then given behavior by implementing the State Pattern to introduce basic behavior with an FSM and evaluate conditions for transitioning between states.

In the next chapter, we'll complete the adventure game by adding health and damage systems for enemies that we'll spawn into the level, implement the attack mechanics with additional weapon types, create a simple quest system for collecting key objects for solving the entryway puzzle, and introduce a new event system for keeping our code loosely coupled.

8
Extending the Adventure Game

In *Chapter 7*, we added polish to the game by applying some simple VFX using post-processing effects (mainly Bloom) with the Universal RP, Shader Graph for a custom 2D shader to make specific parts of a sprite glow, 2D lights to highlight the player, and the Trail Renderer component for a quick VFX win on our bullet sprite.

We then moved away from the player to give some much-needed attention to the enemy NPCs in the game by creating configurable enemies using a ScriptableObject architecture and introducing changing behaviors based on the state pattern as implemented via a simple FSM.

With the base functionality in place for the playable character and enemies with behaviors (mostly) set up, we can now move on to spawning enemies that attack the player and vice versa with a reusable health and damage system.

In this chapter, we're going to cover the following main topics:

- Health and inflicting damage
- Updating the player and enemy to use health
- Enemy wave spawner

By the end of this chapter, you'll be able to spawn enemies with simple health and damage systems – that will also be applicable for the player or any damageable objects in the game!

Technical requirements

To follow along in this chapter with the same artwork created for the project in the book, download the assets from the following GitHub repository:

```
https://github.com/PacktPublishing/Unity-2022-by-Example
```

To follow along with your own artwork, you'll need to create similar artwork using Adobe Photoshop, or a graphics program that can export layered Photoshop PSD/PSB files (e.g., Gimp, MediBang Paint, Krita, and Affinity Photo).

Health and inflicting damage

At this point in our project, we have implemented much of what we specified in our GDD (*Chapter 4*, *Table 4.1*) for our player character and enemies, but an essential system is still missing – health and damage.

In the coming sections, we'll not only tackle adding health to the player and enemies with a reusable component but also finish the enemy's attack behavior for inflicting damage on the player. The player can already fire a weapon that shoots a projectile, so we'll add to the `Bullet` object we previously made in *Chapter 6*, so that it can inflict damage, too.

Health system

We'll develop a `HealthSystem` component to create a reusable component for `Player`, enemy, and other objects (e.g., in the environment – think a destructible crate). This health system will track health, take damage and/or heal, and can be added to any object. The remainder of the setup will include creating Interfaces that tie the system together and make the whole thing operate in an abstract way (i.e., reusable, extensible, maintainable).

Like before, to clearly understand how we'll create the health system, we'll utilize a UML diagram based on the concepts we just described in the preceding paragraph. We can always refer back to this diagram at any point if needed.

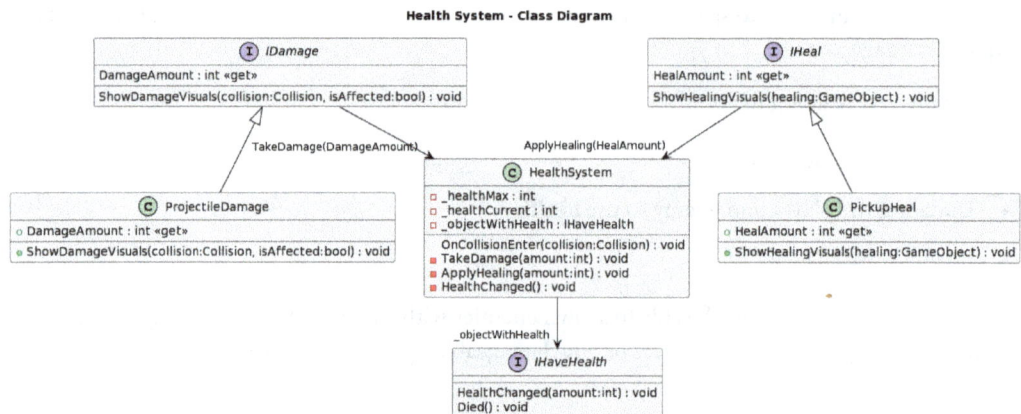

Figure 8.1 – Health system class diagram

Now, let's dive into it! Create a new script in the Assets/Scripts folder and name it HealthSystem. This, being central to the UML diagram, will be the largest class we'll be creating for the health system because, well, it *is* the health system! Since it will consist of the most code we'll be writing in this section, we'll break it up by first just declaring the member variables and a Unity message event, then proceed to flesh it out as we build up the system's functionality:

```
using UnityEngine;
public class HealthSystem: MonoBehaviour
{
    [SerializeField] private int _healthMax;
    private int _healthCurrent;

    private void OnTriggerEnter2D(Collision collision)
    {
        // UNDONE: Test for a collision with a component
        // that can damage us.
        // UNDONE: HandleDamageCollision();

        // UNDONE: Test for a collision with a component
        // that can heal us.
        // UNDONE: HandleHealCollision();
    }
}
```

Task list (IDE)

You've undoubtedly seen // UNDONE: in the preceding code and may have wondered what that's all about. Well, we can use some tokens in our comments that our IDE will pick up, and it will generate a task list based on them! This feature helps us locate incomplete tasks and highlight things that need attention. Additionally, you can create your own custom tokens, which can be super useful for your specific needs (but you'll need to check your specific IDE for support).

Common tokens include the following:

```
// TODO:
// UNDONE:
// HACK:
```

As you may be able to infer from the code we have so far – and if you've read the code comments – HealthSystem functions through collisions with other objects in the scene, such as projectiles from enemies or healing pickups. The other objects can either damage or heal the object with the HealthSystem component on it, and that will depend on the interface the object inherits from (from the UML diagram: IDamage and IHeal). Even the ground as an object can potentially damage the player if, let's say, the player's velocity is above a certain threshold when colliding.

Here's the breakdown for the code added so far, plus the required additions:

- Variable declarations will enable the core function of HealthSystem:

    - _healthMax will specify the maximum health value for the object (i.e., the object we're giving health by adding the HealthSystem component).

    - _healthCurrent will specify the current health of the object. As the object is either damaged or healed, this value will decrease or increase respectively. When the object is created in the scene, we should set the current health to the _healthMax value, which we'll do now by adding the Awake() Unity message event as follows:

    ```
    private void Awake()
    {
        _healthCurrent = _healthMax;
    }
    ```

    - Note here that the max health value (_healthMax is the serialized stored value) will be saved within the Object with Health Prefab, but imagine that you could also use a ScriptableObject, database, cloud-obtained (JSON) data, or even Unity's own **Remote Config** (as part of **Unity Gaming Services**) so that we can change the max health value dynamically (and, with some of these methods, at any time without being dependent on distributing a new build of the game).

> **Additional reading | Unity Gaming Services**
> Remote Config: https://unity.com/products/remote-config

- OnTriggerEnter2D() is where the magic happens! Collisions with other objects drive the health system, so we'll use this Unity message event to handle the interactions when triggered from the physics system.

    Let's add both damage and healing methods first, and then we'll add the evaluations that call these methods next:

> **Note | Physical interactions**
>
> To have the objects physically respond to each other when colliding, use a **Collider** instance on the object that does not have **IsTrigger** enabled and respond to the collision using the `OnCollisionEnter2D()` Unity message event. For an object to not have a physical response when colliding, use a **Collider** instance with the **IsTrigger** field enabled and respond to the collision using the `OnTriggerEnter2D()` Unity message event.

- `HandleDamageCollision()`, as the name implies, handles when we collide with another object that can damage us. We'll pass in the collision parameter from when `OnTriggerEnter2D()` is called and, a future addition, the object that caused the damage (which is the object inheriting from the `IDamage` interface).

- Add the following method to the `HealthSystem` class:

```
internal void HandleDamageCollision
    (Collider2D collision, IDamage damage)
{
    // UNDONE:  TakeDamage(amount);
}
```

- Okay, yes, we've added another unfinished method to be called here that we'll again add to later: `TakeDamage()`. We're keeping the approach simple, taking one baby step at a time to build out the health system. The simple UML diagram may set an expectation that this will be easy. Well, it is – provided we tackle the implementation one step at a time.

- Our future selves will again tackle implementing the `TakeDamage()` method in the coming *Taking damage – IDamage interface* section when we also add the required `IDamage` interface.

- `HandleHealCollision()` is similar to the method for handling damage. We'll add one for handling object healing. However, we'll omit to pass in the `collision` object as a parameter this time; we'll handle things a bit differently for healing compared to taking damage (as you'll see in the coming *Healing – IHeal interface* section).

- Add the following method to the `HealthSystem` class:

```
internal void HandleHealCollision(IHeal heal)
{
    // UNDONE:  ApplyHealing(amount);
}
```

- Okay, we're piling on the *undone* methods to implement here, right?! Just like for `TakeDamage()`, we'll have a method called for healing and changing the health value of the affected object.

And that's the core of the `HealthSystem` class laid out. Let's go ahead and add the required interfaces that our new health system classes inherit from and make things actually function!

Interfaces required!

There's still more work to do before we add a `HealthSystem` component to any of our GameObjects, so let's sort out and evaluate these damage and heal collisions.

The first part we'll need interfaces for is the evaluation of the object that collides with the object that has health (e.g., a `Player` or enemy object with the `HealthSystem` component added). We'll determine whether the colliding object can damage or heal – we're replacing this first UNDONE token in our task list:

```
// UNDONE: Test for a collision with a component
// that can damage us.
```

From the UML diagram, we can see the object we want to evaluate is either inherited from `IDamage` or `IHeal`. Using interfaces to inherit from ensures that the required members exist in our classes that provide the intended functionality (i.e., the class must satisfy the "contract" of the interface).

Standard naming conventions dictate that when naming an interface, it should begin with the letter "*I*," which offers us an opportunity to be a bit clever, or even memorable, in our naming and bring more readability to our codebase. So, for an object that can damage another object, we'll name the interface `IDamage` – as in *I damage [an object]*. In the *Healing – IHeal interface* section, we'll add additional interfaces for `IHaveHealth` and `IHeal` – see what we did there. Not that I'm being terribly clever here because *IDamage*, *IDamageable*, and *ITakeDamage* are pretty common interface names for game code for the same naming reasoning.

Now, let's update the `OnTriggerEnter2D()` method with the following `if` statements that include the interfaces (that we'll create just after this step):

```
private void OnTriggerEnter2D(Collision collision)
{
    // Test for a collision with a component that can
    // damage us.
    if (collision.
        TryGetComponent<IDamage>(out var damage))
    {
        HandleDamageCollision(collision, damage);
    }

    // Test for a collision with a component that can
    // heal us.
    else if (collision.
        TryGetComponent<IHeal>(out var heal))
```

```
        {
            HandleHealCollision(heal);
        }
    }
}
```

Let's have a look at the following two evaluations and how they'll provide the desired functionality for our health system:

- `collision.TryGetComponent<IDamage>()`: If you haven't realized by now, I'm a fan of the *try get component pattern* – we can fail gracefully if the component doesn't exist on the object we're testing. And if it does exist, we conveniently have an `out` parameter that returns the component. Simples!

> **Optimization note**
>
> When using the `TryGetComponent` method, the method does not allocate memory on the heap when it doesn't find a component. When it does find a component, it allocates memory only for the return value, not the component itself. This can be very beneficial for improving performance and reducing garbage collection, unlike the `GetComponent` methods, which can generate garbage and allocate more memory – both negatively impacting performance. By utilizing `TryGetComponent` instead, you can avoid unnecessary memory allocations and keep your game running smooth as butter.

So, what's occurring here is, if the object we collided with has a component that inherits from `IDamage` on it, as in, *I damage this object*, then return the component and pass it into `HandleDamageCollision()` as a parameter along with the collision object itself. Peeking ahead a bit and referring back to the UML diagram, we can see that we'll be implementing a `ProjectileDamage` component (e.g., on `Bullet`) that inherits from `IDamage`.

- `collision.TryGetComponent<IHeal>()`: Ditto here. If the component we collided with has a component that inherits from `IHeal` on it, as in, *I heal this object*, then return the component and pass it into `HandleHealCollision()`. Again, peeking back at the UML diagram, we can see that we'll be implementing a `PickupHeal` component (e.g., on a Water Diamond) that inherits from `IHeal`.

Currently, if the object collided does neither damage nor heal, then we simply ignore the collision (of course, when colliding with anything, the obvious choice here is to do a camera shake!).

Now that we have an implementation of the interface, we need to actually create them. These interfaces aren't going to write themselves, so let's start with `IDamage`.

Taking damage – IDamage interface

We've already seen the UML diagram (*Figure 8.1*) and the `HandleDamageCollision()` code, where the `IDamage` interface is implemented but hasn't been defined yet. I lied above, too: they can write themselves (at least partially) if we use the IDE's refactoring tools – in `OnTriggerEnter2D()`, `IDamage` will have a red squiggly underline. Right-clicking on the word (or clicking anywhere on it and pressing *Alt + Enter* or *Ctrl + .* depending on your IDE) and selecting **Generate interface 'IDamage' in a new file** will generate the following:

```
internal interface IDamage
{
}
```

If you decided not to use refactoring tools (why not?), create a new C# script in the `Assets/Scripts/Interfaces` folder and name it `IDamage`. Even if you did use the refactoring tools, you'll likely still need to move the generated script into the `Assets/Scripts/Interfaces` folder to keep things tidy.

Now, we'll need a field for specifying the value for how much the object inheriting from `IDamage` will damage the object with health, so add a `DamageAmount` variable declaration like so:

```
int DamageAmount { get; }
```

Remember, all members of an interface are public, so there is no need to add the accessor. We'll set the property to be a getter only, though; we'll only want the value to be read by other classes (no modification outside of the class inheriting the interface – all nice and hidden, how we like it).

Now that we have our `IDamage` interface. We can use it for making projectiles and, well, just about any other object that hurt the player, by subtracting health when a collision occurs between the object and `HealthSystem`.

Let's fix up the `HandleDamageCollision()` method now back in our `HealthSystem` class by removing the `// UNDONE:` token comment and using `DamageAmount` for the parameter:

```
internal void HandleDamageCollision
    (Collider2D collision, IDamage damage)
{
    TakeDamage(damage.DamageAmount);
}
```

This sets us up nicely to have our IDE's refactoring tools generate the `TakeDamage()` method for us, so let's continue by doing just that. You know the drill: red squiggly line, etc., etc., then select **Generate method 'TakeDamage.'**.

And here we are:

```
private void TakeDamage(int amount)
{
    // UNDONE: Subtract from current health.
    // UNDONE: HealthChanged();
}
```

I'm so sorry, more *undone* comments! Please don't fret; with this very temporary instruction, we'll fix up the code straight away. Fill in the method with the following statements:

```
private void TakeDamage(int amount)
{
    _healthCurrent = Mathf.Max(_healthCurrent - amount,
        0);
    HealthChanged();
}
```

The first thing we do is update the current health value for the amount of damage received by subtracting amount from _healthCurrent. We're getting some help from the Mathf.Max() function here so that the current health value will never dip below zero (keeping things positive).

> **Additional reading | Unity documentation**
> `Mathf.Max():https://docs.unity3d.com/2022.3/Documentation/`
> `ScriptReference/Mathf.Max.html`

Let's uncomment our placeholder for the HealthChanged() method. Go ahead and create an empty method block for it, but completing this method will be a job for our future selves when we actually have some objects set up to enact the health change:

```
private void HealthChanged()
{
    // UNDONE: If current health is greater than zero,
    // notify the object with health.

    // UNDONE: If current health is zero, the object
    // with health dies/is destroyed.
}
```

We'll call this method anytime the health value has changed so we can evaluate the current health of the object and *do stuff* accordingly – such as notify other classes about the health value of the object changing or die/destroy if health reaches zero.

So, let's get an object set up now that will inflict damage on the player.

ProjectileDamage component

Create a new script called `ProjectileDamage` in the `Assets/Scripts` folder – this is a component we'll add to our `Bullet` Prefab. To ensure this component will cause damage to our health system, it will implement the `IDamage` interface:

```
using UnityEngine;
public class ProjectileDamage : MonoBehaviour, IDamage
{
    public int DamageAmount => _damageAmount;
    [SerializeField] private int _damageAmount = 5;
}
```

The public `DamageAmount` `int` variable is required to be declared to satisfy the `IDamage` interface contract – it's also necessary for the health system to obtain the value for the amount of damage this projectile causes! `DamageAmount` is a public property because all interface members are public and cannot contain fields. Because C# properties are not serialized by Unity, to assign a value in the **Inspector** window (i.e., save a damage amount value within our `Bullet` Prefab), we'll encapsulate a private `_damageAmount` variable and decorate it with the `[SerializeField]` attribute. If you haven't already guessed, this is the structure we'll continue with throughout the remainder of the book.

When you've finished saving the script, open up the `Bullet` Prefab in **Prefab Edit Mode** (double-click on it in the **Project** window) and add `ProjectileDamage` to the root GameObject, as seen in the following figure:

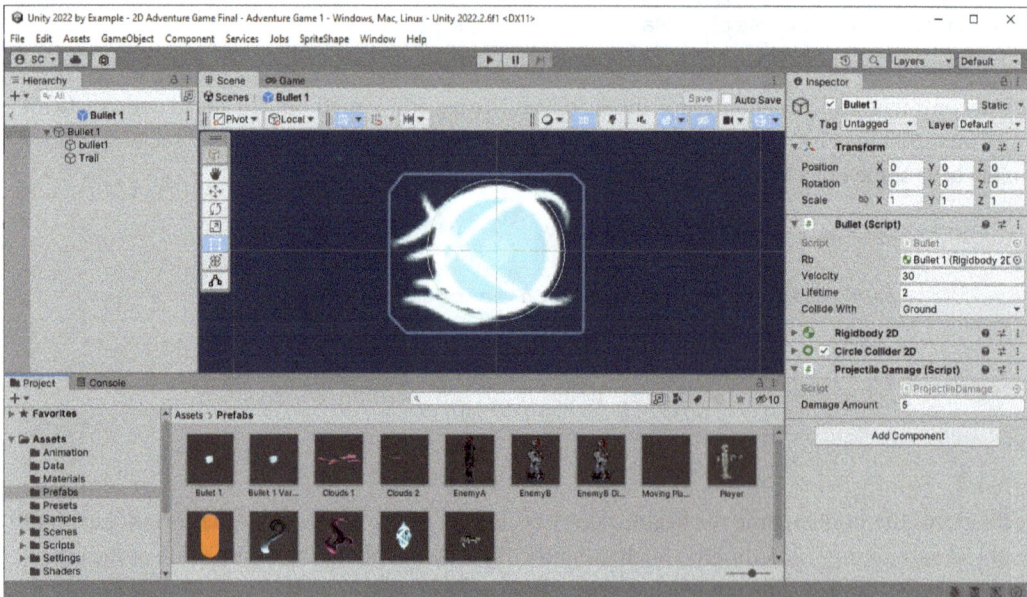

Figure 8.2 – Bullet Prefab ProjectileDamage component

When we declared _damageAmount in the preceding code, we set a default value of 5, which you can already see as the assigned value in the **Inspector** window.

Okay, we can damage stuff now; great! But that's not really fair if objects such as our player cannot also have a chance to heal.

Healing – IHeal interface

We've taken care of the left side of the UML diagram (*Figure 8.1*) for our health system, so now, similar to IDamage, we need to define the IHeal interface to take care of the right side. We'll also create a component to add to objects that can apply healing.

Back in the OnTriggerEnter2D() method now, let's repeat the steps performed for creating the IDamage interface but for IHeal:

1. Create the IHeal interface script (have your IDE generate it) in the Assets/Scripts/Interfaces folder and add the HealAmount variable for, you guessed it, specifying the amount of healing power:

```
internal interface IHeal
{
    int HealAmount { get; }
}
```

2. Update the placeholder HandlHealCollision() method, specifying the HealAmount value from the interface as the parameter when calling ApplyHealing():

```
private void HandleHealCollision(IHeal heal)
{
    ApplyHealing(heal.HealAmount);
}
```

3. Lastly, create the ApplyHealing() method:

```
private void ApplyHealing(int amount)
{
    _healthCurrent = Mathf.Min(_healthCurrent + amount,
        _healthMax);
    HealthChanged();
}
```

The first thing we do is update the current health value for the amount of healing received by adding amount to _healthCurrent. Just like when we took damage, we're getting some help from a Mathf function here again, but it's Mathf.Min() this time, so that the current health value will never exceed the object's maximum health value (no cheating).

> **Additional reading | Unity documentation**
> `Mathf.Min:` `https://docs.unity3d.com/2022.3/Documentation/`
> `ScriptReference/Mathf.Min.html`

And, like with damage, we'll create a healing component that will add health when a collision occurs.

PickupHeal component

Create a new script called `PickupHeal` in the `Assets/Scripts` folder:

```
using UnityEngine;
public class PickupHeal : MonoBehaviour, IHeal
{
    public int HealAmount => _healAmount;
    [SerializeField] private int _healAmount = 10;
}
```

Same as with the `ProjectileDamage` script, we can see that we've inherited from `IHeal` (to ensure healing with our health system) and implemented the `IHeal` interface by defining `HealAmount`. We've also encapsulated `HealAmount` by declaring a private `_healAmount` variable serialized so we can set the value in the **Inspector** window (note a default value of `10` is also assigned here).

> **Note on code architecture**
> If you find yourself needing many different types for damage or heal components, you can create a new base class for each that implements the interface, so you aren't repeating yourself with the event and method to invoke. The current implementation suits our current needs, so you can also leave off here, or challenge yourself to create both a `DamageBase` and `HealBase` abstract class that `ProjectileDamage` and `PickupHeal` inherit from, respectively.

`PickupHeal` is a component we'll add to… hmmm… what object will we add it to? Let's consult our GDD (*Chapter 4, Table 4.2*):

What is a buff mechanic for the player in the adventure game?	The player will be able to collect energy shards (water diamonds) scattered throughout the environment that, when a certain quantity has been collected, will give a power-up state to all of the weapons (increasing damage dealt).

Table 8.1 – Adding a buff to the GDD

Sounds great. Well, actually, I think we can do better. Let's revise; after all, the GDD is a living document:

What is a buff mechanic for the player in the adventure game?	Players can gather energy shards (water diamonds) as they explore the game world. The player can later use the collected energy to power up weapons (increase damage dealt) or heal the player, offering more strategic options for the player while navigating the game's challenges.

Table 8.2 – Revising the buff in the GDD

Better! Allowing players to choose between using energy shards to power up weapons or to heal themselves creates a risk-versus-reward strategy in the game. To make the mechanic meaningful to the player, a game designer must consider the approach carefully, especially if the choice is only sometimes clear!

For now, let's just create an object that will heal the player when picked up.

Creating a water diamond pickup

Let's use the water diamond from the artwork provided at the following link: `https://github.com/PacktPublishing/Unity-2022-by-Example/tree/main/ch8/Art-Assets`.

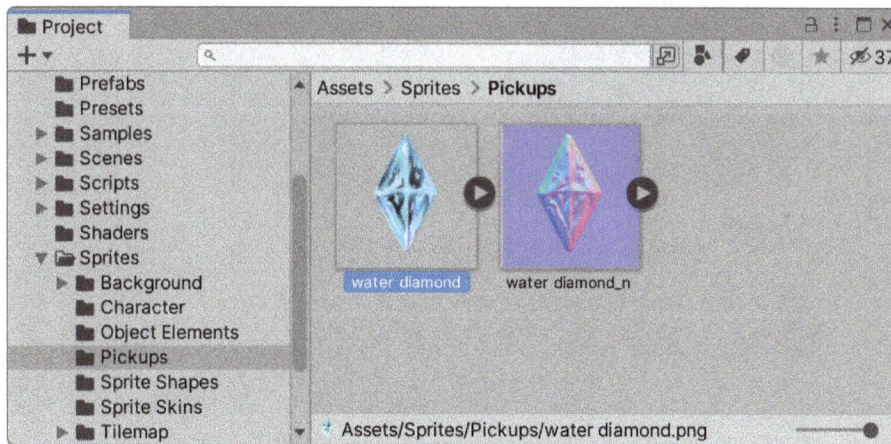

Figure 8.3 – Water diamond artwork

To create our water diamond pickup Prefab based on the water diamond artwork, and as a refresher, follow these steps:

1. Import the artwork to the `Assets/Sprites/Pickups` folder. (I love how this water diamond art came out, by the way; nice job, Nica!)

2. In the **Project** window, select the water diamond sprite and, in the **Inspector** window, use **Sprite Editor** to set the provided normal map as the secondary texture.

3. Drag the water diamond sprite into the **Hierarchy** window and parent it to an empty game object (graphics should be parented to a base object), then name the parent `Water Diamond (Heal)` (remember, you can easily parent a GameObject in the **Hierarchy** window by right-clicking on it and selecting **Create Empty Parent**).

4. Add a `CapsuleCollider2D` instance to the parent object to enable physics (precisely, collision detection).

5. Use the **Edit Collider** button to enable resizing of the collider around the diamond shape, as seen in the following figure.

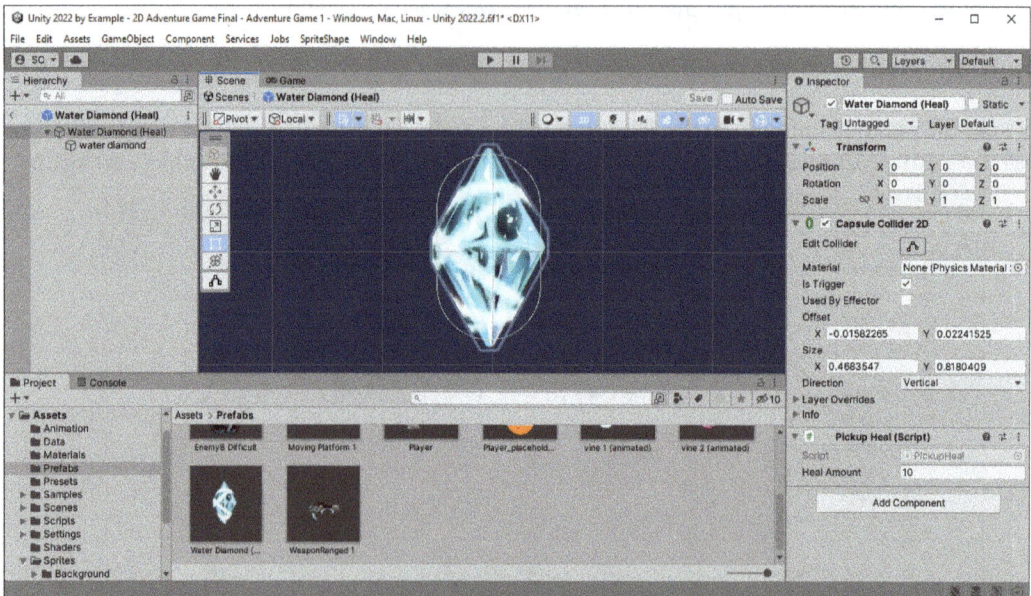

Figure 8.4 – Water diamond healing Prefab

6. Add the `PickupHeal` component to the Prefab root (also seen in *Figure 8.4*).

7. Lastly, drag the parent object from the **Hierarchy** window to the **Project** window to create a Prefab asset in the `Assets/Prefabs` folder.

For a quick hack to bump up the visuals, you can cheat a bit (as indie game devs, it's all about cheating to save time anywhere we can) and reuse the material we created in *Chapter 7*, for the bullet. Follow these steps:

1. In the `Assets/Materials` folder, duplicate the `Bullet 1` material and rename it `Water Diamond 1`.

2. Assign the water diamond sprite to the `MainTex` channel but leave the **Emission** channel set to the `bullet1_emission` map.

3. Assign this new material to the **Sprite Renderer Material** field on the **Water Diamond 1** Prefab's graphic.

4. Adjust the HDR color and intensity for desired visuals.

Our pickup Prefab is looking good. Yay!

We need to add one last bit of functionality for this Prefab to be a pickup for the player. It has to disappear once "collected," and we can do that by revisiting an existing reusable component we previously made – along with a new one.

Composition for pickup behavior

As previously introduced, being able to bring multiple objects together to achieve a desired behavior or functionality is a form of **composition**. To state another way, we'll combine two or more reusable components, each responsible for a specific aspect of an object's behavior or appearance. This will allow us to create a new behavior directly in the editor without introducing new code – this is especially valuable for designers on your team where experimentation can generate new ideas independently.

Let's start by adding the `TriggeredEvent` component to the root `gameObject` of the **Water Diamond (Heal)** Prefab. We made this component back in *Chapter 4*, section *Triggering actions in the level*. Now we have an `OnTriggered` event that fires when the player collides with the pickup object. We want to destroy the pickup, but there isn't a built-in way to do that. However, it's easily solved.

Destroyer component

To destroy the pickup, all we need is an additional component serving as a single-purpose but reusable component. This component will be added to the existing composition for the pickup behavior. To clarify, we need to destroy the pickup object by calling the `TriggeredEvent.OnTriggered` event. So, first, make a new script named `Destroyer` in the `Assets/Scripts` folder:

```
using UnityEngine;
public class Destroyer : MonoBehaviour
{
    public float Delay = 0f;
    public void DestroyMe()
    {
        if (Delay > 0)
            Invoke(nameof(DestroyNow), Delay);
        else
            DestroyNow();
    }
    private void DestroyNow() => Destroy(gameObject);
}
```

Simple. A public method that can be called that destroys the object – `DestroyMe()`. I've added the option for setting a delay before destroying the object – yes, I'm confident this is a typical enough use case, it literally only took seconds to add, and I'll defend that I'm not violating the YAGNI principle!

> YAGNI | "You ain't gonna need it"
>
> This principle states that a programmer should only add functionality if necessary.
>
> Another one I like is **DRY** (**don't repeat yourself**), which is simply directed at reducing repetition (a basic example would be if you find yourself writing the same code more than once, extract it to a method or abstraction).

Go ahead and add `Destroyer` to the root of the `Water Diamond (Heal)` Prefab so we can wire things up in the **Inspector** window:

1. Set the execution state dropdown to **Runtime Only**.

2. Drag the `Destroyer` component to the object field (using its title area).

3. In the function selection dropdown, select **Destroyer | DestroyMe( )**.

4. I'd leave **Delay** at 0 for this pickup, but for other behavior, you may want to adjust (see, you have that option!). And, yes, although `Destroy()` has a second parameter to delay destroying the object, the `Destroyer` class serves as a general example for introducing an execution delay.

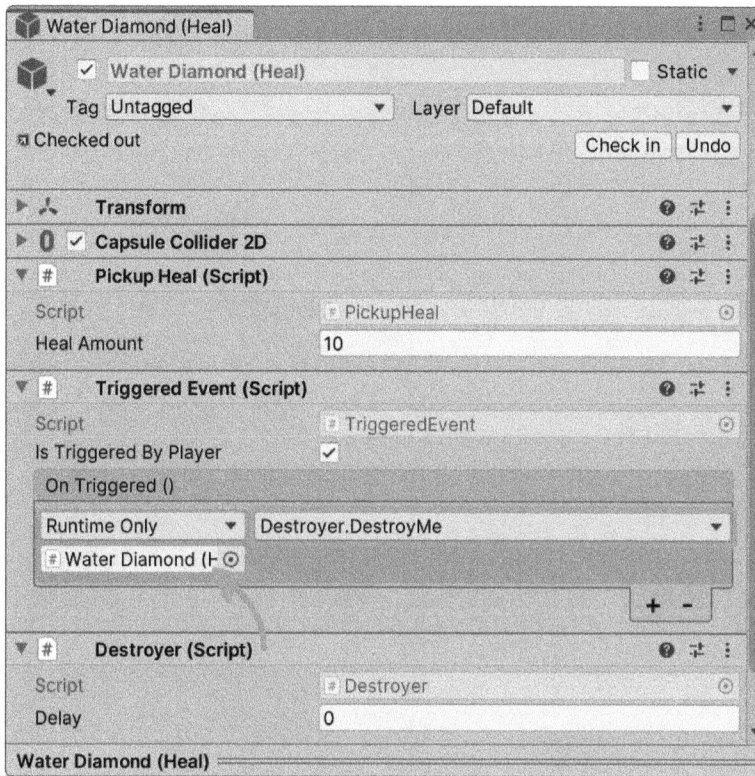

Figure 8.5 – Assign DestroyMe to OnTriggered (UnityEvent)

Okay, so let's talk about what objects can affect the health of other objects – right now, it's like the Wild West out there, with everything able to damage everything else. We can't have any of that going on, so let's put on a mask to start getting everyone to behave properly – a LayerMask, that is.

Controlling what damages/heals what

The problem we have to solve here is finding a simple way to specify what objects can damage other objects – this is essential for reinforcing the game's design. We've solved this problem before by using a **tag** and also a **LayerMask**. My preference, generally, is to use tags only when comparing a single type of object in code and using a LayerMask for sorting out several different types of objects, with the added bonus of a LayerMask being designer friendly since the assignment is made in the **Inspector** window.

What damages

The time to evaluate the objects is at the time of the collision, so we will update the *handle collision* methods in HealthSystem accordingly. But first, we need to define the LayerMask variables in the right places, starting with damage.

Add a DamageMask declaration to the IDamage interface:

```
internal interface IDamage
{
    int DamageAmount { get; }
    LayerMask DamageMask { get; }
}
```

Now, add the following variables so we can use DamageMask and satisfy the existing contract we have in ProjectileDamage for the interface implementation:

```
public class ProjectileDamage : MonoBehaviour, IDamage
{
    public LayerMask DamageMask => _damageMask;
    [SerializeField] private LayerMask _damageMask;
    ...
```

Here is the encapsulation pattern again (told you!) for the _damageMask variable.

We can now revise the HandleDamageCollision() method in HealthSystem to implement the mask check:

```
internal void HandleDamageCollision
    (Collider2D collision, IDamage damage)
{
    if (damage.DamageMask
        & (1 << gameObject.layer)) != 0)
    {
        TakeDamage(damage.DamageAmount);
    }
}
```

This if statement should look a bit familiar; it's the same *is this object's layer in the LayerMask?* evaluation we used for our Bullet back in *Chapter 6*. So, if the ProjectileDamage.DamageMask includes the object with health's layer, only then will TakeDamage() be called.

Damage bad. Heal good. Let's do the same for what can heal.

What heals

We're going to repeat what we did for what damages, so add the LayerMask variable HealMask to the IHeal interface:

```
internal interface IHeal
{
    int HealAmount { get; }
```

```
        LayerMask HealMask { get; }
}
```

Implement the updated `IHeal` interface contract in `PickupHeal` – encapsulating the _ `healMask` variable:

```
public class PickupHeal : MonoBehaviour, IHeal
{
    public LayerMask HealMask => _healMask;
    [SerializeField] private LayerMask _healMask;
    ...
```

As an exception to repeating what we did in the `HandleDamageCollision()` method, let's not repeat ourselves here by also adding the layer mask check code to `HealthSystem`. `HandleHealCollision()`; instead, let's extract the layer mask evaluation to a method, and we'll give it a nice, easy-to-understand name: `IsLayerInLayerMask()` (using `Is` to start the name with makes it obvious this will return a bool `true` or `false` value, no?):

```
        private bool IsLayerInLayerMask(
            int layer,
        LayerMask mask)
                => (mask & (1 << layer)) != 0;
```

> **Utility methods (C#)**
>
> You'll likely need this `IsLayerInLayerMask()` check in other classes, so consider creating a new static class for utility methods such as this one that can be used from anywhere in the code base.
>
> Or, for action on specific types, consider adding an **extension method**: `https://learn. microsoft.com/en-us/dotnet/csharp/programming-guide/classes- and-structs/extension-methods`.

Let's update `HandleHealCollision()` and use our new utility LayerMask check method:

```
        private void HandleHealCollision(IHeal heal)
        {
            if (IsLayerInLayerMask(
                gameObject.layer, heal.HealMask))
            {
                ApplyHealing(heal.HealAmount);
            }
        }
```

Better readability for the win! Don't forget to return to the `HandleDamageCollision()` method to refactor the LayerMask evaluation to use the new `IsLayerInLayerMask()` method too!

> **Optimization note | Physics 2D**
>
> We can also control the physics interactions between objects by specifying what collisions are processed by their layer using the physics Layer Collision Matrix (**Edit** | **Project Settings** | **Physics 2D**, select the **Layer Collision Matrix** tab).

In this section, we created our health system that any object in the game can use for receiving damage and healing – it also provides a method for handling the final death/destruction of the object. You learned how to use interfaces to tie everything together in an extensible way by not relying on concrete class references.

We haven't added our new `HealthSystem` to any of our game's objects yet. Let's do that in the next section, starting with the player.

Updating the player and enemy to use health

Not only `Player` and enemy objects but any object can be set up to use `HealthSystem`. It's barely an inconvenience; in fact, the object simply needs to implement the `IHaveHealth` interface.

Assigning the object with health – IHaveHealth interface

Back in the health system UML diagram (*Figure 8.1*), we see at the bottom that the object having health will implement the `IHaveHealth` interface (again, some meaningful naming here). Create a new file named `IHaveHealth` in the `Assets/Scripts/Interfaces` folder:

```
internal interface IHaveHealth
{
    void HealthChanged(int amount);
    void Died();
}
```

We don't yet have a class for the `Player` object, only `PlayerController`. We don't want to add health to something named `controller` because it wouldn't make sense considering the single-responsibility principle – and the controller's only concern is movement. Let's fix that now by creating a script named `Player` in the `Assets/Scripts` folder:

```
using UnityEngine;
public class Player : MonoBehaviour, IHaveHealth
{
}
```

Make sure it implements IHaveHealth. You can use the IDE's refactoring tools again here. IHaveHealth should have the ever-so-helpful red squiggly underline – so, use the IDE refactoring to *implement interface* on it, and you will get the following:

```
public class Player : MonoBehaviour, IHaveHealth
{
    public void HealthChanged(int amount)
    {
        throw new System.NotImplementedException();
    }

    public void Died()
    {
        throw new System.NotImplementedException();
    }
}
```

Until we decide what actions to perform for these methods, we'll leave it as is for now. You will get a reminder in the **console** anytime these are called because throw indicates an exception has occurred while the program is running – in this case, NotImplementedException: The method or operation is not implemented.

> **Caution when using throw**
>
> Be warned, however: throwing these exceptions will cause program execution in the calling method to stop – meaning any statements that follow HealthChanged() will not be executed! If unsure, replace the throw statements with something such as Debug.LogError("Player.HealthChanged() has not been implemented!");.

Do exactly the same for Enemy:

1. Create a new Enemy script in the Assets/Scripts folder.

2. Add IHaveHealth to the class declaration.

3. Implement the IHaveHealth interface methods.

The last step is to add the components to their respective objects: adding Player to the Player Prefab and Enemy to both enemy Prefabs. Of course, they also both get HealthSystem added, as seen here:

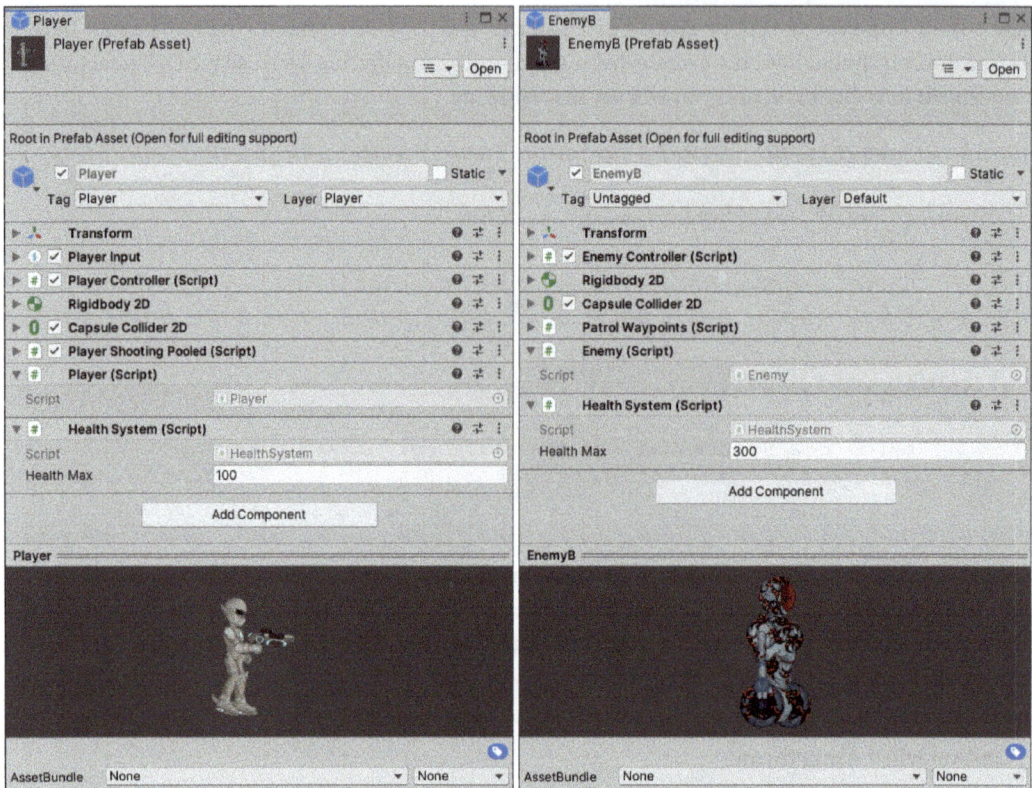

Figure 8.6 – Player and Enemy Prefabs with health

Initial values for max health are just initial values for testing. Playtesting will determine what values they'll eventually land on, depending on difficulty and balanced gameplay. You got this!

Now that we have objects with health, we have one final part of our `HealthSystem` in need of completion – processing.

Process changes to health

To finish up our fully functioning `HealthSystem`, we just need to process the changes to health for our objects that, yeah, have health. Back in our `HealthSystem` class, add the variable that will hold the reference to the object with health and get the object reference in `Awake()` using a `GetComponent()` call:

```
public class HealthSystem : MonoBehaviour
{
    ...
```

```
    private IHaveHealth _objectWithHealth;

    private void Awake()
    {
        ...
        _objectWithHealth = GetComponent<IHaveHealth>();
    }
```

As our future selves now, we'll revisit the `HealthChanged()` method and squash those final lingering UNDONE token comments! We'll use a null check (`if` statement) to ensure we have a sibling component on this GameObject that implements the `IHaveHealth` interface. We'll give ourselves a warning in the console if we don't (and use a `return` statement as a sort of cancellation not to execute any code that follows), and proceed to process the health change otherwise:

```
    private void HealthChanged()
    {
        if (_objectWithHealth == null)
        {
            Debug.LogWarning($"HealthSystem on " +
                $"'{gameObject.name}' requires a " +
                $"sibling component that inherits from " +
                $"IHaveHealth!", gameObject);
            return;
        }

        if (_healthCurrent > 0)
            _objectWithHealth.HealthChanged
                (_healthCurrent);
        else
            _objectWithHealth.Died();
    }
```

With that, our health system is complete! It allows the addition of health to any object and gives the ability for any object to cause damage or heal without any concrete class references! Interfaces for the win!

We covered much territory creating the health system and wrote lots of code back and forth in several classes, so don't forget that you can always refer to the completed project code for this chapter on the book's GitHub repo here: `https://github.com/PacktPublishing/Unity-2022-by-Example`.

Keeping with our composition pattern, let's quickly look at how we can set up the ability to easily add different behaviors (i.e., components) when interactions with `HealthSystem` occur.

Adding behavior with UnityEvent

We've used `UnityEvent` before, for the `TriggeredEvent` component back in *Chapter 4*. It's flexible, in that listeners can be registered by code or assigned in the **Inspector** window (you already know I'm a fan of this one), so it will be a perfect use case for our needs here.

Only a few additions are required to add a `UnityEvent` instance that will be invoked when we handle the collisions for `IDamage` and `IHeal`. Let's start by adding a method declaration to the interfaces:

```
internal interface IDamage
{
    ...
    void DoDamage(Collider2D collision, bool isAffected);
}

internal interface IHeal
{
    ...
    void DoHeal(GameObject healedObject);
}
```

As you can see in the differences in the declarations, we'll change things up just a bit with each implementation. `DoDamage()` will pass two parameters for collision and whether the object is affected by the collision (as in, did it just collide or was it affected by the damage?). We can use this bool to alter things such as the visual effect (e.g., small versus a sizeable varying particle effect), where `DoHeal()` will just pass in the object that is being healed.

Now let's implement the changes to the interfaces, starting with damage in the `ProjectileDamage` class. Add the `UnityEvent` and `DoDamage()` methods:

```
public class ProjectileDamage : MonoBehaviour, IDamage
{
    ...
    public UnityEvent<Collider2D, bool> OnDamageEvent;

    public void DoDamage(Collider2D collision,
        bool isAffected)
            => OnDamageEvent?.Invoke(collision,
                isAffected);
}
```

Here are the specifics of the implementation coded in the preceding snippet:

- OnDamageEvent: Declare as a UnityEvent instance with two parameters. Collider2D takes the collision object that can be used to get the intersection position between the objects. And the isAffected value indicates whether or not the damage was applied as a result of the collision - this is from the layer mask evaluation, as we'll see in a minute.

  Take notice that we won't use the event keyword here because it is a UnityEvent instance – it's not a delegate type but a serializable class. Otherwise, always use the event keyword for your events to enforce the event pattern, where only the implementing class should invoke!

- DoDamage(): This is the public method called from HandleDamageCollision() when the interaction occurs, and its sole responsibility is to invoke the UnityEvent instance (passing the parameters).

And now do the same for PickupHeal – implement the changes with the interface:

```
public class PickupHeal : MonoBehaviour, IHeal
{
    ...
    public UnityEvent<GameObject> OnHealEvent;

    public void DoHeal(GameObject healedObject)
        => OnHealEvent?.Invoke(healedObject);
}
```

Here is the explanation for these code changes:

- OnHealEvent: Declared as a UnityEvent instance with one parameter. The GameObject instance is just the object that is affected by the healing. The usage could simply be getting the object's transform position for instantiating an object or particle effect.

- DoHeal(): Just like the damage method, this is a public method called from HandleHealCollision() when the interaction occurs and is also solely responsible for invoking the UnityEvent instance (passing the parameter).

The final step is to add the public Do calls to the HealthSystem.OnTriggerEnter2D() method. Update HandleDamageCollision() like so:

```
    internal void HandleDamageCollision
        (Collider2D collision, IDamage damage)
    {
        var isAffected = IsLayerInLayerMask(
            gameObject.layer, damage.DamageMask);
```

```
        damage.DoDamage(collision, isAffected);
        if (isAffected)
            TakeDamage(damage.DamageAmount);
    }
```

We introduced a local bool variable, `isAffected`, to get the `IsLayerInLayerMask()` result – we can then use the variable in place of calling `IsLayerInLayerMask()` multiple times.

We can then just call `DoDamage()` and only call `TakeDamage()` if the object is affected by the damaging object.

Now, update `HandleHealCollision()` like so:

```
    private void HandleHealCollision(IHeal heal)
    {
        if (IsLayerInLayerMask(gameObject.layer,
            heal.HealMask))
        {
            heal.DoHeal(gameObject);
            ApplyHealing(heal.HealAmount);
        }
    }
```

Unlike with damage, we don't care about the impact of the damaging object having an effect or not. We'll process healing if it's in `HealMask`. We just need to call the public `DoHeal()` – passing in the object being healed – and we're done!

Now that we have an event exposed on the `ProjectileDamage` and `PickupHeal` components, let's refactor an earlier composition for destroying the water diamond pickup. Hence, we have an example of its usage.

Re-composition for Destroyer

With `PickupHeal` now having a `UnityEvent` instance triggered when the collision occurs, we can improve the composition for destroying the water diamond object when it's collected. We previously used the `TriggeredEvent` component, but now we need to assign the `Destroyer.DestroyMe()` function to the `OnHealEvent` function selection drop-down menu.

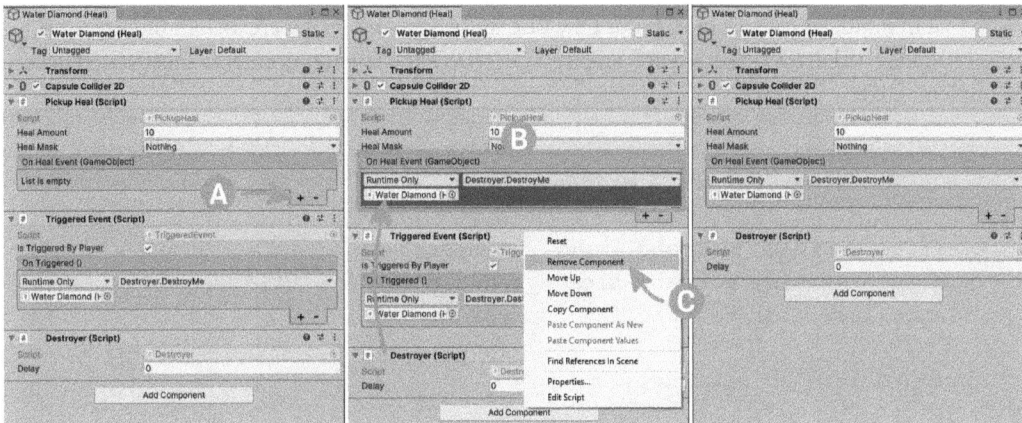

Figure 8.7 – Revising Destroyer

Referring to *Figure 8.7*, let's walk through this change:

1. (*A*) – Click the little + icon on the `OnHealEvent` tab to add a new listener to the list.

2. (*B*) – Drag the `Destroyer` component to the object field (using its title area).

    *   In the function selection drop-down menu, select **Destroyer | DestroyMe( )**.

3. (*C*) – Right-click on the `TriggeredEvent` title area to bring up the context menu and select **Remove Component**.

You end up with an **Inspector** window that looks like the right-most image – you are done. Easy-peasy.

Also, we don't need to worry about `IsTriggeredByPlayer` from the `TriggeredEvent` component anymore since `DoHeal()` will only be called if the `HealMask` check is satisfied.

In this section, we have created a fully implemented health system, and that's a game changer (yes, bad pun). Again, you learned the power of interfaces and how we can quickly add functionality to existing systems. We also practiced composition by refactoring some reusable components to explore a different approach to destroying the heal pickup object.

In the next section, let's put the health system through its paces by having a bunch of pesky enemies to contend with as we introduce a wave spawner.

Enemy wave spawner

A **wave spawner** may sound scary, but it's just a straightforward script. We need to instantiate a new enemy from a given position and on a fixed (or random) time interval. We'll also ensure things don't get out of hand by limiting the number of enemies spawned.

So, with that in mind, let's have a look at our new `EnemySpawner` script – create it in the `Assets/Scripts` folder – and see whether you can point out where the few requirements I just stated have been implemented:

```
using UnityEngine;
public class EnemySpawner : MonoBehaviour
{
    [SerializeField] private Enemy _enemyPrefab;
    [SerializeField] private float _spawnInterval = 5f;
    [SerializeField] private int _maxSpawned = 3;

    private int _objectCount = 0;

    private void Start()
        => InvokeRepeating(
            nameof(SpawnEnemy), 0f, _spawnInterval);

    private void SpawnEnemy()
    {
        if (_objectCount < _maxSpawned)
        {
            var enemy = Instantiate(_enemyPrefab,
                transform.position, Quaternion.identity);
            enemy.Init(DestroyedCallback);
            _objectCount++;
        }
    }
    public void DestroyedCallback() => _objectCount--;
}
```

Let's break this class down – a lot of this should be looking very familiar by now:

1. Declare a variable for the enemy Prefab that will be spawned – we use the `Enemy` type here instead of `GameObject` because when we reference `_enemyPrefab` later, we'll be referencing the Enemy class directly and won't need to do `GetComponent()`.

2. Declare a variable for `_spawnInterval`, which will be the delay before spawning the next enemy.

3. Declare a variable for `_maxSpawned`, which will be the total number of enemies onscreen – from this spawner – at the same time.

4. Declare a variable for `_objectCount`, which keeps track of how many enemies are currently spawned.

5. Create the `Start()` method – here we'll simply use `InvokeRepeating()` to repeatedly call `SpawnEnemy()` at the specified spawn interval (`_spawnInterval`).

6. Create the `SpawnEnemy()` method – we first check to see whether we've already instantiated our `_maxSpawned` amount of enemies, and, if not, `Instantiate()` a new Enemy.

   I.   We create a new (implicit declaration using `var`) local enemy variable – returned from the `Instantiate()` call – so that we can call `Init()` and pass in a callback parameter (as a pseudo constructor). This is in place of what would usually be the C# constructor (objects created with the `new` keyword, which, if you remember, we cannot do with `MonoBehaviour`).

   II. Increment the number of spawned objects with `_objectCount++`.

7. Define the `DestroyedCallback()` method passed into the `Enemy.Init()` call so that, when the enemy object is destroyed, the currently spawned enemy count can be decreased – resulting in the spawner instantiating another enemy to maintain the `_maxSpawned` count.

> **Don't forget about object pooling!**
>
> Note that if we have waves and waves of many enemies, we do want to optimize this by introducing object pooling. Refer back to *Chapter 6*.

We'll have to tie `DestroyedCallback` to the Enemy class because it's passed to the instantiated enemy object via the `enemy.Init()` call. Let's add everything to support that now; it's not much, so open the Enemy script and add the following:

```
public class Enemy : MonoBehaviour, IHaveHealth
{
    private event UnityAction _onDestroyed;

    internal void Init(UnityAction destroyedCallback)
        => _onDestroyed = destroyedCallback;

    private void OnDestroy()
        => _onDestroyed?.Invoke();
    ...
```

Here we have a `UnityAction` instance that we'll use to invoke the callback when the enemy object is destroyed – you've seen all this before.

We just need to actually destroy the enemy object, and we do that when the object dies, as dictated by the `IhaveHealth`-interface-implemented `Died()` method:

```
public void Died() => Destroy(gameObject);
```

A note about `DestroyedCallback` and why we don't have to *unregister from the event* when enemy is destroyed: the responsibility is being flipped here since `EnemySpawner` is not holding a reference to the instantiated enemy object. You only need to unregister (or `RemoveListener`) from events that can become invalid references.

Let's set up a Prefab we can reuse as a preconfigured enemy spawner.

Creating the enemy spawner Prefab

Go ahead and, in your current open scene in Unity, create a new empty GameObject in the **Hierarchy** window and name it `EnemyB Spawner 1` – we can have different Prefabs for different enemy spawning behaviors. Make sure to place it right at ground level in your environment because the spawner's transform position will be used as the enemy instantiation point. Add the `EnemySpawner` component to the `EnemyB Spawner 1` object and drag in the `EnemyB` Prefab from the `Assets/Prefabs` folder to the **Enemy Prefab** field, as seen in *Figure 8.8*.

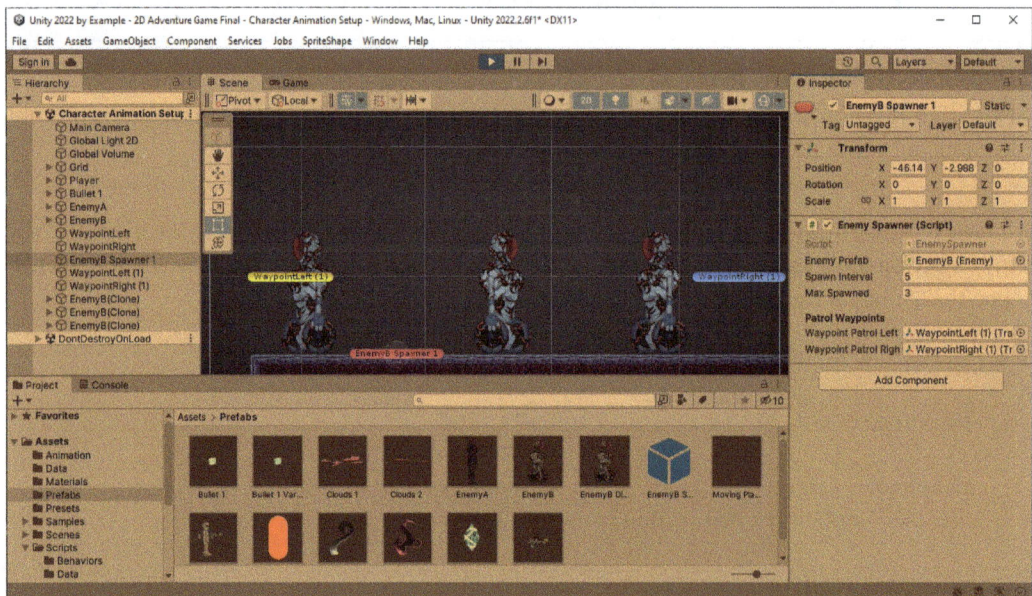

Figure 8.8 – Enemy spawner setup

Finish up by dragging the `EnemyB Spawner 1` object in the **Hierarchy** window to the `Assets/Prefabs` folder. Yay!

If you playtest the enemy wave spawner now, they'll just spawn on top of one another, be disorganized, and not go anywhere – in other words, not acting like robots at all. Let's integrate spawning with the patrolling behavior to keep things orderly.

Integrating spawning with patrol behavior

If you haven't picked up on it already, because we've paid attention to the project's code structure and followed good programming practices from the beginning, maintaining and extending the code to add new functionality has been simple and straightforward. Integrating our new wave spawner to work with the existing patrol behavior will also be quick work. We'll just need to add a few things to set it up.

First things first, let's see if the instantiated Enemy B Prefab has a patrol behavior and, if so, SetWaypoints(). To do that, let's modify the SpawnEnemy() method in the EnemySpawner class. Here you can see we're again using TryGetComponent() to fail gracefully if the component doesn't exist on the enemy:

```
private void SpawnEnemy()
{
    if (_objectCount < _maxSpawned)
    {
        ...
        if (enemy.TryGetComponent
            <IBehaviorPatrolWaypoints>(out var patrol))
                patrol.SetWaypoints(
                    _waypointPatrolLeft,
                    _waypointPatrolRight);
    }
}
```

In the PatrolWaypoints class, add the following SetWaypoints() method (or, again, use the IDE's refactoring tools to generate it) to allow setting the left and right waypoint private variables externally (encapsulation at work):

```
public void SetWaypoints(
    Transform left,
    Transform right)
{
    _waypointPatrolLeft = left;
    _waypointPatrolRight = right;
}
```

And, don't forget: we'll need to add a SetWaypoints() method declaration to our IBehaviorPatrolWaypoints interface so that it's accessible from the reference in the EnemySpawner class:

```
public interface IBehaviorPatrolWaypoints
{
    ...
    void SetWaypoints(Transform left, Transform right);
}
```

Seriously, that's it – three simple additions get the patrolling behavior all wired up to our enemy spawner. You'll just have to add two empty GameObjects to the scene for the patrolling path of this enemy spawner, one for left and one for right, and assign them in the **Inspector** window, as seen in the enemy spawner setup from *Figure 8.8*.

However, you'll notice we have a problem (you playtested, right?). Enemies won't be able to pass each other as they patrol between the waypoints – a simple problem to solve by removing their physics interactions with each other. First, we'll need a layer to set the enemy objects to… how about Enemy?

Using the **Layers** drop-down menu in the top-right corner of the **Editor** window, select **Edit Layers…**.

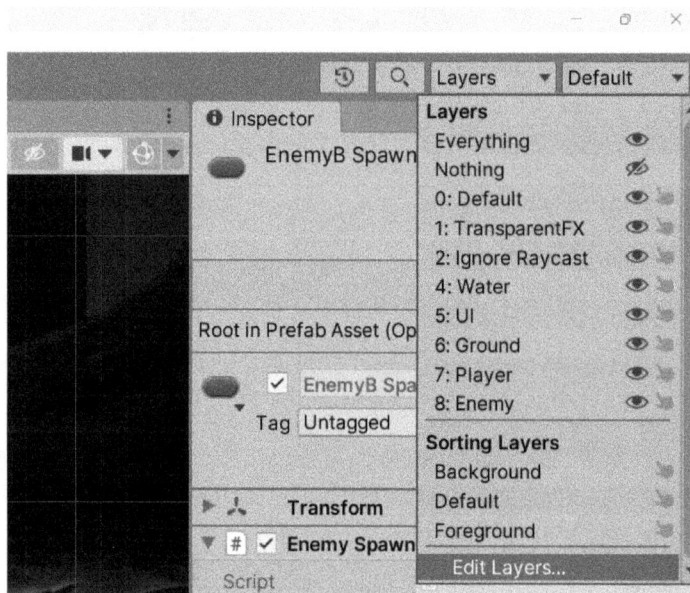

Figure 8.9 – Add an enemy layer

Under the **Layers** sections, in the first empty **User Layer** field, simply type in Enemy to add it.

Now we can control the physics interactions between enemy objects by specifying what collisions are processed by their layer. Using the **Physics Layer Collision Matrix** (**Edit | Project Settings | Physics 2D**; select the **Layer Collision Matrix** tab), we're showing that we disabled physics interactions between all objects set to the Enemy layer (uncheck **Enemy/Enemy**):

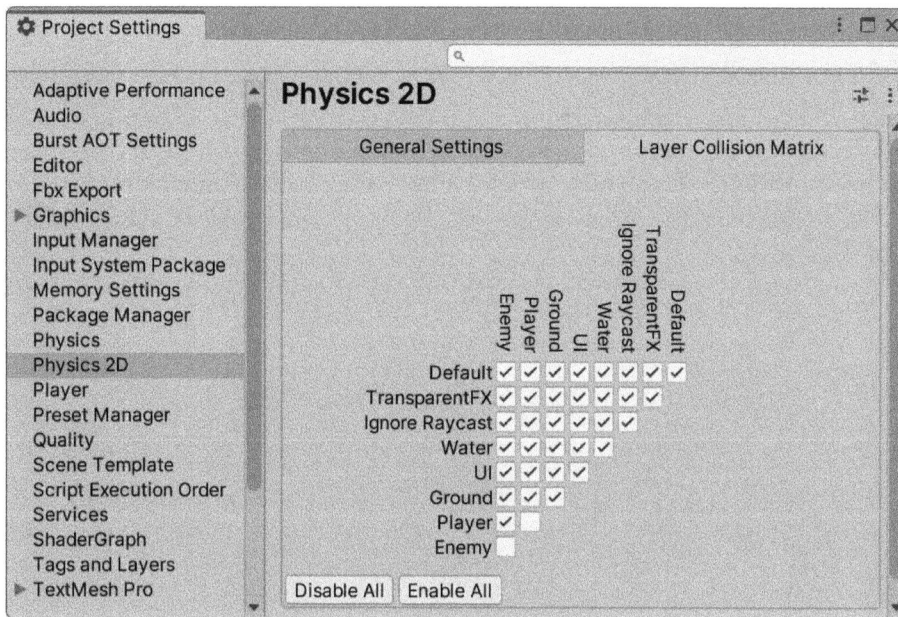

Figure 8.10 – Physics 2D layer collision matrix

When you playtest the enemy wave spawner in the scene now, the spawned enemies will patrol past one another. Awesome!

In this section, we created a wave spawner to instantiate new enemies on a fixed time interval and integrated it with the existing patrol behavior. We finished up by resolving physics interactions between enemy objects to allow them to pass each other while patrolling.

Summary

In this chapter, we covered the implementation of a health system that any object in the game can use to receive damage, heal, and handle the final death/destruction of the object. This system was designed using interfaces to tie everything together in an extensible way without relying on concrete class references – the flexibility of interfaces allowed for the quick addition of new functionality to our existing code.

We continued by creating a wave spawner that instantiates new enemies on a fixed time interval and integrates with the existing patrol behavior. This allows for adding in more complex enemy behavior, which adds new challenges to the game. Additionally, we discussed how to disable physics interactions between objects, which allowed patrolling enemies to pass each other in this case.

Finally, we further explored composition by refactoring some of our reusable components to explore a different approach to destroying the heal pickup object. Through examples such as this, the importance of good programming practices and the use of interfaces to build flexible and extensible systems was highlighted.

In the next chapter, we'll complete the adventure game by creating a simple quest system for collecting key objects for solving the entryway puzzle. We will also introduce a new event system for keeping our code loosely coupled.

9

Completing the Adventure Game

In *Chapter 8*, we started by creating a flexible health system that can be added to any object to give it health, take damage, and heal. The system is extensible, meaning the things that can deal with damage and apply healing can be anything without the need to modify the HealthSystem class because we used interfaces to implement the behavior (not concrete class types). With objects now able to take damage, we continued by updating Player and enemy objects to use health – so, we have the semblance of a real game in the making.

We continued by creating a wave spawner that instantiates new enemies on a fixed time interval and integrates with the existing patrol behavior. This allows us to add more complex enemy behavior, which adds new challenges to the game.

Finally, we further explored composition by refactoring some of our reusable components to explore a different approach to destroying the heal pickup object. The importance of good programming practices and interfaces to build flexible and extensible systems was highlighted through examples.

In this chapter, we'll complete the adventure game by creating a simple quest system for collecting key pieces in the level that are required for solving an entryway security puzzle lock on the habitat station located on the planet's surface. We will also introduce a new global event system for keeping our code loosely coupled. The event system will efficiently manage communication between various quest system components throughout the code base, so we'll tackle that first.

In this chapter, we're going to cover the following main topics:

- Creating an event system in C# to tie things together loosely
- Creating a quest system for a collecting keys mission
- Solving the key puzzle and winning the game

By the end of this chapter, you'll be able to create a quest system that is integrated across different classes while being loosely coupled (that is, reducing dependency by not having external (concrete) class references) and scalable via the use of a new reusable global event system we'll also create. You'll also be able to integrate and customize a puzzle system for your use.

Technical requirements

To follow along in this chapter while using the same artwork that was created for the project in this book, download the assets from the GitHub link provided in this section.

To follow along with your own artwork, you'll need to create similar artwork using Adobe Photoshop. Alternatively, you'll need a graphics program that can export layered Photoshop PSD/PSB files (e.g., Gimp, MediBang Paint, Krita, and Affinity Photo).

You can download the complete project from GitHub at `https://github.com/PacktPublishing/Unity-2022-by-Example`.

Creating an event system in C# to tie things together loosely

We won't need a UML diagram here as the design is quite simple. We'll use a `Dictionary` collection (a special kind of C# collection) to hold the name of an event that we'll assign the event's callback handlers. The added callback handlers will all be invoked when the event is triggered. Although I say this is simple, I didn't introduce it earlier because a few programming concepts still needed to be covered first.

The new event system

Since a UML diagram won't illustrate the functionality of `EventSystem` very well in this case, I've decided to create the following diagram as an introduction to the implementation (see *Figure 9.1*):

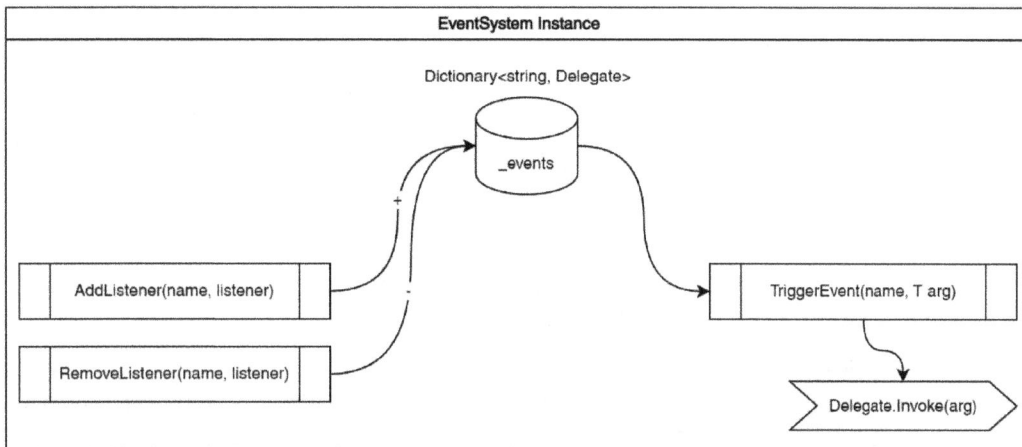

Figure 9.1 – EventSystem diagram

Before looking at the following code, do a quick mental exercise to see if you can visualize what the code should look like from this diagram.

Given this diagram, here is the class template that makes up the basis for our new event system, minus some details we haven't covered yet; is it similar to what you anticipated seeing?

```csharp
using UnityEngine;
using UnityEngine.Events;
using System;
using System.Collections.Generic;

public class EventSystem : MonoBehaviour
{
    private Dictionary<string, Delegate> _events
        = new();

    public void AddListener<T>(
        string eventName, UnityAction<T> listener)
    {
        // UNDONE: Subscribe handler to event name.
    }

    public void RemoveListener<T>(
        string eventName, UnityAction<T> listener)
    {
        // UNDONE: Unsubscribe handler from event name.
```

```
    }

    public void TriggerEvent<T>(string eventName, T arg)
    {
        // UNDONE: Invoke handlers for event name.
    }
}
```

You didn't necessarily know what the `Dictionary` declaration would be – we haven't covered it yet – but you at least knew that we needed a declaration for it, right?

We have a few additional things to unpack here and introduce some new C# items, including `Dictionary`, `Delegate`, and generic types (as you've probably noticed, `<T>` and T sprinkled throughout the code):

- `Dictionary<string, Delegate>`: A C# `Dictionary` is a collection type that contains a list of **keys** and **values**. This differs from a C# `List`, a collection type that contains a single list of a type.

  Let's put this into some game development terms… a C# `Dictionary` is kind of like a magic bag where you can store lots of items and find anything you want, so long as it has a unique name. You can keep track of all the cool stuff your player will need on their adventure, keeping your game organized and awesome!

  - **Key**: From our declaration, the first type is a `string` type, which indicates the key type for the item in the dictionary. We'll set the key to identify the specific event by name, and the event handler methods will be *added to it*.

  - **Value**: From our declaration, the second type is a `Delegate` type and is the value being stored for the specific key item. We'll set the value to a delegate type, such as `UnityAction`, that can have method handlers added to it (note that here, `delegate` is the base type for `UnityAction`). The event can be triggered later by the event's name, invoking all the assigned event handler methods. Simples!

  - **new()**: Note that when we declare our `Dictionary`, we can assign a new empty dictionary by specifying just the new keyword. This is new for C# in Unity 2022.2 since previously, you'd have to specifically restate `Dictionary` and the types for *key* and *value* again here – but you'd still have to do this for serialized variables assignable in the **Inspector** view!

Dictionary (C#)

`Dictionary` represents a collection of keys and values. It has similar properties and methods compared to a **List** collection type. For additional reading, go to `https://learn.microsoft.com/en-us/dotnet/api/system.collections.generic.dictionary-2?view=net-8.0`.

- `AddListener()`: We'll call this method when we add an event to `EventSystem`. The method signature has the following parameters:

  - `<T>`: In C#, there is a programming concept called **generic**, where `T` represents *any type* as a generic type parameter. This allows you to define a method, class, parameter, or something else that can work with any type specified by the calling code, such as `string`, `int`, `float`, `Vector3`, `GameObject`, and so on. This allows reuse across different parts of the code without the need to use specific types and repeat the same code to support those types. It may sound a bit complicated this early on, but I hope it is understandable through the preceding example.

  - Our use case for a generic type is to implement an event delegate that can pass in any type as an argument. This will become clearer as we work through the rest of `EventSystem`.

  - `eventName`: A `string` type that uniquely identifies the event that's been *registered* with `EventSystem`. It's how we add additional listeners to a specific event and how we know what event to trigger.

  - `listener`: `UnityAction<T>`, which is the event delegate. `UnityAction` should be familiar since it was first introduced in *Chapter 3*, as an event listener that was used to update the UI. The difference here is that we are adding a generic parameter of the `T` type, which means the handler method can pass a single argument of any type when invoked.

- `RemoveListener()`: This is used when removing an event from `EventSystem`. The method signature is the same as that for `AddListener()`, so we don't need to repeat ourselves here. However, there will be one glaringly obvious difference: we'll be removing a method handler instead of adding one.

- `TriggerEvent()`: We'll call this method when we want to invoke the specific event for which method handlers were added as listeners (via `AddListeners()`, of course). The method signature has the following parameters:

  - `eventName`: Again, this is the event name string we'll specify for invoking the method handlers (that is, the added listeners).

  - `T arg`: The parameter value we'll pass into the invoked methods that will be handled so that they can receive and process any required data. Again, being a generic type, `T`, we'll be able to pass in an argument of any type (nice, right?).

In the preceding code, I said *event name*, but we won't do something as silly as using string literals for the event names throughout the code base. Instead, let's create an `EventConstants` class that we'll use for the event names. This would be similar to how we added a `Tags` class when referencing tags assigned to our objects in the game through code (for example, `Player`).

Let's define our first event name constant by creating the following `EventConstants` class, saved in the `Assets/Scripts/Systems` folder:

```
public class EventConstants
{
    public const string MyFirstEvent = "MyFirstEvent";
}
```

The event name will be the *constant* (figuratively and literally) that loosely ties things together, so we don't require concrete class references between classes that would observe and respond to events. Now, we can independently test our classes – via **unit testing** – without ensuring references to other classes exist – for example, `Player` doesn't need to specify a reference to `UIManager`, `Enemy`, and so on. We can also move a class to a new project without worrying about bringing in unnecessary classes.

Now that we've learned about the event system template, as well as how we'll specify events, let's dig into the specifics of the implemented methods: `AddListener()`, `RemoveListener()`, and `TriggerEvent()`.

Event management methods

First, let's look at the `AddListener()` method. To recap, `AddListener()` is responsible for adding an event to the dictionary of events if it doesn't already exist (that is, registering the event) and adding the listener to the event's handler method `delegate`, like so:

```
public void AddListener<T>(string eventName,
    UnityAction<T> listener)
{
    if (!_events.ContainsKey(eventName))
        _events.Add(eventName, null);

    _events[eventName] = (UnityAction<T>)_events[eventName]
        + listener;
}
```

Let's break down the code here:

- `ContainsKey()`: The very first thing we'll need to do when adding an event to our dictionary of events is to see if we've already registered one of the same event names (pretty logical, right?). A C# `Dictionary` provides several methods here – similar to a C# `List` in many ways – for achieving this base functionality that we can leverage.

    Specifically, we use `ContainsKey()` and pass in `eventName`. This will return a Boolean value of `true` if the key already exists in the dictionary and `false` if not, so we can use an `if` statement to evaluate and take the proper action.

- `Add()`: Working with the result returned from `ContainsKey()`, we'll go ahead and add the event – using the event name as the key – to the dictionary using the `Add()` method (again, similar to C# `List`).

  Note that for the value parameter – as in `Add(key, value)` – we're specifying `null` for the delegate. That's simply because we'll assign the listener to the delegate in the following line, where it doesn't matter whether this event was just added.

- `_events[eventName]`: We can get the value for a specific event from the `_events` dictionary by simply specifying the key in square braces (this is similar to how we'd return the value of an array by specifying the index (`int`) in square braces).

- `(UnityAction<T>)`: Placing a type in parenthesis before another type is called casting in C#. Since `_events[eventName]` is a `Delegate` type, and `UnityAction` has delegate as its base type, we can operate on the event dictionary value as `UnityAction` via a type conversion by casting it (an explicit conversion). Here, again, `<T>` indicates we will use a generic for a parameter.

Casting (C#)

C# is a strongly typed language (or statically typed) at compile time, so once a variable is declared, it cannot be declared again as a different type. To overcome this, should you need to copy a value into a variable of another type, C# provides various type conversion operations, and casting is one of them.

Here's some additional reading: `https://learn.microsoft.com/en-us/dotnet/csharp/programming-guide/types/casting-and-type-conversions`.

- `+ listener`: `UnityAction` allows you to add additional listeners; using the + (add) operator will enable us to do just that.

  Note that just the + operator is the correct syntax to add a listener to a delegate that may be `null` or already has listeners, unlike previously, all the way back in *Chapter 3*, where we used the `+=` operator to add an event handler method when subscribing to events.

Now, let's look at the opposite. For `RemoveListener()`, we want to remove an event listener when it's no longer needed – especially when an object is destroyed – so that we don't try invoking an invalid handler method reference. This is generally just good practice – for some listener types, not doing so can cause memory leaks:

```
public void RemoveListener<T>(string eventName,
    UnityAction<T> listener)
{
    if (_events.ContainsKey(eventName))
        _events[eventName] =
            (UnityAction<T>)_events[eventName] - listener;
}
```

We don't need to break down the code here again because it's essentially the same as `AddListener()`. The primary differences are as follows:

- We don't need to add the event in case it doesn't exist; we only care if it does exist so that we can remove a listening handler method

- In place of the + operator, we use the - operator to remove the specified listener handler method

Lastly, we have the `TriggerEvent()` method. The naming of these methods has been self-explanatory so far, so it should be no surprise what's next. However, we are doing something different here to retrieve the dictionary's value for the `eventName` key that's been provided to invoke the added handler methods:

```
public void TriggerEvent<T>(string eventName, T arg)
{
    if (_events.TryGetValue(eventName, out Delegate del))
        (del as UnityAction<T>)?.Invoke(arg);
}
```

Here is how things work:

- `_events.TryGetValue()`: You may find this syntax familiar since we've already used `TryGetComponent()` a few times. Very simply, we try to get the value for the specified `eventName` key, and if it exists, we return the value as an `out` parameter called `del`. The `if` evaluation will short-circuit if it's not found, so `Invoke()` will only be called if a value is returned.

- `(del as UnityAction<T>)`: We introduce the as operator keyword here. Similar to how we used `()` to cast to another type, we can also use `as` for type conversion. Specifically, `as` is a good choice when working with a nullable type, which we want to do here because we're also using the null-conditional operator (`?.`) so that we won't erroneously try and call `Invoke()` on a null delegate.

The as operator (C#)

The `as` operator converts objects to a different type but returns `null` if the conversion fails instead of throwing an exception like other type conversion techniques would.

Here's some additional reading: `https://learn.microsoft.com/en-us/dotnet/csharp/language-reference/operators/type-testing-and-cast#as-operator`.

- `Invoke(arg)`: This is how we tie back into the generic parameter specified by T in the `UnityAction<T>` declaration for our listener delegate. The last part, bringing it full circle, will be the method signature for the handler method receiving the argument. This should all make sense when we get to the *Creating a quest system for a collecting keys mission* section, when we put the event system through its paces. Promise!

Next, so that we don't overlook an essential detail for how we will access the event system from our classes that need to work with events, let's make the event system a Singleton – an actual Singleton!

Enforcing a Singleton instance

Our EventSystem will be a **Singleton** instance so that it can be accessed from anywhere in our code. Still, we haven't implemented a full Singleton pattern yet where only a single instance is guaranteed to exist. We'll sort this out by enforcing the pattern and destroying any want-to-be duplicates.

We saw a rudimentary implementation back in *Chapter 3*, where we just set the instance to a static variable without even attempting to seek and destroy additional instances being introduced (bad us). Here we go. Add the following Instance public static variable declaration and the new Awake() method code:

```
public class EventSystem : MonoBehaviour
{
    public static EventSystem Instance { get; private set; }
    ...
    private void Awake()
    {
        if (Instance == null)
            Instance = this;
        else
            Destroy(gameObject);

        DontDestroyOnLoad(gameObject);
    }
    ...
```

The magic happens in the Awake() method, where we previously just did this:

```
private void Awake() => Instance = this;
```

We can now check if our static Instance has been assigned yet; if not, assign it. However, if Instance was already assigned, a want-to-be duplicate will be added to our **Scene Hierarchy**. Let's destroy it right away!

Our EventManager will need to hang around all the time to respond to registering and triggering events at any time, and from any class, throughout the life of our game, so we'll need a way to keep it persistent.

Sometimes, the coding process feels like a game: find something, destroy that, keep this. So, if you're not having fun, you're doing something wrong!

> **EventSystem | Complete code**
>
> To view the complete code for the `EventSystem` class and all the event-related code in this chapter, visit this book's GitHub repository: `https://github.com/PacktPublishing/Unity-2022-by-Example/tree/main/ch9/`.

Singleton persistence

Unity provides a way to keep particular objects persistent, no matter the current scene, survive loading new scenes, and more, and that is by using `DontDestroyOnLoad`. Simply calling it and passing the `GameObject` object you want to persist as the target parameter is all that's needed. In our case, in `Awake()`, we're passing in `gameObject`, which represents the current object the component (script) is attached to.

> **Additional reading | Unity documentation**
>
> `DontDestroyOnLoad`: `https://docs.unity3d.com/2022.3/Documentation/ScriptReference/Object.DontDestroyOnLoad.html`.

Systems GameObject

Objects specified as `DontDestroyOnLoad` must be in the root of the **Scene Hierarchy**, so we'll create a new GameObject in our game's level scene and name it *Systems*, then add the `EventSystem` component. I would place this at the very top of the scene hierarchy to maintain some visible order of dependency for things – it's entirely up to you how you choose to organize things here. Also, remember that you can add empty objects with the sole purpose of having organizational headers (don't forget to set the tag to `EditorOnly`), as seen in the `Systems` object referenced in *Figure 9.2*:

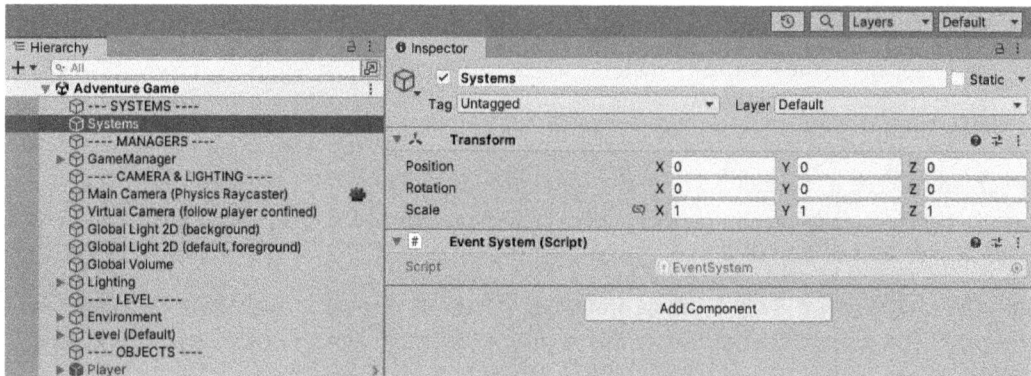

Figure 9.2 – Systems root GameObject

In this section, we learned how to create a global event system that allows our classes to remain loosely coupled because we don't require type references of one class in another to respond to triggered actions. If the `EventSystem` component's usage is still unclear, don't fret – in the next section, we'll cover an example straight away by creating a simple quest system based on it.

Creating a quest system for a collecting keys mission

Now that we have our super decoupled global event system, we'll immediately put it to good use. Referring again to our GDD, we know that the player, at some point, has to collect some key pieces to solve a puzzle so that they can advance:

What is the secondary game mechanic for the adventure game?	The player will search the environment for hidden parts of a key. The pieces will need to be combined correctly as input to gain access to the entryway of the habitat station.

Table 9.1 – GDD quest reference

A typical game system that can support "collecting a certain number of items" is a quest system. A possibly more simplified approach, but also a system-based approach to solving this problem, could be a basic inventory system. A quest system, however, will offer additional opportunities to provide a more complete example – especially for implementing the event system.

Let's have a quick look at what the player will be searching for during their quest and interacting with to regain entry to the habitat station constructed on the planet. The Kryk'zylx technology is a mystery to us, but suffice it to say, they like a good challenge when securing their bases. Both the key pieces that the evil plant entity has scattered and the entryway security puzzle lock with the missing pieces can be seen in the following figure:

Figure 9.3 – Keys and security puzzle art

Our mission now is to code the quest system and flesh out the game mechanics required to implement it. So, let's get started!

The quest system

I'll provide the code template for a `QuestSystem` class, similar to how the event system was introduced. However, I'll give the full implementation this time because you'll find it shares a similar design to `EventSystem`. I also won't provide a diagram this time. I'm going to save it as a challenge for you to create one on your own at the end of this section.

Create a new script named `QuestSystem` in the `Assets/Scripts/Systems` folder with the following code:

```
using UnityEngine;
using System.Collections.Generic;

public class QuestSystem : MonoBehaviour
{
    private Dictionary<string, bool> _quests = new();

    public void StartQuest(string questName)
    {
        if (!_quests.ContainsKey(questName))
            _quests.Add(questName, false);
    }

    public void CompleteQuest(string questName)
    {
        if (_quests.ContainsKey(questName))
            _quests[questName] = true;
    }
}
```

Okay, no, you're not experiencing déjà vu. The basic concept for the `QuestSystem` code is a similar pattern to `EventSystem`. At its core, there's also a C# `Dictionary`.

Here's the breakdown:

- `Dictionary<string, bool>`: The declaration for the `_quests` dictionary will hold a `string` key to identify the quest's name (similar to the event name), and the value for the entry will be of the `bool` type as an indication of whether the quest has been completed (that is, `true` equals completed).

- `StartQuest()`: Starting a quest means adding it to the dictionary if it's not already added. Quests will be identified by a unique `questName` that's passed in as the only argument. Simple. As. That.

- `CompleteQuest()`: As stated earlier, we're going to use the `_quest` value as the quest completion indicator, so if the specified quest exists in the dictionary, then we'll assign `true` for its value (`bool` has a default value of `false`, which is why we didn't need to assign this anywhere).

We'll be repeating ourselves a bit here again with how `QuestSystem` will be accessed when required since we'll be using a Singleton pattern again. Primarily, however, we will decouple references to `QuestSystem` whenever possible by going through the global event system (that is, no tightly coupled objects).

Add the following declarations for the public static `Instance` property and the *Singleton management code* to the `Awake()` method, like so:

```
public static QuestSystem Instance { get; private set; }

private void Awake()
{
    if (Instance == null)
        Instance = this;
    else
        Destroy(gameObject);

    DontDestroyOnLoad(gameObject);
}
```

Using the Singleton pattern solves a big problem for obtaining references to our core systems, but it also comes with some disadvantages compared to more complex patterns (for example, service locator patterns). We're going to address one of these shortcomings now before proceeding further.

We have a potential problem with our systems because we require the `EventSystem` instance to be available to every other system when the game starts – being the core system loosely coupling everything together and playing nice. With that being the case, we must ensure it gets initialized first.

Script Execution Order

To set the execution order for which script event functions are run (for example, the `Awake()` message event), Unity provides a **Script Execution Order** assignment in **Project Settings**. Quite simply, you can set a lower-ordered number for scripts you want to be initialized first and specifically before others that would rely on it being initialized.

> **Additional reading | Unity documentation**
>
> Script Execution Order settings: `https://docs.unity3d.com/2022.3/Documentation/Manual/class-MonoManager.html`.

For our usage here, we require `EventSystem` to be run before `QuestSystem`.

1. Open **Edit | Project Settings… | Script Execution Order**.

2. If they're not already in the list, add these two scripts to the list using the little plus (+) button at the bottom right.

3. Then, click and drag them into the position indicated:

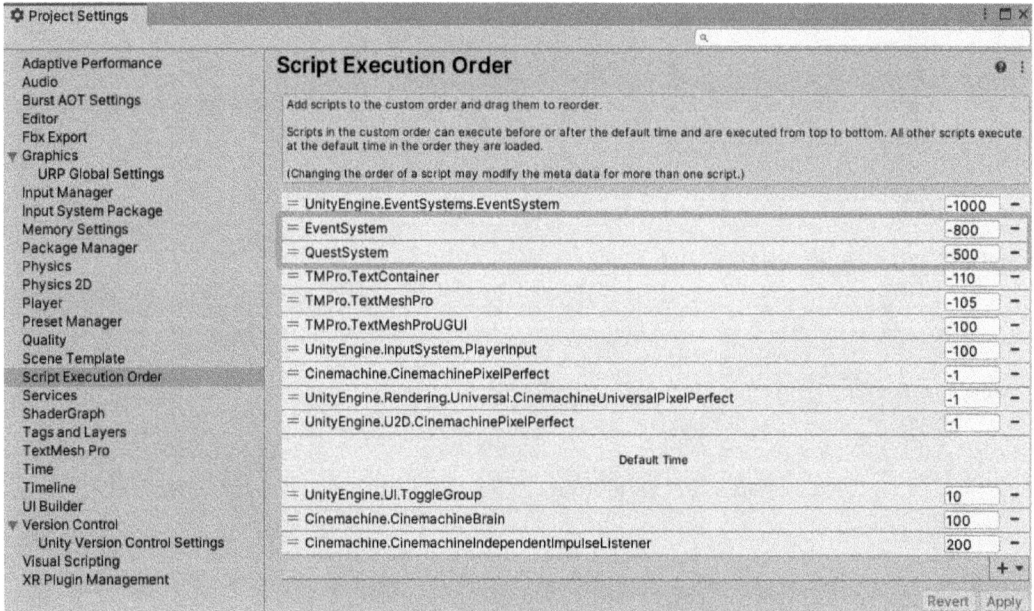

Figure 9.4 – Script Execution Order in Project Settings

4. Click **Apply** when you're finished.

Additional reading | Unity documentation

Unity also provides a code solution for specifying the execution order of scripts using an attribute. The attribute is `[DefaultExecutionOrder(int)]`, and you can decorate the class declaration and set the order value.

For example, adding this attribute to the `QuestSystem` class declaration would look like this:

```
[DefaultExecutionOrder(-500)]
public class QuestSystem : MonoBehaviour
{ …
```

And speaking of execution, make sure to add the `QuestSystem` component to the **Systems** GameObject in the scene (as a sibling of `EventSystem`).

Now, what's a quest system without any quests?!

The quest

Your mission, should you choose to accept it, will be to find and collect the missing pieces that are required to solve the habitat entryway security puzzle lock. The quest object representing the actual quest with its requirements is arguably the essential part of any quest system. As you saw in the *Script Execution Order* section, keeping a list of the active quests is quite simple – we add a quest to a list and set a variable to `true` when it's completed. Easy-peasy.

The next part of the quest system we'll cover will be the quest code. Before we even get started, however, we will ensure that any quest we define will have a unique identifier that is easily assignable (even in the **Inspector** view).

Name consistency

The problem we'll need to solve now is how to guarantee consistency for quest names without using magic strings, reference the quest names from different parts of the code, and have them selectable in the **Inspector** view.

We saw how to use `enum` when the simple **FSM** was introduced in *Chapter 7*. Well, we're going to use enum similarly now, but instead of states for an FSM, we're going to use it for unique quest name identifiers; again, as a reminder, we're not going to rely on string literals! This is similar to how we've used string constants before (for example, event names and tags), but being an `enum` type over a constant means we have some added benefit, including when it comes to selecting from a list of available quests in the **Inspector** view!

Create a new script named *QuestNames* in the `Assets/Scripts/Quests` folder. Replace all of the default script template code with the following code:

```
public enum QuestNames
{
    // Quest name, unique ID.
    CollectKeysQuest = 10
}
```

This is where we'll define any new quest we add to the game. We've populated the `CollectKeysQuest` name and given it a unique ID of `10` (this is any arbitrary number that's not already being used) – be sure to follow this pattern, assigning the name and ID, when adding additional quest names.

We'll see how the quest names will be declared and referenced as we proceed with our coding mission to complete this quest system!

Quest base class

Yup, that's right, another base class means we will use more OOP design here! And since we covered this already in *Chapter 2*, it's barely an inconvenience for you to bang this out, right? Our specific quests will then derive from the new quest base class we'll write next, providing a template for consistency in implementing quests.

We've already defined the quest names. A quest name will need to be a part of every quest, so let's start by setting up the base class for our quests with that.

Create a new script named `QuestBase` in the `Assets/Scripts/Quests` folder:

```
using UnityEngine;

public abstract class QuestBase : MonoBehaviour
{
    public QuestNames QuestName => _questName;
    [SerializeField]
    private QuestNames _questName;
    ...
```

The first thing to note is that the `QuestBase` class is declared as `public abstract class`. This means we won't be able to use this class directly – we cannot add an abstract class to a `GameObject` object in the **Scene Hierarchy**, even with it being a **MonoBehaviour** type. Instead, we will have to create a new class that inherits from `QuestBase` (that is, this is only the *base template*).

Second, we can see the declaration for `QuestName`, which uses the `QuestNames` enum type we previously wrote. We have encapsulated a `private _questName` variable and decorated it with the `[SerializeField]` attribute so that it can be assigned in the **Inspector** view. Spoiler alert: the following screenshot shows the dropdown field the enum type draws in the **Inspector** view:

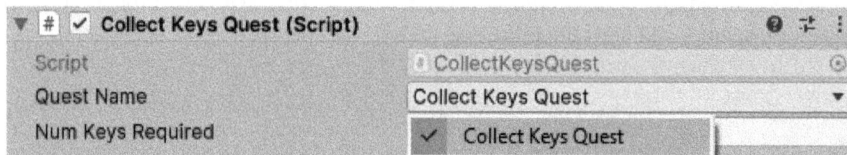

Figure 9.5 – The Quest Name Inspector view assignment dropdown

The `_questName` value is made available to other classes via the `public QuestName` property – the expression body (`=>`) declares the property as a getter only (as in, you won't be able to assign a value).

So far, so good. Still, this was a bit of a review since nothing we've covered here is new. The same goes for the first two methods we'll declare for the `QuestBase` abstract class. We'll use `virtual` methods for both `StartQuest()` and `QuestCompleted()` because the inheriting class may need to override the provided base functionality:

```
...
public virtual void StartQuest()
    => QuestSystem.Instance.StartQuest(QuestName.ToString());

protected virtual void QuestCompleted(string questName)
    => Debug.Log($"Quest '{questName}' completed!");
...
```

The code may be self-explanatory at this point, but it always helps to explain. Let's break it down:

- `public virtual void StartQuest()`: This method has a `public` accessor because it is intended to be called from an external class to trigger the start of the quest. We call the quest system's `StartQuest()` method (via its Singleton instance) and pass the unique quest identifier (that is, the `enum`-based quest name) to add it to the active quests `Dictionary` for later reference.

  We use `virtual` in the method signature in case the inheriting class would need to do more than start the quest.

- `protected virtual QuestCompleted()`: This method has a `protected` accessor, so it can only be called within the class and by derived classes – not an external class because the logic to determine quest completion should be evaluated on the specific quest's requirements and not some external factors.

  Again, we use `virtual` in the method signature in case the inheriting class would need to do something else (such as implementing different logic or behavior for different types of quests, such as side quests). For now, we're simply logging a message to the console that the quest was completed.

Now for the fun part! We'll be leaning on the event system to implement the following function: listening for when the quest has been completed. This is why we didn't declare the `QuestCompleted()` method as `public` – `QuestCompleted()` is the handler method that's passed into the event system as the listener:

```
...
private void OnEnable() => AddListeners();
private void OnDisable() => RemoveListeners();

protected virtual void AddListeners()
    => EventSystem.Instance.AddListener<string>(
```

```
            EventConstants.OnQuestCompleted, QuestCompleted);

    protected virtual void RemoveListeners()
        => EventSystem.Instance.RemoveListener<string>(
            EventConstants.OnQuestCompleted, QuestCompleted);
}
```

Here are just a few points in need of explanation:

- `OnEnable()`, `OnDisable()`: We add and remove the event system listeners for the base class, respectively. `AddListeners()` and `RemoveListeners()` can be overridden in the inheriting class if additional listeners need to be added for the specific quest.

- `AddListeners()`: Here, we add a listener to the event system for quest completion. We'll use a string argument to pass the quest's name as the parameter to the handler method.

- `RemoveListeners()`: We just need to remove the listener added in `AddListeners()` – remember, removing event listeners is always good practice!

  Note that even though it doesn't make sense that simply removing a listener would require an argument for passing a parameter, we still need it because the delegate definition must match the method signature of the handler method that was added as the listener.

As the last step, add the string constant for `EventConstants.OnQuestCompleted` to the `EventConstants` script. Or, in your IDE, *OnQuestCompleted* should have that red squiggly line indicating the definition was not found. Go ahead and use your IDE's refactoring tools to generate the variable (but ensure it is consistent with the other event name constants).

With that, `QuestBase` has been completed! Now, prepare to make a specific quest class derived from the base class.

Collect keys quest

We call the missing security puzzle key pieces that need to be collected *keys*, and three will be required to complete our quest. These are simple requirements for a simple quest, but we still need a way to not only explicitly declare the requirement but evaluate and communicate completion, too.

We'll start by creating a new script called `CollectKeysQuest` in the `Assets/Scripts/Quests` folder and inherit from `QuestBase` instead of `MonoBehaviour`:

```
using UnityEngine;

public class CollectKeysQuest : QuestBase
{
    [SerializeField] private int _numKeysRequired = 3;
    private int _keysCollected = 0;

    ...
```

We also declare the number of keys required to be collected to complete this quest. _numKeysRequired will be `private` so that no other class has access to it, but we'll use the `[SerializeField]` attribute to set this value in the **Inspector** view (although we'll set a default value of 3).

_keysCollected will keep track of the number of keys the player has collected (`private` – it's no one else's business), and we'll increment that value via the event system (so handy):

```
...
protected override void AddListeners()
{
    base.AddListeners();
    EventSystem.Instance.AddListener<bool>(
        EventConstants.OnKeyCollected, KeyCollected);
}

protected override void RemoveListeners()
{
    base.RemoveListeners();
    EventSystem.Instance.RemoveListener<bool>(
        EventConstants.OnKeyCollected, KeyCollected);
}

private void KeyCollected(bool arg0)
{
    _keysCollected++;
    // UNDONE: Evaluate quest completion.
}
}
```

Okay, it is a familiar pattern, but let's clarify it still:

- `override AddListeners()`: For this specific quest, we need to listen for when the player has collected a key. Using the event system, we don't need to know anything about the `Player` object or even the script implemented to collect the key; we observe that a key was collected and handled with the `KeyCollected()` method. We use the `override` keyword here because we need more functionality than the base class alone provides:

  - `base.AddListeners()`: Speaking of what the base class provides, we still need it! We can still ensure the base class methods that are overridden by using the `base` keyword – for accessing base class members within the derived class.

- `override RemoveListeners()`: You got it – maintaining good practice removing what listeners were added:

  - `base.RemoveListeners()`: The same as for `base.AddListeners()`

- `KeyCollected()`: I promise we won't have a breadcrumb trail of UNDONE tokens to follow again! But, for now, we'll just increment the `_keysCollected` variable when the `OnKeyCollected` event is triggered:

  - `bool arg0`: Our `EventSystem` requires us to pass an argument with the handler event. We don't need to pass any parameters for the key collected event, but we must still declare something! A bool value being the smallest type, I feel, is the least evil we can do here.

  - `EventConstants.OnKeyCollected`: Be sure to add it to `EventCostants` to resolve the missing definition error.

Excellent – we have a quest! Now, if only we could be given the quest! We don't have snarky NPCs in our game standing on street corners, just giving out quests to any weary traveler passing through, so we'll leverage some reusable components we already have to trigger the start of our quest.

Quest giver GameObject

We don't need a new script for giving the quest to the player; we just need to start *Collect Keys Quest*. We can easily do that by creating a trigger volume in the environment, using our ever-so-useful and reusable `TriggerEvent` component, and simply calling the publicly declared `StartQuest()` method on a `CollectKeysQuest` component.

Here's what it'll look like in the **Scene** view and **Inspector** view:

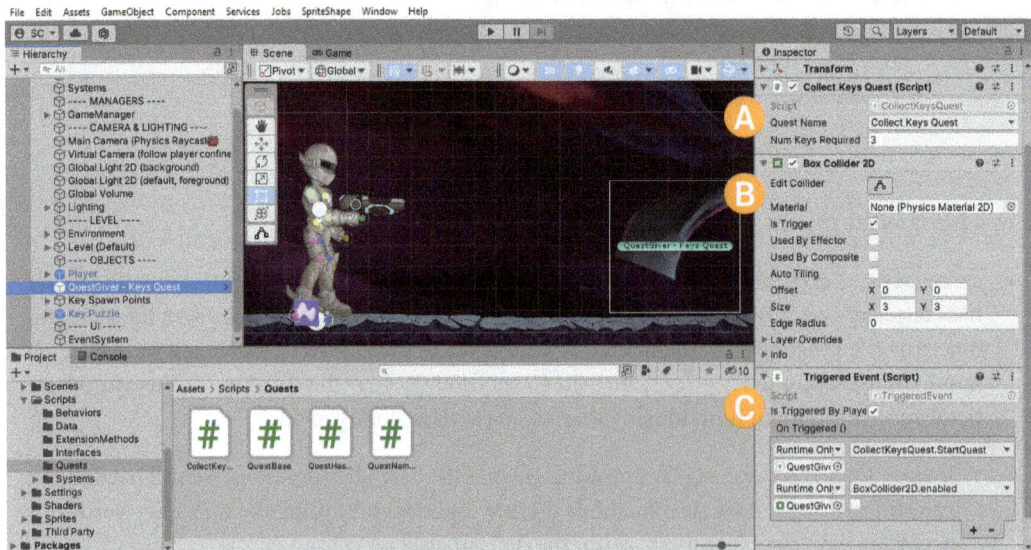

Figure 9.6 – Quest giver object in the scene

Have a go at constructing the object and components that make up the *quest giver* object and place it in the level near where the player starts.

Did you manage to assemble it successfully? Let's review the steps:

1. Create a new GameObject in the scene and name it `Quest Giver - Keys Quest`, then position it in the level near where the player starts (refer to *Figure 9.6* for an example).

2. (*A*) Add the `CollectKeysQuest` script to the new object. *Collect Keys Quest* should be selected in the **Quest Name** dropdown since it's our only quest, and set the number of keys required to three.

3. (*B*) Add a `BoxCollider2D` component next (use the **Add Component** button) and be sure to set **Is Trigger** to `true` because we don't want the player to interact with this GameObject physically; we only want to use it as a trigger volume. Adjust the collider's size to guarantee the player will collide with it.

4. (*C*) Add a `TriggeredEvent` component and wire up the methods that are called when `OnTriggered()` is invoked.

5. Click the little plus (+) button and add two events:

    I.   Drag the `CollectKeysQuest` component to the first event field and select the `CollectKeysQuest.StartQuest()` method from the drop-down list.

    II. Drag `BoxCollider2D` to the second event field, select `BoxCollider2D.enabled` from the dropdown, and leave the checkbox unticked (that is, set **Enabled** to `false`; this will disable the collider and prevent triggering *start the quest* additional times).

    III. Ensure only the player can trigger the quest start by ticking off `IsTriggeredByPlayer`.

6. As the final step, save the **Quest Giver** object as a `Prefab` object in the `Assets/Prefabs` folder (by dragging it from the hierarchy into the **Project** window). That way, we can quickly drop a quest giver into the level later.

Quest accepted! You wouldn't start a quest that was impossible to complete, would you? Now, let's see how we complete *Collect Keys Quest*.

Quest completion

We've defined a specific quest, we've given the quest to the player, and now we must progress on the quest to complete it. Progress is made when we collect keys in the game level (you know, kind of like the collection game we completed as the first project in this book).

In the `CollectKeysQuest` class, as you know, the `KeyCollected()` method is called when the `OnKeyCollected` event is triggered. We left it at only incrementing the variable, keeping track of the number of keys collected. Let's finish it up by evaluating whether the quest requirements have been satisfied to complete the quest.

Complete the `KeyCollected` method by adding the `if` block, like so:

```
private void KeyCollected(int keyId)
{
    _keysCollected++;

    if (_keysCollected >= _numKeysRequired)
    {
        QuestSystem.Instance.CompleteQuest(
            QuestName.ToString());
        EventSystem.Instance.TriggerEvent(
            EventConstants.OnQuestCompleted, QuestName);
    }
}
```

A very simple `if` statement checks whether the incremented `_keysCollected` variable is greater than or equal to (`>=`) the number of keys required to complete the quest, as defined by the `_numKeysRequired` variable (set in the **Inspector** view). If the statement evaluates to true, then we'll do the following:

- Call `CompleteQuest()` on the `QuestSystem` Singleton instance and pass in the quest name as the parameter. We need to use `ToString()` on the `QuestName` variable because it is an enum value, and internally, it is stored as an `int` type; we need a `string` type as the argument.

- Trigger the `OnQuestCompleted` event via the `EventSystem` Singleton instance and, again, pass in `QuestName` as a parameter. This is also a `string` type.

The *UNDONE token quest* is complete. Yay!

Quest event constants added

We need to add some event name constants to support what we just added for the quest system and *Collect Keys Quest*. As a recap, here's what the `EventConstants` script should look like now. I have added some comments for some quick organization to keep things tidy:

```
public class EventConstants
{
    public const string OnMyEvent = "OnMyEvent";

    // QuestSystem events.
    public const string OnQuestCompleted = "OnQuestCompleted";

    // Quests' events.
    public const string OnKeyCollected = "OnKeyCollected";
}
```

Our quest system has come together nicely, but we're still missing one final bit of functionality, and that is being able to know whether or not a quest that was started (that is, added to the `_quests` dictionary) has been completed (that is, the requirements of the quest mission have been satisfied, and `CompleteQuest()` was called).

Quest status

To know the quest's status, go ahead and add the following `IsQuestComplete()` method to the `QuestSystem` class:

```
public bool IsQuestComplete(string questName)
{
    if (_quests.TryGetValue(questName, out bool status))
        return status;

    return false;
}
```

You should recognize a familiar pattern here with `TryGetValue()` (as a method available in `Dictionary`) – we've covered a similar pattern with `TryGetComponent()` several times. `TryGetValue()` will return `true` if the value exists in the `_quests` dictionary. Then, set the returned value to the `out` variable, `status`, for immediate consumption by the following `if` block code. This is extremely convenient, and, as I've espoused, I'm a big fan of this pattern (even going so far as to replicate this pattern in my code).

We'll continue with short-circuiting the method by immediately returning the quest's `status` from the `TryGetValue()` call. Otherwise, we return `false`, indicating a default condition of the quest not being complete due to it not even existing in the dictionary.

Now that we can query the status of a specific quest from our `QuestSystem` Singleton instance, we can create a reusable component to use in our game to respond accordingly.

Quest completed component

We'll need something to check for quest completion when the player gets to the habitat entryway. This will allow us to show the security puzzle lock if the quest has been completed or provide some other action if not.

Create a new script named `QuestHasCompleted` in the `Assets/Scripts/Quests` folder:

```
using UnityEngine;
using UnityEngine.Events;

public class QuestHasCompleted : MonoBehaviour
{
    public QuestNames QuestName;
```

```
public UnityEvent OnQuestComplete;
public UnityEvent OnQuestIncomplete;

public void CheckQuestComplete()
{
    if (QuestSystem.Instance.
        IsQuestComplete(QuestName.ToString()))
    {
        OnQuestComplete?.Invoke();
        return;
    }

    OnQuestIncomplete?.Invoke();
}
}
```

This isn't anything we haven't seen before. The QuestName field will allow us to specify which quest we want to know the completion status for in the **Inspector** view. The UnityEvent fields will enable us to set actions for complete and incomplete statuses. Simples.

Here's how an object based on this new component will look in the **Inspector** view:

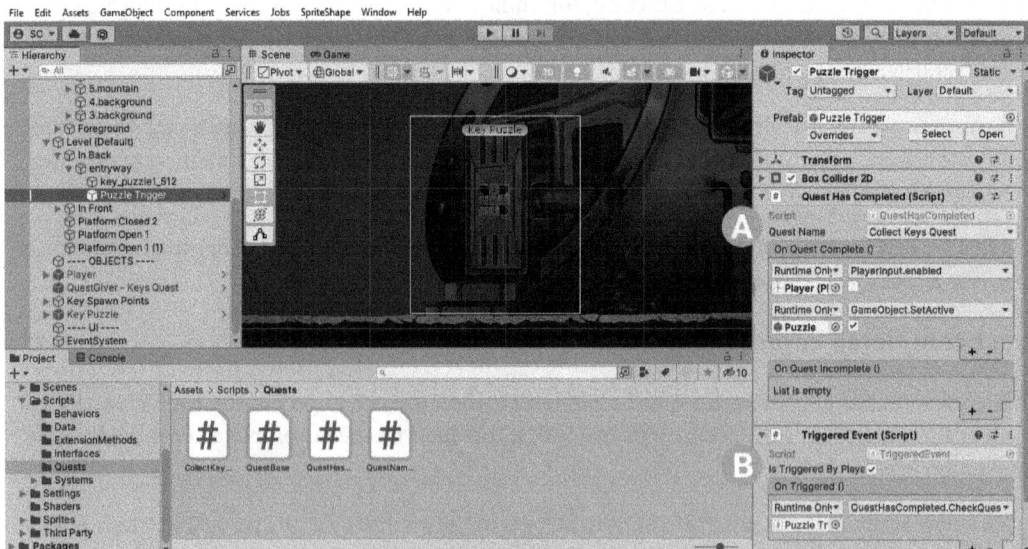

Figure 9.7 – Puzzle trigger object setup

Let's create this *puzzle trigger* object now and wire it up so that when the player enters the trigger volume, and if *Collect Keys Quest* has been completed, it will disable PlayerInput (ensuring the player can no longer move) and show **Puzzle** (solving the entryway security puzzle lock is the player's goal for this level).

Here's how we construct a *Has the quest been completed?* object:

1. Create a new empty GameObject in the **Hierarchy** We'll name this `Puzzle Trigger` because we want to show the puzzle when the player reaches the habitat entryway.

2. (*A*) Add a `QuestHasCompleted` component and select **Collect Keys Quest** from the dropdown of available quest names.

3. Assign the following events for `OnQuestComplete()` (click the plus (+) button to add a new one):

   I. Assign **Player** from the root of the **Scene Hierarchy**, select `PlayerInput.enabled` from the dropdown, and leave the checkbox unticked. Disabling `PlayerInput` in this way will mean no input will be processed to enact with the player – we don't want the player character to move any longer because it would now be time to solve the puzzle.

   II. The following assignment is for our future selves; we'll use a reference to the **Key Puzzle** object and make it active (display it) so that the player can work on solving it – **Key Puzzle** will be added in the last section of this chapter, so we'll revisit this assignment then.

4. (*B*) Add a `TriggeredEvent` component and assign the `QuestHasCompleted` component to the `OnTriggered` event, then select `QuestHasCompleted.CheckQuestComplete()` from the dropdown. Ensure **IsTriggeredByPlayer** is ticked.

5. Don't forget to add a `BoxCollider2D` component and set **Is Trigger** to `true`; the size will be adjusted accordingly concerning the habitat entryway so that the player will interact with it when reaching it.

For placement, if you haven't taken the liberty of adding the habitat entryway to your level yet, now's the time! We haven't explored every nuance of your level design while bringing everything together – naturally, you've been doing your level design homework:

1. Place the habitat entryway sprite *entryway* from the `Assets/Sprites/Object Elements` folder into the environment and use the `Background` **Sorting Layer**. You can organize it in the **Scene Hierarchy** using what we previously established as the **Level (Default) | In Back** structure.

2. Place the **Puzzle Trigger** object at the entryway and set the collider size accordingly (refer to *Figure 9.7* for an example).

As we saw in the preceding code and description, the `QuestHasCompleted` component has an event that is triggered when we check the status for both complete and incomplete quest statuses. We already populated the `OnQuestComplete()` event but also have an `OnQuestIncomplete()` event. We won't be assigning any function here for now, but imagine that we could display a dialog to our player stating that the quest's requirements haven't been met for them to proceed.

> **Challenge | Quest system diagram**
>
> Create a basic object reference diagram for the quest system while using the event system diagram in *Figure 9.1* as a reference. Don't be so concerned with the shapes you choose to represent each part, but do be consistent and use the same shape for the same object type. If you want to challenge yourself further, create a UML diagram!

Everything is in place for us to fully execute a quest's life cycle. Now, we need to collect those pesky key pieces!

Collecting keys

We've tackled player collectible items a few different ways now, so collectible key items will be a familiar concept. The only difference is that we will now rely on `EventSystem` to trigger a *key collected* event. As we already know, the `CollectKeysQuest` quest is listening for `OnKeyCollected` being invoked, so this is where we'll implement triggering it.

As you could probably also guess, we're going to use a `Prefab` object for the puzzle piece keys (via **Prefab Variant**). So, that will require a component to implement the key collection behavior.

KeyItem component

First, let's prepare the artwork we'll use to make the collectible key pieces. For a visual reference, here's what we'll be working with for creating the key pieces and a placeholder security puzzle lock on the habitat entryway door (if it makes you nervous, try disregarding the encroaching plant entity's vines):

Figure 9.8 – Key variants and entryway security puzzle lock

Art assets

To follow along while using the same artwork that was created for this project, download the assets from this book's GitHub repository: `https://github.com/PacktPublishing/Unity-2022-by-Example/tree/main/ch9/Art-Assets`.

Import the art assets for the security puzzle lock, the puzzle placeholder image, and the individual key pieces into `Assets/Sprites/Puzzle`. For the provided key pieces artwork, I've set the following properties for the desired size and placement:

- **Pixels Per Unit**: `500`

- **Pivot: Custom**, then adjust the X and Y values so that the pivot is located in the center of the art:

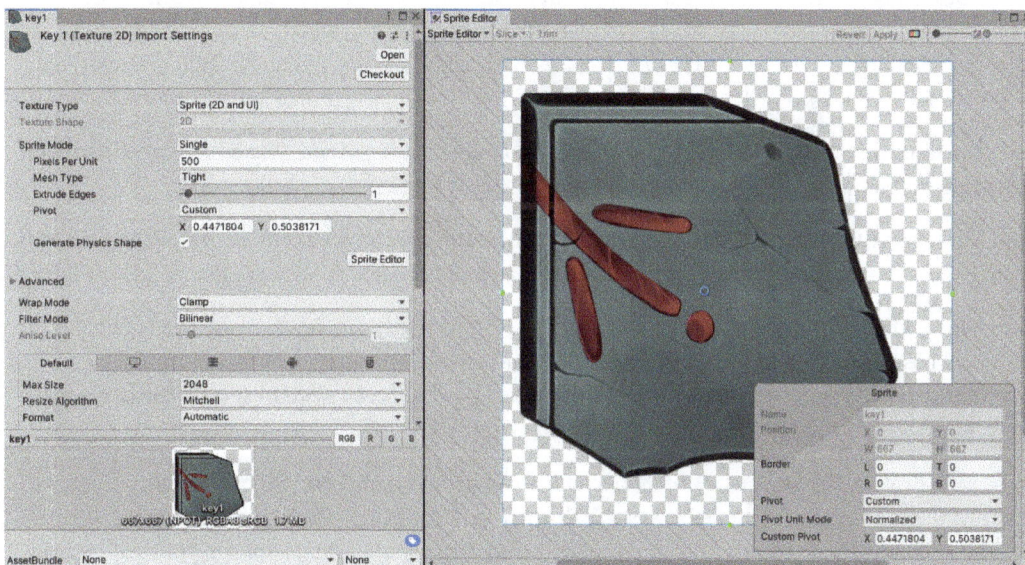

Figure 9.9 – Key pieces art – import settings

Now, let's write the code for the collectible key item so that we'll be ready to assemble the Prefabs for each key piece.

Create a new script named `KeyItem` in the `Assets/Scripts` folder with the following code:

```
using UnityEngine;

[RequireComponent (typeof(Collider2D))]
public class KeyItem : MonoBehaviour
{
    private void OnTriggerEnter2D(Collider2D collision)
    {
```

```
        if (collision.CompareTag(Tags.Player))
        {
            EventSystem.Instance.TriggerEvent(
                EventConstants.OnKeyCollected, false);
            Destroy(gameObject);
        }
    }
}
```

Finally, yes – there is where we specifically trigger the OnKeyCollected event!

This script for collecting items may be even more straightforward than the ones that came before it. We use OnTriggerEnter2D() to detect when the player enters the trigger volume. Does this mean we need a collider component as a sibling on the GameObject to which we attach the KeyItem script? Yes, that's right – we need a Collider2D object with **Is Trigger** enabled. When we assemble our key Prefabs, you'll see we'll use CircleCollider2D, keeping it nice and efficient.

Also, note that we're comparing the tag of the collision object – using our Tags.Player constant – to ensure that only the player interacts with the trigger volume.

Let's give the event system TriggerEvent() call some additional attention because our event system requires a second parameter to be passed in as an event argument. We have to pass in *something*.

As explained when the handler method was introduced, a bool value is the smallest data type in C# (1 byte); we'll pass in false as the type's value – we can also take the liberty of this meaning, "*True or false, will I pass an argument value for this event? False.*" We don't even need to specify the bool type – as in TriggerEvent<bool>() – because the type can be inferred from the argument's value.

Lastly, we'll destroy the key piece object because we collected it and no longer need it in the level. We're just using Destroy(gameObject) directly here for brevity. Still, for consistency, if you'd like, you should be able to hook up the reusable Destroyer component for yourself by now (challenge accepted?).

Now, it's time to construct the collectible key item Prefab!

KeyItem Prefab and variants

Creating Prefabs for reusable items in our project is second nature by now. Let's breeze through the steps to create a new collectible key item **Prefab**; then, we'll make **variants** that have all three key pieces and their individual artwork:

1. Drag the key1 Sprite (which we previously imported) from the Assets/Sprites/Puzzle folder in the **Project** window into the **Scene Hierarchy**.

2. Double-click it to focus on it in the **Scene** view (it should be positioned at (0, 0, 0) if not, **Reset** the **Transform** option in the **Inspector** view so that it is – we don't want any offsets being saved in the Prefab).

3. Rename key1 to simply key (while selected at the top of the **Inspector** view, click it a second time, or use the *F2/Enter* key) – this will make sense later when we create the additional key variants.

4. Right-click on it and select **Create Empty Parent**, then rename it Key1.

5. Add a CircleCollider2D sibling component to the Key1 object and enable **Is Trigger**. Set the **Radius** value so that its hitbox is slightly larger than the puzzle piece Sprite (refer to *Figure 9.10*).

6. Add the KeyItem script to the parent Key1 object – there's nothing to configure; all the behavior is handled in code.

7. Drag Key1 from the **Hierarchy** to the Assets/Prefabs folder in the **Project** window to make it a Prefab.

Now, to create the additional key piece Prefabs as Prefab variants, follow these steps:

1. Select Key1 in the **Hierarchy** and press *Ctrl/Cmd + D* twice to make two duplicates of it.

2. Rename the duplicates Key2 and Key3, respectively.

3. For Key2, on the child key object, change the SpriteRenderer **Sprite** field to key2. Then, do the same for Key3, and set the **Sprite** field to key3.

4. Drag Key2 from the **Hierarchy** into the Assets/Prefabs folder. Then, in the **Create Prefab for Variant?** dialog, click the **Prefab Variant** button (we'll share all the same configurations as the Key1 Prefab but override the **Sprite** field for the key piece variants). Then, do the same for Key3.

> **Tip | Prefab Edit Mode | Variants**
> Note that when opening **Prefab Variant** in **Prefab Edit Mode**, all overrides are indicated by a blue indicator along the left edge of the **Inspector** view.

Here are the key Prefabs and variants we just created:

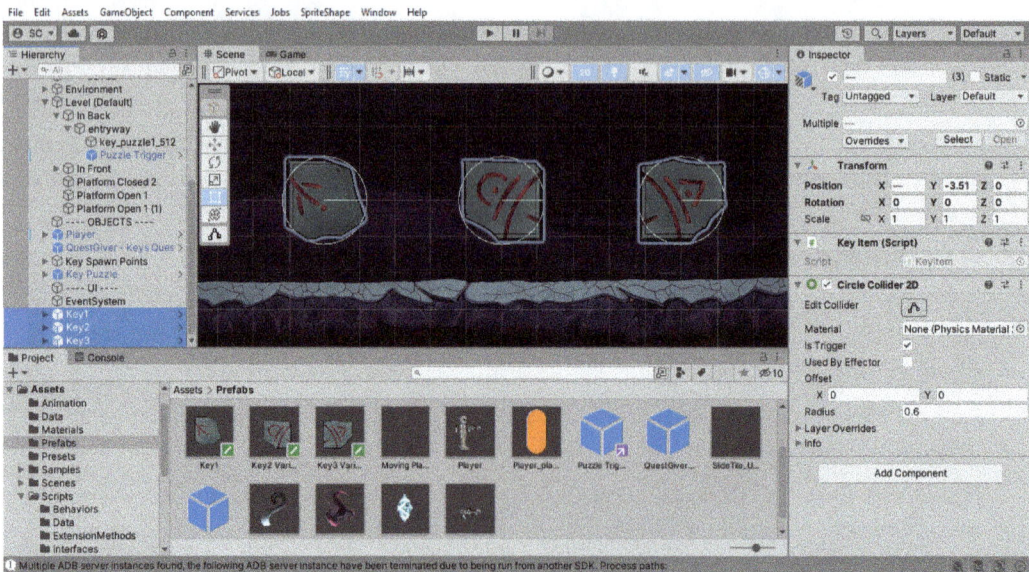

Figure 9.10 – KeyItem pieces Prefabs

The player can now collect the key pieces to complete the quest, but how should we place them in the level? First things first, delete the `Key1`, `Key2`, and `Key3` objects from the scene; we're going to spawn them in.

QuestSystem | Complete code

To view the complete code for the `QuestSystem` class and all the quest-related code in this chapter, visit this book's GitHub repository: `https://github.com/PacktPublishing/Unity-2022-by-Example/tree/main/ch9/`.

Key instantiator – Randomness

Let's explore an easy way to implement some basic random instantiation of objects so that the game is different every time it's played – addressing replayability is good for player engagement!

Randomness in game design

You'll come across the topic of randomness in games in your game development journey – and not just for card games! The role of randomness in game design is appropriate for this entire chapter – heck, this entire book! So, keeping that in mind, this will be one of the most basic examples of how you can add randomness to a simple mechanic with effective results. In the level design, we will instantiate the three keys at random locations identified by a larger number of spawn points.

The structure of the code will be to take an array of KeyItem (objects) (that is, the key pieces) and an array of Transform (positions) (that is, points placed throughout the level) as input and then output (that is, instantiate) the objects in order at the next randomly selected spawn point (being sure not to reuse any of the spawn points).

Did you visualize what the code could look like? Let's see. Let's walk through creating the code for each part.

Start by creating a new script named KeyInstantiator in the Assets/Scripts folder.

We'll declare the arrays that hold the key objects and the spawn points first:

```
using UnityEngine;
using System.Collections.Generic;

public class KeyInstantiator : MonoBehaviour
{
    [SerializeField] private KeyItem[] _keyPrefabs;
    [SerializeField] private Transform[] _spawnPoints;
    ...
```

Here, we declared two arrays:

- KeyItem[] _keyPrefabs: This has been serialized so that it can be assigned in the **Inspector** view; we'll assign the objects to be instantiated (the key puzzle piece Prefabs). We are declaring the type as KeyItem instead of a more generic GameObject because of the following reasons:

  - We only want Prefabs that include the KeyItem component to be assignable to the array.

  - While referencing an item in the collection, we will consume the item as the KeyItem type and avoid making a GetComponent<KeyItem>() call.

- Transform[] _spawnPoints; This has been serialized so that it can be assigned in the **Inspector** view; we'll assign the GameObjects that are placed throughout the level where key pieces can potentially spawn. With game design in mind, be sure to place more than three in a level so that we're not just randomizing what key piece appears at the same three positions – your limit will be the design or your level (and I'd be sure to place one of them where you happen to have several infected robots patrolling).

Array (C#)

An array is a type that's declared with opposing square braces ([]) and represents a collection of items of that type. Items in the array are addressed by their index value, which starts at zero (for example, _spawnPoints[0] is the first Transform stored in the collection).

Here's some additional reading: https://learn.microsoft.com/en-us/dotnet/csharp/programming-guide/arrays/.

Next, we'll add a List type to work with the currently available spawn points from the _spawnPoints array. Why are we doing this? We already have an array type, and now a List type? Yes. Arrays in C# are not easy to work with if we want to resize them (that is, remove an item), but a List type is.

Declare the following List type and add the Start() method with the spawn points assignment:

```
private List<Transform> _availablePoints;

private void Start()
{
    _availablePoints = new List<Transform>(_spawnPoints);
    ...
```

We declared a List type and initialized it in Start() with the _spawnPoints value:

- List<Transform> _availablePoints: This is the private member variable because we'll just work with the points inside the class. We'll use this to determine the points available to instantiate a key.

- Start(): We'll scatter the key pieces throughout the level when the game starts... so we'll use the MonoBehaviour-provided Unity message event Start() for that, yeah.

Now comes the fun part – random position instantiation! Add the following foreach loop to the Start() method's implementation:

```
        foreach (var item in _keyPrefabs)
        {
            var randomIndex = Random.Range(
                0, _availablePoints.Count);

            Instantiate(item,
                _availablePoints[randomIndex].position,
                Quaternion.identity);

            _availablePoints.RemoveAt(randomIndex);
        }
    }
}
```

Okay, let's do a breakdown of this final section:

- foreach (var item in _keyPrefabs): We're using foreach to iterate the available KeyItem Prefabs assigned in the **Inspector** view (all three of our key Prefab/variant pieces).

- `Random.Range(0, _availablePoints.Count)`: A bit of magic? No, this is Unity's **Random Number Generator (RNG)** hard at work returning a random number in a range that is inclusive of the first number and exclusive of the second number (only valid when using `int` values, however, like in our use case; otherwise, it's inclusive of the second number when using `float` values). We don't want to use the `Count` value because array indexes are zero-based (so we'd have to specify `Count - 1` otherwise):

  Note that we obtain `_availablePoints.Count` every iteration. That's because, only two lines below, we're removing the randomly selected point from `List` by the `randomIndex` value that's returned, so it's not used again.

> **Additional reading | Unity documentation**
> `Random.Range()`: `https://docs.unity3d.com/2022.3/Documentation/ScriptReference/Random.Range.html`.

- `Instantiate()`: We spawn the current `item` in the array into the scene at the randomly selected spawn point position, `randomIndex`, with zero rotation (that is, `Quaternion.Identity`).
- `RemoveAt(ramdomIndex)`: A C# `List` type provides a remove method that not only deletes an item from the collection but also resizes it. Hence, the `Count` property reflects the number of items that remain.

It sounds like a lot when breaking it down like this, but it's a short and sweet random placement script.

The last step is, of course, setting the instantiator up in our scene so that the key pieces can be placed in the game level.

Instantiator scene object setup

Here's what we'll do to set up the new `KeyInstantiator` script on an object in our scene so that our keys are randomly spawned for the player to find:

1. Create a **Key Instantiator** object as a child of **GameManager**. To that, add the `KeyInstantiator` component.
2. In the **Inspector** view, drag the key Prefabs from the `Assets/Prefabs` folder in the **Project** window to the **KeyPrefabs** field.
3. Then, after placing – more than three! – GameObjects in the level that represent the possible spawn position of a key piece, select them in the **Hierarchy** and drag them to the **SpawnPoints** field in the **Inspector** view (don't forget that you can lock the **Inspector** window so that it doesn't change when you're selecting the objects you want to assign).

Figure 9.11 shows the result of the preceding steps:

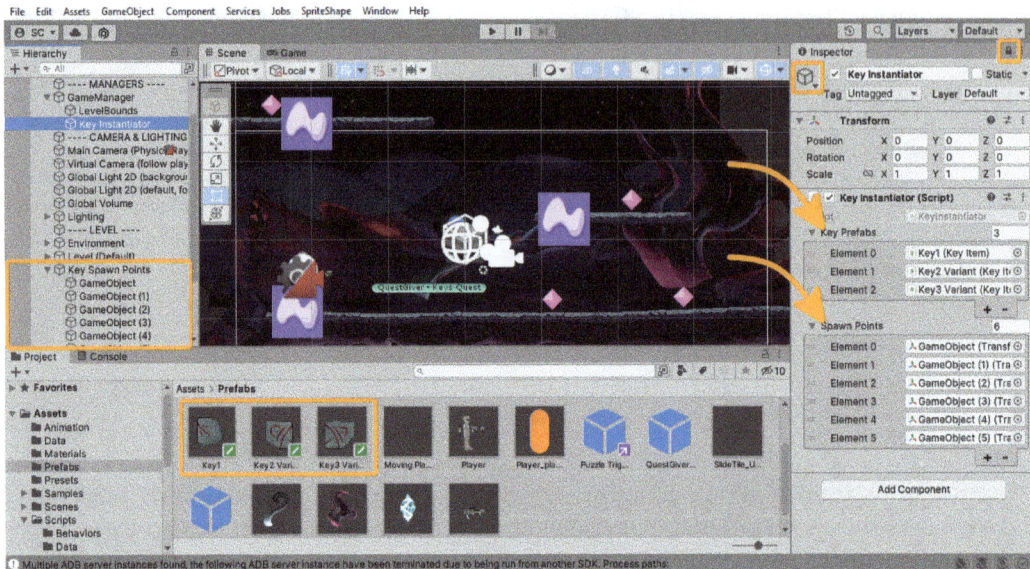

Figure 9.11 – Key instantiator object setup

You may have noticed the pink diamonds in the preceding figure. I've assigned a pink diamond icon (assignable at the top of the **Inspector** view) to the key spawn point objects so that they are easy to find in the **Scene** view while I'm working on the level design. I've also grouped the key spawn points under a parent **Key Spawn Points** object in the **Hierarchy**.

With that, we now have key pieces randomly placed in our level when the game starts. Next, we'll sort out a drawing issue before moving on to solving the key puzzle!

Instantiated Sprite drawing order

The key pieces will be instantiated on the `Default` **Sorting Layer**, and the **Order in Layer** value will be 0 for the default values. Because we've set up our environment with *default objects* originating "in the center of the depth layers," we can be confident that the instantiated pieces will not be obstructed. This can be compensated for when instantiated if needed, but this can simply be avoided if we take some care in the environment layout.

In this section, we learned how to create a quest system, make new quests with unique properties and requirements for completion, assign a quest to the player, and query a quest's status to advance the gameplay. We also saw how to utilize the event system to build other systems on top of its foundation.

In the next section, we'll integrate a sliding puzzle as the habitat entryway's security lock system, where solving it wins the game.

Solving the key puzzle and winning the game

We could spend a whole lot of time here designing a novel puzzle for the security lock system. Still, that falls outside the scope of this book and would not provide the learning opportunity I want to cover – that is, using third-party assets in your game. This isn't to say that you shouldn't strive to introduce a new and original idea in your game – any way you can differentiate your game and offer players a remarkable and unique experience is time well spent!

We will use the well-known sliding tile puzzle for the habitat entryway's security puzzle lock.

Sliding tile puzzle

I've only briefly mentioned the **Unity Asset Store**, but I want to bring some well-deserved attention to it now. The Unity Asset Store contains a wealth of assets that both Unity and third parties provide. You can find just about anything you would need for your games, including pre-made systems, frameworks, characters, animations, 2D and 3D art assets, music and sound effects, VFX, and more, in almost every genre and style you can imagine.

However, integrating solutions and assets from varying vendors isn't always trivial. This section will be dedicated to both the value of leveraging existing pre-made assets and identifying some issues you may need to work through to have them functioning in your project. You may be thinking that I've alluded to a level of quality for third-party assets that indicates they are not good – while that may be true in some rare cases (buyer beware, as usual), issues can arise that have nothing to do with the asset and all to do with the changes and advances in technologies that comes with newer Unity version releases. Rather than continuing in abstract terms, let's move forward with the specific example I'll provide.

For reference, here is the sliding tile puzzle we'll create with the tiles already scrambled (refer to *Figure 9.3* for the unscrambled version):

Figure 9.12 – Sliding puzzle scrambled tiles

As mentioned previously, we're going to leverage an existing asset from the Unity Asset Store to rapidly incorporate a sliding puzzle feature that will include everything necessary to slice an image into tiles, scramble the tiles, respond to user input for sliding the tiles, and calculate when the puzzle has been solved. If we had to develop on our own, all of these requirements would take significantly more time to create, code, debug, and test. Unity Asset Store assets generally come with the benefit of dozens of developers (or hundreds in some cases) using the assets in their projects and reporting back bugs and discrepancies to the asset developer for further improvement – and you directly benefit from others' efforts.

Without further ado, we'll be using a free asset called **Sliding Tile Puzzle Game** by Hyper Luminal Games (`https://assetstore.unity.com/packages/templates/packs/sliding-tile-puzzle-game-41798`):

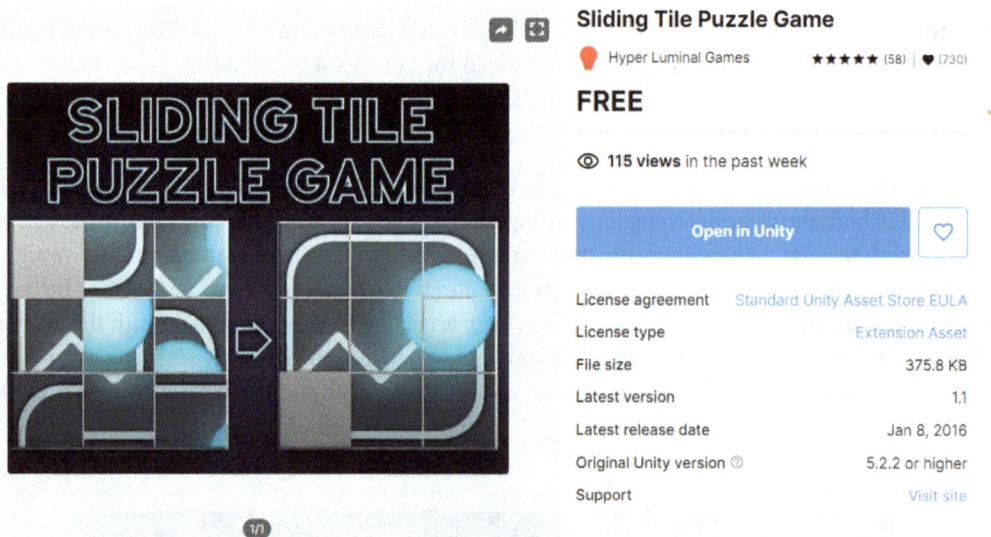

Figure 9.13 – Sliding Tile Puzzle Game

Several steps will be required to use the asset in our Unity 2022 project due to changes in the **Unity Scripting API**, in addition to changes required to support **Universal RP** (**URP**), the *render pipeline* we've based the project on, since the time the asset was released (way back in 2016, for Unity 5).

Let's get started by obtaining the asset and importing it into the project.

Importing the sliding tile puzzle asset

After going through the process of purchasing an asset or clicking **Add to My Assets**, you have the option of using the **Open in Unity** button from within the Unity Asset Store. Alternatively, you may open the **Package Manager** window in the Unity Editor, select **My Assets** from the **Packages** dropdown, find the asset in the list, click **Download**, and then click **Import**.

In my experience, most assets are designed to be imported into your existing project. Still, some assets are provided as *complete projects* and, therefore, cannot be imported directly into your project. The sliding tile puzzle asset requires an additional step to import into your project because it is provided as a complete project, and it's best not to overwrite any of your current project settings!

Let's walk through the steps; this won't take long:

1. Importing the sliding tile puzzle asset from the **Asset Store** page or **Package Manager** will produce the following **Importing a complete project** warning dialog:

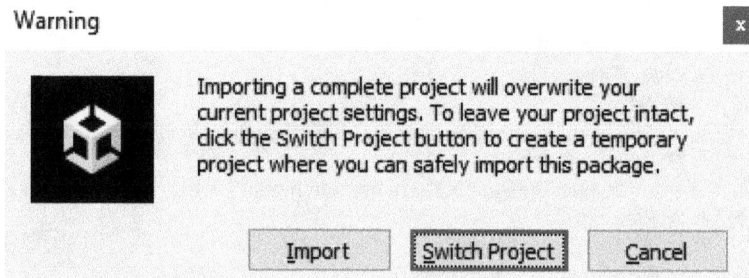

Figure 9.14 – The Importing complete project dialog

2. Click the **Switch Project** button to create a temporary project the asset will be imported into. Unity will automatically generate a temporary project name. We'll delete this project when we're finished extracting the asset from it, so don't worry.

3. When Unity has opened and everything has finished importing, the **Sliding Tile Puzzle Game** asset will be located in the `Assets/HyperLuminal` folder (found in the **Project** window).

4. Right-click on the `SlidingTilePuzzle` folder and select **Export Package...**. This will open an **Exporting package** dialog, where you can change what's included in the export. We want everything, so click the **Export...** button in the bottom-right corner.

5. When the file save window opens, enter `SlidingTilePuzzle` for the `.unitypackage` filename and pick an easily accessible folder; we will be importing from that same folder shortly.

6. Close Unity. You will be prompted with a **Keep Project?** dialog. You can safely click the **Forget** button since we no longer need it:

    Note that even though you confirmed forgetting it, you may still see an erroneous project show up in Unity Hub that represents this temporary project. If so, remove it and delete the project folder.

7. Now, back in our game's project, let's import the saved `.unitypackage` file by going to **Assets | Import Package | Custom Package...** and selecting `SlidingTilePuzzle.unitypackage` from the location we saved it in earlier:

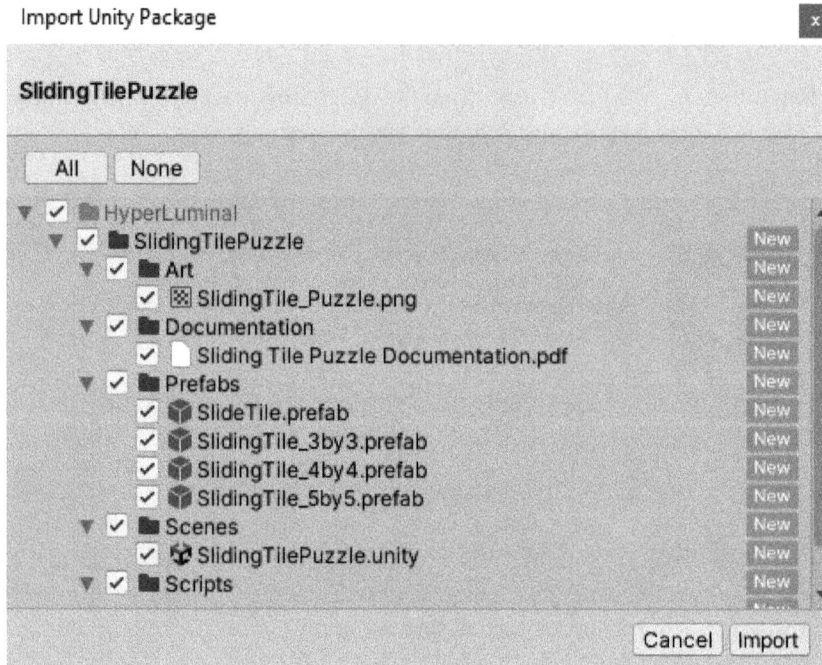

Figure 9.15 – The Import Unity Package dialog

8. All the items should be marked as **New** unless we've previously imported the package. Go ahead and click **Import** so that we can continue integrating the sliding puzzle.

Tip | Project organization

Asset vendors make their own choices on the location of their assets in the project folder structure, so importing many third-party assets can get a bit messy and unorganized. To maintain some level of sanity in your project files, I suggest placing all third-party assets under this `Assets/Third Party` folder.

For example, in *Figure 9.15*, you can see *SlidingTilePuzzle*, which we will import to the `Assets/HyperLuminal` folder. Please create a new `Assets/Third Party` folder and move the `HyperLuminal` folder as a child of it. Mischief managed.

Now, let's update the asset so that it works with our Unity 2022 project requirements.

Updating the puzzle tile shader

Our first order of business is to update from the built-in legacy renderer to the URP renderer. In Unity 5, we only had the built-in renderer, so it makes sense that we'd need to convert items related to the render for this asset.

If you were to open the `SlidingTilePuzzle` example scene in the `Assets/Third Party/HyperLuminal/SlidingTilePuzzle/Scenes` folder and enter **Play Mode** now, you'd get a giant pink square. Pink (or magenta) is the color Unity uses to represent a material or shader error – unless you've specifically made an object this color, seeing it in your **Scene** or **Game** view is generally not good.

The asset takes a Prefab approach for the type of sliding puzzle you want to make and provides Prefabs for 3x3, 4x4, and 5x5 sliding puzzles – we will make a 3x3 sliding puzzle and work with that specific 3x3 Prefab. Follow these steps to update the renderer-specific issues:

1. In the `Assets/Third Party/HyperLuminal/SlidingTilePuzzle/Prefabs` folder, duplicate the `SlidingTile_3by3` Prefab and rename it `SlidingTile_3by3_URP`.
2. Double-click on `SlidingTile_3by3_URP` to enter **Prefab Edit Mode**.
3. Find the `ST_Puzzle Display` component and change the *Puzzle Shader* from `Mobile/Unlit` to `Packages/Universal RP/Shaders/2D/Sprite-Unlit-Default` – you must drag this shader in from the **Project** window after expanding `Packages`.
4. Save the Prefab.

We have one more update to make for the renderer issues, and that's on the puzzle tile itself, but first, we need a new **Universal RP Sprite** material. Follow these steps to update the asset:

1. Create a new *material* named `PuzzleTile 1` in the `Assets/Materials` folder and select it.
2. In the **Shader** drop-down menu at the top of the **Inspector** view, select the **Universal Render Pipeline/2D/Sprite-Unlit-Default** shader.
3. In the `Assets/Third Party/HyperLuminal/SlidingTilePuzzle/Prefabs` folder, we also have the `SlideTile` Prefab. Duplicate it and rename it `SlideTile_URP`.
4. Double-click on `SlideTile_URP` to enter **Prefab Edit Mode**.
5. Find the `MeshRenderer` component and change **Material** at `Element 0` to `PuzzleTile 1`.
6. Save the Prefab.

Now that the rendering issues of the `Sliding Tile Puzzle` asset have been updated, we are ready to work within our **Lit 2D (URP)** scenes!

We only have one more issue with updating, and that's for the interactivity – that is, being able to slide the tiles from player input.

Updating the input system

The new input system is called new because… well, it's new. Unity 5 only had the legacy `InputManager` system; of course, since we're using the new input system in our project, we need to make some changes to support it. All the updates for the renderer were done in the editor, but now, we'll have to change some code.

Make the following code changes:

1. Find the `ST_PuzzleTile` script in the `Assets/Third Party/HyperLuminal/ SlidingTilePuzzle/Scripts` folder and open it in your IDE.

2. Add a new `using` statement at the top of the script. We'll need this `UnityEngine` namespace to support the code changes:

    ```
    using UnityEngine.EventSystems;
    ```

3. Add the `IPointerClickHandler` interface to the class definition to support the new input system's `pointer click` method:

    ```
    public class ST_PuzzleTile : MonoBehaviour, IPointerClickHandler
    ```

4. At the bottom of the script, replace `void OnMouseDown()` with the following:

    ```
    public void OnPointerClick(PointerEventData eventData)
    ```

5. Save the script and return to the Unity Editor; you're done.

Unfortunately, you can no longer test the sliding puzzle using the vendor's `SlidingTilePuzzle` example scene as-is.

To test the updates and the final update requirements for every scene you want to use the sliding puzzle in, follow these steps:

1. Create a new scene (**File | New Scene** or *Ctrl*/*Cmd* + *N*) and select the **Lit 2D (URP)** new scene template.

2. Add a `PhysicsRaycaster` component to **Main Camera**.

3. Add `UI Event System` to the scene. Ensure you update the `StandaloneInputModule` component for the new input system.

4. Add the `SlidingTile_3by3_URP` Prefab to the scene.

5. Enter **Play Mode**. Test. Enjoy.

And that completes all the updates required to modernize *Sliding Tile Puzzle* for the URP render and the new input system. Not too bad.

As you will see, the Unity Asset Store is a great resource with many talented asset publishers. It can help to save you (a lot of) time building game prototypes quickly, polishing your games, and creating vertical slices early (for seeking a publisher or investors in your project).

We will, however, need one additional change to integrate the sliding puzzle into our gameplay. And that is adding an event for when the security puzzle lock has been solved so that we can trigger an appropriate action – a logical one at this point would be granting the player access to the habitat station. Open sesame!

Adding an event for completion

We won't be covering any new ground with this task. We've added events in several different ways already. Our choice this time will be a `UnityEvent` event so that we can set the triggered handlers in the **Inspector** view.

Find the `ST_PuzzleDisplay` script in the `Assets/Third Party/HyperLuminal/SlidingTilePuzzle/Scripts` folder and open it in your IDE. Then, modify the script with the following additions:

1. Add a new `using` statement at the top of the script as required for declaring our event:

   ```
   using UnityEngine.Events;
   ```

2. Add a `public` `UnityEvent` event to be triggered when the puzzle is completed:

   ```
   public class ST_PuzzleDisplay : MonoBehaviour
   {
       ...
       public UnityEvent OnPuzzleComplete;
   ```

3. Modify the `CheckForComplete()` method by adding the `OnPuzzleComplete` invocation line within the `if(Complete)` block:

   ```
   public IEnumerator CheckForComplete()
   {
       ...
       // if we are still complete then all the tiles are correct.
       if(Complete)
       {
           Debug.Log("Puzzle Complete!");
           OnPuzzleComplete?.Invoke();
       }
       ...
   ```

We now have a convenient way to respond to the entryway security puzzle lock being solved/completed.

When we set up the `QuestHasCompleted` component on the habitat entryway, we left unfinished business for our future selves. We are now our future selves, so we can complete the `OnQuestComplete` event assignment and show our security puzzle lock to the player.

Setting up a new puzzle prefab

Yup, that's right – we'll need another Prefab for the sliding puzzle lock. We'll set that up now so that it's shown to the player – with our image – when all three keys have been collected and they reach the habitat entryway.

Follow these steps to create the new Prefab:

1. Drag the `SlidingTile_3by3_URP` Prefab into the **Hierarchy** (ensure its position is at (0, 0, 0)).

2. Right-click on it and select **Create Empty Parent**, then rename the parent object to `Key Puzzle Lock` – you'll end up with the actual puzzle as the child (yes, this is our standard approach to Prefab structure, if you still weren't sure).

3. Find the `key_puzzle1-complete_512` image in the `Assets/Sprites/Puzzle` folder (this is one of the images we imported in the previous *KeyItem component* section) and change its import settings in the **Inspector** view to the following:

    I.   **Texture Type: Default**

    II.   **Texture Shape: 2D**

    III. **Advanced**: Don't enable any options

    IV.   **Wrap Mode: Clamp**

    V.   **Filter Mode: Bilinear**

    VI.   **Max Size:** `512`

4. Find the `ST_PuzzleDisplay` component and assign the following:

    I.   *Puzzle Image*: `key_puzzle1-complete_512`

    II.   *Puzzle Scale*: 0.7, 0.7, 0.7:

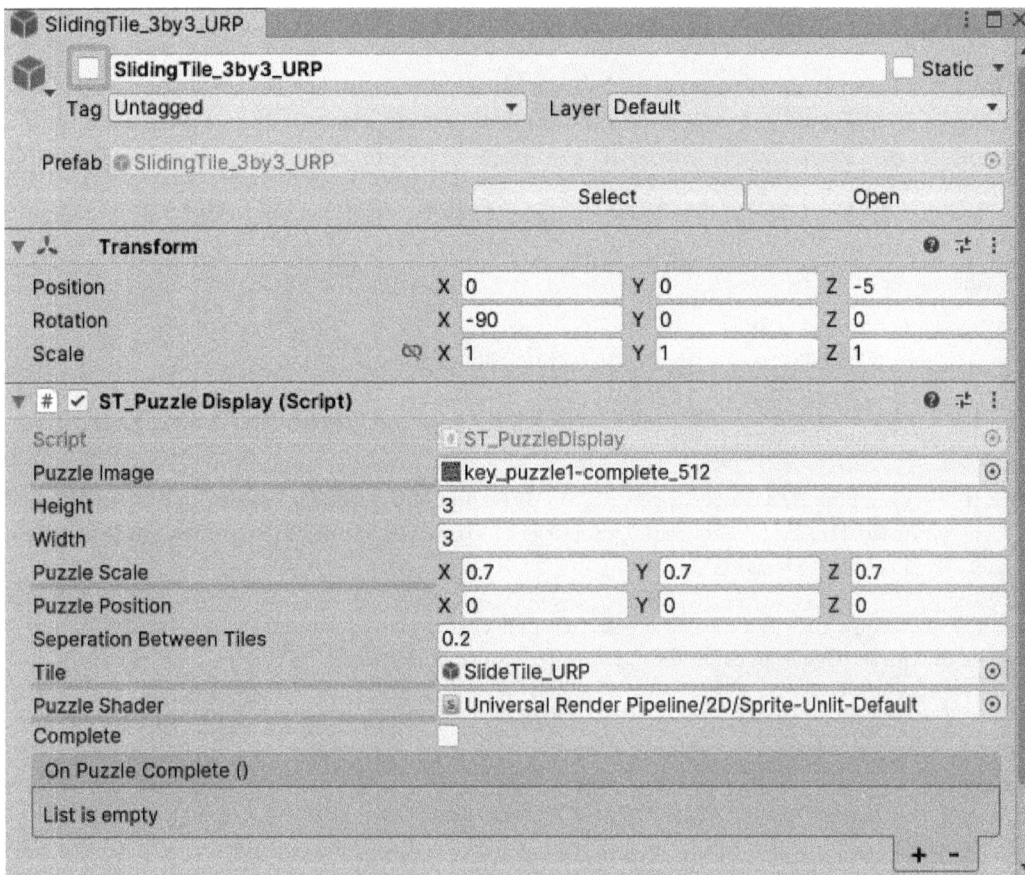

Figure 9.16 – Sliding puzzle configuration

5. Now, disable the `SlidingTile_3by3_URP` object. Yes, you heard that right; we're going to disable the object with the `ST_PuzzleDisplay` component on it because we only want the puzzle to show and do its trick when it's triggered by the player reaching the habitat entryway with all three keys collected.

6. Drag the **Key Puzzle Lock** object from the **Hierarchy** to the `Assets/Prefabs` folder to create the puzzle lock Prefab.

We now have our customized security puzzle lock ready to go! The final setup ensures it shows in place when required at the habitat entryway location.

Adding the puzzle lock to the entryway

We already set up our **Puzzle Trigger** at the location of the habitat entryway. Now, let's add the security puzzle lock to its location so that it shows and is interactable in place at the entryway door. With everything already set up and ready to go, it's a simple two-step process:

1. Position the **Key Puzzle Lock** Prefab in the level nearby, slightly above, or right at the habitat entryway door. You can reference *Figure 9.12* for where I placed it: at the door and just above the player character's head.

2. Show the sliding puzzle from the **Puzzle Trigger** object's `QuestHasCompleted` component's `OnQuestComplete()` event by doing the following:

    I.   Add a new entry for `OnQuestComplete()` as an additional action to `PlayerInput.enabled` by clicking the little plus (+) button.

    II. Assign the `SlidingTile_3by3_URP` child object of **Key Puzzle Lock**. Then, select `GameObject | SetActive(bool)` from the function dropdown, and tick the checkbox to pass `true`.

That's it! You can now fully test the quest life cycle, collect three keys scattered throughout the level to complete the quest, trigger an event for a completed quest, and solve an in-game sliding puzzle. Yay!

All that remains is what we need to do when beating the entryway puzzle lock – that is, winning.

Winning

When the security puzzle lock is solved, we gain entry to the habitat station. The `OnPuzzleComplete()` `UnityEvent` event we just added to the `Sliding Tile Puzzle Game` code gets triggered, so this is our opportunity to do something for a win state.

My game design plan, and plan for this book's projects, is to continue gameplay within the habitat station interior in *Chapter 10*, where things will start having more depth. I don't mean that figuratively; the project the next chapter introduces will be a 3D **first-person shooter** (**FPS**) game. For now, we'll implement a nice cinematic fade to black and *To be continued....* But how do we solve the problem of implementing a fade-out and on-screen text sequence without figuring out a bunch of synchronized linear timing code? Let's see.

Timeline

Timeline is not a new feature in Unity 2022. It's been around since Unity 2017, but I feel it's an underrated core feature that doesn't get nearly the attention it deserves for game developers. For cinematic content creators, however, it was a game changer since Timeline allows the easy creation of linear sequences affecting almost any object in the scene.

Timeline is based on two elements that work in tandem: a **Timeline** file-based asset and a **Playable Director** component. An important thing to note is that a Timeline instance is scene-based.

Additional reading | Unity documentation

Timeline: `https://docs.unity3d.com/Packages/com.unity.timeline@1.8/ manual/index.html`.

We will leverage the pure power of Timeline's simplicity to affect a set of objects linearly to knock out this ending fade-out. Let's start by creating our Timeline instance:

1. First, create a location in our project to store Timeline assets. Create a new folder called `Assets/ Timelines`.

2. Staying within the **Project** window, in the folder we just created, create a new Timeline named `Ending Timeline` by right-clicking and choosing **Create | Timeline**.

3. We can now open the **Timeline** window by double-clicking the **Ending Timeline** asset in the **Project** window.

4. If you're not used to working in a timescale of **Frames**, use the gear icon in the top-right corner of the **Timeline** window to select **Seconds** (as seen in *Figure 9.17*).

5. Finish the Timeline creation by dragging the **Ending Timeline** asset into the **Scene Hierarchy** to make the scene-based instance.

We'll use just two UI widgets to achieve the fade-out effect and title while following these steps:

1. In the **Scene Hierarchy**, select **Create | UI | Panel**. This will add a `Canvas` component to the root of the **Hierarchy**, and **Panel** will be parented to it.

    Just what we need! **Panel** will scale an `Image` component to fill the screen. We don't need an image per se; just the default **Background** sprite will do nicely.

2. Right-click **Panel** and select **UI | Text – TextMeshPro** (select **Import TMP Essentials** if prompted) to add a text widget parented to it, and set these values:

    *   `RectTransform`: **Width** = `400`

    *   `TextMeshPro - Text (UI)`: **Text Input** = `To be continued…`

Lastly, we'll bring these UI widgets into Timeline to set up the sequence. Follow these steps to add the widgets and define their respective sequence:

1. Ensure that **Ending Timeline** is viewable in the **Timeline** window. If the **Timeline** window is not open, you can open it from **Window | Sequencing | Timeline**, then click on the **Endling Timeline** instance in the hierarchy.

2. Drag **Panel** into the **Timeline** window's **Track List** section (the left-hand side column, as seen in *Figure 9.17*). When prompted, click **Add Animation Track**.

    Record keys for the **Panel** `Image` **Color** field to animate the value from 0 (transparent) to 255 (opaque) over 1.5 seconds:

    I.   Click the red *Start recording* circle button on the **Panel (Animator)** track, then set **Image Color** to black (RGB Hexadecimal = 000000) with **Color alpha (A)** set to 0 to set the first keyframe.

    II.   Scrub **Timeline Playhead** to 1.5 seconds, then set **Color alpha (A)** to 255 to set the second keyframe.

    III. Stop recording.

3. Drag **Text (TMP)** into the **Track List** section of the **Timeline** window and select **Add Activation Track** when prompted.

    We want the text to show a heartbeat after the fade has completed, so drag the **Active** clip to the right in the timeline to 1.8 seconds. The GameObject for the **Text (TMP)** binding will only be active within the range of the clip in the timeline and deactivated outside of it. We're done with the text. Easy-peasy.

To preview the results of the timeline sequence in the **Game** view, click the **Play** button or press the spacebar:

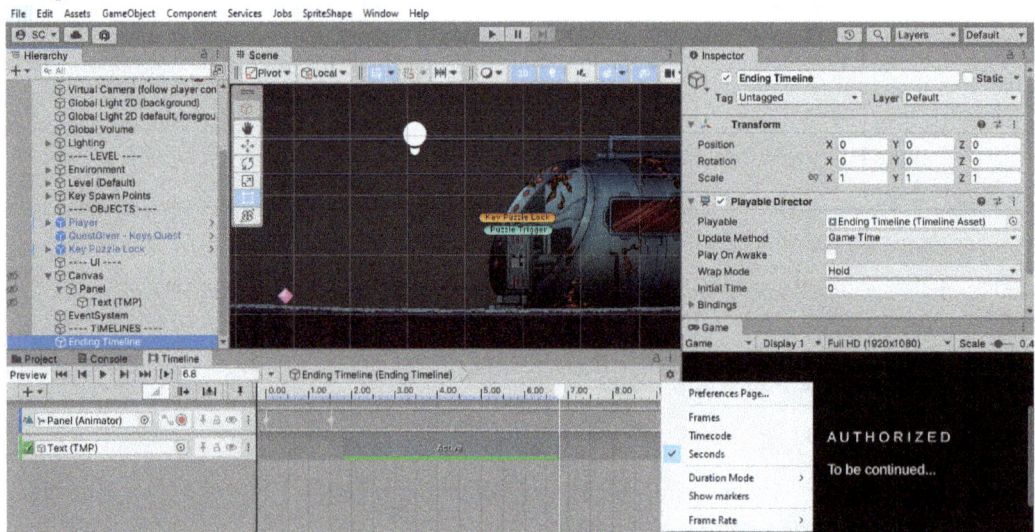

Figure 9.17 – Timeline ending fade out title

If you enter **Play Mode** now, you'll see that our ending fade-out and title will already be shown, and the title will then disappear after a few seconds. We can fix this behavior by changing the settings on the **Ending Timeline** instance – a `PlayableDirector` component was added to the Timeline instance when we dragged it into the scene.

Select the **Ending Timeline** object in the scene and change the `PlayableDirector` component values in the **Inspector** view:

- `PlayOnAwake = false`
- `WrapMode = Hold`

We also want to ensure the UI widgets don't have a visible state at design time in the editor. That way, they are not shown until we trigger the Timeline to play. Make the following changes to the UI widget values in the scene:

1. For the UI **Panel**, set the `Image` **Color** alpha channel value to 0.
2. For the UI **Text (TMP)** object, deactivate the object (untick the checkbox at the top of the **Inspector** view).

Since we already recorded the color values and active state in the Timeline, we can safely set these values in the scene without affecting their sequenced behavior.

The only thing left to do now is to activate our ending sequence when the sliding puzzle is complete. Find the **Key Puzzle Lock** | `SlidingTile_3by3_URP` object in the hierarchy and, on the `ST_PuzzleDisplay` component, add an action to `OnPuzzleComplete:` Runtime Only, **Ending Timeline** | `PlayableDirector.Play()`.

To be continued – or end of Act I, whichever you decide. Either way, this was just the equivalent of a tip-of-the-iceberg introduction to Timeline – it is an extremely powerful cutscenes, cinematics, and gameplay or audio sequences tool!

In this section, you learned how to bring a third-party asset in from the Unity Asset Store, resolved to update the asset for use in Unity 2022, and extended the vendor-provided code with our own to trigger an event. You also received an introduction to Unity's Timeline feature and created a simple but effective cinematic fade to black.

Summary

In this chapter, you learned how to create an efficient and convenient global event system to build out a quest system with components made in a flexible, scalable, and more maintainable way than with directly coupled classes. You also learned how to create a specific quest with unique requirements needing to be met before setting the quest state as completed while also learning how to introduce randomness for collecting required items.

We continued by learning how to import a third-party sliding puzzle asset from the Unity Asset Store and upgrade it for compatibility with Unity 2022 and the URP renderer while also extending upon the code to integrate it into our game code. We finished solving the security puzzle lock and winning the game with a cinematic Timeline sequence for fading to black.

In the next chapter, we'll create a 3D FPS to continue the 2D adventure game directly. We will pick up right where we left off and enter the interior habitat environment, where we will learn about gray boxing to flesh out a playable 3D-level design with **ProBuilder** rapidly. We'll also leverage another asset from the Unity Asset Store – but this time one provided directly by Unity – for our FPS character controller. We'll also look at code reuse by converting some of our existing 2D components for use in 3D.

Part 4:
3D Game Design

In this part, you will learn about 3D games in the FPS genre, 3D art techniques for environment design using Unity's in-Editor 3D toolset, and how to convert 2D systems to 3D systems while covering the Physics API methods in C# code. You will also learn about creating a 3D FPS player character using Unity's Starter Asset and player mechanics by adding reusable components from previous lessons. You will learn about introducing tactics using which players can fend off infected robots in a fully fleshed-out 3D environment with beautiful lighting and optimized performance. You will also learn about sound design, an audio manager, and audio-player components that will help immerse players into the game.

This part includes the following chapters:

- *Chapter 10, Creating a 3D First Person Shooter (FPS)*
- *Chapter 11, Continuing the FPS Game*
- *Chapter 12, Enhancing the FPS Game with Audio*

10

Creating a 3D First Person Shooter (FPS)

In *Chapter 9*, we created a global event system, a quest system, and components that were all loosely coupled and worked together to provide versatile quests for the player to complete. With the systems in place, we made quick work of being able to implement collecting items to notify, update, and satisfy a specific quest requirement. The components created are extensible and reusable for creating any number of quests.

We then imported and fixed the rendering for the **Universal Render Pipeline** (**URP**), and refactored the code of a third-party sliding puzzle asset as an example of how to leverage the Unity Asset Store for our games. We finished by setting up the puzzle in the scene with new artwork and triggering an event when the puzzle was solved that ended the level with a fade-to-black sequence.

This chapter will pick up right where the previous 2D adventure game left off by introducing a 3D FPS game for the habitat interior we just entered.

In this chapter, we're going to cover the following main topics:

- Designing for 3D while continuing the GDD
- Greyboxing a 3D environment with ProBuilder and Prefabs
- Creating an FPS player character with the Unity Starter Asset
- Refactoring environment interactions to 3D API methods
- Code reuse in practice – Adding premade components to the player

By the end of this chapter, you'll be able to design and build a greybox 3D environment from a set of modular parts we'll make right within the Unity Editor, quickly add an FPS character controller, and reuse and refactor 2D code for a 3D project.

Technical requirements

You can download the complete project on GitHub at `https://github.com/PacktPublishing/Unity-2022-by-Example`.

Designing for 3D while continuing the GDD

Level design for a 2D game is more straightforward because players navigate in only two dimensions. In contrast, 3D games involve that extra dimension – depth – in gameplay, contributing to more complex level design. In 2D, the screen space is represented by the **X** and **Y** coordinates. In 3D, the floor plane is represented by **X** and **Z** (depth), with **Y** still being used for the vertical axis – Unity's 3D coordinate system is defined as a **Y-up environment**.

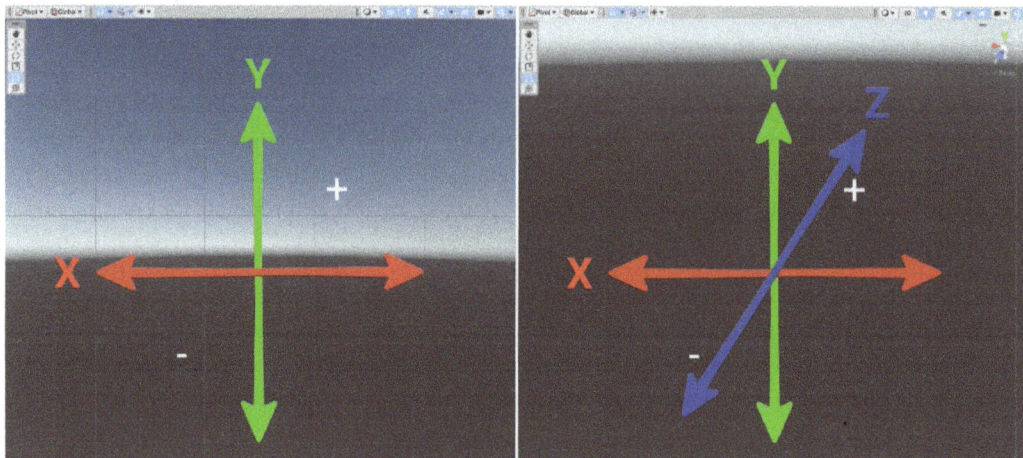

Figure 10.1 – 2D versus 3D coordinates

As we already know from our previous 2D work, the Z axis still exists but is represented straight on with the camera – either in the front or in the back – and only applies in some cases when layering objects in the scene.

Moving on, we'll attempt to simplify the 3D design process using a modular approach, but it all starts with the game design again. Let's review some gameplay changes to the GDD for a new 3D FPS project introduced in this chapter.

Now is our chance to update our blueprint of gameplay mechanics to reflect our evolved vision of the *Outer World* 3D FPS game. We can then ensure that all aspects of the production of the habitat's interior-level design will align with this new experience.

In *Table 10.1*, you can see that I have updated the relevant gameplay sections:

Describe the gameplay, the core loop, and the progression	Make your way to the central control system, peeking around corners and down long corridors to restore operations while dealing with a damaged power suit that must be recharged along the way. Beware of lingering infected maintenance robots!
What is the core game mechanic for the collection game?	With a first-person perspective, the player will navigate the environment, recharge their power suit (health), and shoot the infected maintenance robots.
What systems need to be implemented to support the game mechanics?	The player movement, a weapon with ammo reloading and shooting capabilities, and a health system with pickup (recharging) and damage capabilities.

Table 10.1 – Updated relevant gameplay sections

Next, we'll need to update the player character and enemy backstories to be relevant, as described in *Table 10.2*:

What is the main character's challenge structure?	The habitat station's environmental controls are offline, and the player character's power suit has lost the ability to sustain the player due to damage. The player must seek *recharges* in the station to survive the journey to the central system while battling infected maintenance robots. The player will confront the evil plant entity boss.
Enemy B: Describe the second enemy in the game and how they drive the story. Who is this enemy?	Type: Maintenance Robot Wheeled Backstory: Robot deployed on pre-colonization missions for habitat maintenance and support. Goals: Maintenance, personnel support Skills: Quick charging Weaknesses: Limited mobility
Boss: Viridian Overmind Describe the level boss and how they drive the story Who is this boss?	A sentient plant entity infecting the central control system of the habitat station compels the player to confront the evil alien.

Table 10.2 – Updated character and enemy bios

This is a new game level, so it will also need to be defined:

Describe the environment the game takes place in. What does it look like, who inhabits it, and what are the points of interest?	The game occurs within a habitation station on a remote planet's surface. The habitat is deceptively large with its many small connecting corridors. There are maintenance robots that roam the station carrying out their autonomous duties. Recharge pickups are conveniently placed throughout the station at the corridor intersections.
Describe the game level(s)	The game level is a modular constructed habitat station interior with many corridors and rooms serving different purposes and a central control system room.

Table 10.3 – Environment and level definitions

And lastly, we'll need to update the input control scheme so that it's relevant for functioning in a 3D space:

Define the input/control methods actions	Keyboard: W, A, S, D keys to move, mouse to aim, the left mouse button to shoot the primary weapon, and the E key to interact.
	Game controller: Left-stick/D-pad to move, right-stick to aim, right-trigger or Y to shoot, and button A to interact.

Table 10.4 – Updated input/control methods actions

With these GDD revisions to the *Outer World* game's conceptual design and gameplay mechanics for 3D, we have a foundation for transitioning into the level design phase – ensuring that the level will align with our overall concept and provide the intended player experience.

Greyboxing a 3D environment with ProBuilder and Prefabs

Creating a 3D environment is similar to creating a 2D environment in some ways since you still need to place things meaningfully for the gameplay. Of course, we must consider the additional dimension and use 3D models instead of 2D image assets.

In Unity Hub, please create a new `Unity 2022` project and use the **3D (URP) Core** template as our starting point. We'll continue to use the URP renderer from the previous 2D projects. This is still an excellent choice for 3D since it will be performant on the broadest range of devices, including mobile platforms. With that, even though it sounds like we may be robbing ourselves of some capability, we're not, since URP is also very capable of producing beautiful 3D visuals.

> **Additional reading | Unity documentation**
> URP overview: `https://docs.unity3d.com/Packages/com.unity.render-`
> `pipelines.universal%4014.0/manual/index.html`

We will speed up the design process by using a technique called **greyboxing** (which may also be referred to as **blockout**) – this will allow us to rough in the level design using simple geometry without getting distracted by details. We'll also be able to playtest the level early on and identify any potential issues with the player navigating the environment and working out general playability issues.

Let's start by looking at what we'll be making.

Habitat interior level

Like in earlier chapters, I'll provide an example-level design (subjecting you again to one of my sketches, although a bit more refined this time) that you can follow along with. The level map includes an entry point – where we entered the habitat station at the end of the previous 2D adventure game – and leads to a central control system where we'll confront the evil plant entity!

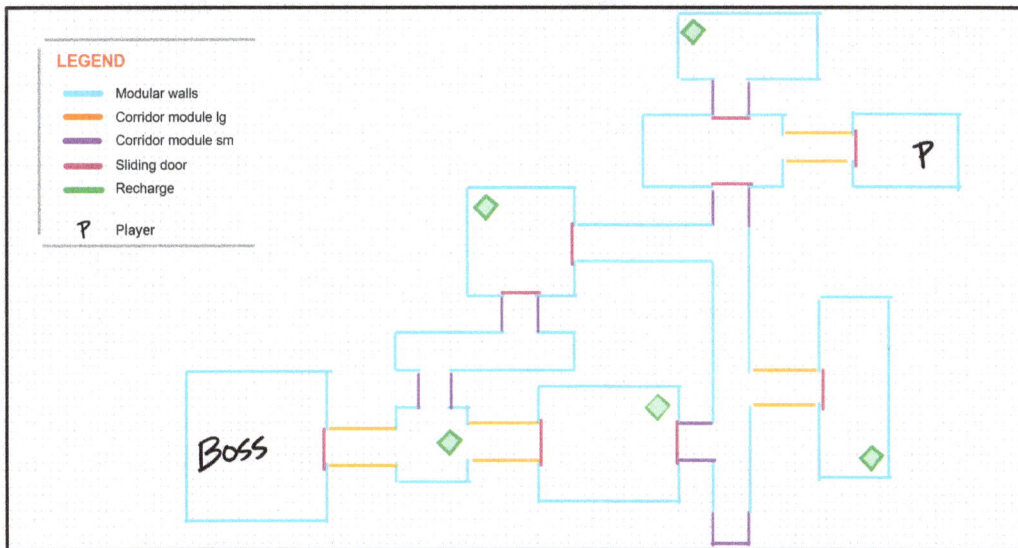

Figure 10.2 – Habitat interior level map sketch

We'll be using a modular approach to designing the level, so following this setup, we'll be able to adapt quickly to different layouts. As you can see, I've added a legend that identifies modules intended to be used – these will be our reusable components serving as the broad strokes to construct and build the level out with. Locations for *recharge* pickups (green), the starting position for the player (**P**), and the location of the boss character are also defined.

Great! Now we have an actionable plan. Next, we can start creating the required 3D modules for our greyboxing kit. If you're thinking, but I have no 3D modeler/artist skills! Don't worry, as this will be a basic introduction – I also believe a game developer should have this as the minimum amount of 3D knowledge.

Installing ProBuilder

Unity provides more built-in tooling that we can use for our greyboxing process in the form of a simple 3D modeling feature called ProBuilder. Like most of Unity's features, we'll install ProBuilder from **Package Manager**. In Unity 2022, tooling is collected into feature sets, so open **Window** | **Package Manager**, select the **3D World Building** feature (including ProBuilder and related tooling), and install it.

After installing the features, we have one additional step to complete the ProBuilder setup. We're using the URP renderer, which requires the supporting Shaders and Materials to be installed:

1. While still in **Package Manager**, find and select **ProBuilder** in the list on the left.

2. On the right side, select **Samples** and click the **Import** button for the **Universal Render Pipeline Support** item.

3. Once the import is complete, go to **Edit** | **Preferences...** | **ProBuilder** and ensure that under **Mesh Settings**, **Material** is set to **ProBuilder Default URP**.

Installing ProBuilder adds a new top menu item named **Tools** that allows opening the **ProBuilder** window and accessing additional functions, exporting, and debugging logging preferences. For our greyboxing, we're going to focus on the object creation process within the **ProBuilder** window, but that's not to say ProBuilder is limited to just creating simple greyboxing objects; it has grown into a full-fledged 3D modeling and texturing product that, depending on the style of your game, could be used for final asset creation too.

> **Additional reading | Unity documentation**
> ProBuilder: `https://docs.unity3d.com/Packages/com.unity.probuilder@5.0/manual/index.html`

Let's open the **ProBuilder** window now by going to **Tools** | **ProBuilder** | **ProBuilder**, and we'll get set up to make the modular building blocks from our sketch's legend.

Modular parts, Prefabs, and Variants

Unlike an exterior environment's level design, where shapes are organic in nature regarding their placement and position in the world, we're making an interior environment for a controlled space. So, we'll be building with modules that will need to *snap* to one another as they are laid out (you know, with the precision an airtight habitat requires). This is generally the approach you'd use for any manufactured structural spaces not found in nature.

Unity handles modular design well by providing some supporting construction options.

Grid snapping

Unity provides a **grid snapping** system that works wonderfully with ProBuilder to simplify our lives for such a task as snapping modules together. Grid snapping allows for precisely positioning a GameObject on X, Y, or Z axis planes, and applies to move, rotate, or scale operations.

Transforming a GameObject – that is, realizing the move, rotate, and scale operations – can also be performed in increments without aligning to predefined gridline snap spacing. Incremental snapping can be performed by holding down the *Ctrl/Cmd* key while using the object transform gizmo in the scene view (**Snap Increment** settings are found on the button to the right of the **Grid Snapping** settings, as seen in *Figure 10.3*).

> **Additional reading | Unity documentation**
>
> Grid snapping: `https://docs.unity3d.com/2022.3/Documentation/Manual/GridSnapping.html`.

Our level sketch is free of any dimensions (I just used a grid for roughly sizing things consistently), so we aren't limited in letting the creativity flow; now that we're making the *precise* modular parts, we need to introduce some unit of measurement to allow for the snapping together of parts. Building modules that need to connect to create our enclosed interior-level design will be pretty tricky if we don't control the size of the details.

To get set up for grid snapping, we first need to activate the **Move** tool and ensure that the handle orientation in the **Tool Settings** overlay is set to **Global** (this is a requirement for the snapping to be enabled in the scene view), as seen in the following screenshot:

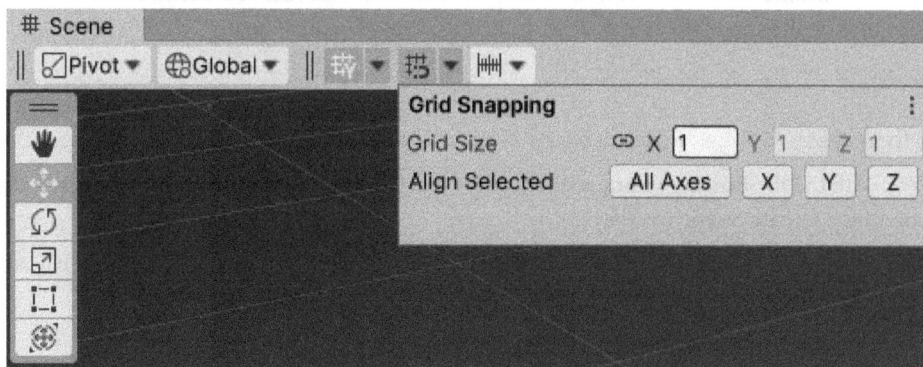

Figure 10.3 – Grid Snapping toolbar

Note that the **Transform** tool snaps the selected GameObject(s) to the grid along the active gizmo axis – using the **Grid Visual** button to the left of the **Grid Snapping** settings, which, as seen in *Figure 10.3*, is currently set to the **Grid Plane** Y-axis.

Let's ensure that the grid snap settings are set up for us to start creating the modular parts for the habitat interior. Referring to *Figure 10.3*, set up the snapping grid with the following steps:

1. Select the **Move** tool from the **Toolbar** overlay in the scene view.

2. Set the handle orientation on the **Tool Settings** overlay to **Global**.

3. Verify **Grid Visual | Grid Plane = Y**.

4. Verify **Grid snapping = enabled**.

5. Set **Grid snapping | Grid size = 2** (default value is **1**).

6. Set **Increment snapping = 0.25**.

I have experience with mechanical drawing and CAD, so I like knowing what dimensions I'm working with – even though, yes, we'll be snapping to the grid units we just defined. ProBuilder's default material also includes a grid texture, but it may still be challenging to establish the required sizes without some visible values. So, let's use the visual dimensions of the objects we'll be building. From the **File** menu, set the **Dimensions** overlay to **Show** by going to **Tools | ProBuilder | Dimensions**.

Grid snapping will be of tremendous help – ensuring consistent sizing of the modular parts – but we can take another step with a simple construction aid.

Construction plane

We will put down a reference plane to assist in creating modules. I've decided the standard module size will be 6 units in length (this is rather arbitrary for how large the base modules are that you'll want to work with; I felt it would be better to work with smaller modules for this design), so we'll base all the dimensions – modules as well as map dimensions – on this standard (and the divisible dimensions of it by the snap unit size of 2; so, hopefully, that all works out).

Let's proceed:

1. In the Assets/Scenes folder, duplicate SampleScene and name it Habitat Interior 1.

2. Now, with the **ProBuilder** window open, select **New Shape** and then click the **Plane** selection icon in the **Create Shape** overlay.

3. Grid snapping and dimensions showing are enabled, so click and drag out a plane that is 6 x 6 units in size.

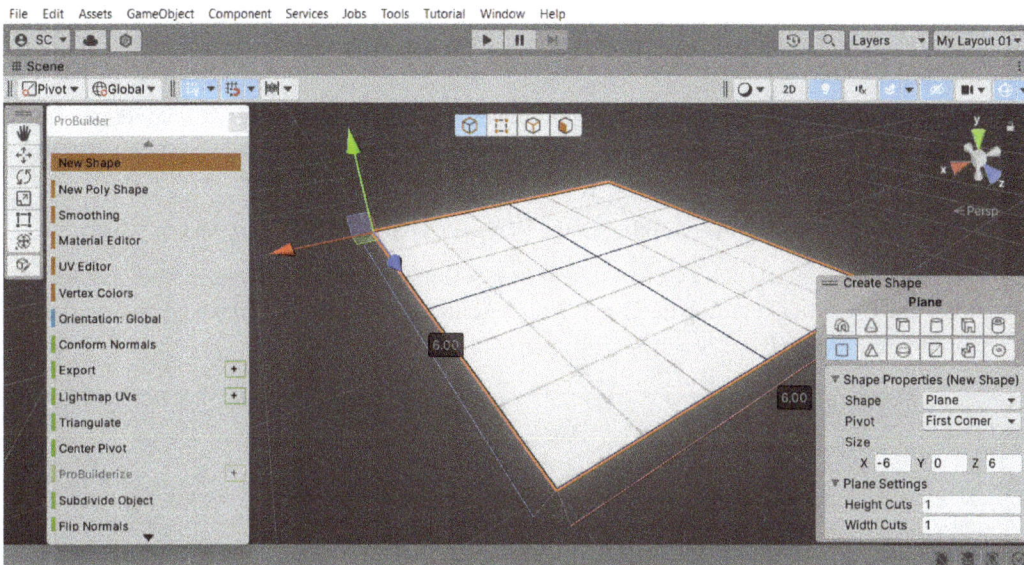

Figure 10.4 – ProBuilder reference plane

4. Click **New Shape** again to stop creating shapes.

5. Now, let's reset the plane position to (0, 0, 0) in world space. With the plane selected, right-click the **Transform** heading in the Inspector and then click **Reset**. This will ensure that all our modular parts' anchor pivots will also be at (0, 0, 0) – giving the correct snap anchor position (such as the location of the move tool gizmo in *Figure 10.4*). I've renamed the plane Module Reference Plane.

Tip | Maximize Unity window

To maximize the currently active window in Unity, as seen with the scene view window in *Figure 10.3*, you can use the keyboard shortcut *Shift + spacebar*.

And now, to make our first modular part!

Making modular parts

Our modular parts will seamlessly snap together in numerous configurations to make the variations we need to create the larger complex space of the habitat station's level design. So, let's start making modular parts for our environment with a wall segment:

1. To make a wall segment, click **New Shape**, but click the **Cube** selection icon this time.

2. Starting at (0, 0, 0) drag out a shape that is the length of the reference plane (6 units) and -2 units in the **Z** direction (away from the reference plane; the blue gizmo axis).

3. Then, move up (in the **Y** axis direction) until the cube is 4 units in height.

As a result of snapping to the grid, it looks like the box is a bit too thick to represent a wall, so let's edit the shape and make it a bit thinner:

1. As seen in *Figure 10.5*, rotate the camera to see the other side of the wall. We can rotate within the scene view by holding down the *Alt/Option* key and then clicking and dragging the mouse pointer.

2. Then click the **Face Selection** option in the **ProBuilder Selection** overlay (*A*).

3. Clicking on the facing side of the wall will now select just the face polygons. While holding down the *Ctrl/Cmd* key, drag the Z-axis handle (blue) of the move transform gizmo until the wall is 1 unit thick (*B*).

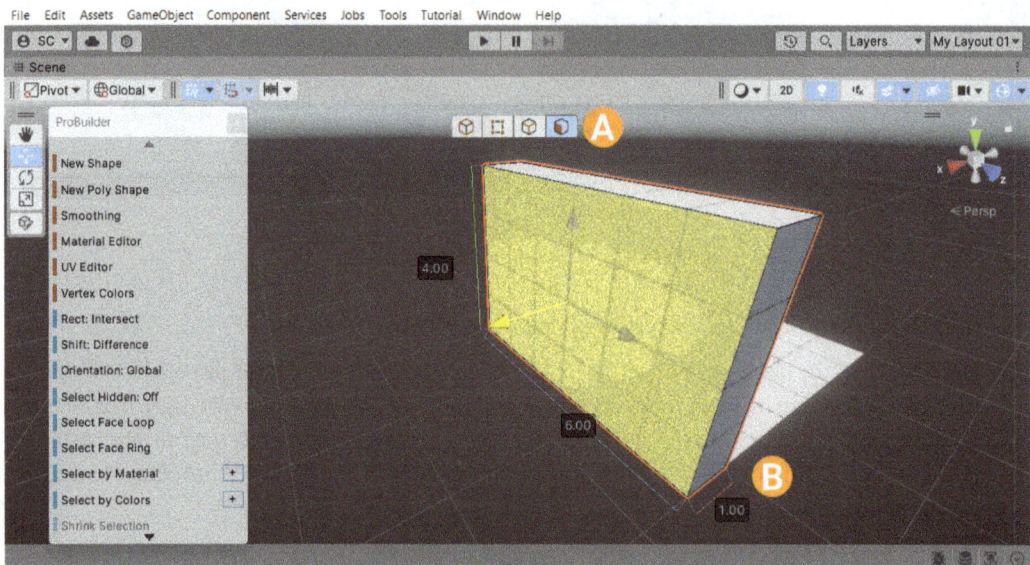

Figure 10.5 – ProBuilder incremental surface snap editing

4. When done editing the face, return to the **Object Selection** mode (*A*).

Greyboxing doesn't necessarily mean 100% void of all details; I generally like to add at least some small elements to my greyboxing kit parts to convey some design aesthetics, so we'll add a few minor details to the wall segment before building up some additional parts.

Let's add some geometry detail to the wall segment:

1. Make a new cube of 2 x 2 units with a height of 4 units. Then, using **Face Selection** and **Incremental Snapping**, edit its size to be 1.5 x 0.25 x 4. Position it at the end of the wall segment to provide an end cap.

2. Duplicate the **End Cap** cube and place it at the other end of the wall segment (like bookends). Note that both ends should be within the reference plane and will therefore intersect the wall segment (this is fine – we want to constrain the parts within our reference size so everything snaps together like Lego bricks).

3. Make a new cube of 6 x 2 units with a height of 2 units, then edit its size and position it to make a sort of connecting bar thing along the length of the wall segment.

The results of our 3D modeling can be seen in the following screenshot:

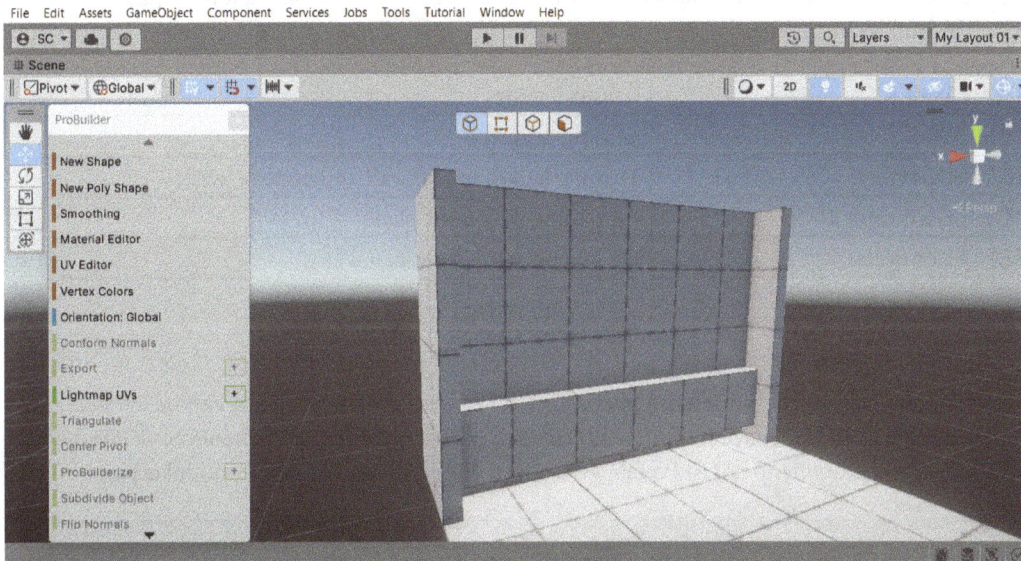

Figure 10.6 – Wall section details added

> **Tip | Lighting**
>
> Note that the shadows on the lighting in the scene have been turned off to focus on the modeling task. You can do this by selecting the **Directional Light** object in the scene hierarchy, then, in the Inspector, find the **Shadows** section and set **Shadow Type** to **No Shadows**.

I would say this wall section is now complete for our first modular kit part. So, we'll, of course, want to make it a Prefab so we can reuse it where needed, and should we need to modify it, all references in the scene will be updated.

Follow these steps to create the wall Prefab:

1. Add a new empty GameObject and reset **Transform** so its position is (0, 0, 0). This will be our parent object and ensure that our modular parts maintain consistency for their snapping point – the individual objects that all make up the module will be transformed by this anchor pivot.

2. Select and drag all the parts of the wall assembly to be a child of the new empty GameObject.

3. Now, rename the parent object to `Wall 1` and drag it to a new `Assets/Prefabs` folder to make the Prefab.

You now have all the tools and knowledge to make the rest of the modular parts needed for our greyboxing kit. That was easy, yay!

What are all the Prefabs we need for our greyboxing kit to produce a complete block out of the level? We can get our answer from the map legend (see *Figure 10.2*). There are only a few to make – minus one for the wall we just made:

(A)	Wall 1	6 units length
(B)	Wall 2	12 units length
(C)	Doorway	18 units length (equal wall, open space, wall)
(D)	Connector 1	6 x 6 units with floor plane
(E)	Connector 2	6 x 12 units with floor plane

Table 10.5 – Greyboxing kit parts

Connector 1 (*D*) will be the same type of connector between rooms so that it can be created as a completed module and not have to be assembled from individual wall parts whenever we need one. It's the same for the *Connector 2* (*E*) Prefab, just longer. Here, we can see all the modular parts:

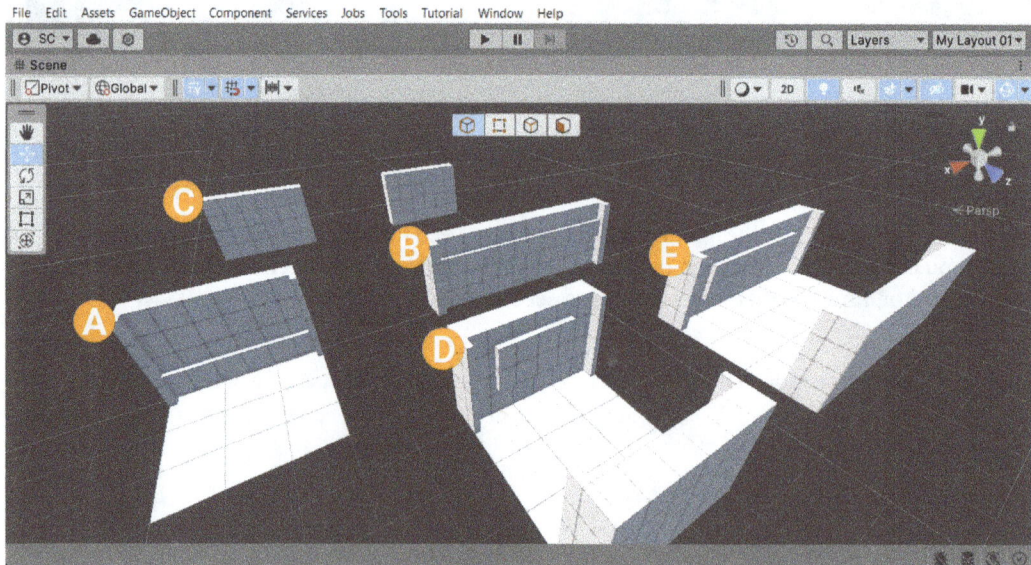

Figure 10.7 – Greyboxing kit Prefabs

Go ahead and create the additional wall, door, and connector modular parts now; I'll wait. Just be sure to always start from the origin of our construction plane to ensure the anchor pivot is at (0, 0, 0) – by creating a new empty GameObject there and parent the cubes to it – so that our kit parts will always snap into place properly. These anchor pivot instructions are based on what's seen in *Figure 10.4* and not as seen in *Figure 10.7*, where the wall parts were moved for better visibility after being created.

With our greyboxing kit complete, we can now start mapping out the level.

Greyboxing the level design

We have our map sketch as a reference, but it would still be too abstract if we start haphazardly dragging in kit Prefabs to try and block it out. To streamline the process a bit more, we can use ProBuilder planes to lay out the map sketch according to our snapping grid. This will ensure that adding the kit modules will be straightforward and quick.

Another drawing aid (we take all the help we can get) is to set the **Grid Visual** setting for **Opacity** to the max value (as seen in *Figure 10.7*) so the grid is fully visible.

Following our sketch and considering the size of our modular kit parts in *Table 10.5*, use the ProBuilder planes to map out the level: walls, doorways, and connectors. ProBuilder will create the planes at **Y, 0** by default, which is right where we want them (floor level). It would then help to have the view locked to a top-down view while we draw out the level. So, using the scene gizmo (top-right corner of the scene view, (*A*) in *Figure 10.8*), click the **Y** (green) handle, click the center cube to set the view to orthographic (no perspective), then click the little lock icon.

Additional reading | Unity documentation

Scene view navigation: `https://docs.unity3d.com/2022.3/Documentation/Manual/SceneViewNavigation.html`

Here is the start of our level map:

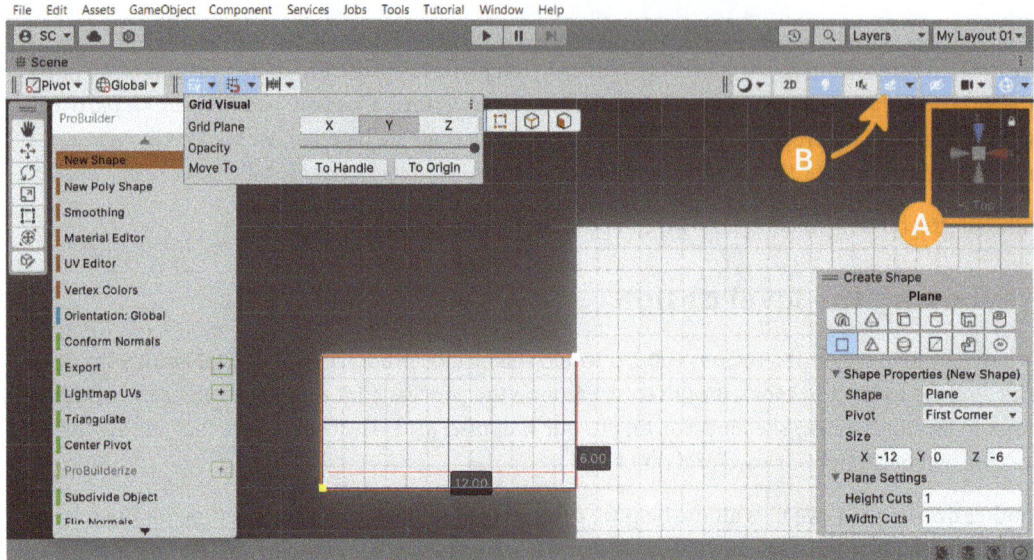

Figure 10.8 – Map layout planes

As you'll notice (and as seen in *Figure 10.8*), the white ProBuilder planes are glowing a bit. This is because the post-processing volume was in the scene from when we duplicated **SampleScene**. If it's a distraction, simply disable the **Global Volume** object in the hierarchy or, toggle off the effects via the **View Options** toolbar (*B*) while mapping out the level.

Here is the completed level map with the entryway at location (*A*), and making our way to the central control system room at location (*B*):

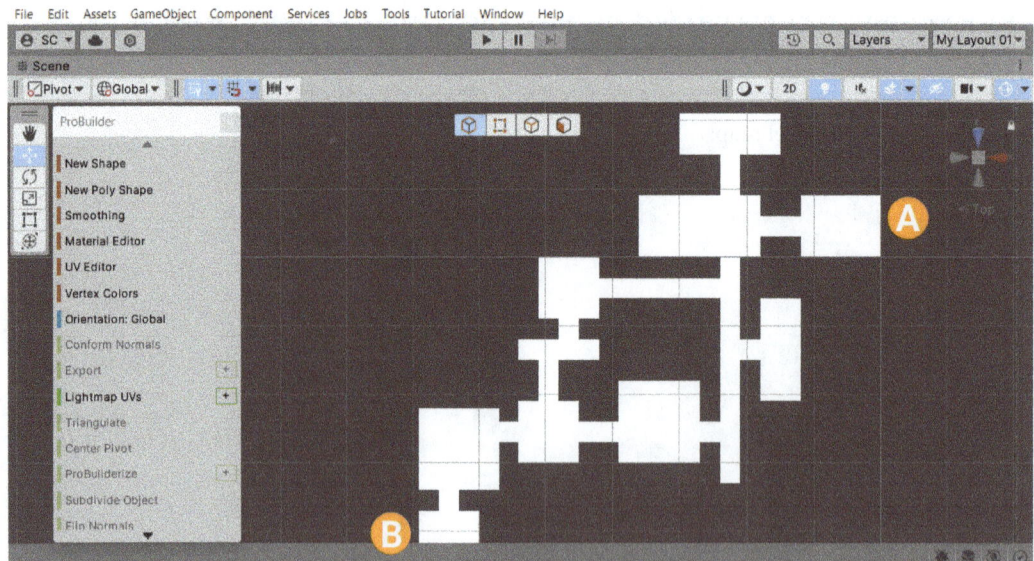

Figure 10.9 – Completed map layout

I'm using the *of greatest importance* definition for the word central here, not the *in the middle of something* definition, just to clear up any possible confusion with the layout of the rooms.

All the hard stuff is done. All we have to do now is to drag our modular kit parts to the edges of the mapped floor planes – with the anchor pivots we ensured are at the origin of the module. We've guaranteed everything should snap in place and to each other. Once you have a modular Prefab in the scene, you can use *Ctrl/Cmd + D* to duplicate it quickly, then move and rotate it into place.

Note that you will have to rotate modules to enclose the perimeter of the rooms and corridors. This is where the anchor pivot also plays a crucial role – the part will rotate at the pivot, ensuring proper snapping will be maintained. You can turn the parts by typing a value in the Inspector or using the **Rotate Transform** tool in the **Toolbar** overlay (hold the *Ctrl/Cmd* key to rotate incrementally and ensure precisely 90 or 180 degrees of rotation).

If we make any mistakes in the size or spacing of the map's rooms and corridors, we'll quickly find out, but rapidly making changes for corrections is trivial with this modular approach. Easy-peasy.

Here are the results of my effort in bringing the interior habitat level together with our modular greyboxing kit:

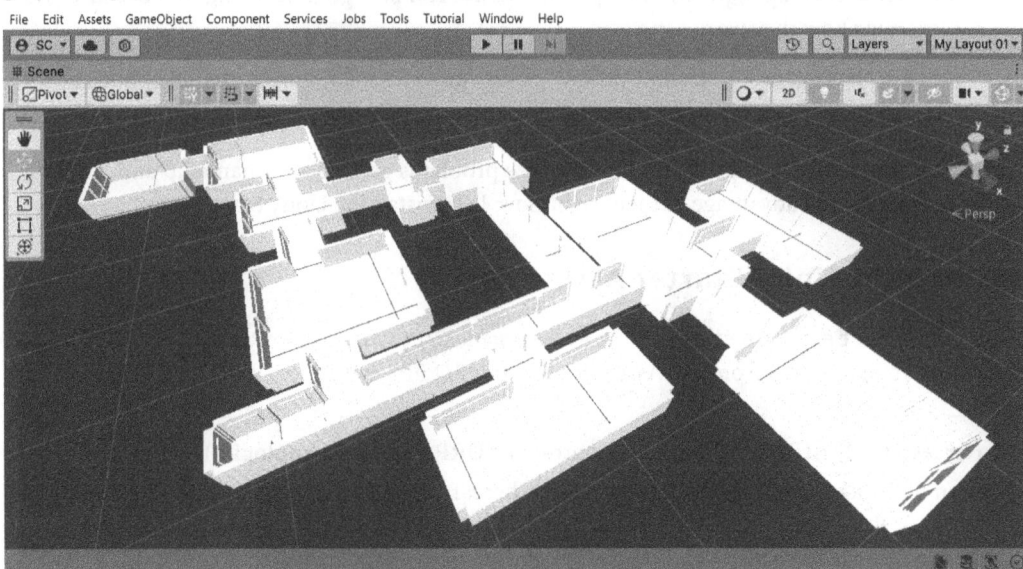

Figure 10.10 – Finished habitat interior greyboxing

In this section, we learned how to block out a 3D environment made of simple ProBuilder model Prefabs used to create a simple greyboxing parts kit. We then discovered how to map out a level design and use Unity's grid snapping system to make fitting everything together quick and easy.

For the next part, we need to playtest what we've built. Rather than coding a player controller from scratch, this time, we'll be leveraging Unity's Starter Assets to construct the player rapidly.

Creating an FPS player character with the Unity Starter Asset

Let's quickly look at some of the general benefits of using prebuilt assets – like Unity's **Starter Asset character controller** – compared to coding ones ourselves:

- They save time and effort – complex systems take time to build and troubleshoot any issues that arise along the way

- They are tested and optimized for performance and generally use best practices – as they're provided by Unity (they know a thing or two about creating components for use in their engine) and widely used by game developers of all levels, these assets will be performant and much less likely to have bugs

- They are built on Unity's `CharacterController` component – these assets are built modularly, ensuring compatibility with other systems (such as camera and combat systems) and assets; they are a great starting foundation for an FPS game

- They offer a learning opportunity – prebuilt assets can be a great learning tool because you can examine how they work

- They have plenty of customization options – Unity provides many customization options out of the box for most player controller uses (especially for an FPS in this case, which is fantastic for us!)

These advantages are pretty significant to minimize problems getting your game up and running quickly; we will take full advantage of them right now. Let's install the Unity Starter Assets.

Installing the Unity Starter Assets

We'll use the `Starter Assets - FirstPerson CharacterController | URP` asset provided (for free) by Unity from the Asset Store for our 3D FPS game.

> **Starter Assets - FirstPerson CharacterController | URP (Unity Technologies)**
> You can find this asset in the Unity Asset Store here: `https://assetstore.unity.com/packages/essentials/starter-assets-first-person-character-controller-urp-196525`.

Installing will be straightforward and similar to how we've already installed assets from the Package Manager. However, since the Starter Assets have some required dependencies, there will be a bit of a hiccup in the process. Don't worry; it's only a minor inconvenience, as you'll soon see by following these steps:

1. Save your scene now (*Ctrl/Cmd + S*) if you haven't already – you'll see why in a few steps.

2. Follow the preceding URL to open the asset in the Unity Asset Store.

3. Sign in if not already signed in.

4. Click the **Add to My Assets** button (accept **Asset Store Terms of Service and EULA**).

5. With the project already open in the Unity Editor, click the **Open in Unity** button that appears at the top of the browser window (you can also always click the **Open in Unity** button on the Asset Store page at any time later) and you'll see the following dialog:

Open Unity Editor?

https://assetstore.unity.com wants to open this application.

☐ Always allow assetstore.unity.com to open links of this type in the associated app

Open Unity Editor   Cancel

Figure 10.11 – Open in Unity dialog

6. Clicking the **Open Unity Editor** button will set the focus to the Unity Editor and open the Package Manager with the **Starter Assets - FirstPerson CharacterController | URP** package already selected (how convenient).

7. Click the **Download** button (top right of the window).

8. When the download has finished, click the **Install** button.

9. The Starter Assets packages require the new **Input System** and **Cinemachine** packages to function (we've covered both already in *Chapter 2* and *Chapter 3*, and I'm glad to see Unity building assets based on them). When you import a Starter Assets package into your project, these two packages will automatically be installed (via the `PackageChecker` script), and we'll be prompted concerning this, as seen in the following screenshot:

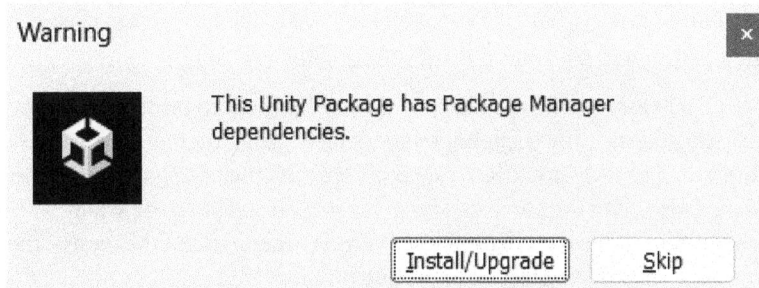

Warning   ×

This Unity Package has Package Manager dependencies.

Install/Upgrade   Skip

Figure 10.12 – Starter Assets dependencies warning

10. Of course, we want to click **Install/Upgrade** so we can actually use the Starter Asset in our project! If you accidentally skip this step, you can still install the new **Input System** package and **Cinemachine** package through the Package Manager. When the **Input System** dependency is installed, however, we'll be presented with this dialog for enabling the native platform backends:

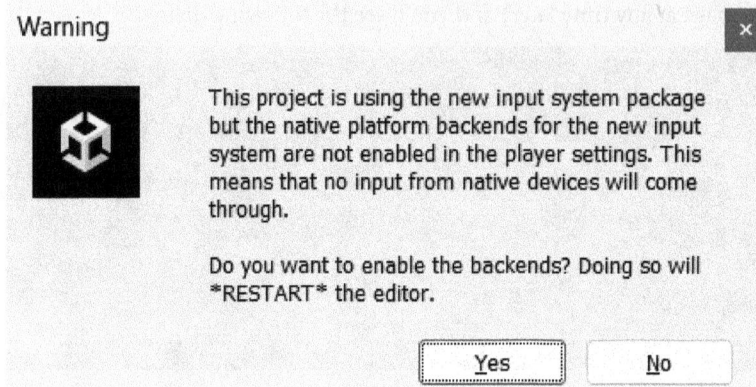

Figure 10.13 – Update native platform backends

11. Click **Yes**, and the Unity Editor will restart (aren't you glad I told you to save your scene in *step 1*?).

12. When the Unity Editor reopens, return to the Package Manager and click the **Install** button again (just a bit annoying but it seems unavoidable with the current dependency requirements). This time, the installation will finish without a hitch.

That takes care of the installation, so let's look at what the Starter Assets provide.

Starter Assets Playground scene

We can quickly test the first-person character controller by opening the provided playground scene. Go to `Assets/StarterAssets/FirstPersonController/Scenes` to find the **Playground** scene. In the **Playground** scene, we have the necessary objects for the first-person character controller, as well as a simple environment we can test the character controller in.

However, you can quickly manipulate the playground environment to provide some additional test geometry with the available parts by snapping them together using their mesh vertexes rather than trying to figure out a snapping grid setup. To snap objects together using their vertexes, select the mesh you want to transform with the **Move** tool active, then press and hold the *V* key to activate vertex snapping. Move to the vertex you want to use – generally a corner point of the mesh – then click and drag to any other vertex on another object. Simple and quite helpful!

> **Additional reading | Unity documentation**
>
> Positioning GameObjects and vertex snapping: `https://docs.unity3d.com/Manual/PositioningGameObjects.html`
>
> Of particular note, you can also use vertex snapping to place an object precisely on the surface of another object. To accomplish this, while already holding down the *V* key, hold down the *Shift + Ctrl/Cmd* keys while moving over the surface you want to snap to.
>
> Don't forget that you can use the *F* key in the scene view window at any time to refocus on the currently selected object – this sets orbiting, panning, and zooming around that object.

Okay, let's look at getting around with the first-person character controller we'll use for our game.

Getting around

First things first, enter Play Mode. Now, you can look around using the mouse, move the player with the *W, A, S, D* keys, hold *Shift* to sprint, and press the *spacebar* to jump. The *move, look, jump,* and *sprint* actions for both keyboard/mouse and gamepad input are established with the `StarterAssets` Input Action Asset (Input Actions) located in the `Assets/StarterAssets/InputSystem` folder.

As seen in the scene **Hierarchy** window in the following screenshot, we have the objects that make up the Starter Assets first-person character controller:

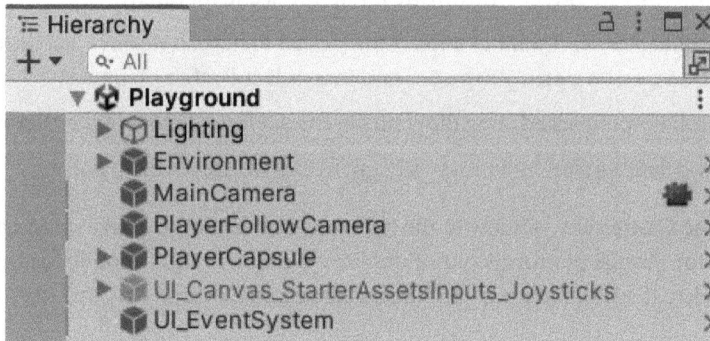

Figure 10.14 – Standard Assets Playground scene Hierarchy

The `PlayerCapsule` object is the primary object (Prefab) holding the modular components that provide the behavior for the first-person character controller. Starting with `PlayerCapsule`, the process flow for the components – from input to transform manipulation – looks like this:

`PlayerCapsule` → [`StarterAssets` (Input Action Asset / Input Actions)] `Player` → `PlayerInput` [SendMessages] → `StarterAssetsInputs` → `FirstPersonController` → `CharacterController`

I encourage you to look into each of the components to gain some familiarity with how things are connected to provide this functionality. We don't actually need to dig into these components to get things working, so I won't be covering it. We will, however, extend upon the base functionality the Starter Assets provide when required.

However, with this brief overview, we should understand what's needed to bring the first-person character controller into our greybox habitat level.

Adding the first-person controller to our level

Now that you've finished having some fun in the Starter Assets `Playground` scene, let's go back to our habitat interior scene to add the player so we can progress further with our game. As we saw in the `Playground` scene hierarchy, a few Prefabs worked together to provide the first-person character controller functionality. So, we'll use these Prefabs too.

Bringing this in is made simple with a provided nested Prefab. Here are the steps we'll follow to bring the controller setup into the scene:

1. In the `Assets/StarterAssets/FirstPersonController/Prefabs` folder, find the `NestedParent_Unpack` Prefab. As the name no doubt indicates, this is a nested Prefab containing everything we need to set up the player.

2. Drag and drop `NestedParent_Unpack` into the scene hierarchy.

3. As the name indicates, we'll want to unpack this nested Prefab, so we have just the child Prefabs. Right-click `NestedParent_Unpack`, and then select **Prefab | Unpack**.

4. With the prefab now unpacked, drag the child objects out into the root of the **Hierarchy** window.

5. You can now delete `NestedParent_Unpack`, and we are ready to go!

As you can see in the **Game** view window in the following screenshot, we have some onscreen input controls intended for mobile platforms out of the box. We won't target mobile for this project, so either disable or delete the `UI_Canvas_StarterAssetsInputs_Joysticks` Prefab object in the hierarchy.

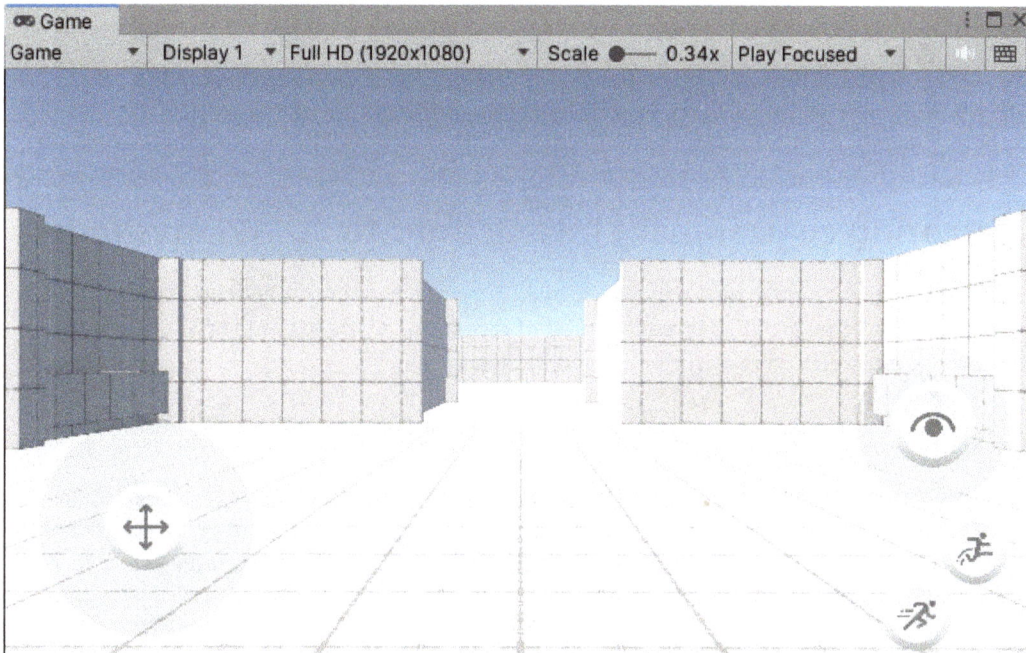

Figure 10.15 – Starter Assets mobile controls

I think it's great that Unity has decided to include mobile support with the Starter Assets; again, this saves us a lot of time if we want to get a mobile game up and running quickly – as well as all the previous benefits noted.

We're now ready to give the level a playtest!

Playtesting the level

To get things started, we'll have to move the `PlayerCapsule` object into a position at the start of our level map, assuming we just entered the habitat station after solving the entryway puzzle. You can see an example of placement in *Figure 10.15*.

The ProBuilder models include **Colliders** for the planes we used to map out the level floor and the cubes we used to make the modular wall Prefabs, so the level is already set! Enter Play Mode and have a look around.

> **Additional reading | Starter Assets documentation**
>
> Additional details for the Starter Assets package can be found in the documentation included with the package. The documentation is in PDF form and can be found in the `Assets/StarterAssets` folder.

In this section, we learned how to rapidly add a first-person character controller to our game's level by leveraging the Unity Starter Assets.

In the next section, we'll revisit some reusable components from the 2D game projects to use in our 3D FPS game project.

Refactoring environment interactions to 3D API methods

From the previous 2D game projects, we have a small library of components built up already, but they are 2D, and we are now in 3D, so it requires a bit of refactoring for us to use them. Let's first revisit the `TriggeredEvent` component we created in *Chapter 4*.

Revisiting the TriggeredEvent component

The Physics 2D API method we used for the `TriggeredEvent` component is `OnTriggerEnter2D()`. Its 3D counterpart simply drops the *2D* part, and so it's just `OnTriggerEnter()` (Unity is 3D under the hood, and it would make sense that only the 2D-specific methods are indicated as such).

Now, considering the above, let's look at how we'll update the `TriggeredEvent` code. Provided that you copied in the `Assets/Scripts/TriggeredEvent.cs` file from the previous 2D project, only a few changes are required. Otherwise, you can first revisit the earlier code in the book or even download the 2D version of the script from the GitHub project repo: `https://github.com/PacktPublishing/Unity-2022-by-Example/tree/main/ch4/Unity%20Project/Assets/Scripts`.

Here are the changes we'll make:

1. The Unity Physics message event from this appears as follows, but we will change it:

    ```
    private void OnTriggerEnter2D(Collider2D collision)
    ```

    And here it is after the change:

    ```
    private void OnTriggerEnter(Collider collision)
    ```

2. Update the `[RequiredComponent]` attribute from the original, which looks like this:

    ```
    [RequireComponent(typeof(Collider2D))]
    ```

    This is what you'll change it to:

    ```
    [RequireComponent(typeof(Collider))]
    ```

    I believe this one is self-explanatory – we don't require the collider to be a `Collider2D` type.

If you save the script now, you'll likely notice we have an error displayed in the Console window with the `Tags` type not being found: **The name 'Tags' does not exist in the current context** (that is, if you didn't already notice your IDE complaining about it with a red squiggly line under `Tags.Player`).

We can resolve this error by copying the `Assets/Scripts/Tags.cs` file again from the 2D project, creating the `Tags.cs` file manually and typing out the constant variable declaration, or using the IDE's refactoring tools to generate `Tags.cs`. Either way, we'll end up with the following:

```
internal class Tags
{
    // Ensure all tags are spelled correctly!
    public const string Player = "Player";
}
```

Concerning our **Player** object having the `Player` tag, the `PlayerCapsule` object from the Unity Starter Assets first-person character controller is already tagged as `Player`. How convenient!

With the `TriggeredEvent` script all sorted, let's implement it now by adding a triggered interaction to the interior habitat level.

Implementing a TriggeredEvent in our level design

We've already seen how we can leverage editor tooling and reusable components to build out environments and add behavior or functionality to our games. This current implementation will be no exception. So, let's see how we can use both ProBuilder and the `TriggeredEvent` component to add a door to the level design that automatically opens when the player gets near it:

1. We'll start by creating a trigger volume using a ProBuilder cube. Using *Figure 10.16* as a reference, draw out a cube just in front of the first doorway of the room the player starts from – and don't forget your *grid snapping*!

2. Once the cube is created, select it, then in the ProBuilder **Tools** window, click **Set Trigger**.

    This will not only change the appearance of the cube – making it a translucent yellow, as seen in *Figure 10.16* – but will also enable **IsTrigger** on the collider. This allows the player to pass into the volume and ensures the `OnTriggerEnter()` message event will be fired.

With that, we have quickly made a trigger volume visible at design time but hidden at runtime – another excellent feature ProBuilder provides.

Let's go ahead and add the refactored `TriggeredEvent` component to our trigger volume ProBuilder cube object now. We can now use the `TriggeredEvent` component to wire up and trigger an animation.

Animating the door opening

Now that we have a `UnityEvent` that can be triggered when the player enters the trigger volume at our doorway, let's add a closed door that will animate open.

First, we'll create the door following these steps:

1. Create a new ProBuilder cube.

2. Use face editing and incremental snapping to make the cube thin (you know, like a door).

3. Place the cube in the doorway to obstruct entry (as if it were closed).

4. Rename the object Door.

5. By assigning a color, let's quickly differentiate the door mesh from our greyboxing kit parts. We can do that using the **Vertex Colors** command in the ProBuilder window. With the **Door** object selected, clicking **Vertex Colors** will open a new window with a default color palette shown. Clicking one of the colors' related **Apply** button will set a new color on the object. Easy-peasy. (Oh… and as seen in *Figure 10.16*, I chose blue.)

We can now add an animation to the door that we can trigger to play from the TriggeredEvent component.

With the **Door** object selected, open the **Animation** window by going to **Window | Animation | Animation** (or *Ctrl/Cmd + 6*) and clicking the **Create** button. This will create an animation asset file (.anim) that we'll name Door-idle and save to a new Assets/Animation folder. In the associated **Animator**, this sets the idle animation as the default state. Since we don't want the door to do anything at all in its idle state, we're done with it. Now, we want to create the animation for opening the door.

Now, follow these steps to create the door-opening animation:

1. In the **Animation** window, with the **Door** object selected, click on the **Animation Clip** dropdown – currently set to **Door-idle** – then select **Create New Clip…**.

2. When prompted, save the new animation clip and name it Door-open.

3. Click the red **Record** button, then scrub the timeline to 0.5 seconds.

4. Using the **Move** tool, move the **Door** into an open position (or change the **Transform** position value in the Inspector for the open direction axis) wide enough for the player to pass through. This will create two keyframes in the timeline: one at the start and one at the current time.

5. Click the **Record** button again now to stop animating.

6. Now, open the **Animator** window by going to **Window | Animation | Animator**. You'll see both the **Door-idle** and **Door-open** nodes (the animations) already there and a **Default State Transition** line from **Entry** to **Door-idle** – precisely what we want. We don't want the door to do anything when we start playing the game.

7. Double-click the **Door-open** animation, then, in the Inspector, uncheck **Loop Time**. We want the door to open and stop playing the animation, not continuously loop back and do it repeatedly.

The last thing to do is wire up `UnityEvent` on the `TriggeredEvent` component:

1. Select the **Cube** trigger volume object in the hierarchy, then in the `TriggeredEvent` component, click the little plus (+) icon to add a new event listener.

2. Drag the **Door** to the **Object** field.

3. In the function selection dropdown, select **Animator > Play (string)**, then type `Door-open` in the provided field – and double-check your spelling! The spelling here must match the name of the animation node in the **Animator**.

The preceding `TriggeredEvent` listener assignment, the **Door-open** animation timeline, the blue **Door** object, and the ProBuilder **Cube** trigger volume object can all be seen in the following screenshot:

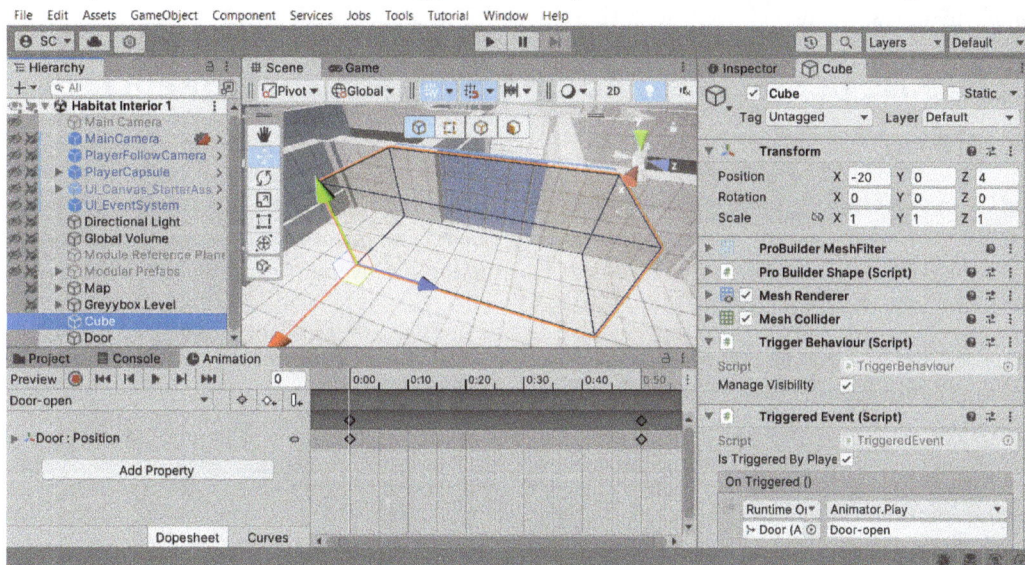

Figure 10.16 – Triggered door open animation

Let's make our triggered/animated door a Prefab as the final step:

1. Parent the **Cube** and **Door** objects in the hierarchy to a new empty GameObject by selecting both objects.

2. Then right-click and choose **Create Empty Parent**.

3. Name the new GameObject `Door_Triggered`.

4. Then drag it from the hierarchy to the `Assets/Prefabs` folder.

Note that you may need to fix the **Door-open** animation since the **Door** object is now a child GameObject whose local position is offset in relation to the parent. If so, select **Door**, open the **Animation** window, and select the **Door-open** animation from the dropdown. You can now manually reset the values for each keyframe in the timeline. Use the **Preview** buttons to play the animation and test to adjust accordingly until resolved (you've got this).

You can now duplicate the `Door_Triggered` Prefab in the scene (*Ctrl*/*Cmd* + *D*), drag it in from the **Project** window, and place it throughout the level where it makes sense for the desired gameplay. For example, having a patrolling enemy behind a closed door is always fun.

In this section, we learned how to easily refactor a 2D API method to reuse some existing code and proceeded to use it and grasped some new learnings for rapidly creating 3D objects with ProBuilder to implement new functionality for the level. Next, we'll look at more code reuse.

Code reuse in practice – Adding premade components to the player

In addition to refactoring some existing code to work in our new project, we can also bring in our existing system code – this could then be considered premade components, ready for use. So, referring back to the GDD now, we will use `HealthSystem` from the 2D adventure game project to decrease and recharge the player's power suit (i.e., health).

If you don't already have the 2D adventure game scripts locally, you can download the project source from the GitHub repo here: `https://github.com/PacktPublishing/Unity-2022-by-Example`

Now, from the 2D adventure game project, copy the following files into the 3D FPS game project (in the same locations):

- From `Assets/Scripts/Systems`
  - `HealthSystem.cs`
- From `Assets/Scripts/Interfaces`
  - `IHeal.cs`
  - `IDamage.cs`
  - `IHaveHealth.cs`

Now that the health system has been added to the project, the first thing to do is refactor any of the 2D API methods and types to the non-2D counterparts just like we did in the previous section, *Refactoring environment interactions to 3D API methods*. This includes changing all references of OnTriggerEnter2D to OnTriggerEnter and Collider2D to Collider.

Now, let's go ahead and add HealthSystem as a component on the **PlayerCapsule** Prefab (again, in the Unity Starter Assets first-person character controller, this is the **Player** object).

To satisfy the GDD requirement of decreasing the player's power suit level (health), let's write some code that extends upon the existing HealthSystem code that will slowly decrease health at an assigned rate. We'll ensure we expose fields in the Inspector for the amount to reduce health by and the rate of decrease.

Constant damage script

Create a new C# script named ConstantDamage in the Assets/Scripts folder and open it for editing. We'll replace the script template code with the following code, which should look familiar since we will implement the IDamage interface. The difference is that we won't be triggering the damage from an object collision (like before with the ProjectileDamage class). Instead, we'll be applying damage directly to the HealthSystem over time.

Let's start with the required implementations for IDamage:

```
using UnityEngine;
using System.Collections;

public class ConstantDamage : MonoBehaviour, IDamage
{
    public LayerMask DamageMask => _damageMask;
    [SerializeField] private LayerMask _damageMask;

    public int DamageAmount => _damageAmount;
    [SerializeField] private int _damageAmount = 1;

    public void DoDamage
        (Collider2D collision, bool isAffected) {}
}
```

A quick breakdown of the code looks like this:

- ConstantDamage : MonoBehaviour, IDamage: This class is inheriting the IDamage interface, which means we'll have to implement the properties and methods defined (the contract): DamageMask, DamageAmount, and DoDamage().

- _damageMask: This encapsulated private variable is serialized and assignable in the Inspector, with the public DamageMask getter (to satisfy the interface contract) referenced by HealthSystem.

  The _damageMask is a LayerMask that determines what object **Layer** can be damaged by this.

- _damageAmount: This encapsulated private variable is serialized and assignable in the Inspector, with the public DamageAmount getter (to satisfy the interface contract) referenced by HealthSystem.TakeDamage().

  This determines the amount of damage that will be applied to the object over time (via **coroutine**).

Now, we'll add the code for applying damage over a set time interval:

```
[SerializeField] private float _damageInterval = 5f;

private void Start()
    => StartCoroutine(ApplyDamageOverTime());

private IEnumerator ApplyDamageOverTime()
{
    var healthSystem = GetComponent<HealthSystem>();
    while (true)
    {
        healthSystem.HandleDamageCollision(null, this);
        yield return new
            WaitForSeconds(_damageInterval);
    }
}
```

And here's a quick breakdown of the preceding code:

- _damageInterval: This is the private member variable that is serialized and assignable in the Inspector.

  It determines the time interval of the damage amount that will be applied to the object (the damage rate).

- StartCoroutine(ApplyDamageOverTime()): This is how we start the coroutine in the Start() Unity message event, which will begin applying damage to the player right away.

- `ApplyDamageOverTime()`: This is our `IEnumerator` coroutine method that uses `while (true)` to loop indefinitely, calling `HandleDamageCollision()` then delaying 5 seconds via `WaitForSeconds()` before looping again.

  - `HandleDamageCollision(null...`: Here, we're passing in a value of `null` for the `Collider` parameter since, well, we don't have a collision occurring. We'll just have to ensure we do a **null check** against this value before acting on it in `DoDamage()`.

- `DoDamage()`: This declaration is required for the interface implementation (to satisfy the contract), but we won't use it now.

Whew, that makes it seem like we have to do so much for only a dozen or so lines of code! And that's the point. We introduce good architecture, patterns, and practices into our projects to accomplish more with less.

Add `ConstantDamage` as a component on the `PlayerCapsule` object (this will be a sibling component of `HealthSystem`) – so that the player's health constantly decreases. Before we can test it out, assign the following values to the fields in the Inspector:

- **DamageMask** = `Player` (you may have to add `Player` to the project's layers list first, using the **Layers** dropdown menu on the right-side area of the Toolbar; then, also assign the `PlayerCapsule` object in the hierarchy to the `Player` layer)
- **DamageAmount** = 1
- **DamageInterval** = 5 (seconds)

Now you can save (*Ctrl/Cmd + S*) and enter **Play Mode** to test it all out. While playtesting, don't forget that you can adjust the `PlayerCapsule FirstPersonController` values under the `Player` header until the movement feels right in the level.

It will be difficult to see if anything is happening right now because we don't have any visual indicators in the scene, and the variable holding the current health value isn't visible in the Inspector. No worries, Unity has a solution for this.

Inspector Debug

The Unity Inspector has a Debug mode that will peek into our components' code and expose private member variable fields as read-only values. While still in Play Mode, go ahead and switch the Inspector into Debug mode by clicking the vertical ellipsis (⋮) **More Items** menu (also known as the kabab menu) button (*A*), then click **Debug** to switch from **Normal** mode, as seen in the following screenshot.

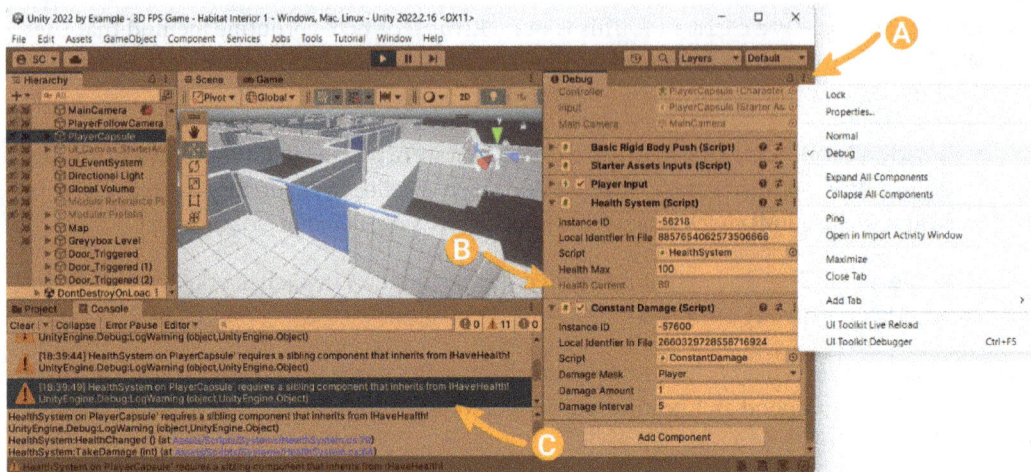

Figure 10.17 - Applying damage to the player

And now, we can see the private `_healthCurrent` variable field as **Health Current** (*B*) on the `HealthSystem` component, which, according to our `ConstantDamage` assignments, will decrease `_healthCurrent` by a value of 1 every 5 seconds. Be sure to switch the Inspector back to Normal Mode when you've finished inspecting the value (you won't generally need to see all the additional debug information).

Additional reading | Unity documentation

Learn more about working in the Inspector: `https://docs.unity3d.com/Manual/InspectorOptions.html`.

Also seen in *Figure 10.17*, the **Console** warning message is (*C*): **HealthSystem on PlayerCapsule' requires a sibling component that inherits from IHaveHealth!**

This console output results from a null check we had implemented in the original `HealthSystem` code when the `HealthChanged()` method is called. It just ensures that an object – that implements the `IHaveHealth` interface – exists before calling `HealthChanged()` or `Died()` methods on it, as follows:

```
private void HealthChanged()
{
    if (_objectWithHealth == null)
    {
        Debug.LogWarning(…
        return;
    …
```

That takes care of the constantly decreasing health of the player. Now, how do we recharge this Kryk'zylx power suit?

Recharging aka healing

We're going to create a pickup that will recharge (heal) the player, and this might be the quickest we're going to add new functionality to our game yet! This is because we will again use previously written code and a similar approach to things we've already made.

From the previous 2D adventure game project, copy in the `PickupHeal.cs` and `Destroyer.cs` scripts. We're going to place pickups for recharging/healing the player throughout the level – which is something we've made before, so maybe try making this on your own first? The only difference is that now we will create a ProBuilder object as a 3D pickup compared to the 2D sprite pickup we previously made.

Don't worry; we're still going to follow the below steps that you can check your work against. So, here we go. To make a 3D object as a healing pickup, follow these steps:

1. Make a ProBuilder object to use as a pickup (a cube, sphere, cone, etc.). With grid snapping enabled, you'll want to use incremental snapping (hold *Ctrl/Cmd*) to make it a suitable size compared to the player.

2. On the new object you've made for the pickup, set **MeshCollider** as **Convex** and **IsTrigger** as **Enabled** (don't use the **SetTrigger** option from ProBuilder because we want the mesh to be visible as is).

3. Use ProBuilder's **Vertex Colors** to set a new color on the pickup object.

4. Add both `PickupHeal` and `Destroyer` as components on the ProBuilder object.

   I.   Assign `Player` as the **HealMask** value.

   II.   Add `Destroyer.DestroyMe` as a listener for `OnHealEvent` (ensure it's set to **Runtime Only** and not **Off**).

5. Rename the object to `Heal Pickup` (or `Recharge Pickup`) and drag it from the hierarchy to the `Assets/Prefabs` folder to make it a reusable Prefab.

How'd you do on your own? With our healing pickup Prefab completed, we can now scatter them throughout the level in a strategic way to create good, challenging gameplay for the player. Enter Play Mode now and verify in the Inspector **Debug mode** that the current health value is restored when a pickup is collected. As always, you'll continue to playtest – or better, enlist others to playtest as you observe – and balance values to provide the best player experience!

In this section, we saw how to reuse existing components to quickly add health system functionality to the player and extend functionality with a new component that applies constant damage. We also learned how to view and debug private member component values in the Inspector, then finished with more code reuse to rapidly create a 3D pickup object that recharges the player's health.

Summary

This chapter first covered a quick addition to our GDD to add specifics for our new 3D FPS game, including considerations for 3D-level design. We used the updated GDD to block out the interior habitat level environment from a greyboxing kit of modular parts that snap together and draw out a level map, all using Unity ProBuilder.

We continued by learning how to rapidly add an FPS player character to our game by leveraging the Unity Starter Assets first-person character controller. We also rapidly added behavior to our player with code reuse for the health system and healing pickups. Code reuse and refactoring 2D components to the 3D API method counterparts also allowed us to quickly make an animated door that opens when the player triggers.

In the next chapter, we'll continue with the environment-level design by replacing the modular greybox kit Prefabs with art assets and decorating with more Unity-provided tooling: Polybrush and Decals. We will then finalize the environment design process by baking the 3D environment lighting to ensure high-performance rendering while comparing it with real-time lighting to evaluate the pros and cons of each.

11
Continuing the FPS Game

In *Chapter 10*, we made some updates to our GDD for our new 3D FPS game. We added specific details about the level design, which helped us to create the habitat's interior-level environment using a graybooxing kit of modular parts that *snap* together. We also used Unity **ProBuilder** to draw out a level map.

We finished by learning how to quickly add an FPS player character to our game by utilizing the **Unity Starter Assets First Person Character Controller**. We added the health system and healing pickups to engage the player's abilities by reusing previous systems and code. We also refactored to 3D API method counterparts to make an animated door that opens when the player triggers it.

In this chapter, we will update and finalize the 3D environment with polished assets, scatter objects to increase realism, and incorporate wear-and-tear effects. Additionally, we will improve the lighting to create an immersive experience for the player and ensure optimal game performance.

In this chapter, we're going to cover the following main topics:

- Decorating the 3D environment
- Immersing the player using Polybrush and Decals
- Lighting design – Probes, Decals, light baking, and performance

By the end of this chapter, you will have the skills to enhance a 3D environment by replacing **Prefabs** and materials, increase realism and immerse the player by decorating the environment using Unity artist tools such as **Polybrush** and **Decals**, and bake lighting along with techniques to overcome limitations for lighting and shadows of dynamic objects in the scene. Additionally, you will gain knowledge on performance considerations related to lighting.

Technical requirements

To follow along with the same artwork that was created for the project in this book, download the assets from GitHub at `https://github.com/PacktPublishing/Unity-2022-by-Example/tree/main/ch11/Art-Assets`.

You can download the complete project from GitHub at `https://github.com/PacktPublishing/Unity-2022-by-Example`.

To follow along with your own 3D artwork, you'll need to create similar artwork using 3D modeling and texturing software (for example, Blender, Maya, 3ds Max, Cinema 4D, ZBrush, Silo, Substance 3D Painter, Quixel Mixer, or 3DCoat).

Decorating the 3D environment

Our 3D FPS game is well on its way to becoming something of a real game, but it must graduate from its graybox environment first if we want it to be attractive to potential players. This section will explore Unity 2022 features for replacing the graybox kit module's Prefabs with polished and textured 3D mesh Prefabs that properly represent our habitat's interior environment.

We'll not only replace the existing graybox Prefabs but also introduce new assets to decorate the environment to make it feel more complete and lived-in. This will be a combination of both 3D models and textures that we'll apply in different ways, again, using Unity's artist feature tooling (specific to **Universal RP**).

Here is an example of what our interior habitat environment – which has gone through the processes outlined in this chapter – will look like:

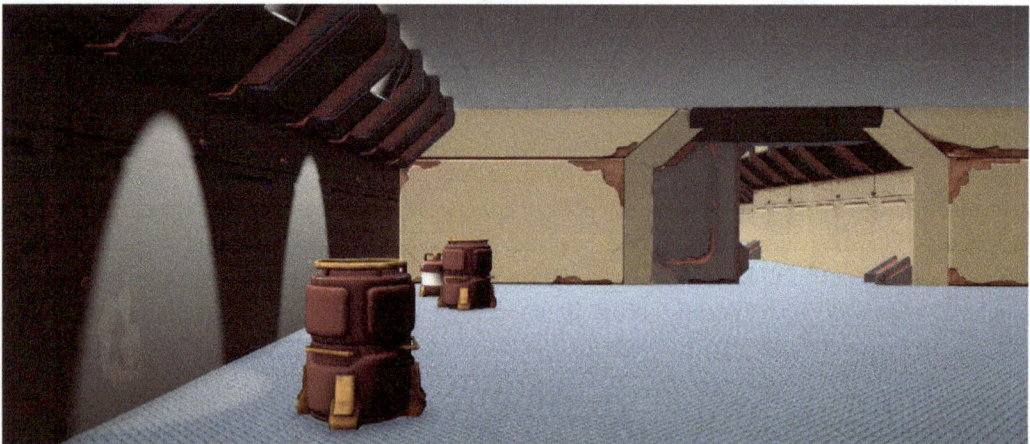

Figure 11.1 – Habitat interior scene lighting

Our first order of business is to replace those boring graybox Prefabs. Having the environment look correct will help with the following steps as we decorate and detail it.

Updating and replacing Prefabs

Replacing Prefabs is more of a structured approach to dealing with the art assets; since we're working with fixed modular parts, the art is in creating polished 3D assets. In this case, we're going to be using 3D assets produced by my friends over at Polypix Studios (`https://polypixstudios.com/`).

> **Note of gratitude**
>
> Miguel Dumars was kind enough to provide select assets from their *Neon Street* and *Stylized Scifi Modular Corridor* Unreal kits for use in this Unity project, and I'm very excited to be working with these assets!

Polypix Studios has permitted the use of the provided game art for learning purposes only; commercial use is strictly prohibited. Polypix Studio's portfolios can be viewed on ArtStation at `https://www.artstation.com/polypixcc` and Unreal Marketplace at `https://www.unrealengine.com/marketplace/en-US/profile/Polypix+Studios`.

To get started, let's import the Polypix artwork.

Importing and reviewing the assets

You should be a pro in the process of importing assets into Unity projects by now, so we won't waste any time detailing each little step. Download the 3D art assets file, `3DArtwork.zip`, from this book's GitHub repository (link in the *Technical requirements* section) to a temporary directory, then import the `.unitypackage` files into your current 3D FPS project.

In *Figure 11.2*, we can see the new modular kit parts in the **Polypix Modular Kit** scene provided in the package:

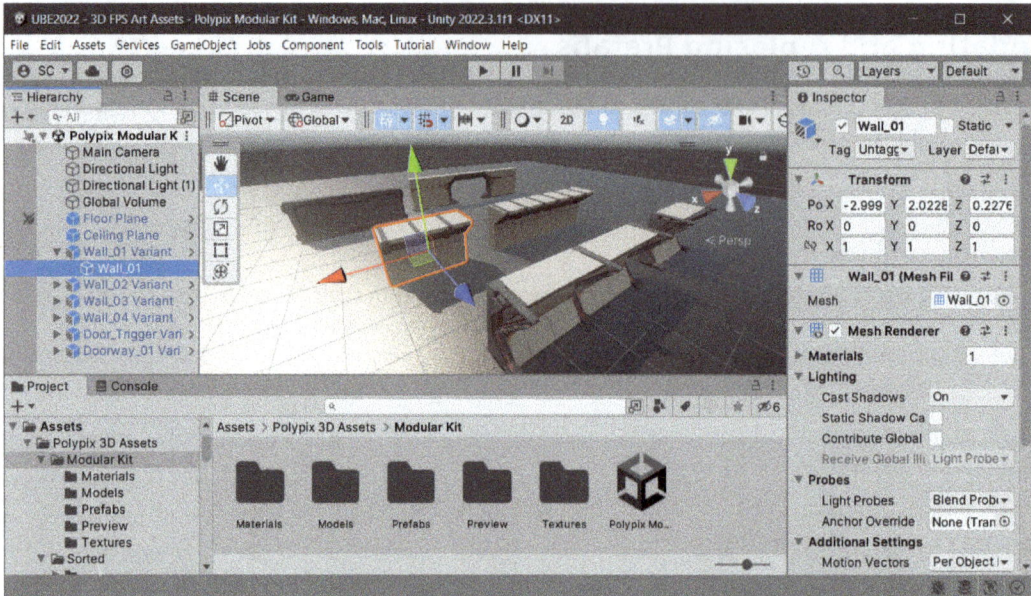

Figure 11.2 – Polypix Studios modular kit

Since this is the first time we're working with third-party 3D assets, let's have a look at the files that were imported into the `Assets/Polypix 3D Assets/Modular Kit` folder (also shown in *Figure 11.2*):

- **Materials**: These materials are applied and shared across 3D models of the same category (walls, doorways, and so on). The same material is shared across several models to keep things more optimized as this reduces draw calls for the renderer (that is, it's less work, resulting in higher FPS).

- **Models**: These are the optimized polygon meshes (vertices, edges, and faces) that make up the 3D geometry representing the shape of the objects.

- **Prefabs** (yes, I know you know what Prefabs are): These are the Polypix models with the materials applied and colliders added as finished objects we can use directly in our game.

- **Textures**: Textures are the image files that are mapped onto the 3D geometry via assignment to the material that's applied to 3D geometry to give it color and details.

We'll mainly be concerned with the Prefabs since we'll use them to replace the initial graybox kit modular parts we made to build the level. In the `/Prefabs` folder, we have the same graybox wall assets but all textured and fancy.

Now, let's see how we can rapidly replace the modular graybox kit Prefabs with this new artwork.

Replacing Prefab instances

With the replacement art assets imported into the project, we can go ahead and start replacing Prefabs. Unity 2022 has introduced some new Prefab workflow features and **Search**, which will greatly help us in this endeavor.

> **Additional reading | Unity Blog**
>
> Unity's new Prefab workflow features go far beyond simple Prefab replacement, transfer overrides, reconnecting Prefabs, and inspecting Prefab Variant relationships. You can read about these additional features in the following Unity Blog article: *What's new for Prefabs in 2022.2?* `https://blog.unity.com/engine-platform/prefabs-whats-new-2022-2`.
>
> Unity Blog is a fantastic resource for learning content! I highly recommend regularly consuming Unity's blog articles to rapidly broaden your knowledge and elevate your understanding of what Unity is capable of.

We will use the following process as much as possible to mass-replace the Prefabs in our level. However, we'll no doubt still have to make some manual adjustments to some of the layout's finer details – due to some changes in the art direction (that may or may not have been my fault).

Some of the Prefabs that we've added behavior to already are examples of Prefabs that we must manually update and cannot simply replace in the scene with new art, but we'll still be able to merely replace the art within the Prefab because we've maintained keeping the *graphics* as separate child objects in the Prefab (as you may recall, we've been using this practice throughout this book thus far, and this is just another example of the benefit of being consistent in terms of our approach).

Let's get to it and make our first Prefab replacement. We'll replace the `Wall 1` graybox Prefab in the scene with the `Wall 1` Prefab from the Polypix `/Prefabs` folder. We'll use both the new **Search** feature as well as the **Prefab Replace** workflow by following these steps:

1. Open **Search** by going to **Window | Search | New Window**, clicking the **Open in Search** button at the top of the **Hierarchy** window, or pressing *Ctrl/Cmd + K*.

2. In the **Search Unity** text box at the top of the **Search** window, type `Wall 1` with quotes to search explicitly for this string (removing the quotes will search for all occurrences; note that **Search** is not case sensitive).

3. Select the **Hierarchy** tab to filter the search to only objects in the open scene **Hierarchy**.

4. Now, select all the items in the results list by clicking on the first item, then scroll to the bottom and hold *Shift* and click the last item.

5.   Right-click and click **Select** (or hit the *Enter* key):

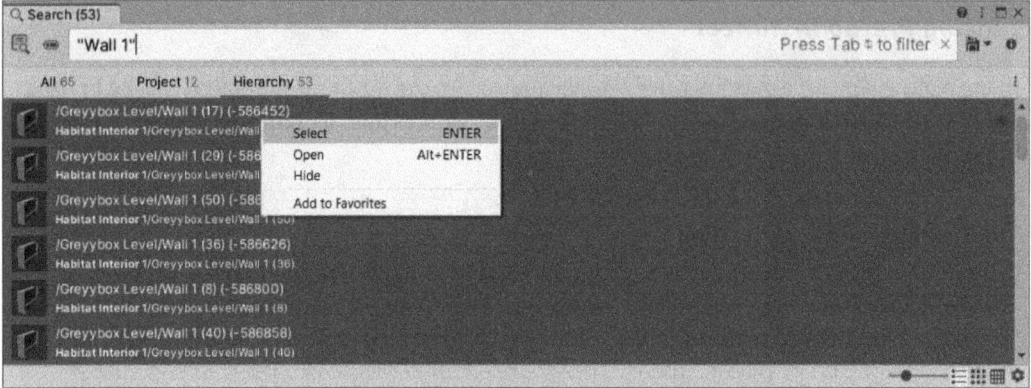

Figure 11.3 – Unity Search for easy scene Prefab selection

With all of the `Wall 1` Prefabs in the scene now selected, we can easily replace them with **Prefab Replace** – which is only a two-step process:

1.   Right-click one of the selected `Wall 1` Prefabs in the **Hierarchy** area, then select **Prefab | Replace…**:

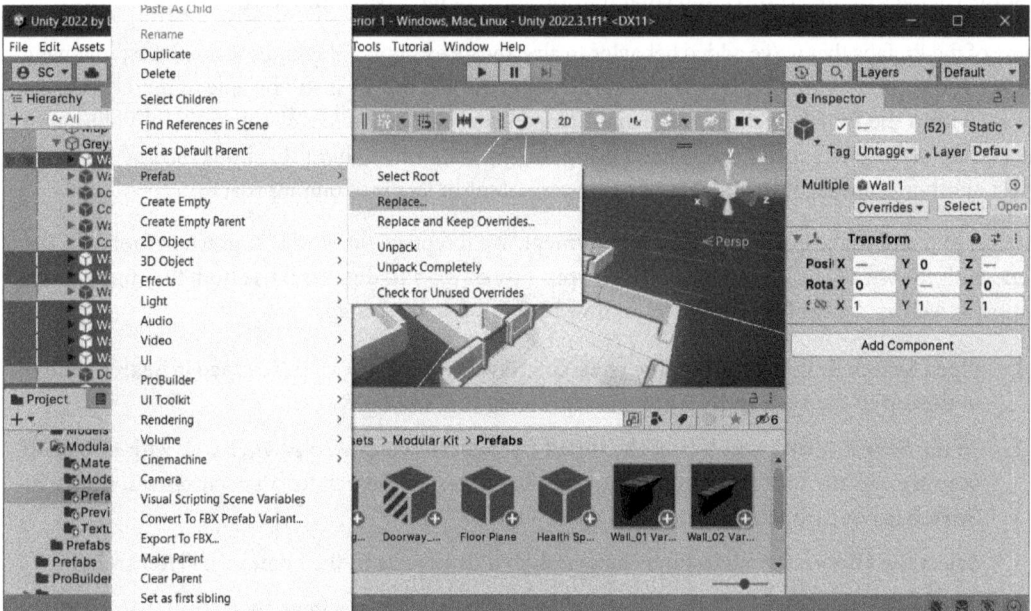

Figure 11.4 – Prefab – replace instance selection

2. Select the replacement Prefab in the `Assets/Polypix 3D Assets/Modular Kit/` `Prefabs` folder (using *Table 11.1* as a reference):

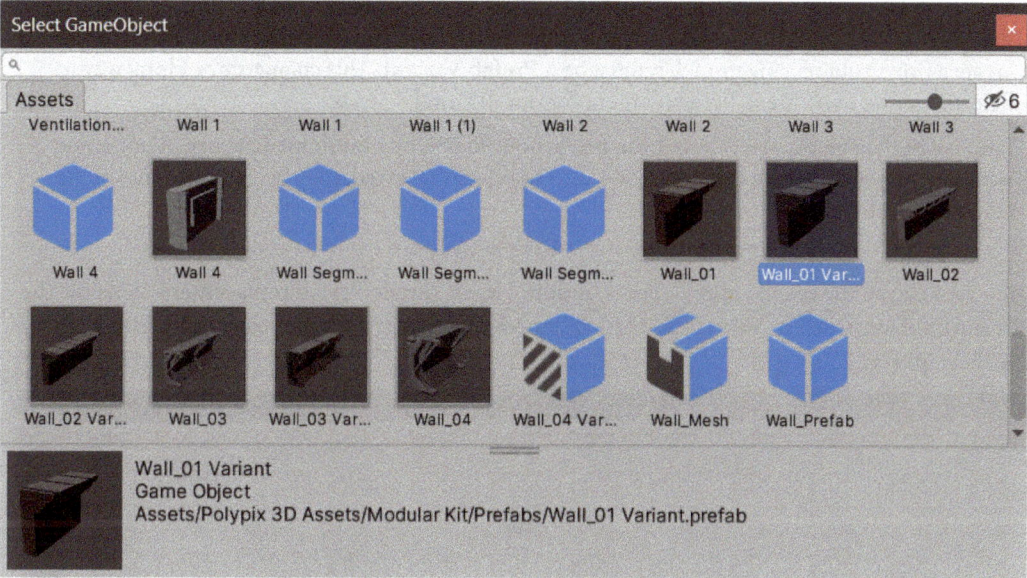

Figure 11.5 – Selecting the replacement Prefab

You can use the replacements indicated in the following table for the graybox kit Prefab and corresponding Polypix imported Prefab as a guide:

Greybox Prefab	Polypix Prefab
Wall 1	Wall_01 Variant
Wall 2	Wall_02 Variant
Wall 4	Wall_04 Variant
Doorway 1	Doorway_01 Variant

Table 11.1 – Replacement graybox Prefabs

> **New to Unity 2022 – 3D models are Prefabs**
>
> You might be wondering why the Polypix replacement Prefabs listed in *Table 11.1* all have *Variant* in their naming. Well, that's because you can no longer make original Prefabs from 3D models (for example, FBX files). Unity now imports 3D models as **Model Prefabs**, so when you create a Prefab out of a 3D model, it must be a **Prefab Variant**. This change goes a long way to help ensure we don't break Prefab content for our 3D model assets.
>
> Since all of the Polypix assets have been pre-produced in Unity for immediate use in our scene with the proper materials, textures, scale, and so on and re-saved, they become variants of the original 3D Model Prefabs.

Figure 11.6 shows the results of my efforts in doing these simple Prefab replacement steps for the walls and doorway listed in *Table 11.1*. The job of updating the artwork is already more than halfway done! Easy-peasy:

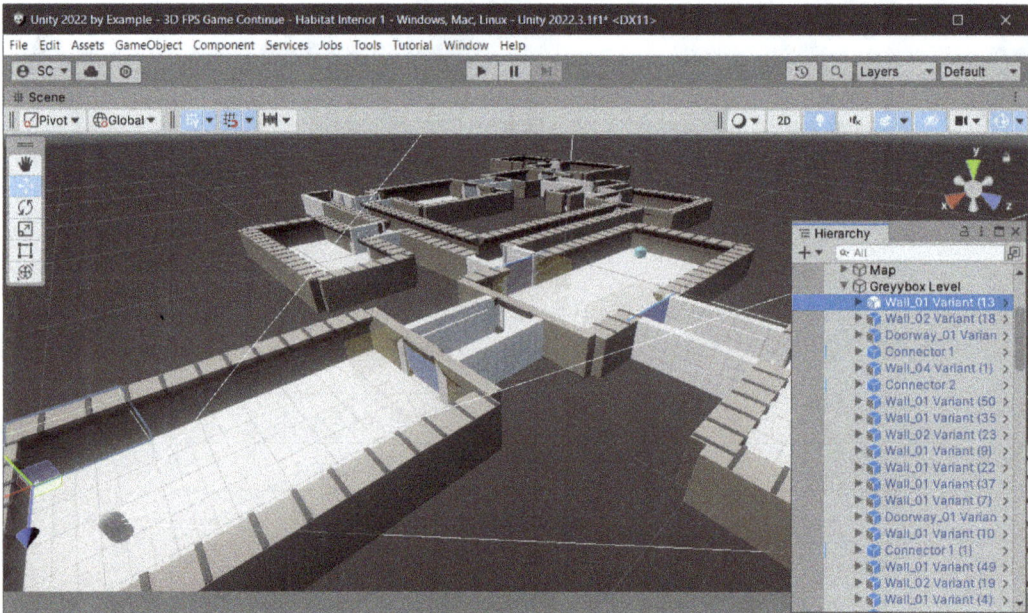

Figure 11.6 – Wall Prefabs replaced in the scene

> **Tip**
>
> Note that in my level design, I found that I could not use all the replacements directly with the finished artwork and instead changed some of the wall modules out for different ones. So, your mileage may vary as well. It's all part of the design process; much like coding, it's an iterative process.

Just repeat the preceding steps to replace all the static graybox modular parts – *static* means these are structural and don't contain any behavior, interactivity, or animation. We'll address the `Door_Triggered` Prefab next since it has animation and won't be directly replaced.

These additional updates on the graybox kit parts will occur in **Prefab Edit Mode**, but we'll still simply replace the *graphics* with the updated artwork.

Updating the existing modular Prefabs

Some Prefabs, such as `Doorway_Trigger`, cannot just be replaced because they have the behavior we added in the form of a collider-triggered animation. Using this as an example workflow, let's update `Doorway_Trigger` by performing the following steps:

1. Open the `Door_Triggered` Prefab in **Prefab Edit Mode** for direct editing.

2. Add the `Door_Trigger` **Variant** Prefab from the Polypix / `Prefabs` folder.

3. Reconnect the animation for the new `Sliding_Door_01` object that replaced the `Door` object's animation – ensure you disable or delete the `Door` object since we're replacing it and don't need it anymore.

4. Update `TriggeredEvent.OnTriggered() Animator.Play(string)` to reference the new `Sliding_Door_01` object that replaced the `Door` object so that our `Door-open` animation will still be functional on the new door mesh. Remember that the string value for the `Animator.Play()` function is the same as the animation's name: *Door-open*.

Figure 11.7 illustrates the refactored setup:

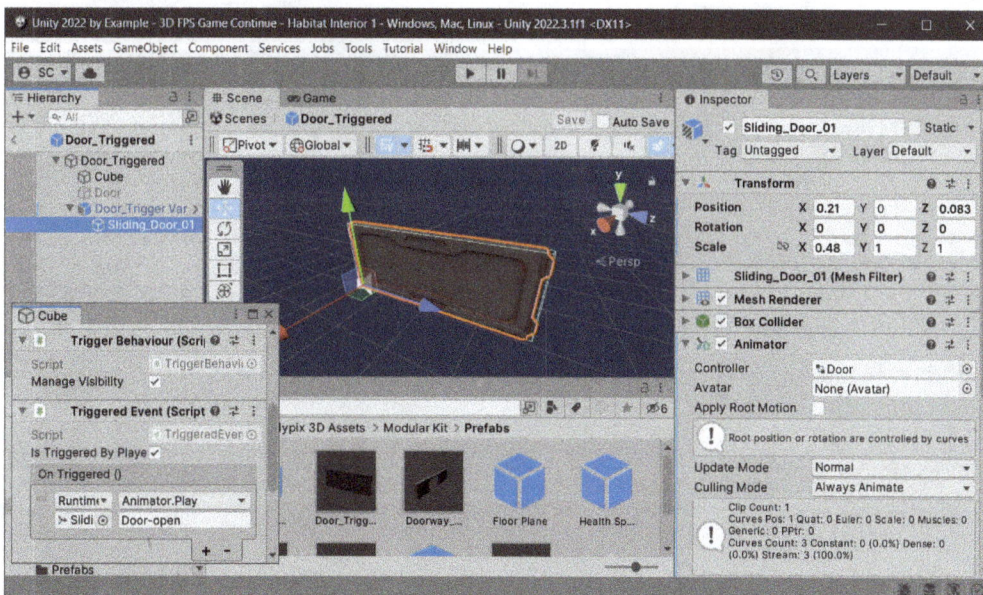

Figure 11.7 – Updating the door trigger Prefab

I believe this leaves only the **connector** Prefabs that need updating. The Polypix assets don't include direct replacement Prefabs for connectors, so just make them yourself! You can create a new connector Prefab by duplicating the original graybox asset and then replacing the child graphics – as **nested Prefabs** – to create the replacement connector Prefabs.

I won't provide the steps for this process since you have all the required knowledge to accomplish this now. Still, if you get stuck, you can always reference the new connector Prefabs in the completed project files in this book's GitHub repository. You've got this!

With that, we have replacement Prefabs from Polypix and some new material textures we can apply.

Applying new materials

While we're doing some art updates, we also have some new materials we can use to update our floor planes' appearance (our **ProBuilder** map layout). Within the Polypix /Materials folder, we have a **FloorPlate** material, so let's assign that to our map to have it become our textured floor.

In *Figure 11.8*, you can see that I have renamed the **Map** root's GameObject in the **Hierarchy** area to **Floors**. With **Floors** selected, pressing *Shift + H* will enter **Isolation View** (*A*) – temporarily hiding all the other objects in the **Hierarchy** area, which makes it easy for us to assign a new material to the selected object by clicking and dragging the **FloorPlate** material (*B*) onto a plane in the **Scene** view area:

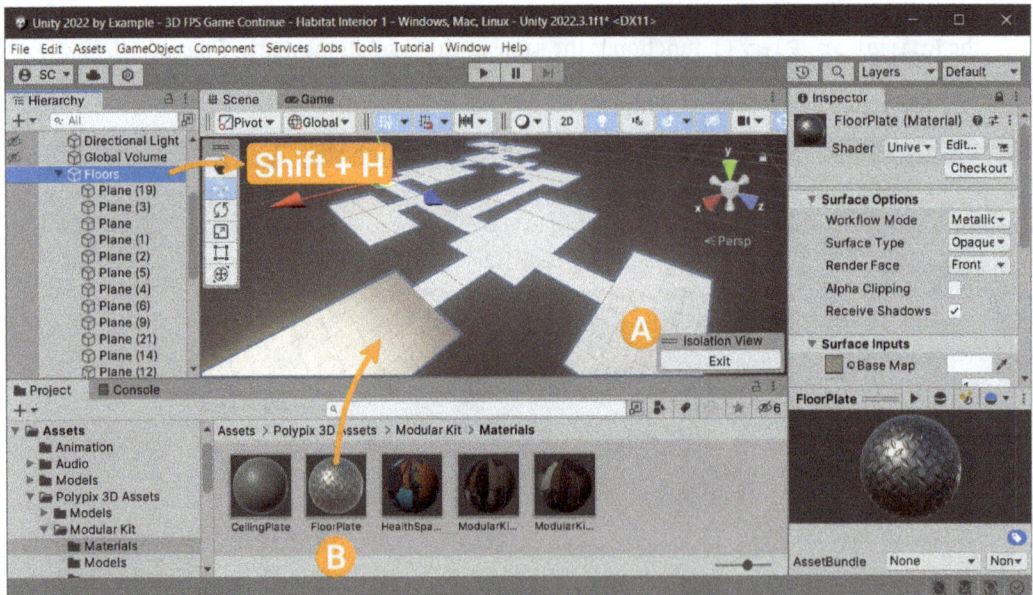

Figure 11.8 – Updating the floor material

Toggle visibility shortcut

Note that you can also use *Alt/Option + Shift + A* to toggle the visibility of the currently selected object.

Alternatively, since these planes are ProBuilder objects, we can open the **ProBuilder Window** area (**Tools | ProBuilder | ProBuilder Window**), select **Material Editor** from the toolbar, and assign the **FloorPlate** material to the next available slot (see *Figure 11.9*):

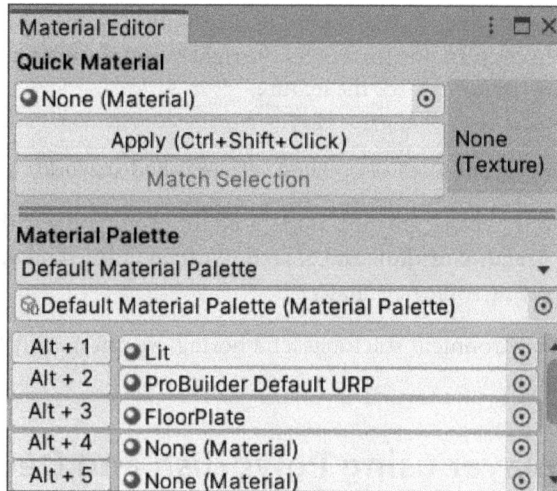

Figure 11.9 – ProBuilder Material Editor assignment

With the material assigned, you can now select all the floor planes in the **Hierarchy** area and press the shortcut key assigned to the **FloorPlate** material in one shot – in our case, that's *Alt + 3*. Simple!

While we're at it with the floor, let's add a ceiling to our habitat interior too! This is quite simple for us now:

1.   Use **ProBuilder** and create a new plane.

2.   Scale it to contain all of the rooms in our level, then position its height at the top of the wall Prefabs – just where a ceiling should be (in my level, this comes out to a Y transform position value of 4).

3.   From the Polypix /Materials folder, assign the **CeilingPlate** material to the ceiling plane.

4.   Lastly, for the texture to be visible from the bottom of the plane, not the top, as is the default and not what we want since our player is below the ceiling and looking up at it, in the **ProBuilder** window, click **Flip Normals**.

Surface normals

Normals – or **surface normals** – depict the direction a polygon's surface faces (that is, its visible side):

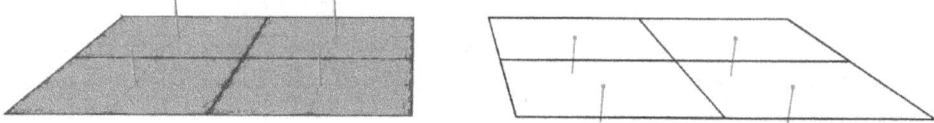

The surface on the left has its normals (the orange lines) pointing up, facing the camera so that we can see the texture. In contrast, the surface on the right has its normals pointing down and away from the camera, so we cannot see the texture.

And with that, we have our graybox level all updated with polished artwork!

In this section, we saw how to replace graybox assets used for prototyping with imported polished artwork using Unity's new **Prefab Workflow** and **Search**. We then finished up the interior 3D habitat environment refresh by applying some new materials.

Next, since our 3D habitat environment still looks a bit boring, we can quickly fix that by scattering some Prefabs about the place.

Immersing the player using Polybrush and Decals

In contrast to the previous structured method of using fixed, modular parts such as Prefab replacements, a more freeform method offers a different perspective to the artistic approach. So, instead of relying on prefabricated objects, we'll create dynamic, spontaneous pieces from scratch (well, randomization) using Polybrush.

Painting objects with Polybrush

Polybrush offers us an unconstrained way to decorate the environment and randomly scatter Prefabs for their placement – so, not modular at all. Considering this, we're only going to cover the **Scatter Objects** feature of Polybrush, but know that it has more capabilities than that.

Additional reading | Unity documentation

You can read more about Polybrush here: `https://unity.com/features/polybrush`.

While preparing for writing, I experienced some issues using Polybrush to paint objects onto ProBuilder'ized meshes, so while waiting for Unity to respond to my bug report concerning this, we'll continue with a workaround, which is simply to enable **Use Additional Vertex Streams** in **Preferences | Polybrush**. Remember this should you run into any problems with ProBuilder objects!

As an alternative, you can mesh'ify the ProBuilder objects by selecting them and going to **Tools |
ProBuilder | Actions | Strip ProBuilder Scripts in Selection**, but this also means you will no longer
be able to edit the mesh with ProBuilder.

Now that we've learned how to avoid issues while painting on specific object meshes, let's paint!

Painting/scattering objects

We'll need some objects to scatter throughout our environment, so if you haven't already, import
`Sci_Fi_Assets.unitypackage` from the GitHub artwork download. You'll want to do it
now, if not earlier.

Now, while referencing *Figure 11.10*, follow these steps to set up the Prefab painting:

1. Open **Polybrush** by going to **Tools | Polybrush | Polybrush Window**.

2. Click the **Scatter Prefabs on meshes** button (*A* in *Figure 11.10*).

3. Drag the `Barrels` and `Trashcan` Prefabs from the `Assets/Polypix 3D Assets/
    Prefabs` folder from the **Project** window to the **Current Palette** section (*B* in *Figure 11.10*):

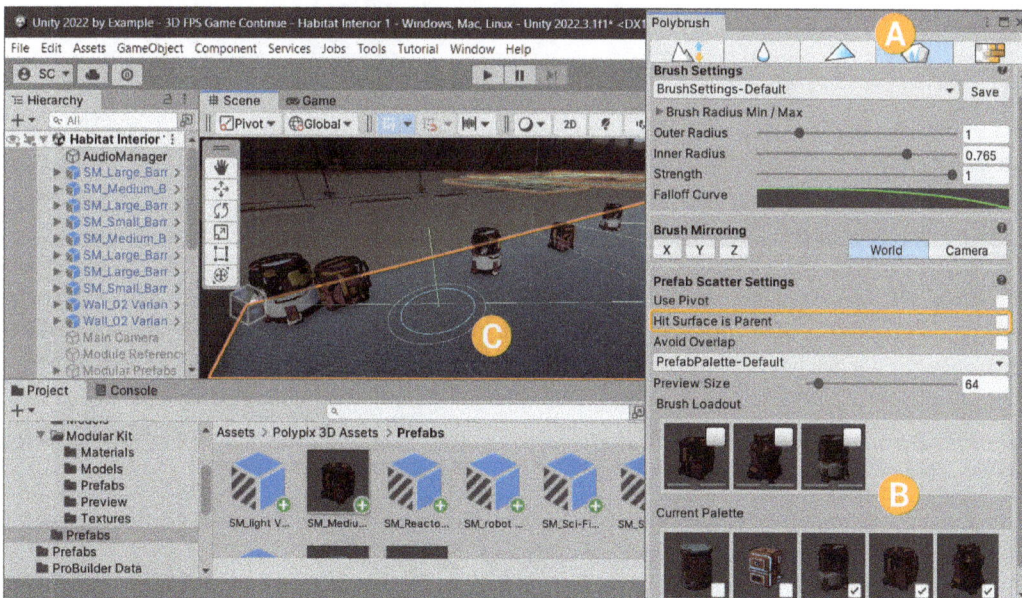

Figure 11.10 – Polybrush Prefab painting setup

By default, **Hit Surface is Parent** is enabled in the **Polybrush Window** area, making all the painted
Prefabs a child of the mesh being painted on. In *Figure 11.10*, you can see that I have disabled **Hit
Surface is Parent** so that all the painted objects can be collected into a single parent **GameObject** in
the **Hierarchy** area (this is my preference, but you may prefer to have the painted objects as children

of the mesh they were painted on). If you're keeping this option disabled, once you've painted objects throughout the environment, ensure you select all the painted objects from the root of the **Hierarchy** area and move them into a new, empty *Scattered Objects* root GameObject.

We can now proceed to decorate the environment by using the assets that were added to the **Current Palette** area (wearing our *interior designer* hat?). First, add the Prefabs for our brush from the **Current Palette** area to the **Brush Loadout** selection (as seen in *B* in *Figure 11.10*) by checking off the item (selecting the item in the **Current Palette** area will also produce a drop-down section below it where you can adjust settings for how the item should be considered for painting).

Painting Prefab objects in the environment is now as simple as hovering the mouse cursor over the floor mesh in the **Scene** view and holding the left mouse button down while dragging in the area you want to have objects scattered – as seen in *C* in *Figure 11.10*. Holding down *Ctrl/Cmd* while painting will work as an eraser and remove objects should you dislike any placements. Have fun!

> **Tip**
> Polybrush works on vertical surfaces too, so you can scatter Prefabs on the walls (just know that you may need to re-orient or position the Prefab's anchor for the objects to paint as intended; I have provided an example Prefab for this with `Exhaust_01 Pb`).

Scattering objects to break up the environment is one part of environment design that can lead to the better immersion of the player, while another is also breaking up visually repeating patterns. We can solve the latter with another Unity artist tool that recently got some love in version 2022.

Surface story with Decals

As my friend Miguel, over at Polypix Studios, would say, *surface story matters*, and I wholeheartedly agree! Unless you're going for a super-pristine futuristic-clean look, you'll want to ensure your environments are grounded in the real world. This means that these environments will communicate their history of usage – wear and tear – through their texture details. Let's say our habitat station's maintenance robots are not very good at janitorial duties, so the environment should be indicated as such. Plus, you know that there should be some indication of the plant entity's effect on the environment!

This is a great opportunity to either refer back to the GDD or expand upon the details that support the story we're telling through the environmental design.

As designers (or developers wearing the hat of a designer), we don't need much in the way of 3D artistic skills to leverage surface story in our environment design. We can use tooling that Unity provides (surprised?) to add surface details to the environment, and we do that with **Decals**.

Additional reading | Unity documentation

You can read more about the Decal renderer feature (URP) here: `https://docs.unity3d.com/Packages/com.unity.render-pipelines.universal%4014.0/manual/renderer-feature-decal.html`.

Before we can use Decals in our project, we have to enable the feature.

Enabling the Decal feature in URP

To enable **Decals** in the **Universal RP** settings, while referring to *Figure 11.11*, follow these steps:

1. In the `Assets/Settings` folder, select the **URP-HighFidelity-Renderer** asset (*A* in *Figure 11.11*). Note that this is the default setting unless you've changed the default **Quality** value in the **Project Settings** area.

2. Click the **Add Renderer Feature** button (*B* in *Figure 11.11*) in the **Inspector** area, then select **Decal** from the list:

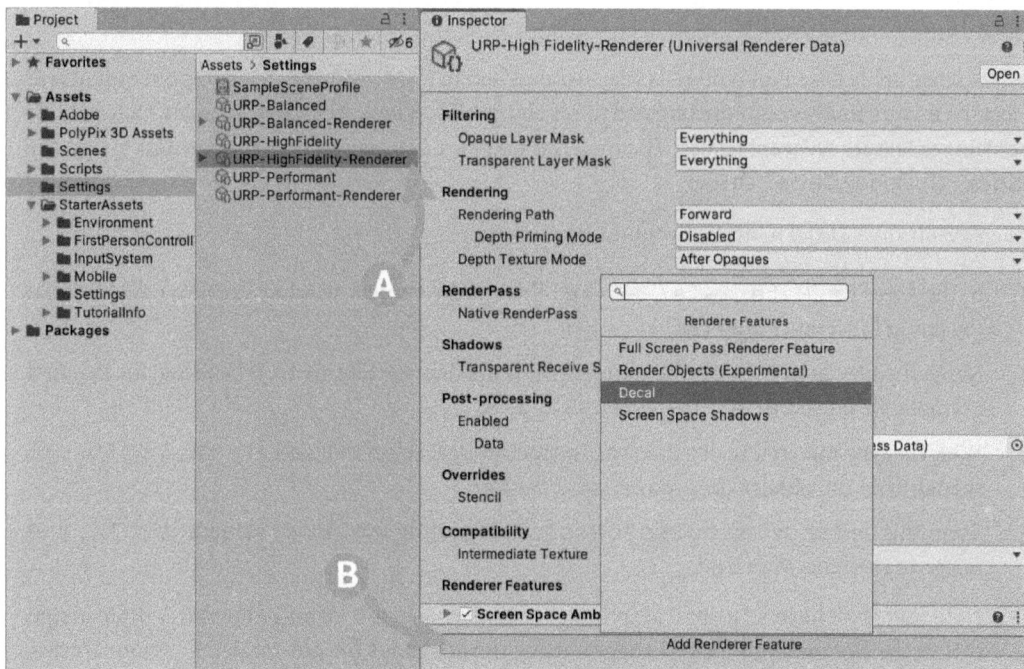

Figure 11.11 – Enabling Decal for Universal RP

Now, we need some decal textures to use for our Decals!

Decal textures

First, we'll need to get some detailed textures into our environment via textures files. I've already sourced some royalty-free textures we can use in our project. Download the `cgtrader_2048986_Damage.zip` file from this chapter's GitHub repository at `https://github.com/PacktPublishing/Unity-2022-by-Example/tree/main/ch11/Art-Assets`.

> **Free decal textures | cgtrader**
>
> The decal textures we're using in the project are from the *Decals Damage 48 Texture* files (royalty-free license) at cgtrader, available at `https://www.cgtrader.com/free-3d-models/textures/decal/decals-damage-48-texture`.

Unzip the file and import the images to a new `Assets/Textures/Decals` folder. While you're creating folders, go ahead and create a new `Assets/Materials/Decals` folder too, since we'll need that for creating decal materials in the next step.

Creating decal materials

The Decal feature is based on materials that are assigned to a **Decal Projector** component that will project the texture images onto the targeted mesh objects. Therefore, we'll have to create materials for each damage texture we want to use. The material setup for Decals requires that we also assign the `Shader Graphs/Decal` shader.

Let's create our first Decal material by following these steps:

1. In the `Assets/Materials/Decals` folder in the **Project** window, create a new **Material** (by going to **Create | Material**).

2. Name the new material with the same name as the damage texture we'll be using; for this first example, we'll use `DecalsDamage0032_1_S`.

3. With the new material selected, in the **Inspector** area, select the `Shader Graphs/Decal` shader from the **Shader** drop-down list at the top.

4. From the `Assets/Textures/Decals` folder, drag the `DecalsDamage0032_1_S` image to the **Base Map** field.

   Note that if we have a Normal Map image included with our damaged textures, then assign that to the **Normal Map** field (I suggest using normal maps for greater detail in your Decals; we just don't, however, have one included in the provided textures).

With that, our first damage Decal material is ready to go! We're almost ready to start applying damage to our environment with the Decal projector component.

Rendering Layers for Decals

Decals are projectors! Due to the nature of a projection in the scene, some objects may pass between the projector and the target mesh(es) and get hit by the projection, producing an undesirable result. So, although we can take a more freeform approach to the art direction here – since we aren't limited to where we can place Decals in the environment – we may still have to control how Decals are projected and protect some objects from receiving the decal texture. Fortunately, using **Rendering Layers**, we can limit what meshes the decal affects.

> **New to Unity 2022**
>
> Rendering Layers | How to use Rendering Layers with Decals: `https://docs.unity3d.com/Packages/com.unity.render-pipelines.universal%4014.0/manual/features/rendering-layers.html#how-to-rendering-layers-decals`.

For example, we will want to have our Decals affect the walls but not the objects we've scattered throughout the environment from our painting with Polybrush! Let's ensure we have Rendering Layers for Decals enabled now. So, let's revisit the **URP-HighFidelity-Renderer** asset (refer to *Figure 11.11*) and enable **Use Rendering Layers** (it's turned off by default):

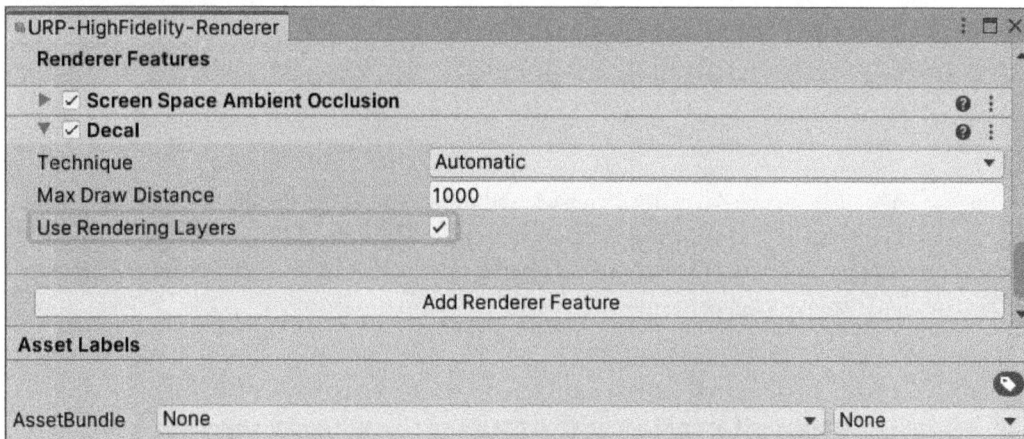

Figure 11.12 – Use Rendering Layers enabled

The next requirement for a functional Decals **Rendering Layers** setup is to ensure a layer is specified for our Decals case. We can do that by editing the **Rendering Layers (3D)** list of layers:

1. Select the **Universal Render Pipeline Global Settings** asset in the root of the `Assets/` folder.

2. As shown in *Figure 11.13*, we'll create our Decal layer by renaming the **Layer 1** slot to `Receive Decals`:

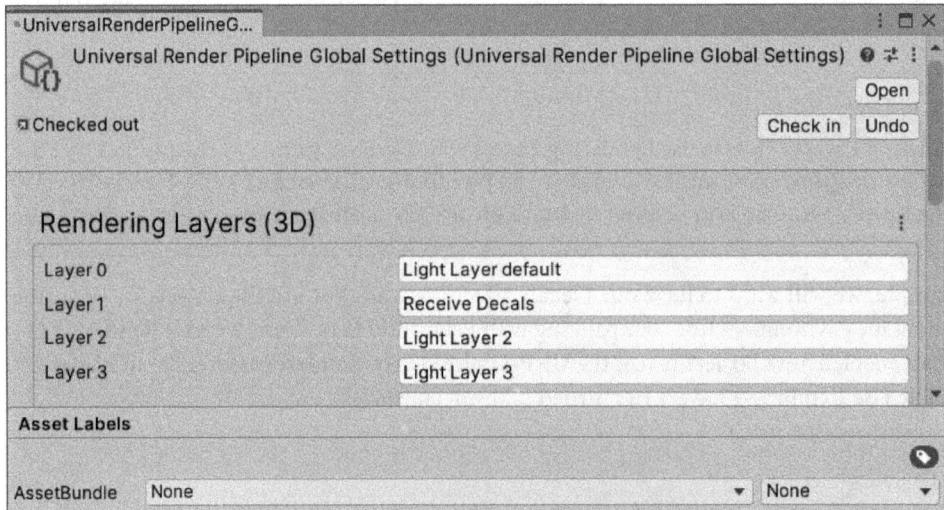

Figure 11.13 – Decal Rendering Layers

Now that we have an assignable layer, we'll go back and modify all of the Prefabs we want to be able to receive decal projections on – ensuring that only objects that have the *Receive Decals* layer selected will show our damage textures. Working with the setup is straightforward – that is, once you wrap your head around the layers being the glue that binds all the parts together.

As shown in *Figure 11.14*, we'll modify the `Wall_02 Variant` Prefab (in **Prefab Mode**) from the `Assets/Polypix 3D Assets/Modular Kit/Prefabs` folder so that it includes the *Receive Decals* layer in the **Rendering Layer Mask** field on the `Mesh Renderer` component:

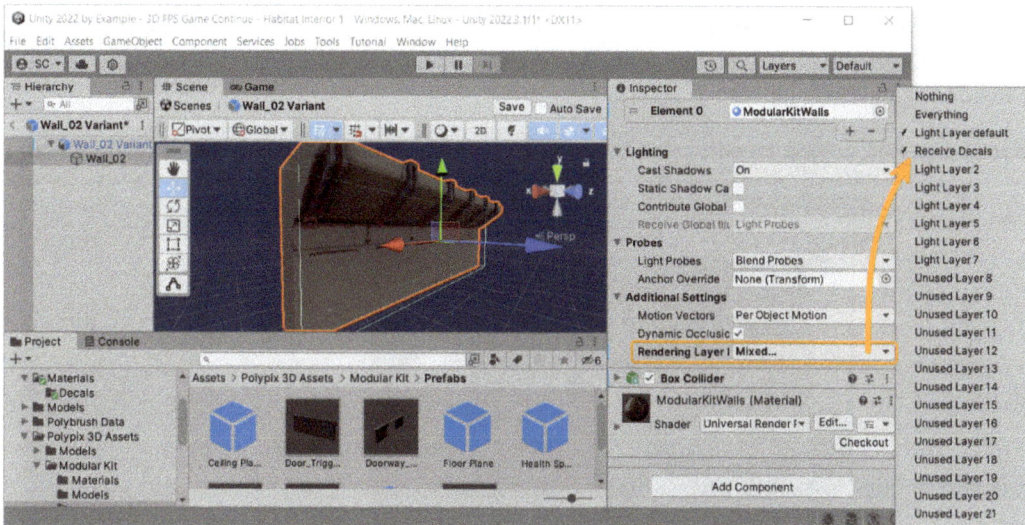

Figure 11.14 – Assigning a decal rendering layer to Prefabs

You'll want to update all the Prefabs you intend to project Decals onto with the *Receive Decals* layer. If your Decals aren't showing where you believe they should be, you'll want to double-check the layer assignments (and, of course, that Decals are enabled in the renderer settings).

Now, we can create our first **Decal Projector** in the scene and start fleshing out the environment's story. Yay!

Using the Decal Projector component

To add a Decal, you can either create a new `Decal Projector` GameObject or add a `Decal Projector` component to an existing object in the scene. Since we don't have any objects in the scene that are appropriate for adding a `Decal Projector` component, we'll create a new one: in the **Hierarchy** window, go to **Create | Rendering | URP Decal Projector**.

As shown in *Figure 11.15*, we must assign the Decal material we previously made for the `DecalsDamage0032_1_S` texture to the **Material** field and select the *Receive Decals* layer in the **Rendering Layers** field:

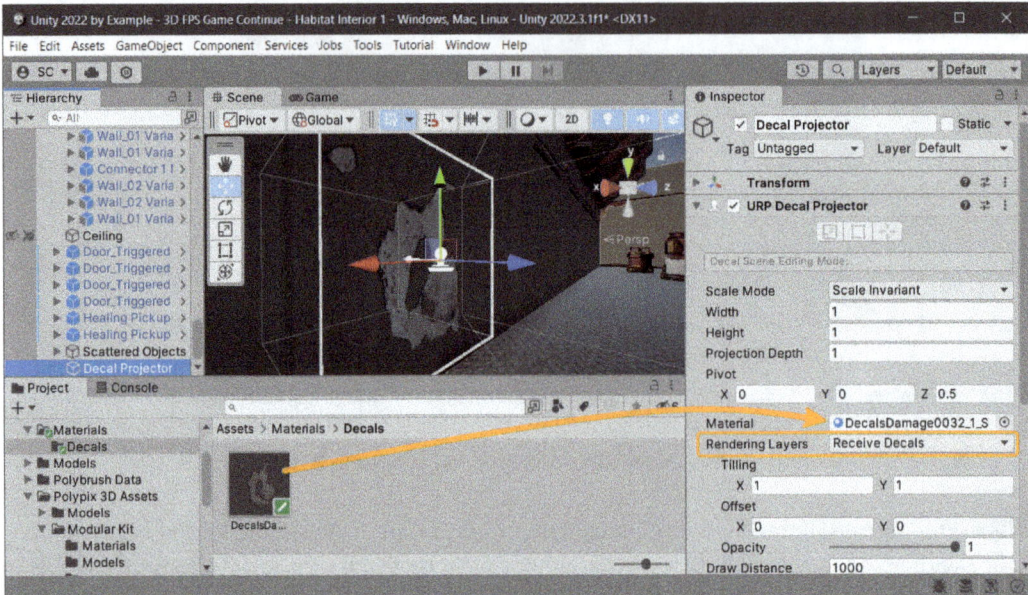

Figure 11.15 – Decal projected onto the wall Prefab

As shown in *Figure 11.15*, I've positioned and rotated the **Decal Projector** *cube gizmo* (which shows its bounds) using the **Transform** tool. The base of the gizmo has a thicker line, and the projection direction is indicated by the white arrow emanating from the pivot point anchor (on the Z-axis).

You can get away with using the **Transform** tool for the initial positioning and rotation of the projection. Still, you'll likely want to use the **Scene** view editing tools to fine-tune the decal further – this requires further exploration and experimentation on your part, so I recommend additional reading.

> **Additional reading | Unity documentation**
>
> Decal Projector component: `https://docs.unity3d.com/Packages/com.unity.render-pipelines.universal%4014.0/manual/renderer-feature-decal.html#decal-projector-component`.

Continue to decorate – ah – design the level environment further and add *surface story* details to sell your environment to the player by utilizing the varying damage textures we have at our disposal. Again, playtesting and feedback are essential here to help guide the art direction.

As you start projecting textures everywhere, you may be thinking, what is this doing to my game's performance?

Decal performance

Well, Unity has got us covered for performance optimization too. If we ensure that the Decals in the scene with the same damage texture use the same *Material*, and if we turn on **Enable GPU Instancing** for the *Material*, Unity will use a technique called **instancing** to make rendering more efficient. Instancing on the GPU can minimize the work Unity has to do by reducing the number of draw calls (which is like drawing all the Decals at once instead of each decal individually).

Furthermore, to reduce the different number of materials we need for Decals (because, again, each *Material* is grouped to draw at once), we can put many damage texture images into one larger image (that is, a texture atlas). Then, the **Decal Projector** component allows us to use **UV offset** properties to select which part of the texture atlas we want to show – our selected individual texture image. This way, we can manage all the decal images more efficiently and keep our games running smoothly.

> **Additional reading | Unity documentation**
>
> Decal Renderer Feature | Performance: `https://docs.unity3d.com/Packages/com.unity.render-pipelines.universal%4014.0/manual/renderer-feature-decal.html#performance`.

In this section, we learned how to polish our 3D environment by replacing graybox Prefabs with finished art assets using Unity's new Prefab workflow and manually updating Prefabs with new graphics while maintaining their existing behavior in Prefab Mode. We continued by polishing the environment design by painting scattered objects with Polybrush and immersing the player in *surface story* by adding Decal textures.

In the next section, we'll continue to add polish to the habitat's interior environment with lighting.

Lighting design – Probes, Decals, light baking, and performance

Let me immediately set some expectations for this section – we won't spend much time on lighting design. It's a subject worthy of an entire book all on its own. So, what we will be covering are some basics any game developer should be familiar with when working with lighting 3D scenes in Unity.

The first thing I would like to address is that Unity has released a new rendering path for **Universal RP (URP)** with the 2022.3 LTS version called **Forward+ Rendering**.

> **Additional reading | Unity documentation**
>
> Forward+ Rendering Path: `https://docs.unity3d.com/Packages/com.unity.render-pipelines.universal%4014.0/manual/rendering/forward-plus-rendering-path.html`.

The **Forward+ Rendering Path** (Forward+) gives several advantages over the previous **Forward Rendering Path** (Forward), but it primarily overcomes the per-object limit on the number of lights that can affect GameObjects in the scene (note that the per-camera limitation does still apply). Let's see how we can update our project URP settings to use **Forward+**.

Setting the URP Forward+ Rendering Path

Now, let's change our URP **Rendering Path** setting to use the **Forward+ Rendering Path**. We can do that in the **Universal RP settings** area (the same asset we added to the **Decal** feature in *Figure 11.11*):

1. In the `Assets/Settings` folder, select the **URP-HighFidelity-Renderer** asset.

2. Under the **Rendering** section, select **Forward+** in the **Rendering Path** dropdown:

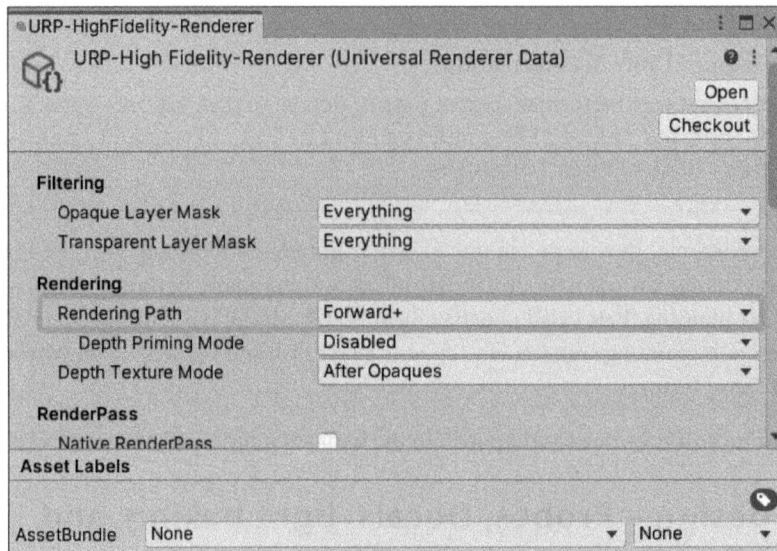

Figure 11.16 – Setting the Forward+ Rendering Path

As with most decisions about the tech stack that you'll be making for your Unity projects, you'll have to experiment and test to see what fits best for the project's direction or target platform (mobile, we're looking at you), and the renderer pipeline choice is no exception. It's pretty important. So, we will be changing this project to the **Forward+ Rendering Path** now. Just know that we may end up reverting this decision in the future should we need to change our lighting approach (for example, from real time to baked lighting; more on this in the *Bake that lighting?* section).

The consensus with performance – concerning **Forward+** over **Forward** – is that gains are realized with greater than six real-time lights due to the overhead **Forward+** introduces when clustering lights (**Forward+** gathers light data into clusters computed in the fragment shader rather than individual light data). That's excellent news for lighting designers because **Forward+ Rendering** gives us precisely what we may need – more real-time lights in a scene!

Speaking of lights, we can get away with adding lighting effects to our scenes, and not touch a single light – this is accomplished with proxy lighting.

Proxy lighting with Decals (yes, Decals)

Decals here refer to *lighting decals* (that is, proxy lighting) and not *texture decals*, which we're already familiar with. To accomplish this effect, **Decal Projector** uses a special shader to modify the emissive color of affected surfaces without any real-time lights being involved.

The *Material* we'll be using for the proxy lighting **Decal Projector** in our scene is based on a **Shader Graph** shader, **Spotlight**, that comes from the **Universal RP** samples.

> **Additional reading | Unity Blog**
>
> You can explore the latest package samples for the Universal Render Pipeline here: `https://blog.unity.com/engine-platform/explore-the-latest-package-samples-for-the-universal-render-pipeline`.
>
> These are not just for **Universal RP**! You should always check the Unity package content to see what samples are provided as examples that supplement the Unity documentation for learning the feature and can give a jump start on implementation.

To import the **Universal RP** samples, follow these steps:

1. Open **Package Manager** by going to **Window | Package Manager**.
2. Ensure you have **In Project** selected in the **Packages** dropdown.
3. Find and select **Universal RP** from the list (left-hand side).
4. On the right-hand side of the window, click the **Samples** tab.
5. For **URP Packages Samples**, click the **Import** button to the right. The sample files will be installed in the `Assets/Samples/Universal RP/14.x.x/URP Package Samples` folder.

Once the samples have been imported, find and open the **ProxyLighting** scene in the /URP Package Samples/Decals/ProxyLighting folder:

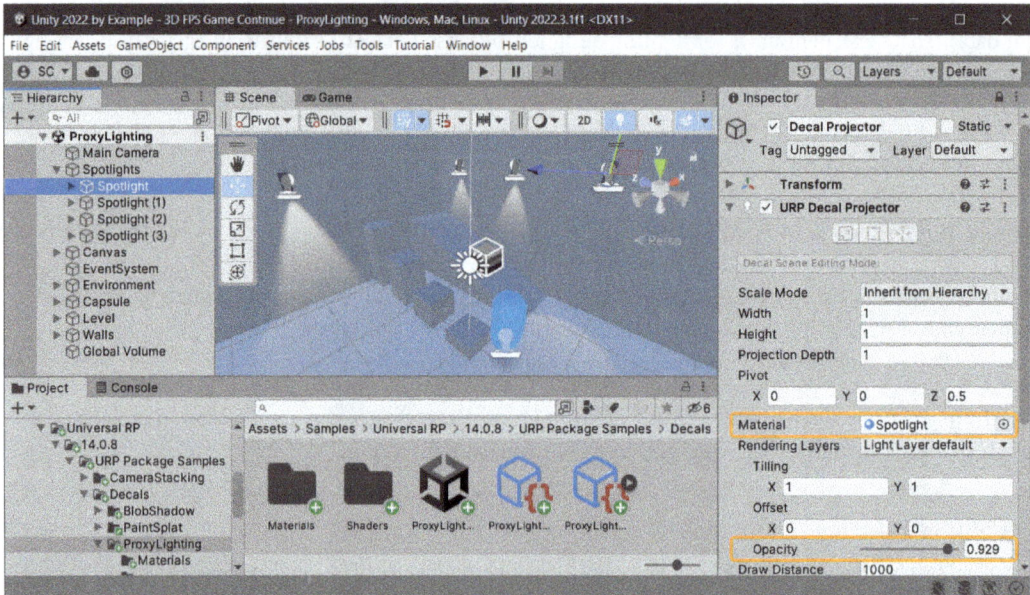

Figure 11.17 – Unity Universal RP ProxyLighting sample

Now, go ahead and find the **Spotlight** object in the **Hierarchy** area, then drag it into the Assets/ Prefabs folder. Name it Decal Spotlight so that we can use it as a *lighting Decal spotlight* in our habitat interior scene. It's already pre-configured with the **Spotlight** material (and the default **Rendering Layer** so that it will affect any object in the scene), so we're all set!

Now, go back to our interior habitat scene and place a **Decal Spotlight** material into a wall overhang (as pictured in *Figure 11.18*), adjust the **Decal Projector** component's **Opacity** setting (refer to *Figure 11.17*) to a lovely lighting value, then duplicate it several times (three times in this example; four in total). Don't worry that they're all sitting on top of each other at the moment.

We could manually place the spotlights further along the wall, or, to make things easier on ourselves, we can use a bit of math in the **Inspector** area's **Transform** field to array the spotlights. Yes, you can use **numeric field expressions** in the **Inspector** area! You can use not only simple arithmetic but trigonometric functions such as cos(a) and sin(a) too!

Additional reading | Unity documentation

Editing Properties | Numeric field expressions: https://docs.unity3d.com/Manual/ EditingValueProperties.html.

Select all the spotlights we just added to the **Hierarchy** area; then, in the **Inspector** area, enter `L(-5.4,-17.3)` in the **Transform Position X** field; `L(a,b)` is a linear ramp expression where the selected objects are distributed between the values.

While you have the expression in the field, you can play around with the values until you have the spotlights positioned where you want (your values may differ from mine based on your wall position and so on).

You should end up with something like this:

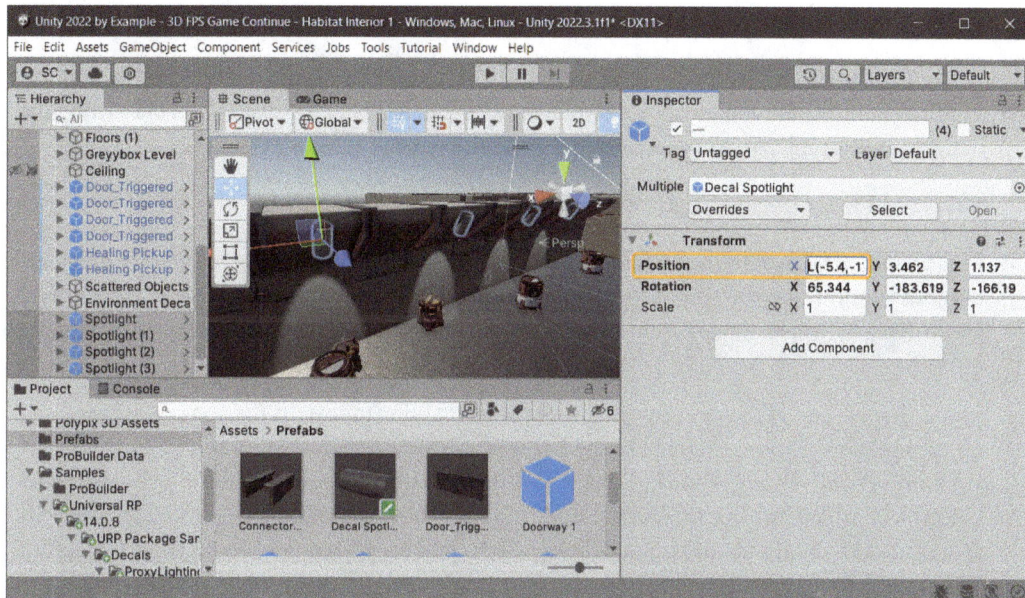

Figure 11.18 – Decal spotlights array on the wall

Pretty cool, right? And remember, no real-time lights are involved! We can save on lighting resources while still enjoying additional lighting effects in our environment design. Just note that **Decal Material** performance considerations are still applicable.

Speaking of performance, we have another approach to lighting that we can consider besides the use of real-time lighting, and that's baked lighting.

Bake that lighting?

350° F (175° C) for 45 minutes… sure!

Using baked lighting in Unity can have several benefits over real-time lighting in your game, but deciding to use one or the other depends greatly on the target platform and the target framerate you've determined the game needs to hit on specific hardware specs (you may have already been introduced to this type of information related to games you'd like to play on your PC or mobile device – system requirements: minimum and recommended).

Let's look at the three main benefits of baked lighting:

- **Performance**: Baking the lighting means all of the light interactions in the scene regarding objects' surfaces are computed ahead of time (in the editor) and saved into lightmap textures files. During runtime, the lightmaps are used to determine the light that's received by the objects' surfaces in the scene, which is much faster than real-time calculations.

- **Quality**: Baked lighting usually produces better lighting fidelity, especially when it comes to indirect bounced light, soft shadows, and more complex light diffusion effects.

- **Lighting complexity**: Baked lighting can handle higher levels of lighting complexity with lots of lights and complex shadow interactions (just be prepared for longer bake times!).

Now, let's consider some disadvantages of baked lighting:

- **Lightmaps**: The size of the generated lightmap texture files can be of concern for both memory usage at runtime and the size of the game on disk (that is, mobile and lower-end platforms). Optimizing lightmaps is generally a balancing act between performance and quality (which, of course, takes time; see *Long iteration time*).

- **Static**: The baked lightmaps only work for objects that don't move in the scene (that is, static objects; they are assigned as such in the **Inspector** area). Dynamic lights and moving objects are more suitable for real-time lighting. Unity does, however, provide a solution for dynamic objects with baked lighting via **Light Probes**, but they come with their own limitations (such as area lights and volumetric lighting not being supported, not working well with some materials, and no real-time reflections), so often, a combination of techniques applied with real-time direct lighting and shadow casting is necessary to obtain the desired results.

- **Additional work**: Additional work must be done in the form of setting objects as static, setting up **Light Probes** for dynamic objects in the scene, a solution for basic shadow needs, and balancing the generated lightmap quality, for starters.

- **Long iteration time**: The process of baking lightmaps can be resource-intensive and, therefore, time-consuming on lower-end development system CPU/GPU hardware, which can cause a significant slowdown in the environment design iteration process.

There's no doubt that lighting an environment with real-time lighting is more accessible. Still, you may not have a choice if you're going to target mobile or lower-end hardware specs (for example, the Nintendo Switch). Baked lighting is generally more performant across devices and platforms compared to real-time lighting.

Our approach so far has been to use the default lighting setup with the **Directional Light** property provided by the **Standard (URP) Scene Template** and create additional lights in the scene – which are *Realtime* by default. Now, we must change our approach because we will target a lower-end platform but still want to target 60 FPS. So, we're going to need baked lighting.

Setting up baked lighting

To set up our baked lighting, let's duplicate our current habitat interior scene with real-time lighting so that we can non-destructively experiment with converting it into baked lighting.

Select your **Habitat Interior** scene in the `Assets/Scenes` folder in the **Project** window, press *Ctrl/Cmd + D* to duplicate, and rename it with a (*baked*) suffix. Let's open the duplicate scene and start the setup:

1. Select **Directional Light** in the **Hierarchy** area and, in the `Light` component's **General** section, select **Baked** in the **Mode** field dropdown instead of **Realtime**.

    Please do the same for any additional lights you may have added to the scene that we want Unity to include in the baking. Note that we won't have to change the **Decal** spotlights we added in the *Proxy lighting with Decals (yes, Decals)* section – remember, these are not lights!

2. We need to let Unity know what objects won't move so that their lighting can be baked. So, for all of the modular Prefabs that make up the walls, the floors, the doorway, and scattered Prefab objects in the scene, select them in the **Hierarchy** area. Then, in the **Inspector** area, enable **Static** (the top right of the window):

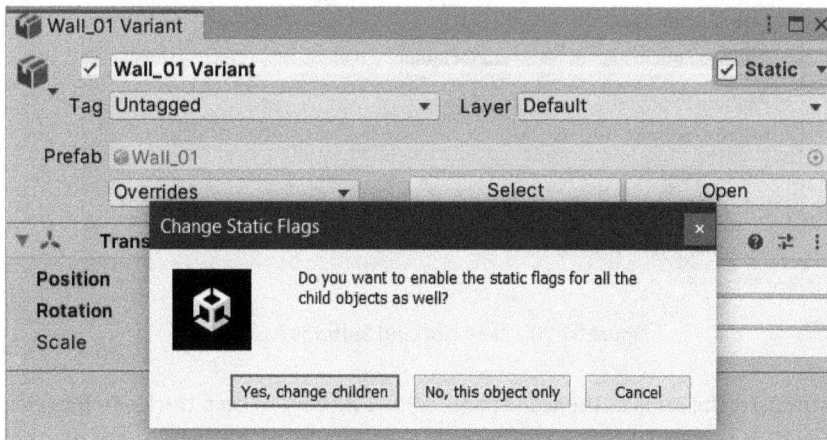

Figure 11.19 – Marking a GameObject as static

3. When you're asked whether you want to mark the child objects as static too, click the **Yes, change children** button.

Now that the objects in the scene have been set up, we will configure **Lightmapping Settings** just before baking.

Lighting settings

Lightmapping Settings configure the light baking calculation and how it applies to the scene. We need to use the **Lighting** window to create a new **Lighting Settings Asset** to store our configuration. Unity will use default read-only light settings to bake the scene lighting until we create an asset.

> **Additional reading | Unity documentation**
> The Lighting window: `https://docs.unity3d.com/2022.3/Documentation/Manual/lighting-window.html`.

To create a new **Lighting Settings Asset**, follow these steps:

1.   Open the **Lighting** window by going to **Window | Rendering | Lighting**.

2.   At the top of the **Lighting** window, under the **Lighting Settings** section, click the **New** button to the right of the **Lighting Settings Asset** field to create a new **Lighting Settings Asset** in the **Project** window; it will be assigned immediately:

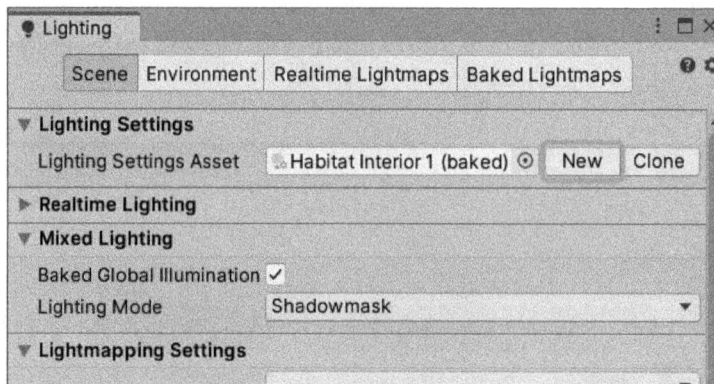

Figure 11.20 – New Lighting Settings Asset

3.   Name the asset the same as the scene name so that it's easy to keep track of – for reference, in *Chapter 10*, we created a new scene named `Habitat Interior 1`; but use the current scene name you're using here.

> **Additional reading | Unity documentation**
> Lighting Settings Asset: `https://docs.unity3d.com/2022.3/Documentation/Manual/class-LightingSettings.html`.

We'll work in the **Scene** tab to optimize our settings for the desired balance of quality, lightmap texture size, and bake time. Using the default values is a good starting baseline for your baking.

Okay, click that **Bake** button! Sadly, no. The actual button to click for baking the lighting is the **Generate Lighting** button, located at the bottom of the **Lighting** window. When started, the lighting process will display a progress indicator at the bottom right of the **Editor** window with an estimated time for completion.

After some time – depending on the power of your system hardware – the generated lighting process will finish, and the **Scene** view will update with the new baked lighting. Inspect the result of the baked lighting in your scene, and, if it's not to your satisfaction, adjust **Lightmapping Settings** and/or both the settings and position or rotation of light sources in the scene and bake again; repeat as needed.

Let's have a gander at the results of our labor by comparing screenshots of the same view with baked and real-time lighting:

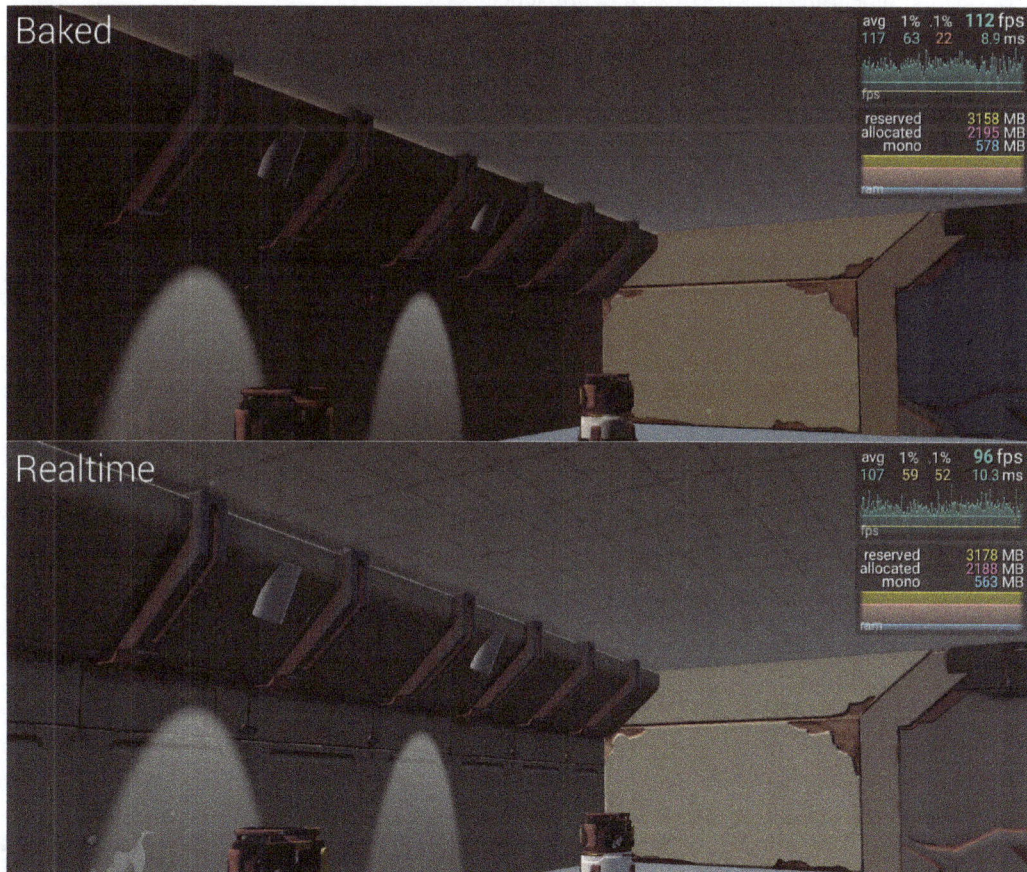

Figure 11.21 – Baked versus Realtime renders

> **Tip**
>
> You'll also want to compare the results of the baked lighting to the real-time lighting. Analyze the quality of the light, the lighting features, memory usage, and, most importantly, the FPS the game is running at compared to your target framerate on the desired platform(s).
>
> You can get some performance gauge while running in **Play** mode in **Unity Editor**, but you'll want to build and test on the target device(s) for the best results. Graphy – Ultimate Stats Monitor & Debugger (`https://github.com/Tayx94/graphy`) is a free tool that is greatly helpful in analyzing your game's performance. Output from Graphy can be seen in the game screenshots in the top-right corner in *Figure 11.21*, where the baked lighting shows a slight FPS benefit over real time (this is in the **Editor** window, so I expect greater gains in a build).

You may not be happy at all with how long the baked lighting generation took with the default **Lightmapper** setting set to **Progressive CPU**. Fortunately, if your hardware supports it, we can improve the calculation time using the GPU-based lightmapper.

> **Tip**
>
> If you notice graphical artifacts appearing on models in your baked lighting, then imported model UVs for baked lightmaps may not be provided. In that case, you can tell Unity to generate them automatically to fix their appearance.
>
> To have Unity generate lightmap UVs, select the offending model in the **Project** window to view **Model Import Settings** in the **Inspector** area. Once you've done this, ensure the **Model** tab is selected, and then, in the **Geometry** section, *enable* the **Generate Lightmap UVs** option.
>
> Unity Documentation | Generating lightmap UVs: `https://docs.unity3d.com/Manual/LightingGiUvs-GeneratingLightmappingUVs.html`.

Now, about that iteration time.

Improving iteration time

Lightmapping Settings allows us to choose between the **Progressive CPU** and **Progressive GPU** lightmappers (the latter still being in preview for 2022.3.1f1). **Progressive GPU** may be much, much, quicker to generate lighting, but it all depends on what GPU (that is, what graphics card) you have installed in your system running Unity. In my example, **Progressive CPU** is estimated to take 3+ hours, whereas **Progressive GPU** (on modest GPU hardware) is estimated to take 40+ minutes.

> **Additional reading | Unity documentation**
>
> The Progressive Lightmapper: `https://docs.unity3d.com/2022.3/Documentation/Manual/progressive-lightmapper.html`.

Now that we've covered baked lighting for static objects in the scene (that is, objects we marked as *Static* because they don't move), we just need to cover a bit about how dynamic or moving objects in the scene can be affected by light when there is no real-time lighting!

Light Probes

Again, Unity has us covered for dynamic object lighting. We can use **Light Probes** to apply our baked lighting to moving objects – as mentioned previously; however, this is additional work compared to real-time lighting (as indicated in the *Additional work* bullet point in the *Bake that lighting?* section). Note that light probes will not affect the objects we marked as **Static** – moving objects should never be marked **Static**.

Light Probes are placed throughout the environment, not only where the moving objects will be but also in areas where lighting changes (especially dramatic changes). The light probes we put in the scene will capture the light information when baking and then use that data to light the moving objects based on their relative position to the probes.

> **Additional reading | Unity documentation**
>
> Light Probes: `https://docs.unity3d.com/2022.3/Documentation/Manual/LightProbes.html`.

Let's place our first **Light Probe Group** in our scene by going to **Create | Light | Light Probe Group**. Then, using the **Transform** tool, move the new **Group** to the center of a room in the level. Light changes occur primarily along the walls in these rooms since the lighting is pretty uniform, so we'll spread out the probes to cover a wider area.

In *Figure 11.22*, you can see that I first used the toggle pickability feature (*A* in *Figure 11.22*) in the **Hierarchy** area to turn off the selection of the wall, floor, and ceiling so that we can work with the probes and not accidentally select anything else in the scene (that would be not very pleasant, after all).

You can now enable **Light Probe Editing** (*B* in *Figure 11.22*) and drag a selection window (*C* in *Figure 11.22*) in the **Scene** view to select the probes on the right-hand side of the group. Use the **Duplicate Selected** button – or the tried-and-true *Ctrl/Cmd + D* shortcut – to duplicate the selected probes, then use the **transform move** tool to position the duplicated probes near the wall:

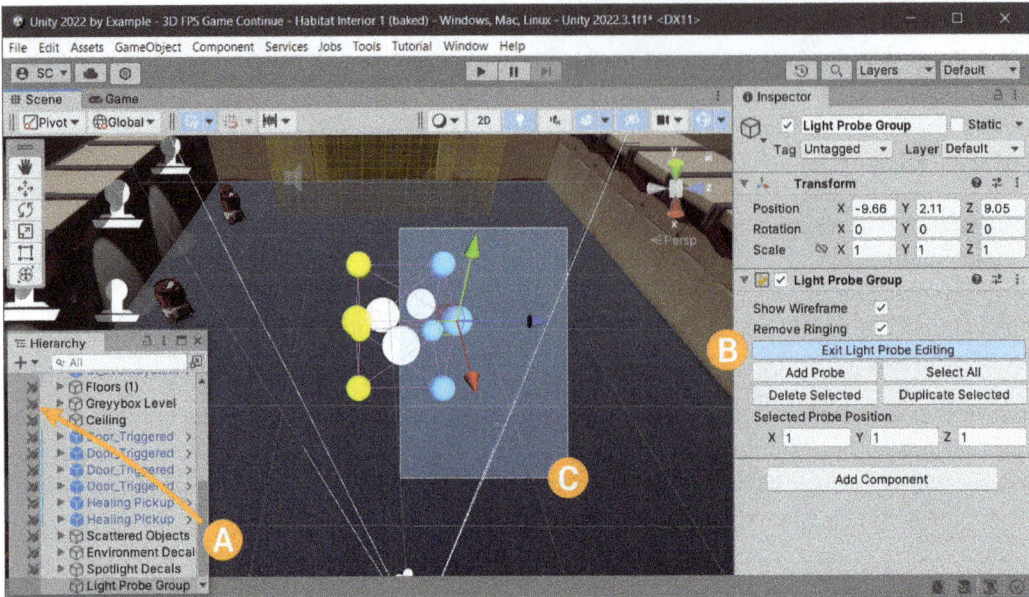

Figure 11.22 – Editing Light Probes in the scene

> **Additional reading | Unity documentation**
>
> Pick and select GameObjects: https://docs.unity3d.com/Manual/ScenePicking.html.
>
> Scene visibility: https://docs.unity3d.com/Manual/SceneVisibility.html.

Repeat the process of selecting, duplicating, and positioning probes until you have something similar to the probe group setup shown in *Figure 11.23*:

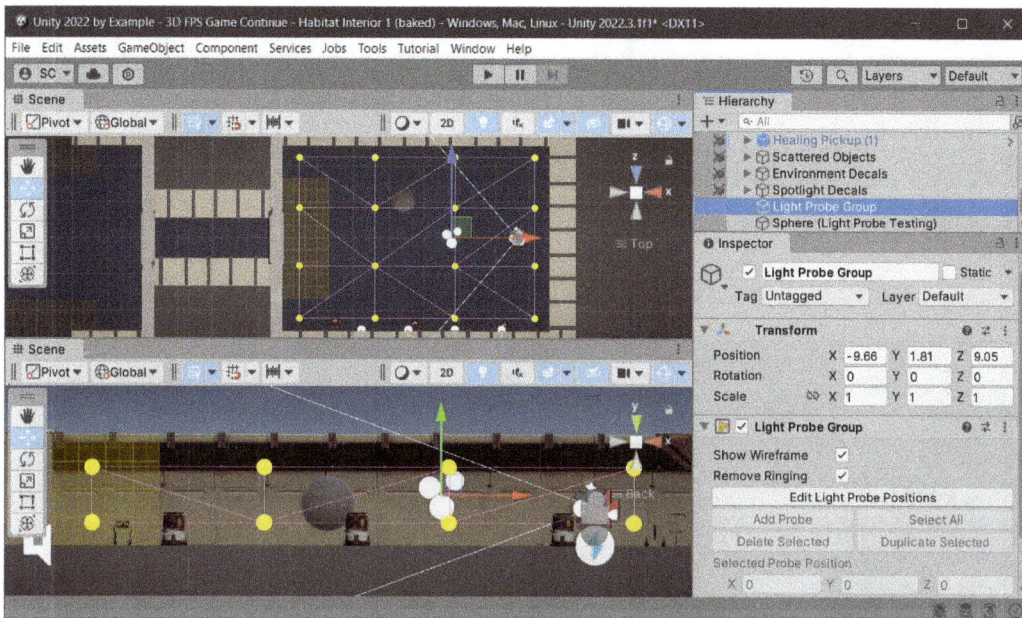

Figure 11.23 – Light Probes placement example

> **Tip**
>
> Remember, in *Figure 11.23*, as uniform as the probes grouping looks in this scene, you won't always have it set up this way because you'll want to position probes where the light changes (again, especially where the light changes dramatically) for the best results.

Once you have the **Light Probe Groups** set up in your level, you'll have to bake (err, generate) the lighting again so that the probes have all the lighting information stored. The probe lighting data will then be used at runtime to light dynamic objects.

You can visualize the effect light probes have on dynamic objects by using the **Light Probe Visualization** setting in the **Lighting** window. When you have a dynamic object selected in the scene, the probes that influence the object will be shown when set to **Only Probes Used By Selection**, with the selected **Sphere** (temporarily added to the scene for the sole purpose of visualizing the Light Probes), as shown in *Figure 11.24*:

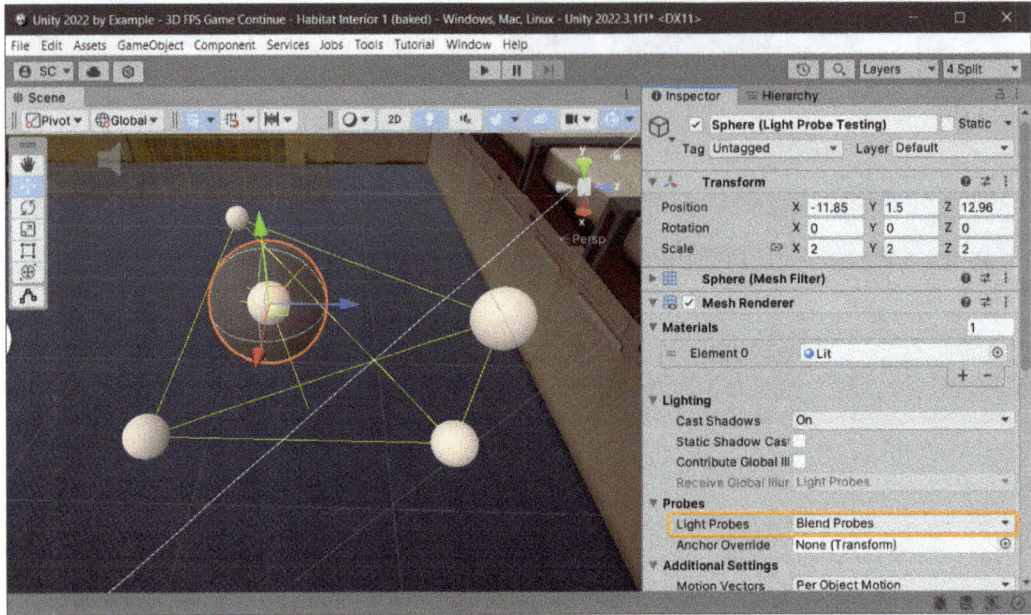

Figure 11.24 – Light Probes visualization selection

> **Tip**
>
> Unity has another useful lighting tool called **Reflection Probes**. Their use can significantly enhance the visual fidelity of scenes by contributing to the quality of both baked and real-time lighting – that is, these probes work with both static and dynamic objects.
>
> As the name indicates, reflection probes add the effect of reflections of shiny things in the environment. I recommend exploring **Reflection Probes** further to enhance the quality of your game's visuals!
>
> Unity documentation | Reflection Probes: `https://docs.unity3d.com/2022.3/Documentation/Manual/class-ReflectionProbe.html`.

There is one additional subject we need to cover regarding baked lighting, and that is shadows for dynamic objects. Baked shadows are a part of baked lighting for static objects – the shadows were baked into the lightmaps. Our lights in the scene were also set to *Baked*, so that doesn't leave any means for providing shadows to our dynamic objects.

We can achieve a semi-dynamic shadow effect for baked lighting in a few different ways. We'll cover two in the following section.

Baked lighting dynamic shadows

The first thing we can do for dynamic object shadows is use **Mixed** mode for lights. **Mixed** mode will bake shadows (just like the **Baked** mode setting) for static objects *and* compute real-time shadows for dynamic objects (just like the **Realtime** mode setting).

Light Probes don't solve this problem because they don't represent any direct lighting and don't produce shadows; they only influence the lighting applied to dynamic objects, so we'll use them in combination with this mixed lighting mode technique.

So, to implement it, simply set **Directional Light** to **Mixed** mode and bake. Done.

The second thing we can do is fake it (yes, fake it – often, the game developer must resort to creative techniques that produce the desired results visually or close to it, but in clever and creative ways behind the scenes). No worries – Unity still has us covered.

Blob shadows

A shadow projector can be used as a more optimized way to represent a simple *blob shadow* on the ground under objects. This technique benefits performance-constrained platforms (for example, mobile games and low-end hardware) where real-time shadow casting can be too performance-heavy.

To use a blob shadow projector, we don't need to look further than the **Universal RP** samples we installed from the **Package Manager**. You can find the *BlobShadow* example scene in the `Assets/ Samples/Universal RP/14.0.8/URP Package Samples/Decals/BlobShadow` folder.

With the scene open, find the **Capsule** object in the **Hierarchy** area. The **BlobShadow** child object contains the **URP Decal Projector** component (this should be familiar to you) with the provided **BlobShadow_Mat** material assigned.

Create a Prefab of it and use it for dynamic objects in your scene that require dynamic shadows in a baked lighting setup. It couldn't be easier!

In this section, we learned how to light an interior environment scene with both real-time and baked lighting and weighed the visual differences and the performance tradeoffs between these approaches. We also learned some techniques to overcome lighting limitations.

With that and the knowledge and technique learned in this chapter, take the time to finish your level's interior environment design to make it appear lived-in – immersive and engaging for the player. Tell a story. Iterate as necessary with feedback from your playtesting group. Most importantly, just have fun!

Summary

In this chapter, we covered the process of transforming a grayboxed prototype environment by importing and replacing modular Prefabs with high-quality 3D model assets using Unity's new 2022 Prefab workflow, preserved Prefabs with additional behavior using Prefab Mode, and applied new materials to finish up the level's structural visuals.

We continued to decorate the environment by painting scattered Prefabs with **Polybrush**, strategically placing wear-and-tear **Decal** projectors for an additional layer of uniqueness and detail, and finalizing our artistic vision with technical implementations of both real-time and baked lighting setups, where we addressed additional considerations for lighting and shadows for dynamics objects while using baked lightmaps.

In the next chapter, we're going to continue enhancing the player experience by adding some much-needed sound design (so far, audio is something we haven't spent any time on). We'll code `AudioManager` and create reusable *audio player* components to make adding music, **sound effects** (**SFX**), and ambient sounds simple and straightforward (even for artists and designers).

12

Enhancing the FPS Game with Audio

In *Chapter 11*, we transformed a gray-boxed prototype environment by replacing modular Prefabs with high-quality 3D models, preserved Prefabs with extra behavior in Prefab Mode, and applied new materials. We also added objects to the environment by painting Prefabs, telling surface stories by placing wear-and-tear decals, and implementing lighting and shadow setups for real-time and baked lighting.

To improve the player experience, we have to work on the sound design aspect of our game. Until now, we haven't given much attention to audio and **sound effects** (**SFX**). Let's change that now by developing an audio manager and reusable audio player components system. This will allow us to easily add audio and SFX to our game, even for designers and artists who may not have much experience with coding.

In this chapter, we're going to cover the following main topics:

- Adding audio using the Audio Mixer
- Building an immersive soundscape with music, SFX, and ambiance
- Enhancing the audio experience with footsteps and reverb zones
- Deeper SOLID refactoring

By the end of this chapter, you'll be able to code an audio manager and its components to play game music and SFX through an Audio Mixer. You will also gain an understanding of the differences between playing 2D and 3D sound and how to reuse audio playing components to create new sound behavior, such as footstep sounds. Additionally, you will learn how to add effects zones to your game's level.

Technical requirements

You can download the complete project on GitHub at `https://github.com/PacktPublishing/Unity-2022-by-Example`.

Adding audio using the Audio Mixer

In any interactive experience, sound is pivotal in enveloping the player in the game world. This section will introduce different ways to play audio using Unity's audio tooling. You'll learn how to elevate your game's audio experience by effectively utilizing music, both 2D and 3D SFX, ambient noise, and reverb zones.

Let's get started by defining what sound design is.

Sound design 101 for games

Good sound design can significantly enhance the player's experience and immersion in the game. Play your favorite game but turn off the sound, and you'll see quite quickly that the experience is not the same!

Several audio concepts come together to form the foundation of sound design in creating the atmosphere for an immersive world. While I'll be introducing sound design components, just know it's more art than technical in bringing all the concepts together – this is the job of a sound designer!

We'll be implementing audio scripts that will enable us to add the following elements of sound design:

- **Music/background music**: Set the mood with music! Also, changes in music at the right moments of gameplay can be impactful to the player.

- **SFX**: Every time the player swings their sword, picks up a coin, bumps into a crate, or defeats an enemy, SFX is crucial for player immersion and enhancing the gameplay experience.

- **Ambience or atmosphere**: The environmental sounds, or background noise, you hear that sets the scene and lets you know you are in a real living place. Generally, 3D audio changes based on the player's position and orientation to an audio source in the environment, which can greatly enhance immersion.

- **Character dialogue**: Yes, dialogue is sound design too. The style of their voice can further define a character. Voice can also be used to drive a story-driven narrative.

- **UI feedback**: *Clicks*, *boops*, and *beeps* on button presses and changes in the UI when the player is interacting provide a satisfying feedback experience that you don't want to omit from your overall sound design.

Gathering all of the preceding sound design elements together into a cohesive audio soundscape for a game experience is another job performed by the sound designer, and it is called audio mixing. **Audio mixing** allows for further engineering of the audio experience and can include additional audio tooling such as filters, effects, and reverb zones.

> **Additional reading | Unity documentation**
>
> When it comes to audio, there are many factors to consider, such as formats, file size, compression quality, looping capabilities, and runtime performance – for starters. While a deep dive into the technical aspects of varying audio formats is not within the scope of this chapter, rest assured that I won't shy away from pointing out crucial format details wherever necessary in the upcoming sections.
>
> You can read more about audio here: `https://docs.unity3d.com/2022.3/Documentation/Manual/Audio.html`.

Now that we understand sound design, we'll put on our *audio developer hat* and add audio to our 3D FPS game.

Adding audio to the game

We'll start by adding an Audio Mixer. Even though I mentioned audio mixing last in the previous section, we'll add the capability to mix our game audio first and patch every audio component through it.

Working with the Audio Mixer

The **Audio Mixer** asset will allow us to set levels for each sound element we define independently – tailoring our game's soundscape. It will also allow us to change levels when required in gameplay and provides a convenient way for the player to adjust levels to their taste via UI controls.

> **Additional reading | Unity documentation**
>
> You can read more about the Audio Mixer here: `https://docs.unity3d.com/2022.3/Documentation/Manual/class-AudioMixer.html`.

To add an Audio Mixer, we'll first have to ensure we have our project open in Unity and have our habitat interior level design scene from *Chapter 10*, open (e.g., `Habitat Interior 1`).

We can now create a new **Audio Mixer** asset named `AudioMixer 1` in a new `Assets/Audio` folder (using the **Create | Audio Mixer** menu). Once you've created **AudioMixer 1**, double-clicking on it (in the **Project** window) will open the **Audio Mixer** window, as shown here:

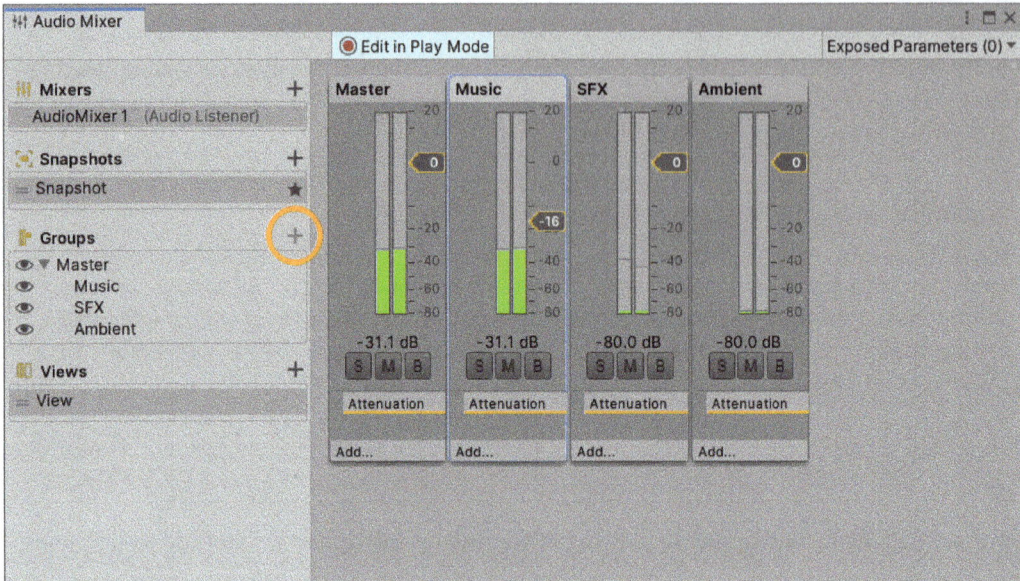

Figure 12.1 – Audio Mixer setup

With the **Audio Mixer** window open, we can now add groups for the channels of audio we'll want to mix the levels for.

Add the following groups (see *Figure 12.1* for reference) by clicking the big plus sign (+) to the right of the **Groups** heading. Note that when the group first appears, the field is highlighted so that you can give it a name right away:

- **Music**: The level at which the game's music will be played is set with this group
- **SFX**: The game's SFX level will be set with this group
- **Ambient** [sound]: All of the environmental ambiance and noise will be set with this group

> **Tip**
> In **Play Mode**, there is a toggle button called **Edit in Play Mode** in the **Audio Mixer** window. This toggle, when enabled, allows us to adjust or mix audio levels while playtesting.

In *Figure 12.1*, you can see I left all the levels at zero except for the **Music** group, which I've brought down to a starting value of -16 to see how loud that sounds during gameplay – it should be lower so as not to drown out the SFX.

Okay, that was easy! We'll continue coding our way through creating the audio system with components for playing the different sound elements in the following sections. It all starts with an audio manager class.

Creating a simple audio manager

Following what should be a recognizable code architecture at this point in the book, introduced in *Chapter 2*, section *SOLID principles and design patterns*, we'll create an audio manager class by following SOLID principles. Specifically, we'll rely on the SOLID **open-closed principle (OCP)** and polymorphism, and also introduce an interface for our different types of audio player components responsible for playing the different kinds of sounds in the game.

Now, let us create a C# script in a new `Assets/Scripts/Audio` folder named `AudioManager`. We'll start by adding the necessary variable declarations:

```
using UnityEngine.Audio;

public class AudioManager : MonoBehaviour
{
    [SerializeField] private AudioMixerGroup _groupMusic;
    [SerializeField] private AudioMixerGroup _groupSFX;
    [SerializeField] private AudioMixerGroup _groupAmbient;

    private AudioSource _audioSource2D, _audioSourceMusic;
}
```

Here, we can see the groundwork for implementing the Audio Mixer groups we configured in the previous section. We've also declared some `AudioSource` variables for sound playback – these will be explained as we add to the code while implementing the play audio functionality in the following *Building an immersive soundscape with music, SFX, and ambiance* subsections.

> **Important note**
>
> I'm just going to assume by now that you know that if we have a class inheriting from `MonoBehaviour`, we need a `using UnityEngine;` statement at the top of the C# script. As you should also know, this `using` statement is already there from the default script template. :)

Again, one of the key functions the `AudioManager` class will handle for us is setting the Audio Mixer group for the type of audio we want to play. This will ensure sound designers can use the Audio Mixer to set the initial audio playback levels of the game (i.e., designing the soundscape).

Now, add the following code to the `AudioManager` class:

```
public enum AudioType { Music, SFX, Ambient };

private AudioMixerGroup
    GetAudioMixerGroup(AudioType audioType)
        => audioType switch
```

```
    {
        AudioType.SFX => _groupSFX,
        AudioType.Music => _groupMusic,
        AudioType.Ambient => _groupAmbient,
    };
```

---

switch expression (C#)

You can read more about pattern-matching expressions using the `switch` keyword here: `https://learn.microsoft.com/en-us/dotnet/csharp/language-reference/operators/switch-expression`.

Here, we've created an `AudioType` enum that we'll use to map an audio player component to an Audio Mixer group. We've added the `GetAudioMixerGroup()` method to get the appropriate Audio Mixer group by using a `switch` expression and passing in an `AudioType` enum. This will all become clearer in the next section when we create our first audio player component.

Discards (C#)

Note that the compiler will generate a warning if the switch expression in the preceding code doesn't handle all possible input values. In that case, we could use the following discard pattern so all possible input values are handled – avoiding a console warning:

```
_ => throw new ArgumentOutOfRangeException(nameof(AudioType),
     $"Not expected audioType value: {audioType}"),
```

The underscore (_) is a placeholder variable that won't hold a value and is not intended to be used.

You can read more about discards here: `https://learn.microsoft.com/en-us/dotnet/csharp/fundamentals/functional/discards`.

With the `AudioManager` class we have coded up until now, we are ready to start making the audio player components! But before we do that, we'll need to have a way to access `AudioManager` to call on its methods. So, we'll continue to use the Singleton pattern for our managers – go ahead and add the required code for it now.

As a reminder, here it is:

```
public static AudioManager Instance { get; private set; }

private void Awake()
{
    if (Instance == null)
        Instance = this;
    else
        Destroy(gameObject);
```

```
      DontDestroyOnLoad(gameObject);
}
```

To add the `AudioManager` script, a class derived from `MonoBehaviour`, to our project, create a new GameObject named `AudioManager` in the **Hierarchy** (drag it to the top) and add the `AudioManager` script to it.

We can now assign the `AudioMixerGroup` fields by clicking and dragging the group from the **Audio Mixer** window or using the field's **Object Picker** window (the small circle icon at the right side of the field), using *Figure 12.2* as a guide.

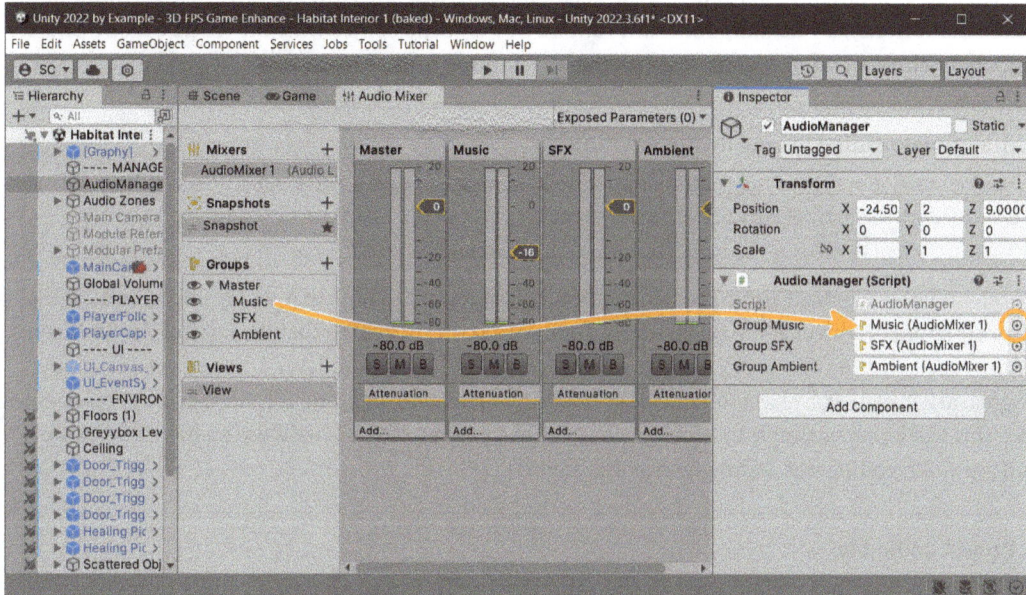

Figure 12.2 – AudioManager mixer groups assignment

Done! Well, we don't actually have any methods to play sound in our audio manager… yet…

The audio manager foundation for the game is in place, so we can now begin writing the individual components to play the different types of audio needed in the game.

Building an immersive soundscape with music, SFX, and ambiance

We'll cover writing audio player components for what I consider the most common scenarios for the type of game audio we'll need. We'll have components for playing music, SFX, and ambient sounds. We'll also cover playing the sound in 2D or 3D space, as required or desired, for each type of audio player component.

Everything that plays sound in Unity requires an `AudioSource` component (think of it as a speaker, and you can have many), and the scene requires a single `AudioListener` (think of it as the microphone). By default, our main camera in the scene comes with the `AudioListener` component already added – so we're good; there is nothing further to do there.

> **Additional reading | Unity documentation**
>
> You can read more about the Audio Listener here: `https://docs.unity3d.com/2022.3/Documentation/Manual/class-AudioListener.html`.

As seen in the `AudioManager` implementation, it's important to note that we are ensuring all audio player components direct audio playback through `AudioManager` so that the correct Audio Mixer group is used. This design is intentional, so anyone adding an audio player component (developer, designer, artist, etc.) won't have to remember to set the correct mixer group for the specific component type. This will be super simple to accomplish because – you guessed it – we'll use an interface (as we'll soon see when coding our audio player components).

> **Project audio files**
>
> The audio files that will be used in the following sections are all either public domain or **Creative Commons Zero** (**CC0**) (`https://creativecommons.org/publicdomain/zero/1.0/deed.en`) and are available from the `Audio-Assets` folder in the GitHub repo, from the individual Unity project assets folders, or from the original download links provided in the corresponding sections.
>
> The GitHub repo audio assets for this chapter can be found here: `https://github.com/PacktPublishing/Unity-2022-by-Example/tree/main/ch12/Audio-Assets`.

Okay, on to our first audio player component. Time to play some sweet music!

Playing music

We'll delve into the technical aspects of incorporating music into your games. A music composer is responsible for creating impactful soundtracks that resonate with your audience and become synonymous with your game's identity. Therefore, I highly recommend partnering with a composer for your projects to achieve this level of quality.

However, if you are looking for a more budget-friendly option, you can use sourced music, where composers create pre-made soundtracks in different genres that can be purchased and used in your project (sometimes, as in our case here, even for free).

But, before we can assign any music clips to an audio player component, we first have to create one.

> **Warning!**
>
> The `AudioPlayerMusic` class introduced next will not follow the SOLID OCP. This is intentional. We will instead use this initially coded approach as an example to be refactored for the remainder of the audio player components. Hopefully, it will serve as a reinforcement learning opportunity to solidify knowledge.

We're starting with playing music, so create a new C# script named `AudioPlayerMusic` in a new `Assets/Scripts/Audio` folder with the following fields and methods declared:

```
public class AudioPlayerMusic : MonoBehaviour
{
    [SerializeField] private AudioClip _musicClip;
    [SerializeField] private bool _playOnStart = true;

    private void Start()
    {
        if (_playOnStart)
            Play();
    }

    private void Play()
        => AudioManager.Instance.PlayMusic(_musicClip);
}
```

Now, let's have a look at what we did here:

- `_musicClip`: An audio clip holds the audio data representing the sound to be played, which supports many popular audio file formats (`.wav`, `.mp3`, and `.ogg` being the most used). In the **Inspector** window, this variable will be assigned a reference to an audio file asset from the **Project** window folder for our game's music to be played.

> **Additional reading | Unity documentation**
>
> You can read more about audio clips here: `https://docs.unity3d.com/2022.3/Documentation/Manual/class-AudioClip.html`.

- _playOnStart: This variable is a simple Boolean flag we'll set in the **Inspector** to tell this audio player component to start playing when the game begins or not.

- Start(): We evaluate the value of _playOnStart in the Start() Unity message event method and call Play() if true. Simple.

- Play(): This is where the magic happens, where we tell AudioManager to start playing our lovely music's audio clip.

As you can see, we call a PlayMusic() method in the AudioManager Singleton instance. However, the method doesn't yet exist. Let's fix this now to complete the playing music functionality.

Adding the PlayMusic() method to AudioManager

Now, add the PlayMusic() method to the AudioManager class as follows:

```
public void PlayMusic(AudioClip clip)
{
    if (_audioSourceMusic == null)
        _audioSourceMusic =
            gameObject.AddComponent<AudioSource>();

    _audioSourceMusic.outputAudioMixerGroup = _groupMusic;
    _audioSourceMusic.clip = clip;
    _audioSourceMusic.spatialBlend = 0f; // 2D
    _audioSourceMusic.loop = true;
    _audioSourceMusic.Play();
}
```

AudioManager will play the music directly by adding AudioSource to the GameObject it resides with:

```
gameObject.AddComponent<AudioSource>();
```

Then, following the architecture we established, set the appropriate mixer group, the music clip passed in as a parameter to the _audioSourceMusic audio source, and the remaining properties set the music to play in 2D space and on an infinite loop.

The _audioSourceMusic properties for playing music are as follows:

- spatialBlend = 0f: Based on the player's position in the environment, we don't want the music to be heard differently, so setting the spatial blend property to 0f makes it a full 2D sound.

> **Tip**
> When adding 2D audio sources to the scene, note that their transform position in world space won't matter; the audio listener always hears them.

- `loop = true`: This one is pretty clear – does `loop` equal `true`? Yup. Then play the audio clip in a loop forever! Otherwise, play it once.

The remaining properties for an `AudioSource` component will be set to their default values, which will all be just fine for background music.

> **Additional reading | Unity documentation**
>
> You can learn more about `AudioSource` here: `https://docs.unity3d.com/2022.3/Documentation/Manual/class-AudioSource.html`.

Finally, the music is played using the `Play()` method on `_audioSourceMusic`. All that remains to get the music playing in our game is to add `AudioPlayerMusic` to our scene.

Implementing AudioPlayerMusic

Implementing our music audio player component means adding the `AudioPlayerMusic` script to our scene (no surprise here).

In the **Inspector**, add `AudioPlayerMusic` as a sibling component to the `AudioManager` component on the `AudioManager` object in the **Hierarchy**. *Figure 12.3* shows our final music audio player component setup.

Figure 12.3 – Adding AudioPlayerMusic to the AudioManager object

We can see in the preceding figure that the audio clip for the music is already assigned to the **Music Clip** field. So, let's add the `Arpent` audio file to our project now and make that assignment.

If you haven't yet imported the audio assets from the project GitHub repo, refer back to the *Building an immersive soundscape with music, SFX, and ambiance* section and do that now. We'll need the audio files for the next step and the following sections.

The music file we'll be using for our game is `Arpent.mp3` and comes from FreePD (`https://freepd.com/music/Arpent.mp3`). The music is licensed CC0, meaning we can use it for free.

Ensure the `Arpent.mp3` file is imported to the `Assets/Audio/Music` folder. Then, with the file selected in the **Project** window, let's adjust the import settings in the **Inspector** window to something appropriate for what is generally a longer music clip – you can see in the **Inspector** window in the following figure that the music file is nearly three minutes long (**2:42.064** to be exact).

Figure 12.4 – Audio clip import Inspector settings

> **Important note**
>
> You'll have noticed by now that when we select assets in the **Project** window, the **Inspector** will adjust to the type of asset chosen, so the import settings will look different.

Settings to note that we've changed here are as follows:

- **Load In Background**: Enabling loading the audio file in the background for lengthy or high-quality music files ensures the game runs smoothly since asynchronous loading (i.e., not blocking the main thread) eliminates frame drops and the possibility of stuttering

- **Load Type**: For longer music files, it's advised to set this to **Streaming** so all of the audio data is not loaded into memory all at once

- **Quality**: A value of around 70% is generally a good balance for music – file size versus quality

> **Audio Clip Import Settings | Unity documentation**
>
> You can learn more about audio clip **Inspector** options here: `https://docs.unity3d.com/2022.3/Documentation/Manual/class-AudioClip.html`.

When you've made these changes, clicking the **Apply** button will save them. Now, return to the `AudioPlayerMusic` component by selecting the `AudioManager` object in the **Inspector**. Click and drag the `Arpent` music file to the **Music Clip** field, and we're done!

Save your scene and enter Play Mode to hear music play when the game starts. If you don't hear anything, you may need to toggle the *mute audio* icon in the **Game** view, as seen here:

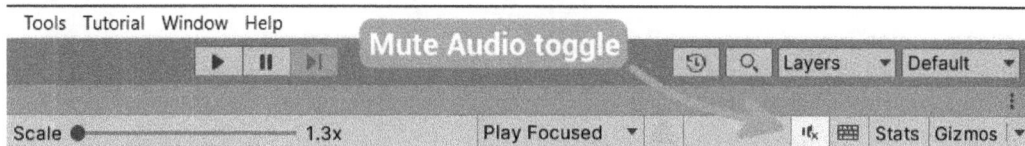

Figure 12.5 – Game mute audio toggle

We've started on our journey of audio immersion and exploration in Unity! Hearing music playing is a nice way of establishing the mood and setting the stage for our gameplay. Game sound designers carefully choose music tracks to create the desired emotional and thematic tone, evoking wonder, mystery, or the unknown to establish an expansive and intriguing atmosphere. Upon hearing the *Arpent* music track, I realized it would be a perfect fit for the sci-fi environment and the player's exploration of the habitat station for this level.

But as great as playing music is, I'm sure you already know that a game is nothing without sound effects!

Playing SFX

Tell me, would a particular series of sci-fi movies (set in a galaxy far away) be complete without its iconic sound of laser swords humming then cutting through the air and clashing? The answer is no… no, it wouldn't.

SFX are the unsung heroes of game development – only 50% of the game is complete without them (no, you cannot argue this). So, what are we waiting for… these SFX aren't going to play themselves!

SOLID refactoring with the IPlaySound interface

We're going to refactor for SOLID principles now. Specifically, `AudioManager` will be closed for modification, adhering to the OCP. For the *open* part, we'll pass in the type of audio player object – this is also the polymorphism part – each with its own implementation (i.e., modification) to play sound, and the audio player types implement an interface: `IPlaySound`. The interface ensures we have a consistent public method to call on each different audio player object type we implement.

We are making an exception to the *closed-for-modification* principle by allowing the use of the `AudioType` enum. This decision is based on making it easier for developers and designers to add audio player components. By simplifying the process and preventing potential mistakes, we aim to save time and eliminate the need for manual assignment of audio groups. The trade-off is that we'll need to modify `AudioManager` whenever a new mixer group is added, which I'm okay with.

With all that explanation out of the way, let's look at the `IPlaySound` interface. Create a new C# script named `IPlaySound` in the `Assets/Scripts/Audio` folder with the following code:

```
using AudioType = AudioManager.AudioType;

public interface IPlaySound
{
    AudioType PlayAudioType { get; }
    void PlaySound(AudioSource source);
}
```

Here is the breakdown of the interface declarations:

- `using AudioType`: Sometimes, we must distinguish a type name shared across namespaces. We defined that `AudioManger.AudioType` and `AudioType` exist in `UnityEngine`. So, we need to tell the script which one we want. To do that, we use an alias: `using AudioType = AudioManager.AudioType;`.

The using alias (C#)

You can learn more about the `using` alias here:
`https://learn.microsoft.com/en-us/dotnet/csharp/language-reference/keywords/using-directive#usingalias`

- PlayAudioType: This property will get the default value we assign from our AudioType enum. We will set AudioMixerGroup based on this value for the specific type of audio player component we're making – and avoid any assignment mistakes in the **Inspector**.

- PlaySound(): Exactly as its name states, the implementing class will use this method to play the audio clip.

And that's all we need for our interface. Our audio manager class can now implement multiple audio players without modification, following the SOLID OCP, by passing each different audio player class by the interface type. The audio player classes that implement the interface will provide unique play functionality.

Let's see our first example by writing the play SFX component. Create a new script named AudioPlayerSFX in the Assets/Scripts/Audio folder with the following code:

```
using AudioType = AudioManager.AudioType;

public class AudioPlayerSFX : MonoBehaviour, IPlaySound
{
    [SerializeField] private AudioClip _audioClip;
    [Range(0f, 1f)]
    [SerializeField] private float _volume = 1f;

    public void Play() =>
        AudioManager.Instance.PlayAudio(this);
}
```

First, we added the IPlaySound interface to the AudioPlayerSFX class declaration to conform to our audio player component design.

Next, we have these core variables and methods that will be common to all the audio player components since they are all considered required for playing sound:

- _audioClip: Every audio player component needs a clip to play. This variable refers to the audio clip asset assigned via the **Inspector** and then played.

- _volume: Not every audio file will have the same levels for its playback, or, for some sounds in the game, they may satisfy the sound design by playing at a lower volume; set with this variable in the **Inspector**.

- Play(): The public method we'll call to start playing the assigned audio clip using the AudioManager Singleton instance's PlayAudio() method (which we'll be adding in the next section).

When calling the audio manager's `PlayAudio()` method, note that we're passing in `this` as the parameter. The `this` keyword refers to the current class instance – in this case, `AudioPlayerSFX`, but the parameter type we will use for declaring `PlayAudio()` will be `IPlaySound`. This is the magic of polymorphism. We'll pass in `this` as the type parameter, representing any number of different classes for unique audio player functionality, but all accessible as `IPlaySound`.

---

this (C#)

You can read more about the `this` keyword for class instances here:
`https://learn.microsoft.com/en-us/dotnet/csharp/language-reference/keywords/this`

Now, implement the `IPlaySound` interface's public properties and methods (remember, interface declarations must be `public`):

```
public AudioType PlayAudioType => AudioType.SFX;

public void PlaySound(AudioSource source)
    => source.PlayOneShot(_audioClip, _volume);
```

Without repeating the interface details again here, let's delve into the assignments:

- `PlayAudioType`: Here is where we pre-assign the `AudioType` value for the Audio Mixer group we want this audio player component to use.

- `PlaySound()`: Here is where we actually play the sound with the audio-playing code specific to this type of audio player component, using the referenced audio source. In the case of SFX, we'll use the `PlayOneShot()` method because it allows playing multiple sounds on a single `AudioSource` component.

  Note that the `AudioSource` component being passed in here will be added to the `AudioManager` object for the 2D SFX. So, there is no need to add an `AudioSource` component to the objects with this audio player component added. You'll see how this works in the next section when we add to `AudioManager`.

Additional reading | Unity documentation

You can read more about `PlayOneShot()` here:
`https://docs.unity3d.com/2022.3/Documentation/ScriptReference/AudioSource.PlayOneShot.html`

To provide some additional clarity to this architecture, now that you've seen the code and how the method calls are set, let's look at a **Unified Modeling Language (UML)** diagram for this `AudioPlayerSFX` and `AudioManager` relationship:

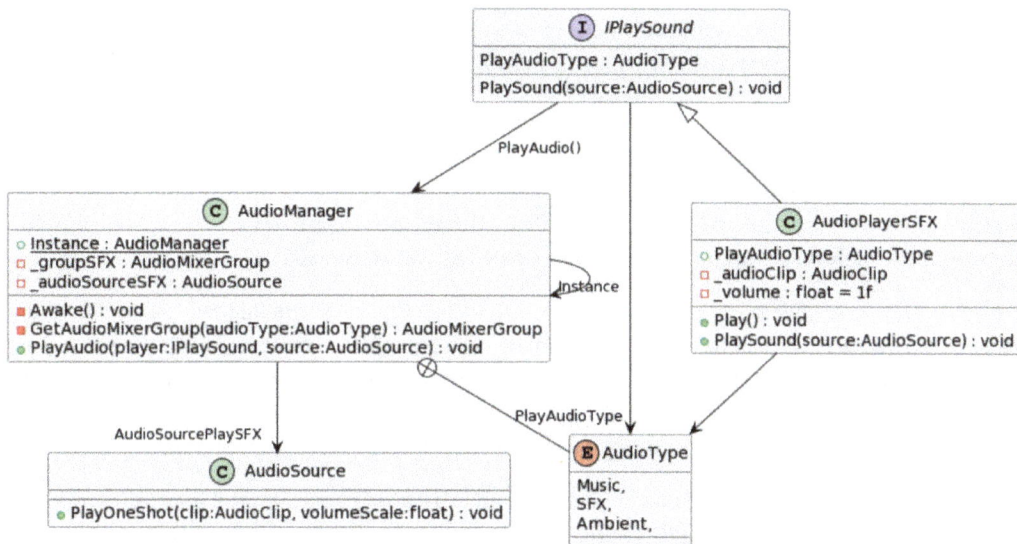

Figure 12.6 – Audio architecture UML class diagram

The code calls on the audio manager – `AudioManager.PlayAudio()` – to assign the correct Audio Mixer group. It then uses the interface to call back to its specific play sound function (with the audio source responsible for playing the sound) – `IPlaySound.PlaySound()`. It'll become even clearer as we implement additional audio player components in the subsequent sections, each with a unique play sound function for playing 3D SFX and ambient sounds.

This finishes up our audio player component for playing SFX, but it's not done yet because it won't play anything with the audio manager's `PlayAudio()` method not existing. Let's solve this problem now by updating our audio manager.

Updating AudioManager

Bringing it all together now, we're adding in the `AudioManager.PlayAudio()` method to set the assigned mixer group and call back to the passed-in component's specific `PlaySound()` functionality to play the audio – in this case, an SFX.

Now, open the `AudioManager` script and add the following method:

```
public void PlayAudio(
    IPlaySound player,
    AudioSource source = null)
{
    if (source == null)
        source = AudioSourcePlaySFX;
```

```
        source.outputAudioMixerGroup =
            GetAudioMixerGroup(player.PlayAudioType);
        player.PlaySound(source);
    }
```

PlayAudio() will be closed for modification so its structure is suitable for any number of audio player components that implement the IPlaySound interface. As such, we've made the second argument in the method signature, source, have a default value of null.

For playing 2D sounds, we're not going to require every object in the scene that will play a sound to have AudioSource attached to it. Having too many AudioSource components in the scene can potentially lead to performance issues related to memory usage, processing overhead, and audio clipping if too many of those audio sources are playing simultaneously (see the *Optimization note* information that follows because Unity limits the number of audio sources that can play simultaneously).

For an audio source with a null value, we're first going to assume it's a 2D sound that will be played – 3D sounds will have their own audio source passed in. We'll use a private property to get an instance of AudioSource for playing the audio clip (using a property getter) instead of using a private method to return (i.e., get) the AudioSource reference in the class requiring it.

Let's go ahead and add the AudioSourcePlaySFX property to AudioManager now so we have a valid audio source component added to the scene for playing our 2D sounds:

```
private AudioSource AudioSourcePlaySFX
{
    get
    {
        if (_audioSource2D == null)
        {
            _audioSource2D = new
                GameObject().AddComponent<AudioSource>();
            _audioSource2D.spatialBlend = 0f; // 2D
        }
        return _audioSource2D;
    }
}
```

For this property, you can see we have a getter defined. It evaluates the private member _ audioSource2D (previously added) to see whether it already has an AudioSource component assigned to it, and if not, uses the AddComponent() method to add one to the scene (as a sibling on the same GameObject as AudioManager).

It then sets the spatialBlend value to ensure 2D playback of the sound and returns the new AudioSource to have its mixer group assigned and then passed back to the audio player component for playing the sound. Whew!

> **Optimization note**
>
> Note that for the `AudioSource` component that plays the SFX sounds, we have a limit to the number of audio clips that can be played simultaneously. `AudioSource` supports 32 voices, and each audio clip consumes 2 voices to play. Using the `PlayOneShot()` method, we can play several audio clips up to this limit, at which point it will start clipping. You may have already guessed that we can support additional audio clips by object pooling the `AudioSource` object via the `AudioManager` component. Refer back to *Chapter 6*.

Great, all our audio playback needs should be satisfied per our architecture at this point. Let's see how to implement playing SFX in our scene now before moving on to making additional audio player components.

Implementing AudioPlaySFX – Unity event

To add our first SFX to the game, let's revisit something you're already familiar with: our health pickup. Having followed along in the previous chapters – such as *Chapter 10*, section *Recharging aka healing* – you should have one in your scene already. Find and select it. Now, let's add the `AudioPlayerSFX` component to the root GameObject. As seen in the following figure, it's added just below the `Destroyer (Script)` component.

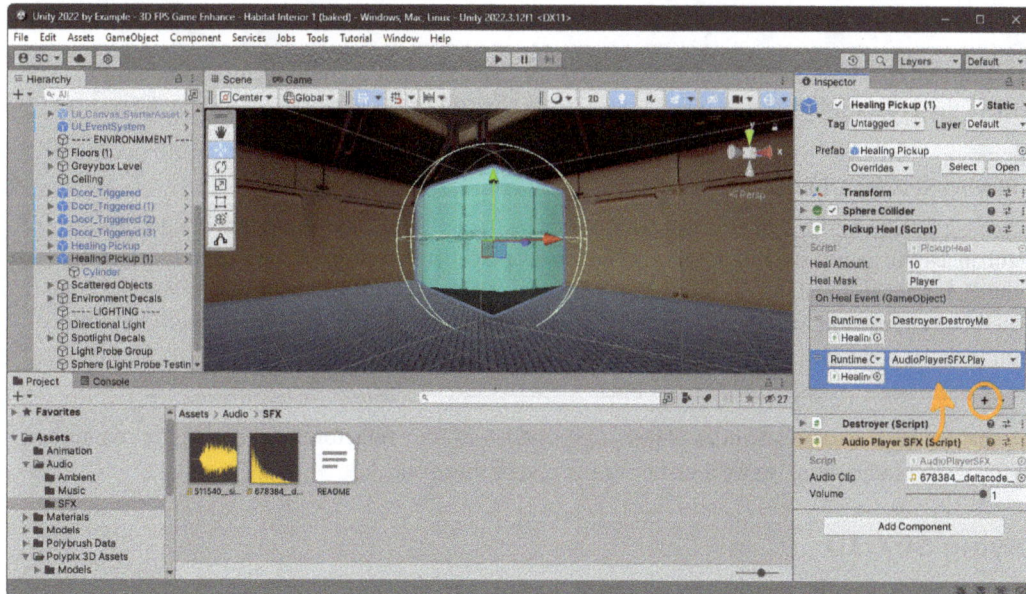

Figure 12.7 – Health pickup SFX

Assign `AudioPlayerSFX` to play a sound when `OnHealEvent` is triggered from the preceding `PickupHeal` component.

1. On the `PickupHeal (Script)` component, click the plus (+) to add a new entry to the `UnityEvent` field.

2. Click and drag the `AudioPlayerSFX` component (by its title bar) to the new entry.

3. In the function selector dropdown, select **AudioPlayerSFX | Play()**.

Okay, it's all wired up, but we're missing something… what sound will we play?!

Assigning the audio clip

The sound file we'll be using for our health pickup is `item-pickup-v1.wav`. It comes from the Freesound website (`https://freesound.org/people/DeltaCode/sounds/678384/`) and is licensed CC0, which means we can use it for free.

Ensure the `item-pickup-v1.wav` file is imported to the `Assets/Audio/SFX` folder. Unlike with the music audio file, we'll use the default import settings values for this sound file. Return to the `AudioPlayerSFX` component on the healing pickup, then click and drag the `item-pickup-v1` file to the **Audio Clip** field. At the top of the **Inspector**, click **Apply All** in the **Prefab | Overrides** dropdown to ensure all the health pickup instances in the level are updated to include playing the SFX.

Playtest the level, adjust the **Volume** field of the `AudioPlayerSFX` component as needed, and take notice of the 2D sound playback. Next, we'll play SFX in 3D space so you can hear the difference.

> **Playing SFX for UI feedback**
>
> Adding SFX to button clicks is also an essential aspect of game sound design. It significantly enhances the game's overall polish and improves player satisfaction. It may seem like a small detail, but it can substantially affect how players perceive and interact with your game UI. For a UI button, use the `AudioPlayerSFX` component and wire up the `Play()` method to the button's `On Click()` event in the **Inspector**.

Now that we have created an audio player component with our refactored approach, the remainder will follow the same setup, so we'll just bang the rest of these out.

Playing SFX 3D

Unlike with the `AudioPlayerSFX` component, where we did not add an `AudioSource` component to the GameObject for playing the 2D sound, we will add one here because we want a "speaker in 3D space" to emanate the sound. The player, via the `AudioListener` component on the `Player` object, will hear the sound as you naturally would in a real-world environment – whichever "ear" is turned toward the source of the sound will hear it. Hearing sounds in 3D space further builds on the player's immersion in your game world, so we definitely want to take advantage of 3D audio!

Noting the preceding information concerning AudioSource, we'll ensure the developer/designer adding a "play 3D sound effect" component to an object configures an audio source for this sound by requiring the AudioSource component to be added as a sibling.

Now, create a new script named AudioPlayerSFX3D in the Assets/Scripts/Audio folder, and start with the following code for the audio source requirements:

```
using AudioType = AudioManager.AudioType;

[RequireComponent(typeof(AudioSource))]
public class AudioPlayerSFX3D : MonoBehaviour, IPlaySound
{
    [SerializeField] private AudioSource _audioSource;

    private void OnValidate()
        => _audioSource = GetComponent<AudioSource>();
}
```

This is what we did here:

- [RequireComponent]: By decorating the class with this attribute, we require that a sibling component exist on this GameObject. Specifically, we need an AudioSource component because this will be a 3D sound.

- OnValidate(): We can use this Unity message event to pre-assign the _audioSource variable with the AudioSource instance added by the RequireComponent attribute. OnValidate() only runs in the Editor and is called when the script is loaded or a value changes in the **Inspector**.

- IPlaySound: Don't forget to add the interface! We're required to implement the interface for all our audio player components.

Additional reading | Unity documentation

You can read more about OnValidate() here:
https://docs.unity3d.com/2022.3/Documentation/
ScriptReference/MonoBehaviour.OnValidate.html

Let's continue by borrowing from the AudioPlayerSFX class; we'll also have _audioClip and _volume serialized private member variables. To those, we'll add an additional field for assigning the 2D to 3D sound ratio with a _blend2Dto3D variable (which we'll use to assign to the AudioSource. spatialBlend property).

So, add the following code:

```
public class AudioPlayerSFX3D : MonoBehaviour, IPlaySound
{
    ...

    [SerializeField] private AudioClip _audioClip;
    [Range(0f, 1f)]
    [SerializeField] private float _volume = 1f;
    [Tooltip("0 = 2D, 1 = 3D"), Range(0f, 1f)]
    [SerializeField] private float _blend2Dto3D;


    ...

    public void Play() =>
        AudioManager.Instance.PlayAudio
            (this, _audioSource);
}
```

The `Play()` method here is only slightly different from the previous `AudioPlayerSFX` implementation of `Play()` in that we're now passing the audio source reference to `AudioManager`. As a reminder, we need to do this because we want to assign playback to the correct Audio Mixer group before playing the sound.

Speaking of playing the sound, now, implement the `IPlaySound` interface public properties and methods:

```
public class AudioPlayerSFX3D : MonoBehaviour, IPlaySound
{
    public AudioType PlayAudioType => AudioType.SFX;

    ...

    public void PlaySound(AudioSource source)
    {
        source.spatialBlend = _blend2Dto3D;
        source.PlayOneShot(_audioClip, _volume);
    }
}
```

We've set `PlayAudioType` to play the sound using the SFX mixer group while the `PlaySound()` method remains largely the same – except we're setting the `source.spatialBlend` value before using `PlayOneShot()` to play the audio clip at the transform position of this object (in 3D space) and at the set volume level.

PlayClipAtPoint() | Unity documentation

Those of you familiar with the Unity scripting API may wonder why I didn't just use the `AudioSource.PlayClipAtPoint(_audioClip, transform.position, _volume)` static method.

Well, the reason falls back to the primary goal of implementing the `AudioManager` class the way we did – and that is to ensure the Audio Mixer groups are being utilized for all the distinct audio player components. While `PlayClipAtPoint()` does play an audio clip at a position in 3D world space, it does not work with the Audio Mixer, which ruled it out as an option.

You can read more about `AudioSource.PlayClipAtPoint()` here: `https://docs.unity3d.com/2022.3/Documentation/ScriptReference/AudioSource.PlayClipAtPoint.html`

`AudioPlayerSFX3D.Play()` now calls `AudioManager.Instance.PlayAudio()` with the added `_audioSource` parameter. The passed-in audio source is modified in `AudioManager` and passed back to the interface implemented `PlaySound()` method to use the audio player component's specific functionality to play the sound.

Note on code architecture

The relationship between `AudioManager` and the `AudioPlayerSFX3D` class may seem cyclical since we've implemented playing the sound with the interface and not within `AudioManager`. This is an okay trade-off in the architecture for me here because I've prioritized composition, ease of use, and eliminating errors for **Inspector** assignments in Unity. Doing things in Unity sometimes means developing novel approaches and compromises to otherwise "standard approaches" to C# OOP software development. I came to terms with that long ago. :)

We'll now follow through with the new 3D audio player component by looking at an implementation again.

Implementing AudioPlaySFX3D – Animation event

Much like how we revisited an object we previously created for playing SFX (the health pickup), we'll do the same here and add a 3D SFX to the door-opening animation we created in *Chapter 10*. The following figure shows the sliding door Prefab – we'll add the 3D sound effect when it slides open.

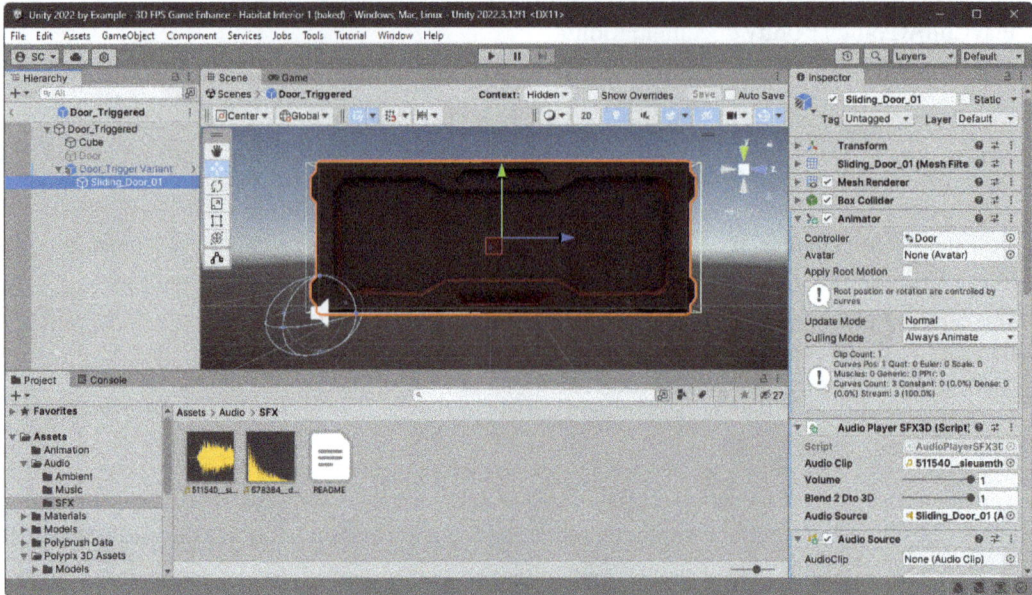

Figure 12.8 – Sliding door 3D SFX

Because the door does not have a component with an exposed `UnityEvent` being invoked, as with the health pickup, we have to trigger the sound playing differently. We'll still use an event, but this event is one we'll add to the door-opening animation directly.

Follow these steps to add the 3D SFX to the sliding door animation:

1. Open the `Door_Triggered` Prefab for editing.

2. Add the `AudioPlayerSFX3D` component to the same object as the `Animator` component. As seen in *Figure 12.8*, it's added just below the `Animator` component on the `Sliding_Door_01` child object.

3. Now, while `Sliding_Door_01` is still selected in the **Hierarchy** window, open the **Animation** window by going to **Window** | **Animation** | **Animation** (or press *Ctrl/Cmd + 6*).

4. Using the following **Animation** window figure as a reference, scrub the timeline a frame or two ahead, then click the **Add Event** button. This will add an **Animation Event** check mark to the timeline and select it (it will be blue when selected and white when not).

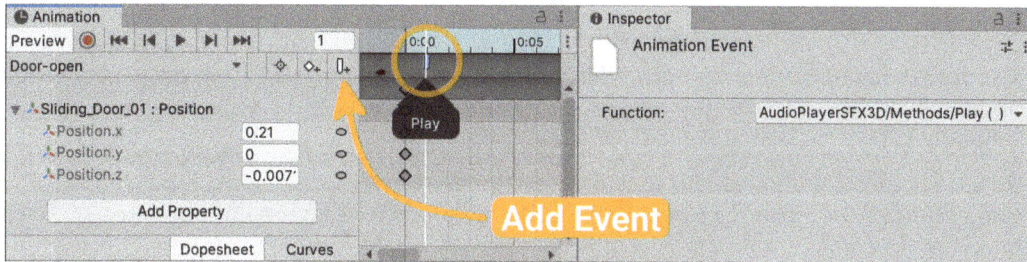

Figure 12.9 – Adding an animation event

When **Animation Event** is selected, the **Inspector** window will show a **Function** drop-down menu so that we can choose a method to invoke for the event.

5. Select **AudioPlayerSFX3D** | **Methods** | **Play()** in the **Function** drop-down.

Animation event | Unity documentation

The `AudioPlaySFX3D` component must be on the same object as the `Animator` component to select `Play()` in the **Function** drop-down from the **Animation** timeline.

Additional reading | *Using Animation Events*:
`https://docs.unity3d.com/2022.3/Documentation/Manual/script-AnimationWindowEvent.html`

---

6. Save the Prefab (or apply overrides) because we're done wiring up the sliding door for playing the 3D SFX!

Hold on a second, don't jump the gun and start entering Play Mode to test it out just yet… we still need to assign the sound the door will play when sliding open!

Assigning the audio clip

I was somehow tempted to grab the sound of the *Star Trek: The Next Generation* Enterprise NCC-1701-D's sliding door opening sound, but yeah, copyright laws and all. ;)

Not to worry, Freesound is to the rescue again! The sound file we'll be using for our sliding doors is `cua-ien-tu-mo.wav` (in Vietnamese, *cửa điện tử mở* or "open electronics"), which you can download here: `https://freesound.org/people/SieuAmThanh/sounds/511540/`. It is also licensed CC0.

Ensure the `cua-ien-tu-mo.wav` file is imported to the `Assets/Audio/SFX` folder and go back to the `AudioPlayerSFX3D` component by opening the `Door_Triggered` Prefab again. Click and drag the `cua-ien-tu-mo` file to the **Audio Clip** field, then click **Save**.

Playtest the level, adjust the **Volume** field of the `AudioPlayerSFX3D` component as needed, and take notice of the 3D sound playback.

> **3D Sound Settings | Unity documentation**
>
> Use the **3D Sound Settings** section of the `AudioSource` component to adjust the sound as desired for the spatial effect you're looking for. The parameters in this section are applied proportionally to the **Spatial Blend** parameter.
>
> You can learn more about audio source properties here: `https://docs.unity3d.com/2022.3/Documentation/Manual/class-AudioSource.html`.

The next audio player component we'll tackle is also for 3D sound – in the next section, let's make an audio player for ambient sounds.

Playing ambient sound

Adding ambient 3D sounds to a game's environment is essential for defining the setting, bringing it to life, and enhancing the player's sensory experience. While adding ambient sound to our environment, let's keep in mind that it's crucial to match the sounds to the setting, layer different sound elements to create the soundscape, and change or vary the sound – interactively or not.

We first need our ambient audio player component to get started adding ambient sounds to our level. Because ambient sound is 3D in nature, we'll again take the same approach as our 3D SFX and rely on an audio source added to our object to produce its sound.

We'll use `Play()` on the `AudioSource` component directly this time because other methods, such as `PlayOneShot()`, cannot trigger looping sounds. However, there are some limitations to using `AudioSource.Play()`. It only allows one audio clip to be played simultaneously with an audio source. Calling `Play()` on an audio source that is already playing will stop the audio source and start the clip again. But we only care about the ability to loop the audio for playing ambient sounds in the environment so these limitations are not an issue here.

Let's take care of setting up the `AudioSource` component right away. Create a new script named `AudioPlayerAmbient` in the `Assets/Scripts/Audio` folder and add the following code:

```
[RequireComponent(typeof(AudioSource))]
public class AudioPlayerAmbient : MonoBehaviour, IPlaySound
{
    [SerializeField] private AudioSource _audioSource;

    private void OnValidate()
        => _audioSource = GetComponent<AudioSource>();
}
```

Next, we'll need the **Inspector** field to assign the audio clip for the ambient sound to play on this object and the required play methods. Add the following code:

```
public class AudioPlayerAmbient : MonoBehaviour, IPlaySound
{
    ...
    [SerializeField] private AudioClip _audioClip;

    private void Start() => Play();

    public void Play() =>
        AudioManager.Instance.PlayAudio(this, _audioSource);
}
```

The specifics related to the methods added to play ambient sounds are as follows:

- `Start()`: Ambient sounds always play! (That's a rule I just came up with.) So, we'll start playing the ambient sound when the level begins. Note that because this is a 3D sound and configured with its specific audio source attached to the 3D object in the environment, the player will only hear it when they are within range.

- `Play()`: Similar to what we did before, we're calling the audio manager `PlayAudio()` method and passing in the ambient sound's `AudioSource` component to modify it and assign the correct Audio Mixer group for playback.

Again, implement the `IPlaySound` interface's public properties and methods so the audio manager knows which Audio Mixer group to use:

```
using AudioType = AudioManager.AudioType;
...
public class AudioPlayerAmbient : MonoBehaviour, IPlaySound
{
    public AudioType PlayAudioType => AudioType.Ambient;
    ...
    public void PlaySound(AudioSource source)
    {
        source.clip = _audioClip;
        source.spatialBlend = 1f;    // 3D
        source.loop = true;
        source.Play();
    }
}
```

This is an ambient sound, so we must set `PlayAudioType` to `Ambient` and then play the audio clip indefinitely for this audio player component's specific `PlaySound()` method functionality, like this:

1. Set `source.clip` to the `_audioClip` value (set in the **Inspector**).

2. Set `spatialBlend = 1f` much like how we enforced music to play in 2D (`0f`) within the `AudioPlayerMusic` component. Here, we'll enforce the audio to play as a fully 3D sound.

3. Set `source.loop` to `true`. This is because ambient sounds continuously loop indefinitely!

4. Play the sound using `source.Play()` because we cannot use `PlayOneShot()` for a looping sound.

Okay, these various audio-playing components have really come together! In the next section, let's continue providing example implementations for each with an ambient sound.

Implementing PlayAmbientSound

First things first, we need to identify something in our environment to use as an example ambient sound implementation. Let's look at the models provided by Polypix Studios for an answer… I see we have a 3D model Prefab asset imported to the `Assets/Polypix 3D Assets/Prefabs` folder called **Ventilation 1**.

Let's use this as a ventilation shaft inlet somewhere on our level. These types of things generally have something running – making noise – that pulls and circulates the air. A perfect environmental audio addition!

> **Art assets**
>
> The art assets used in this section are available from the GitHub project files repo. The `Ventilation 1` Prefab, in particular, is available from the `Art-Assets` folder included in the `3DArtwork.zip` file (or from the Unity project files directly): `https://github.com/PacktPublishing/Unity-2022-by-Example/tree/main/ch11/Art-Assets`.

Add the `Ventilation 1` Prefab from the `Assets/Polypix 3D Assets/Prefabs` folder to the level somewhere – by dragging it in from the **Project** window – to give a lovely ambiance to one of the habitation station's rooms. In *Figure 12.10*, I've added it to a rather sparse-looking room that needs some interest.

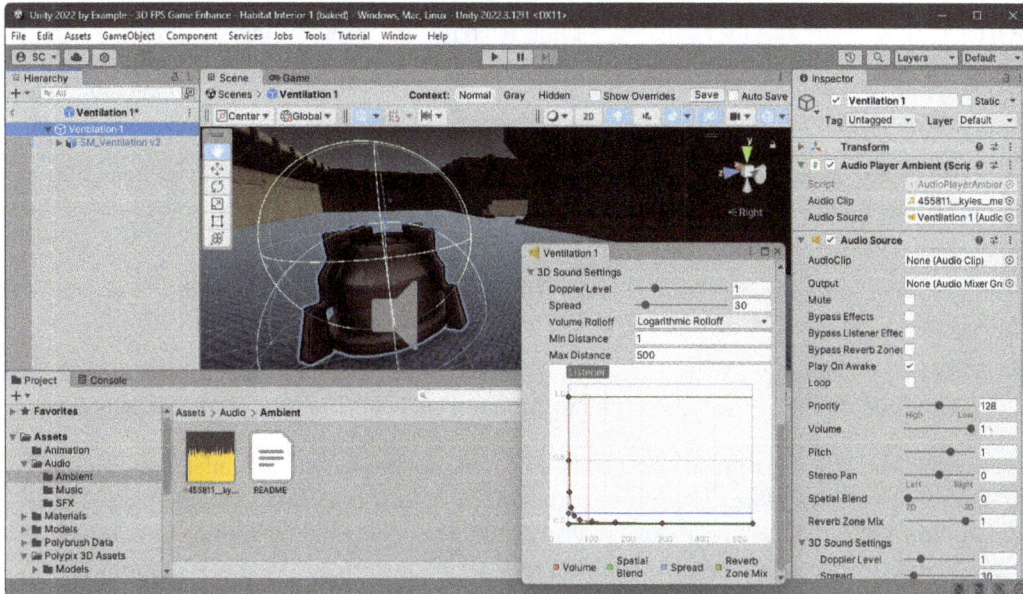

Figure 12.10 – Ventilation ambient sound

With the ventilation shaft inlet added to the scene, let's add the ambient 3D sound audio player component.

1. Open the `Ventilation 1` Prefab in Prefab Edit Mode.

2. Use your preferred method to add the `AudioPlayerAmbient` component to the root object.

3. An `AudioSource` component will automatically be added when we add the audio player component (due to the `RequireComponent` attribute).

4. Unlike the previous audio player components, where we provided fields for **Volume**, **Spatial Blend**, and so on, we'll configure the sound playing properties directly on the `AudioSource` component to tailor the sound to the proper environmental audio effect for its location.

Before we continue configuring the audio source, it would help to have a sound to play so we can make the proper adjustments.

The sound file we'll use for our ventilation shaft inlet is `metro-subway-hallway-corner-noise-heavy-ventilation-rumble.flac` (no, I didn't name it!). It comes from Freesound (`https://freesound.org/people/kyles/sounds/455811/`) and is licensed CC0 (still free to use).

Ensure the `metro-subway-hallway-corner-noise-heavy-ventilation-rumble.flac` file is imported to the `Assets/Audio/Ambient` folder. We'll make a few changes to the import settings values for this sound file.

Referring back to *Figure 12.4* for an import settings reference, set the following:

- **Force To Mono** = true, **Normalize** = true: Disregard the L/R stereo channels of the sound file (combine them) and set the audio levels to a normalized value.

- **Load In Background** = true: Loads the larger-sized audio file without causing the main thread to lag. Sound will start playing once the file is loaded (which may not be at the start of the scene being loaded).

- **Load Type** = **Streaming**: Decode audio with minimal memory use from disk and using a separate CPU thread.

- **Quality** = 30: Balance playback quality and file size by adjusting the compression slider for a compressed clip. Keep the file small for distribution while maintaining playback quality.

Return to the `AudioPlayerAmbient` component on the ventilation shaft inlet, then click and drag the `metro-subway-hallway-corner-noise-heavy-ventilation-rumble` file to the **Audio Clip** field. Finish by saving the `Ventilation 1` Prefab (by clicking **Save** in Prefab Edit Mode, or at the top of the **Inspector**, click **Apply All** in the **Prefab | Overrides** dropdown).

Before we playtest the level with the ambient sound playing, let's review the 3D sound settings that affect how ambient sound can be heard in the environment.

3D ambient sound settings

It's time to put our sound designer hat on! When we added the audio player component to the ventilation shaft inlet Prefab, I mentioned we'd tailor the sound directly using the audio source properties.

Referring to the **Ventilation 1** object's property window, shown in *Figure 12.10*, I've adjusted the **Spread** value from its default value to 270 to influence how the sound is distributed in 3D space. Lower values create more directional sound (heard only in front of the source), whereas higher values produce more omnidirectional sound (can be heard from wider angles). A value of 360 would make it seem like the sound is coming from everywhere around the listener.

Tip | Open the Properties window

In the **Inspector**, right-clicking on the **AudioSource** component title, then clicking **Properties…** will open it in a floating **Properties** window. You can do this for objects, components, and file assets from most windows in the Unity Editor.

Playtest the level, adjust the **Spread** value of the audio source as desired, and take notice of the ambient sound's playback.

In this section, we learned how to create components to play audio for different auditory experiences and implement them on objects in our environment. In the next section, we'll add to immersion in the environment by adding footstep sounds to the player.

Enhancing the audio experience with footsteps and reverb zones

Creating an immersive and enjoyable gaming experience requires paying close attention to sound design. Even seemingly simple elements, such as footstep sounds, play a significant role in grounding characters in their environment and conveying a sense of physicality and presence.

With footsteps in particular, it's crucial to have a variety of footstep sounds to randomize them and adjust tempo or cadence based on the player's speed to prevent auditory fatigue and repetition (you could even take it so far as to have different sounds for each surface type the player walks on).

Optimizing how continuously randomized audio clips are played back is also essential to ensure the process doesn't negatively impact the game's performance.

With our footstep sounds implementation, we will address all these factors. Considering the audio player components and manager code we've already written, the implementation is more straightforward than you may think.

Reusing audio player code

Nothing could be simpler than starting with one of our existing audio player components to play the footstep sounds! So, we'll rely on `AudioPlayerSFX` for the footsteps and not create footstep sounds with a new direct implementation of the `IPlaySound` interface. I mean, we could, but we'd just be duplicating the functionality that `AudioPlayerSFX` already provides. For those of you keeping track, the cards in play here are code reuse, **don't repeat yourself (DRY)**, **keep it simple, silly (KISS)**, and let's also claim the single responsibility card.

This time, instead of requiring a built-in component provided by the Unity game engine, we'll require our own `AudioPlayerSFX` component to be added to the GameObject that we're adding a new play footstep sounds component to.

Let's see this implementation by creating a new `AudioPlayerFootsteps` script in the `Assets/Scripts/Audio` folder with the following initial code:

```
[RequireComponent(typeof(AudioPlayerSFX))]
public class AudioPlayerFootsteps : MonoBehaviour
{
    [SerializeField]
    private float _walkInterval = 0.5f;

    private AudioPlayerSFX _playerSFX;
    private float _timerStep;

    private void OnValidate()
```

```
        => _playerSFX = GetComponent<AudioPlayerSFX>();

    private void Start()
        => _timerStep = _walkInterval;
}
```

Remember, the `RequireComponent` attribute forces a composition pattern by combining components to implement the desired functionality. So, as mentioned previously, we've required the `AudioPlayerSFX` component and pre-assigned the reference to the `_playerSFX` variable in the `OnValidate()` method.

We also declare these variables:

- `_walkInterval`: This variable value should match the cadence, or speed, of the player walking – in other words, the time between the footstep sound playing

- `_timerStep`: This variable will hold the current interval for footstep sounds to play (spoiler alert: we'll have different values assigned based on whether the player is walking or sprinting)

Okay, we have the base boilerplate code in place. Let's add our footstep audio clips list variable and an update loop to play a random footstep sound from the list at the assigned time interval:

```
public class AudioPlayerFootsteps : MonoBehaviour
{
    ...

    [SerializeField]
    private AudioClip[] _footstepSounds;

    ...

    private void Update()
    {
        float currentStepInterval = _walkInterval;

        _timerStep -= Time.deltaTime;
        if (_timerStep <= 0)
        {
            _playerSFX.Play(_footstepSounds[
                Random.Range(0, _footstepSounds.Length)]);
            _timerStep = currentStepInterval;
        }
    }
}
```

And here, we've done precisely what's needed:

- `_footstepSounds`: Here, we have our `AudioClip[]` array of footstep sound file assets that will be randomly selected and played at the set time interval. Refer to *Figure 12.11* for a sneak peek at this **Inspector** assignment.

- `Update()`: The update loop will keep our `_timeStep` variable current with its new time. Then, we'll evaluate to see whether `_timeStep` has expired to play the following randomly selected clip. We finish up by resetting `_timeStep` to delay the next play at intervals.

Looking good! Optionally, we can do a bit better for code readability, and that would be with a quick refactor extracting the code that gets a random footstep sound into a local function, like so:

```
AudioClip GetRandomFootstepClip()
    => _footstepSounds[
        Random.Range(0, _footstepSounds.Count)];
```

The line of code for playing the SFX would now look like this:

```
if (_timerStep <= 0)
{
    _playerSFX.Play(GetRandomFootstepClip());
```

It's, again, a quick, optional refactor that doesn't change the functionality. Still, one that makes for better code readability – and anything that helps code readability is worth the little bit of extra work for clarity (for someone else looking at your code or yourself in six months), and if it requires a bit more, add a code comment!

> **Complete code for the audio player components**
>
> Don't forget, at any time, if you need to see the completed code for these sections, you can find it at the GitHub repo here:
>
> `https://github.com/PacktPublishing/Unity-2022-by-Example/tree/main/ch12/Unity-Project/Assets/Scripts/Audio`

The `AudioPlayerFootsteps` component doesn't play sound itself. As we know, it uses `PlaySoundSFX.Play()` and passes an audio clip as the parameter. The only problem is that `Play()` doesn't currently accept a parameter! Let's fix that now by adding a method overload.

Adding a method overload to AudioPlayerSFX

So, now, we'll overload the `Play()` method in `AudioPlayerSFX` to add the required `AudioClip` parameter for the current audio clip for the footstep we want to play (well, that was randomly selected for us to play).

In `AudioPlayerSFX`, add the following method with its code:

```
public void Play(AudioClip clip)
{
    _audioClip = clip;
    AudioManager.Instance.PlayAudio(this);
}
```

We already have a `Play()` method without a parameter, so we're overloading the `Play()` method now by declaring another with the same method name but with a different method signature (because we've added the `AudioClip` parameter). Now, when `Play()` is called, the one that matches the method signature will be the specific method executed – either with the passed-in audio clip or without.

> **Tip**
>
> Avoid frequent clip changes: It's important to remember that repeatedly changing an `AudioSource` clip can be less efficient than using multiple audio sources or the `AudioSource.PlayOneShot()` method. In such cases, it's recommended to use the `PlayOneShot()` method, as it allows you to play a clip without changing the audio source's main clip.

We can now move on to adding the footstep sounds to our player character in the next section.

Implementing AudioPlayerFootsteps

Okay, this will be our fifth implementation of an audio player component, so there's no reason to delay. Let's go right ahead and add to `PlayerCapsule`. Since we'll be requiring some of the other components already on the `PlayerCapsule` object in the next section (such as `CharacterController` and `PlayerInput`), we'll want to add `AudioPlayerFootsteps` right to the root of this object.

Now that we've added `AudioPlayerFootsteps`, you should have seen `AudioPLayerSFX` added, thanks to the `RequireComponent` attribute, so that takes care of the ability to play sound through the SFX Audio Mixer channel. All that's left to do to get something playing is to add the sound files to the `_footstepSounds` array in the **Inspector**.

The sound files we'll use this time come from the Unity Asset Store. We'll use `Classic Footstep SFX (Free)` by Matthew Anett (`https://assetstore.unity.com/packages/audio/sound-fx/classic-footstep-sfx-173668`). From its name, you can see that we can use it freely in our projects. Yay!

Ensure the `Classic Footstep SFX (Free)` package is imported to your project. By default, this will be the `Assets/Classic Footstep SFX` folder. Unlike with the other audio files, since this is a package, the import settings are already set for us by the author, so we're all set to start using them.

Follow these steps to assign the provided footstep sound files to the `AudioPlayerFootsteps` component's `_footstepSounds` array field:

1. Return to the `AudioPlayerFootsteps` component on `PlayerCapsule` and lock the **Inspector** window (using the little *lock* icon at the top-right of the **Inspector** tab). By locking the window, no matter what we select, the **Inspector** will stay on this window, which is essential for selecting multiple objects in the **Project** window to be assigned to a component field.

2. With the **Inspector** window locked, go to the `Assets/Classic Footstep SFX/Floor` folder in the **Project** window.

3. While within the **Project** window folder, you can press *Ctrl/Cmd + A* to select all the sound files or click on the first sound file at the top (e.g., `Floor_step0`), then hold *Shift* and click on the last sound file, or any range of files to select them.

4. With the sound files selected, click and drag (from anywhere in the selection) to the **Footstep Sounds** field label, and you'll see the mouse cursor change from an arrow to an arrow with a box and a plus sign. Releasing the mouse button while hovering over the field name will populate the array with all the sound files in the selection.

5. You can unlock the **Inspector** window now and check the assignment of the sound files by expanding the array (with the arrow to the left of the **Footstep Sounds** field label) as seen in *Figure 12.11*. Note that I've only shown five assigned for brevity in the figure.

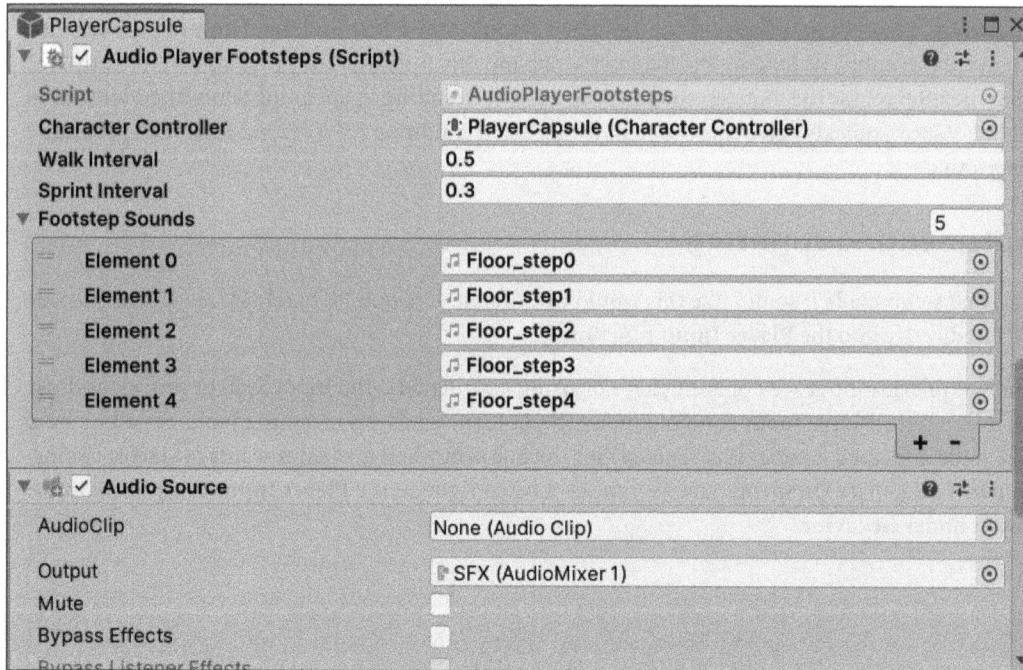

Figure 12.11 – AudioPlayerFootsteps component assignments

If you try to use the `AudioPlayerFootsteps` code as it sits now, you're going to be hearing footsteps all the darn time because there are no conditional statements to tell us when we should not play footsteps! That just won't do, so let's modify `AudioPlayerFootsteps` with a conditional check.

First, we need a reference to the player character controller... so add a new field to `AudioPlayerFootsteps` to hold the reference (you'll have to set the reference in the **Inspector** for this field manually, so don't forget to – you inevitably will fail, I know I do):

```
[SerializeField]
private CharacterController _characterController;
```

We can now evaluate "Is the player grounded or moving?" by adding the following `if` statement:

```
private void Update()
{
    if (!_characterController.isGrounded
        || _characterController.velocity.magnitude <= 0)
        return;
    ...
```

Using `return`, we short-circuit the `Update()` method so it doesn't continue and, therefore, doesn't play any footstep sounds. Groovy.

At this stage, playtesting will give us our footstep sounds only while walking. However, we can also have the player sprint by holding the *Shift* key while moving forward. The problem we currently face is that the interval for the footstep sounds is consistent, resulting in sprinting sounding identical to walking. We can quickly solve this problem by adding a condition for a second interval, specifically for sprinting.

Implementing sprinting

Okay, that setup made it sound like this would be difficult. It is not. It's barely inconvenient because we'll tap directly into the **Player Input** functionality.

When we provide input – by pressing keys or using a gamepad – the Input System processes those keys through the **Player Input** component, which, in turn, sends out message events. So, all we have to do is add a method handler (i.e., listener) for the one sent when the player wants to start sprinting. In *Figure 12.12*, that's **OnSprint**, and we can see it listed right on the **Player Input** component in the box just under **Behavior**.

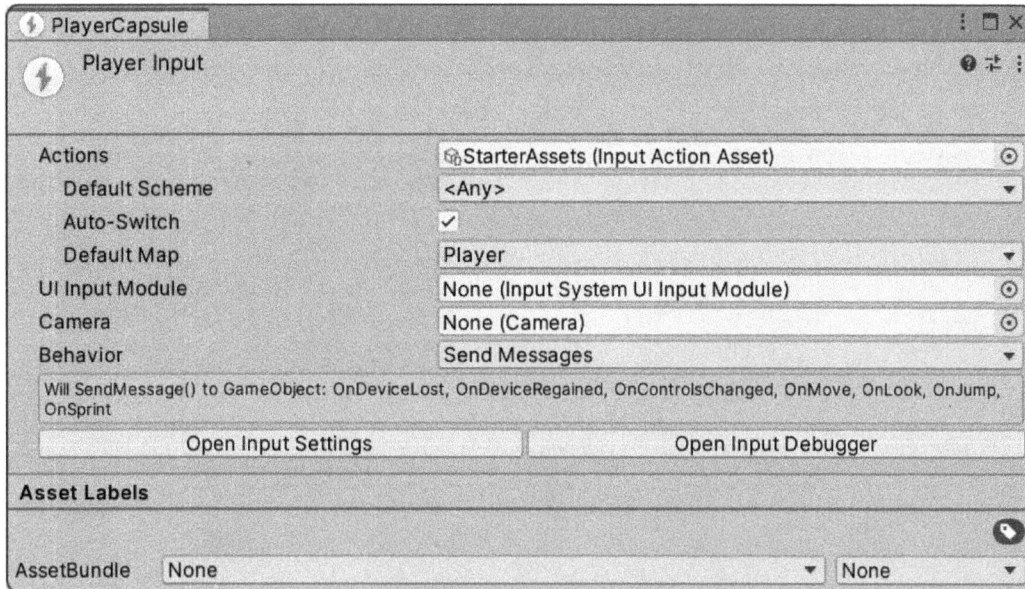

Figure 12.12 – Player Input SendMessage() to GameObject list

Adding the ability to respond to the player sprinting is now just a matter of updating the `AudioPlayerFootsteps` class with the required code. Add the following:

```
using UnityEngine.InputSystem;
public class AudioPlayerFootsteps : MonoBehaviour
{
    ...
    [SerializeField] private float _sprintInterval = 0.3f;
    private bool _isSprinting;
    ...
    public void OnSprint(InputValue value)
        => _isSprinting = value.isPressed;
}
```

We add a new variable to assign the time interval for sprinting in the **Inspector** with _sprintInterval and a related Boolean variable to evaluate whether the player is currently sprinting or not, with _isSprinting. Then, OnSprint() will set _isSprinting to true or false depending on value.isPressed passed in from the Input System.

All we need to do now is assign the proper step interval to the `_currentStepInterval` variable that will be used along with the `_timerStep` variable for playing footsteps at the correct time interval. Do so by making the following changes to the `Update()` method:

```
private void Update()
{
    if (!_characterController.isGrounded
        || _characterController.velocity.magnitude <= 0)
        return;

    float currentStepInterval =
        _isSprinting ? _sprintInterval : _walkInterval;
    ...
```

And that wraps up footstep sounds! Playtesting now should produce footsteps matching the player while walking or sprinting. This is precisely the attention to detail that sets your games apart from the field – players notice and appreciate an indie developer's efforts with things like this.

One additional attention to immersive sound details in your level that you can quickly add to the production value of your game soundscape is reverb zones, and the next section will show the simple steps to add one.

Adding reverb zones

Reverb zones simulate the acoustics of sound in different spaces, whether echoing in a large warehouse or dampening in a small storage room, adding depth to the soundscape and enhancing the game's atmosphere as the player moves between different areas in a level.

Additionally, simulating the acoustic properties can assist in storytelling and subtly guide player emotions and expectations, which is just another tool game designers can use to build more immersive and believable gaming experiences.

Find some key areas in your level where you feel the acoustics would be affected by the scale of the space. Using *Figure 12.13*, we'll use the following example to add a reverb zone to one of the larger rooms of the level:

1. Add an empty GameObject to the scene and position it in the center of a large room.

2. Rename the GameObject `Reverb Zone` (optionally, parent it to an object to organize all zones in the **Hierarchy**).

3. Add an `AudioReverbZone` component to the `Reverb Zone` object.

> **Additional reading | Unity documentation**
>
> You can read more about reverb zones here:
> `https://docs.unity3d.com/2022.3/Documentation/Manual/class-AudioReverbZone.html`

4. Choose an appropriate **ReverbPreset** option. Or, select **User** and customize the property sliders underneath to your desired effect. In the following figure, you can see I chose the **Hangar** preset.

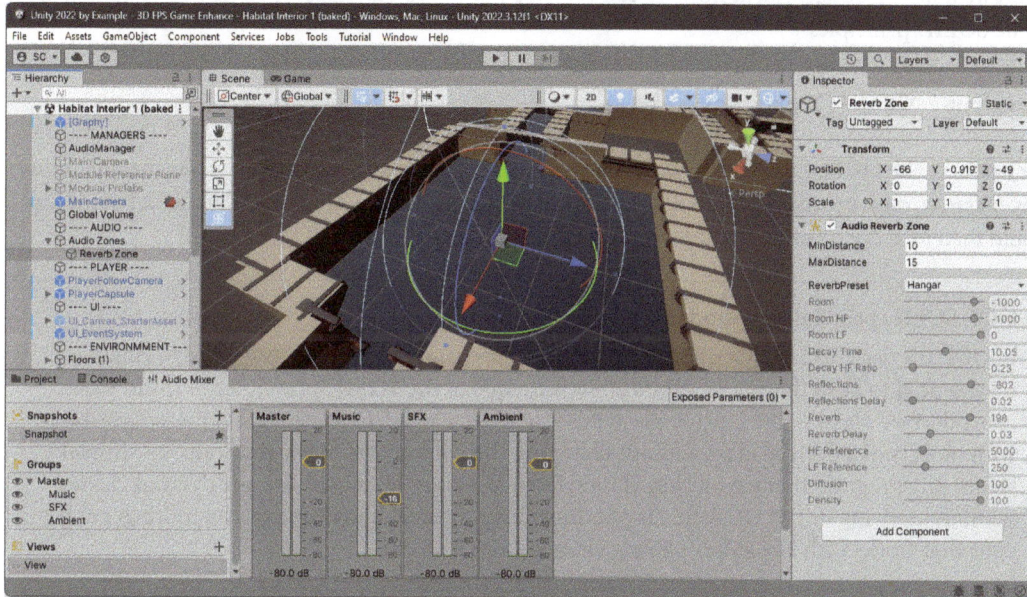

Figure 12.13 – Reverb zone placement and settings

And that's all that's required to add the reverb zone to our level – playtest and adjust the zone properties to your liking. However, we'll want to ensure that the music playback isn't affected by the reverb zone (that wouldn't make sense for the music to be affected by the environmental changes, would it?).

We can do this very easily in our code – we *have to* because we don't have a resident `AudioSource` sitting on a GameObject in our scene for playing music; we're adding it via code.

So, in our `AudioManager` script, in the `PlayMusic()` method, simply add this line to the list of audio source music settings we assign:

```
_audioSourceMusic.bypassReverbZones = true;
```

That's it! Our music playback will no longer be affected by the reverb zone, while the remainder of the playing sounds, especially the footstep sounds, will be.

This brings us to the end of adding audio to our game. In this section, we learned how to create varying audio playback components that route through an audio manager class to establish the rules for how sound is played with this system.

We now have an audio toolset at our disposal to handle most use cases for the different types of audio playback a game requires. You can now revisit the earlier projects in the book to add your own sound design and level up the gameplay experience for the player. Have fun while wearing your sound designer hat!

Next, we'll have a quick look at how we can refactor our audio player components to better adhere to the SOLID principles.

Deeper SOLID refactoring

We can take the OCP of the SOLID principles further by using an abstract base class that all audio player classes derive from. This can add further required implementations and default behavior and would then be a classic **object-oriented programming (OOP)** inheritance example.

> **OOP SOLID principles reminder**
>
> In OOP, the derived class inherits all base class members and can also add its members. However, it's essential to keep the *Liskov substitution* principle of the SOLID principles in mind when using derived classes. The *L* principle states that objects of a base class should be replaceable with objects of a derived class without changing the correctness of the program. In simpler terms, any program that uses a base class reference should be able to use any derived classes without knowing it. For OOP, this is polymorphism, which allows us to write more general code that works with any audio-playing class.

Let's take a look at how we might write an abstract base class that the audio-playing classes in our project could inherit from:

```
using AudioType = AudioManager.AudioType;

public abstract class AudioPlayerBase : MonoBehaviour, IPlaySound
{
    [SerializeField] protected AudioClip _audioClip;

    public abstract AudioType PlayAudioType { get; }

    public virtual void PlaySound(AudioSource source)
        => source.PlayOneShot(_audioClip);

    public virtual void Play()
        => AudioManager.Instance.PlayAudio(this);
}
```

Here is what we did:

- `_audioClip`: The field is marked protected, which means inherited classes can access it. Makes things pretty easy, eh?

- `abstract PlayAudioType` (property): This abstract property ensures that each derived class will define its own `AudioType` value. This is a good use of an abstract property because it enforces each audio-playing class to specify its own `AudioType` value.

- `PlaySound()`: The method provides a default implementation for playing a sound that all inheriting classes can share. Because we included the `virtual` keyword in the method signature, if a derived class needs a different behavior, it can override this method. We must also define `PlaySound()` – and `PlayAudioType` – because we must implement the `IPlaySound` interface!

- `Play()`: The method provides a way to play the specified `AudioClip`. Here, we're delegating the play functionality to the `AudioManager` instance, adhering to our pattern and assigning the `AudioMixerGroup` group accordingly for the `AudioType` value. Note it's also `virtual`, so it can be overridden by inheriting classes if a different behavior is needed.

The `AudioPlayerBase` abstract base class provides a good foundation for creating specific audio-playing classes – as many different kinds as needed. We've enforced a consistent interface for our classes to provide default implementations for behavior while allowing for customization by the inheriting classes. Yay!

Now that you've seen how we can approach an abstract base class that all our audio-playing classes can derive from, go ahead and try the refactor yourself! The project code on the GitHub link for the chapter has already provided an example of a refactored `AudioPlayerSFX3D_Derived` component that inherits from the `AudioPlayerBase` abstract base class, so you can use that as a reference (give it an honest try, no peeking).

> **AudioPlayerBase completed code**
>
> To view the completed code for the `AudioPlayerBase` class and an example of an inherited class, visit the GitHub repo here:
> `https://github.com/PacktPublishing/Unity-2022-by-Example/tree/main/ch12/Unity-Project/Assets/Scripts/Audio`

In this section, we learned how to refactor our code to be more aligned with the SOLID programming principles. While our code is pretty complete already (I'm sure someone could always think of more types of audio player components to extend), this refactor makes it even easier to extend upon the code to add additional audio player component types without having to modify the manager class.

Summary

In this "audio chapter," we extensively covered adding audio to our game by introducing an audio manager and individual reusable audio player components. We used the audio manager as a fool-proof implementation for designers and developers alike to target the appropriate mixer group for setting playback levels for the game's soundscape.

We continued by coding individual audio components for sound playback of the different types of audio common in most games – music, SFX, and ambient sound. We created these reusable audio player components following SOLID programming principles so we can extend additional types of audio components without modifying the manager class. We finished with a footsteps sound example showing how we can compose components for a simple implementation and also quickly add an environmental audio effect with reverb zones.

In the next chapter, we'll continue to flesh out the game by adding an intelligent non-player character (NPC). We'll accomplish this by, again, refactoring our previous 2D code for 3D usage as well as introducing cutting-edge technology for dynamic enemies utilizing sensors, a behavior tree, and machine learning (AI) techniques.

Part 5:
Enhancing and Finishing Games

Game developers can achieve a remarkable gameplay experience through advanced AI-based interactions using **NavMesh**, **Sensors** – both visually via Raycast and audio – implemented with a **behavior tree**, and Unity's machine learning solution **ML-Agents**. By refactoring code from the 2D game, you will gain an understanding of the difference between implementing interactions in 2D and 3D.

You will learn how to easily convert the FPS game to an immersive **mixed-reality** (**MR**) experience with the **XR Interaction Toolkit**. In no time, you can create a sense of presence in a mix of virtual and real-world environments for players as they understand how to spawn and interact with objects to facilitate gameplay mechanics. You will take away an understanding of how to maintain smooth gameplay while targeting required devices with a simple but immersive game. Winning the game after battling and defeating a boss room encounter will bring all the systems together and provide you with a comprehensive understanding of a working game structure.

Unity Gaming Services is essential for commercial game developers who want to earn revenue from releasing a game while operating it as a **game as a service** (**GaaS**). Unity LiveOps services analyze data to respond to player demand and provide dynamic player content with **Remote Config**, while a feature-rich version control system, **Unity Version Control**, and **Build Automation** streamline the DevOps process. Additionally, you will learn how to distribute your game worldwide using leading game distribution storefronts.

This part includes the following chapters:

- *Chapter 13, Implementing AI with Sensors, Behavior Trees, and ML-Agents*
- *Chapter 14, Entering Mixed Reality with the XR Interaction Toolkit*
- *Chapter 15, Finishing Games with Commercial Viability*

Implementing AI with Sensors, Behavior Trees, and ML-Agents

In *Chapter 12*, we dove into all the required details for approaching adding audio to our games. We did this by introducing an audio manager and individual reusable audio player components so that designers and developers can easily add different types of game audio to create a sound experience that encompasses our players. We enforced good coding practices to ensure we're writing maintainable code, with an emphasis on reusability and extensibility, to simplify the challenges in our daily game developer lives.

Now that we have addressed the sound design of our game, we can continue to finish out the enemy **non-player character** (**NPC**) mechanics for our FPS game's level by implementing some basic **artificial intelligence** (**AI**). We'll accomplish adding simple AI navigation by reusing and refactoring our previous 2D components and code to 3D. We'll continue to discuss the elevation and sophistication of NPC systems with **behavior trees** and **machine learning** (**ML**) tools.

In this chapter, we're going to cover the following main topics.

- Refactoring the 2D enemy systems to 3D with NavMesh
- Dynamic enemies with sensors and behavior trees
- Introducing ML with ML-Agents

By the end of this chapter, you'll be able to create a remarkable gameplay experience for your players by implementing AI-based interactions with NPCs, seamlessly navigating the level environment, and executing a set of behaviors on their own. You will better understand the differences between 2D and 3D implementation because we'll revisit the 2D methods and refactor for 3D while making improvements in the process.

Technical requirements

You can download the complete project from GitHub at https://github.com/ PacktPublishing/Unity-2022-by-Example.

Refactoring the 2D enemy systems to 3D with NavMesh

In this chapter, the goal is to guide you through the comprehensive process of bringing this hovering adversary to life, turned against us by the evil plant entity invading our systems. Its mission is to patrol the corridors of the habitat station to prevent the player from eradicating the infestation and getting back to Kryk'zylx normalcy (whatever that is).

Figure 13.1 – Enemy hover bot on patrol

Way back in *Chapter 8*, for our 2D game, we solved the problem of enemy NPC navigation by using a simple patrol waypoint behavior where the enemy robot moves between two waypoints in 2D space – a left and a right.

Well, we're going to do something similar here. However, because we're now working in 3D space and have a more complex floor plan to navigate, we'll still set up a patrol path using waypoints. Still, we'll now use Unity's **AI Navigation** package and its **NavMesh** components to accomplish the task of navigating between them. NPC navigation is crucial for creating immersive game environments, and Unity's updated 3D NavMesh system, introduced in a Unity 2022.3 release, offers an efficient solution.

> **Additional reading | Unity documentation**
>
> Information about the AI Navigation package can be found here: `https://docs.unity3d.com/Packages/com.unity.ai.navigation%401.1/manual/`

Revising and refactoring previously programmed components to save on development time is something we've discussed a few times already since the inception of the 3D FPS project, and there will be no exception now. There is no sense starting from scratch every time; let's rely on the assets we already have in our bag of tricks and revise where required (ultimately, adding to our bag).

We'll need a bit of a refresher for the basic component structure to understand where 3D-related revisions, specifically for NavMesh refactoring, will be required. So, here we go. Let's review the following **Unified Modeling Language** (**UML**) diagram representing the behavior responsible for moving the NPC in the 2D project:

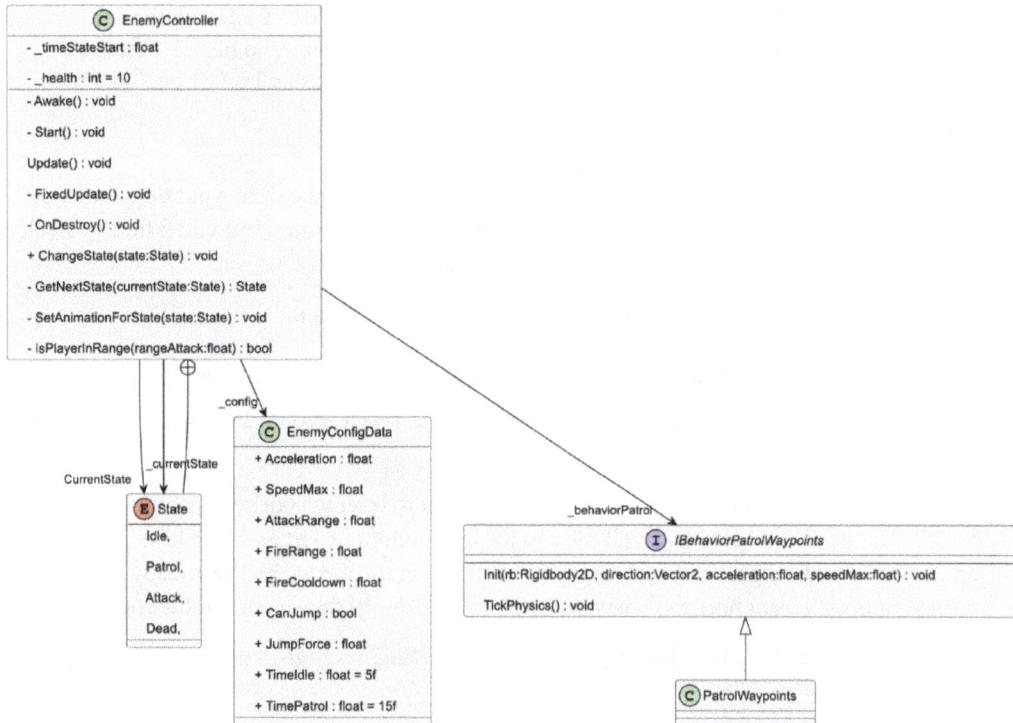

Figure 13.2 – 2D patrol waypoints reference UML

Interestingly enough, we're going to reuse about 96% of what we see in the preceding diagram (no, I didn't do the math; don't hold me to that percentage). It's primarily the `PatrolWaypoints` class that is in need of revising because it's the implementation of the `IBehaviorPatrolWaypoints` interface that actually moves the enemy NPCs between the waypoints. We simply need a behavior implementation for the NavMesh instead – because we did the work upfront, following good programming practices, to extend functionality without modifying the implementing class.

First things first, let's gather the 2D scripts for reuse.

Importing scripts from the 2D project

We currently require the scripts from the previous 2D adventure game project. We'll need several files this time, unlike the previous time we reused and refactored the 2D components for 3D in *Chapter 10*, in the *Refactoring environment interactions to 3D API methods* section, where we only needed a few files. So, for the scripts necessary for the 3D FPS project, let's use the Unity package exporter to obtain the scripts along with their dependencies.

> **Download scripts**
>
> Alternatively, download `patrol-behavior-and-dependencies.zip`, with the already exported scripts and dependencies required, from the book's GitHub repo here: `https://github.com/PacktPublishing/Unity-2022-by-Example/tree/main/ch13/Script-Assets`.

For a review of the process, you can revisit *Chapter 9*, where we exported the sliding puzzle game assets. This process certainly isn't one of the most fun parts of developing games, but you'll find yourself doing these types of things from time to time. Let's get to it.

Open the previous 2D adventure game project in Unity now and follow these steps:

1. Go to the `Assets/Scripts/Behaviors` folder in the **Project** window.

2. Right-click on the `PatrolWaypoints.cs` file and select **Export Package…**, as seen in *Figure 13.3 (A)*.

   The 2D patrolling waypoints behavior is what we want to leverage our reuse of for our enemy bot NPC in the 3D FPS game, so we'll use it as the basis for our export and grab its dependencies and other related scripts in the process. Don't worry – in the next step, I've already sorted out what scripts we'll want here to save you the time and stress of figuring it out for yourself.

3. Ensure you've selected the following files in the **Exporting package** dialog:

   - `Assets/Scripts/:`

     - `Bullet.cs`

     - `Enemy.cs`

     - `EnemyController.cs`

     - `PlayerShootingPooled.cs`

     - `ProjectileBase.cs`

     - `ProjectileDamage.cs`

     - `WeaponRanged.cs`

- `Assets/Scripts/Behaviors/`:

  - `PatrolWaypoints.cs`

- `Assets/Scripts/Data/`:

  - `EnemyConfigData.cs`

- `Assets/Scripts/ExtensionMethods/`:

  - `LayerMaskExtensions.cs`

- `Assets/Scripts/Interfaces/`:

  - `IBehaviorAttack.cs`

  - `IBehaviorPatrolWaypoints.cs`

- `IWeapon.cs`:

  - `IWeaponLaser.cs`

- `Assets/Scripts/Systems/`:

  - `EventSystem Example/`:

  - `ExampleListener.cs`

  - `ExampleTrigger.cs`

  - `EventSystem.cs`

You can see from the list of dependencies selected that we'll be using some of these later to fast-track the implementation of projectile shooting and events.

4. Once you have the files selected, click the **Export…** button and save the exported package file (with any name you choose) to a location where we can easily find it to import into our current 3D FPS project.

5. Now, open the 3D FPS project, if not already, so we can import the files.

    As shown in *Figure 13.3 (B)*, we'll be presented with the **Import Unity Package** dialog when we drag and drop the exported package from your system's file manager to the 3D FPS game's **Project** window.

6.   Click **Import**.

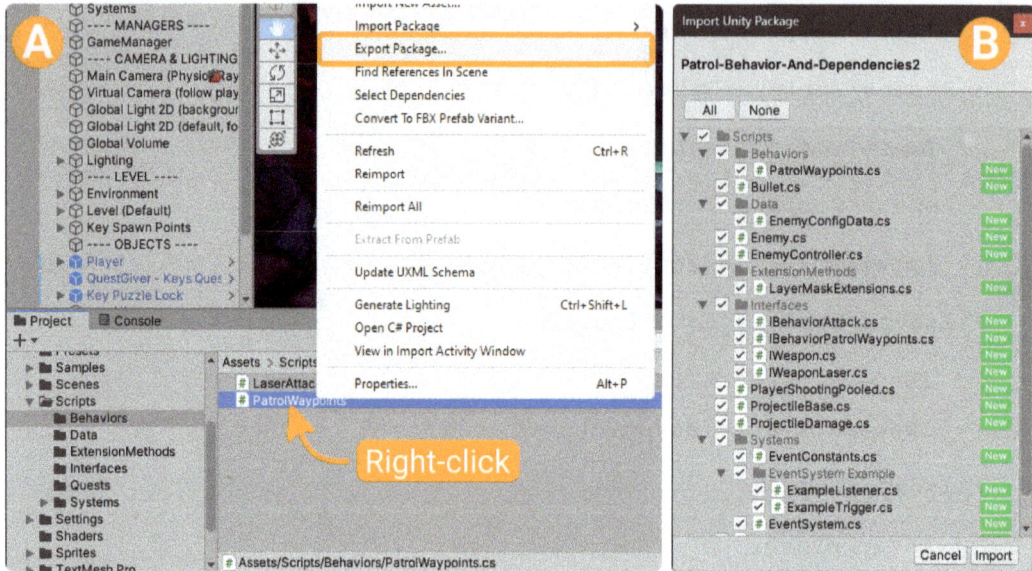

Figure 13.3 – 2D project export and 3D project import

You'll likely notice after the import finishes that we immediately have a console error:

```
Assets\Scripts\ProjectileDamage.cs(4,48): error CS0535:
'ProjectileDamage' does not implement interface member 'IDamage.
DoDamage(Collider, bool)'
```

We'll first – quickly and easily – fix this compiler error while the remainder of the changes will be focused on the refactoring of the 2D to 3D functionality. So, to fix the error, we simply need to change `Collider2D` to `Collider` in the `ProjectileDamage` class's `DoDamage()` method signature to properly implement the `IDamage` interface as defined in this 3D project.

Open `Assets/Scripts/ProjectileDamage.cs` in your IDE and change both the `UnityEvent` declaration and the `DoDamage()` method signature to the following:

```
public UnityEvent<Collider, bool> OnDamageEvent;


public void DoDamage(Collider collision, bool isAffected)
    => OnDamageEvent?.Invoke(collision, isAffected);
```

Simple.

Moving on with the remaining refactoring will require an additional step to ensure we can reference the required 3D navigation types. This simply means we must first add the **AI Navigation** package to our project.

Open **Package Manager** from **Window | Package Manager**. Ensure that **Unity Registry** is selected from the **Packages** menu drop-down list to set the context. Then, referring to *Figure 13.4*, enter nav in the search box to filter the available packages and select **AI Navigation** from the list view. Finish by clicking the **Install** button.

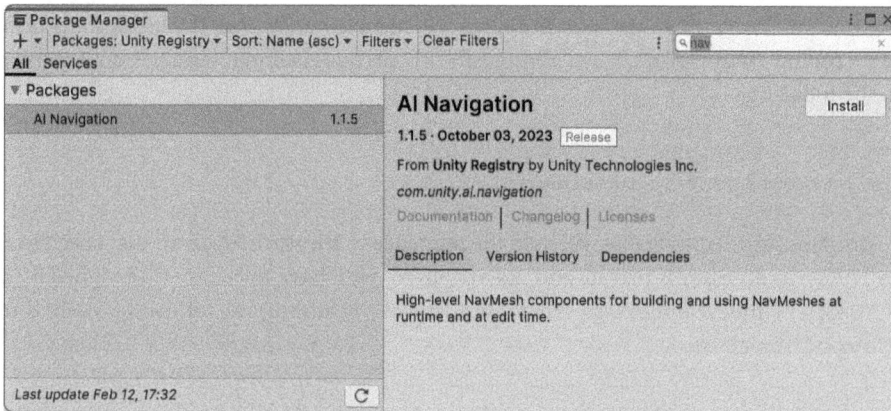

Figure 13.4 – Installing the AI Navigation package

Okay, now we're ready to move on updating our code reuse to use the NavMesh components.

Refactoring the PatrolWaypoints class for NavMesh

Unity's NavMesh system greatly simplifies setting up complex 3D waypoint navigation where NPCs can move more naturally and interact in believable ways, which contributes to enhancing the game's immersion for our players. The modular, adaptable code design (i.e., SOLID principles and composition) we adhere to also makes switching between 2D and 3D and adding or swapping components easy. That's why we'll be able to quickly make only a few updates to our existing 2D components to have the 3D FPS game up and running with patrolling enemy bot NPCs!

Primarily, as indicated earlier in the UML diagram, the most significant revision will be with the patrol waypoints behavior class. We'll actually end up replacing most of it, so open the newly imported `Assets/Scripts/Behaviors/PatrolWaypoints` script in your IDE now so we can get started. And the first thing we'll start with – beginning at the top of the script and working our way down – is the waypoint variables.

Updating the waypoint variables

We're simply going to replace the old transform WaypointPatrolLeft and WaypointPatrolRight variable declarations with the following single list of waypoint transforms declaration – serialized with the [SerializeField] attribute so they're assignable in the Inspector:

```
[Header("Patrol Waypoints")]
[SerializeField] private List<Transform> _waypoints;
```

Okay, off to a good start; we'll now be able to add any number of waypoints to the _waypoints list for the enemy bot NPC to navigate. We'll tackle the remaining variables next.

We'll replace the movement-related private member variables with only the following two – _waypointCurrentIndex and _navMeshAgent – to hold the reference to NavMeshAgent and to keep track of the current waypoint the agent will head towards from the list of transforms we add to the preceding _waypoints list. Remove the 2D private member variables now and replace them with the following:

```
private int _waypointCurrentIndex = 0;
private NavMeshAgent _navMeshAgent;
```

Since the _navMeshAgent variable will hold the reference to NavMeshAgent that's used to move our enemy NPC – its accessor is private and not serialized, we'll have to ensure it's assigned by using the Unity Awake() lifecycle message event. Do that now by adding the following method to the PatrolWaypoints class:

```
private void Awake()
{

    _navMeshAgent = GetComponent<NavMeshAgent>();

    Debug.Assert(_navMeshAgent != null,
        $"[{nameof(PatrolWaypoints)} NavMesh agent is null!]",
            gameObject);
}
```

We'll be adding the NavMeshAgent component to our enemy patrol bot when we set up its Prefab in the *Configuring the enemy NavMesh Agent (Prefab)* section. So, if we forget to add the component, GetComponent() will return a null assignment to the _navMeshAgent variable, and that just won't do! To make ourselves aware of this potential forgetfulness, we'll use a Debug.Assert() statement – if the condition is not met, an error message will be logged to the Unity console.

Additional reading | Unity documentation

Information about Debug.Assert can be found here: https://docs.unity3d.com/2022.3/Documentation/ScriptReference/Debug.Assert.html.

That takes care of the variables! We can now move down to our first method to update, which involves getting things set up for the new patrolling behavior.

Initializing the patrolling behavior

Now, let's look at the Init() method. First, we'll change the method signature, specifically the parameters, because we no longer need the values for the rigid body nor the direction, as NavMeshAgent will now take care of these. We're just going to leave the acceleration and speed parameters since they still apply.

Modify the Init() method signature to the following:

```
public void Init(float acceleration, float speed) { }
```

For the contents of the Init() method, we're going to remove all the previous movement code and replace it with the following – for working with NavMeshAgent:

```
public void Init(float acceleration, float speedMax)
{
    _navMeshAgent.acceleration = acceleration;
    _navMeshAgent.speed = speedMax;

    if (_waypoints.Count > 0)
    {
        _navMeshAgent.SetDestination(_waypoints[0].position);
    }
    _waypointCurrentIndex = 0;
}
```

A quick breakdown of the code looks like this:

- _navMeshAgent.acceleration: We assign the acceleration that will be passed in from our enemy configuration data to determine how quickly the agent will reach its maximum speed.

- _navMeshAgent.speed: We assign the speed passed in from our enemy configuration data to set the maximum speed the agent will travel along its path to the current waypoint.

- _waypoints.Count > 0: A quick insurance check that we indeed have a list of waypoints to work with before attempting to get the position of one.

- _navMeshAgent.SetDestination: When the behavior is initialized, what waypoint will the NavMesh Agent travel to? This one. The index value of 0 addresses the first waypoint that was added to the _waypoints list.

- _waypointCurrentIndex: Again, specifying the first waypoint in the _waypoints list by its index value of 0 as the current waypoint the agent will travel to.

To quickly summarize what we did here, just as before in the 2D code, the Init() method is used to initialize the movement parameters for the NavMesh Agent. It is specifically designed to initiate navigation between waypoints as part of the implemented enemy NPC patrolling behavior.

NavMeshAgent parameters can be a vital part of game balancing. They allow developers to adjust NPC behavior and game difficulty to match the design – creating challenging areas or guiding gameplay experiences. Hint: you'll be adjusting these values later as you playtest the game level.

With the initialization of the behavior completed, we can move down further in the class to revise the required methods.

Revising the update methods

The `UpdateDirection()` method wasn't technically responsible for moving the object in the previous 2D implementation (the `UpdateVelocity()` method was). Still, to keep within the same logical naming, I've decided to reuse it here.

The previous `UpdateDirection()` code was simply responsible for facing (flipping) the patrolling sprite object in the direction of movement – since the 2D game is a side-view, we don't really have a rotation. Well, in a 3D perspective game, we certainly do have object rotation to contend with. Thankfully, the `NavMeshAgent` component will automatically calculate and apply the object's rotation while pathfinding its way between waypoints.

> **Tip**
>
> To have more control over the agent's rotation, you'll have to manage it manually. Keep in mind that if you choose to, you'll need to disable the NavMesh Agent's automatic rotation by setting `_navMeshAgent.updateRotation = false`.

Let's see what the new `UpdateDirection()` method looks like now. Replace the code with the following:

```
private void UpdateDirection()
{
    if (!_navMeshAgent.pathPending
        && _navMeshAgent.remainingDistance
            <= _navMeshAgent.stoppingDistance)
    {
        MoveToNextWaypoint();
    }
}
```

As we can see here, for the `UpdateDirection()` method, we're now dealing directly with `NavMeshAgent`. We're evaluating the current values for the progression using the `pathPending` and `remainingDistance` properties of the agent heading to the currently set destination waypoint. Precisely, we'll execute a call to `MoveToNextWaypoint()` to set moving to the next waypoint when the agent's current remaining distance is less than the stopping distance.

> **Important note**
>
> In this example, we rely on the `NavMeshAgent` component's default values for its stopping distance, obstacle avoidance, and path-finding values.

As you can see, we've added a new method to handle setting the next waypoint from the list of available waypoints for when the agent arrives at the currently assigned waypoint. If you're following along with the code directly, you can automatically use your IDE's refactoring tools to generate the new `MoveToNextWaypoint()` method. Either way, with the new method created, add the following code to it:

```
private void MoveToNextWaypoint()
{
    if (_waypoints == null
        || _waypoints.Count == 0)
            return;

    _waypointCurrentIndex =
        (_waypointCurrentIndex + 1) % _waypoints.Count;
    _navMeshAgent.SetDestination
        (_waypoints[_waypointCurrentIndex].position);
}
```

Nothing too tricky is happening here; it's just a simple *null check* and looping through the waypoints. Even so, a simple breakdown of these lines looks like this:

- `if (_waypoints == null || _waypoints.Count == 0)`: Doing a double-check on the available waypoints in the list here, we're ensuring it's not `null`, and the count for the number of waypoints in the list is not 0 (zero) before continuing to execute statements to assign the next waypoint for the agent to travel to.

- `_(_waypointCurrentIndex + 1) % _waypoints.Count`: The modulus operator (or remainder operator) ensures we loop back to the *beginning of the list* index when the `_waypointsCurrentIndex` value equals the list count.

- `_navMeshAgent.SetDestination`: This is the `NavMeshAgent` method we call to tell the agent to start heading toward the passed-in position – the index assigned in the preceding line.

> **Modulus or remainder operator % (C#)**
>
> A typical shorthand way of using the modulus operator is to guarantee a seamless infinite loop of a list of items by ensuring that the current item index progresses linearly through each list item and only wraps around to 0 upon reaching the end of the list.
>
> Additional reading can be found here: `https://learn.microsoft.com/en-us/dotnet/csharp/language-reference/operators/arithmetic-operators`.

In this section, we've sorted out the update methods to work with the new NavMesh system. Next, we'll still work with methods, but instead of revising, we'll remove ones that are no longer needed.

Cleaning up unused methods

Still working within the `PatrolWaypoints` class, we have some no longer needed methods. Instead of just leaving them in the class, let's clean them up – by removing them – so if we ever need to revisit the class later, we won't be confused by methods with zero references.

Remove the following methods from the `PatrolWaypoints` class now:

- `UpdateVelocity()`: The velocity is now calculated internally by `NavMeshAgent`.
- `SetWaypoints(Transform left, Transform right)`: We won't be spawning the patrolling enemy NPC NavMesh agents in this example. Consider this an opportunity to add a patrolling enemy spawner to your game later!
- Oh, and give "update `TickPhysics()` " a quick fix by removing the now invalid call to the removed `UpdateVelocity()` method, leaving simply this expression:

```
public void TickPhysics() => UpdateDirection();
```

Methods removed; check! To compare your revisions to the completed `PatrolWaypoints` class, you can find it here: `https://github.com/PacktPublishing/Unity-2022-by-Example/blob/main/ch13/Unity-Project/Assets/Scripts/Behaviors/PatrolWaypoints.cs`.

Updated the behavior interface? Next refactoring item to check!

Updating the behavior interface

This will be another quick section (very quick). The only changes we need to make for the `IBehaviorPatrolWaypoints` interface – you've probably already noticed the errors from saving the previous changes to the `PatrolWaypoints` class – are directly related to the implementing class's changes.

Update the `IBehaviorPatrolWaypoints` code to the following to resolve the issues:

```
public interface IBehaviorPatrolWaypoints
{
    void Init(float acceleration, float speedMax);
    void TickPhysics();
}
```

Lastly, in addition to the error we just resolved, we have a few additional ones to clean up before we can call this process entirely done.

Resolving remaining console errors

For the last bit of housekeeping on the 2D code refactoring, when we changed the behavior `PatrolWaypoints.Init()` method's signature in the `PatrolWaypoints` class, we caused an error in the implementing classes. So, we'll simply remove the unnecessary parameter variables to fix it.

Open the `EnemyController.cs` script and update the portion of the `if` block body shown to the following:

```
private void Awake()
{
    ...
    // Get behaviors and initialize.
    if (TryGetComponent<IBehaviorPatrolWaypoints>(
        out _behaviorPatrol))
    {
        _behaviorPatrol.Init(
            _config.Acceleration,
            _config.SpeedMax);
    }
    ...
```

For the `_behaviorPatrol.Init()` call, the only `config` variables we must pass now are the acceleration and speed values. Easy-peasy.

> **Completed 2D to 3D refactored code**
>
> It wasn't a ton of changes, but it certainly may have seemed that way as we stepped through each one at a time and touched multiple scripts. If it left you a bit fuzzy on the final state of the code, you can refer to the completed scripts in the GitHub repo here: `https://github.com/PacktPublishing/Unity-2022-by-Example/tree/main/ch13/Unity-Project/Assets/Scripts`.

Also, since we're no longer using a rigid body to move the enemy NPC between waypoints, we can remove all references to `Rigidbody2D` in the `EnemyController` class, which includes removing the following lines:

```
// In class EnemyController, delete:
  private Rigidbody2D _rb;
// In Awake(), delete:
  _rb = GetComponent<Rigidbody2D>();
// In FixedUpdate(), delete:
  else
      _rb.velocity = Vector2.zero;
```

Whew, we are done! We didn't even break a sweat with all the 2D to 3D refactoring we completed in this section, right!? We are nearly there to test out our game's new NavMesh setup for the 3D FPS enemy patrols. Two steps remain, including configuring the patrolling enemy Prefab next, and they're also no sweat for us to complete now.

Configuring the enemy NavMesh Agent (Prefab)

Setting up a NavMesh Agent for waypoint patrol requires technical and visual setup in the Unity Editor, as well as careful consideration of the bot's behavior for optimal player experience – nothing we haven't done before. But, because there are several steps involved in the process, let's first break it down into high-level tasks:

1. **Designing the enemy bot**: We'll select the 3D model to use for the NavMesh Agent and create the Prefab asset.

2. **Adding the enemy base type and controller components**: We'll add the required components to give the enemy object properties and state.

3. **Configuring the patrolling behavior**: We'll add the behavior component that implements patrolling waypoints.

4. **Adjusting NavMesh agent for the environment**: We'll set the NavMesh Agent type and properties for the enemy bot.

5. **Testing and tweaking settings**: We'll playtest and adjust values for the desired gameplay.

Not so bad, only five steps! Here we go; let's knock out the first step.

Designing the enemy bot

This will be super easy because we'll leverage more Polypix Studios 3D models provided for our game's use. We have a lovely hovering camera drone (armed with some type of weaponry) available in the assets we previously added to the 3D FPS project in *Chapter 11*, which will be perfect for a patrolling enemy!

Figure 13.5 – Patrol hover bot reporting for duty!

Here are the steps we'll follow to build out the enemy patrol bot. Use *Figure 13.6* as an object hierarchy reference while completing the following steps:

1. Create an empty GameObject in the Hierarchy, reset its transform, and rename it `Enemy Hover Bot A`.

2. In the `Assets/Polypix 3D Assets/Prefabs` folder, find the `SM_Camera_Drone Variant` Prefab and parent it to the new `Enemy Hover Bot A` object – seen in *Figure 13.5*.

    Remember that it's always preferred to maintain the graphics as a child of the root object to make future changes easily, should you want or need to.

3. Set the following transform values for the **SM_Camera_Drone Variant** object:

    -   **Position**: `(0, 0, 0)`
    -   **Scale**: `(1.5, 1.5, 1.5)`
    -   Then, for the child `SM_Camera_Drone` object, set its Y-position value to `0.6` so that it's a good height off the ground – this is a hovering patrol bot! We change the child mesh object because we want the parent object anchor to maintain its current position.

4. Back to the `Enemy Hover Bot A` root object, add the following components to it:

    -   `Enemy`
    -   `EnemyController`
    -   `PatrolWaypoints`
    -   `NavMeshAgent`

> **Additional reading | Unity documentation**
>
> More information about NavMesh Agent can be found here: `https://docs.unity3d.com/Packages/com.unity.ai.navigation%401.1/manual/NavMeshAgent.html`.

5. Now, make a Prefab by dragging `Enemy Hover Bot A` from the Hierarchy to the `Assets/Prefabs` folder.

The results of these setup steps can be seen in *Figure 13.6*. Note that for assigning the waypoints to the `PatrolWaypoints` component, we'll add the waypoints to the scene in the *Adding waypoints to the level and testing* section.

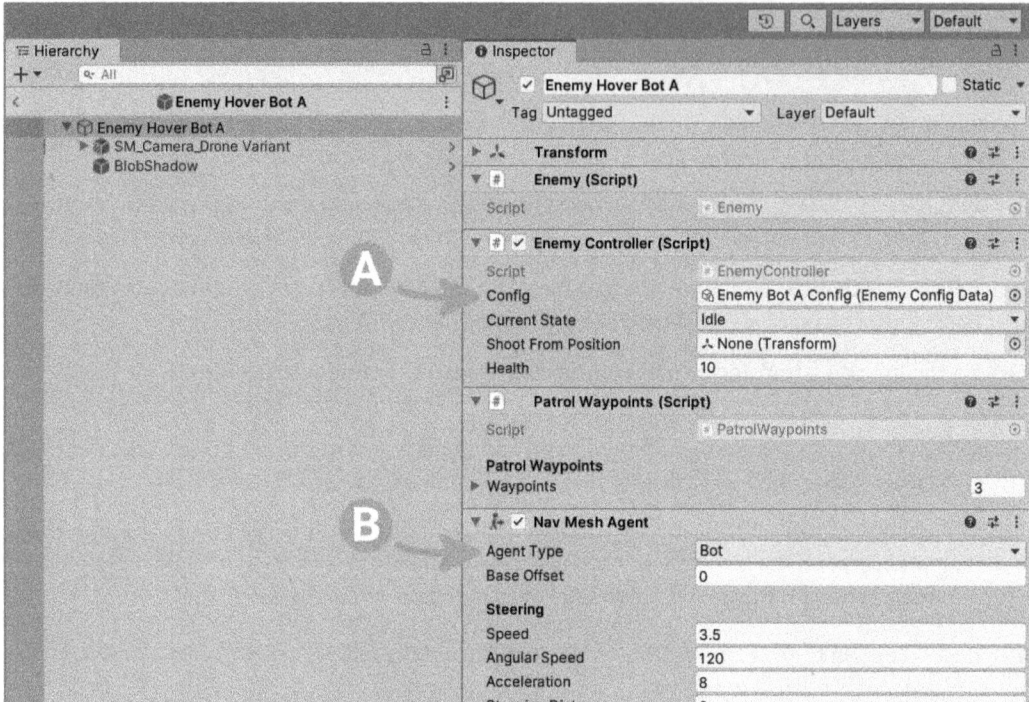

Figure 13.6 – Enemy hover bot setup

That was just the first part. The second part of the setup requires the enemy configuration data that we'll assign to the EnemyController component's **Config** field, seen as (*A*) in *Figure 13.6*.

Configuring the enemy bot properties

If you remember, in *Chapter 7*, we're using a ScriptableObject asset to hold varying configurations to have different property values (for example, to change the enemy's difficulty).

Following these steps, we'll create a new enemy configuration asset to assign right now:

1. Create a new Assets/Data folder in the **Project** window.

2. While within the new folder, use the **Create** menu and select **ScriptableObjects | EnemyConfigData**.

3. Rename the newly created file asset Enemy Bot A Config.

4. Set the following starting values:

   - **Acceleration** = 10

   - **SpeedMax** = 4

- **TimeIdle** = 0 (correct, we won't have the patrol bot idle)

- **TimePatrol** = Infinity (yes, **infinity** is a valid float value!)

5. Finally, assign the Enemy Bot A Config asset to the Enemy Bot A Prefab's Enemy component's **Config** field and save.

Bonus activity

Add a blob shadow projector to the enemy bot and light probes to the scene in the area of the level where the bot will be patrolling. This will maintain the high-quality visual fidelity of the game. You can refer to *Chapter 11* as a reminder of how we implemented these features.

And that's all there is to creating the NavMesh Agent enemy patrol bot Prefab! Only two tasks remain to get the bot on patrol: defining the navigation surface and setting the waypoints.

Baking the NavMesh Surface

Now that our NavMesh Agent enemy bot Prefab is ready and equipped with all the necessary components, baking only the desired patrol areas is the next crucial step. We accomplish this with a NavMesh Surface component. By selectively defining the patrol areas in our game's level, we can influence our level design for more strategic gameplay.

Additional reading | Unity documentation

More information about the NavMesh Surface can be found here: https://docs.unity3d. com/Packages/com.unity.ai.navigation%401.1/manual/NavMeshSurface. html.

Using *Figure 13.7* as a reference, you can see I've decided to have the enemy hover bot patrol the main corridors, indicated by the blue surfaces, connecting the two major sections of the habitat station. We use layers to accomplish selectively baking NavMesh surfaces. Comparatively, we've previously utilized layers in a physics context to identify and limit interactions – so, layers have different purposes in Unity.

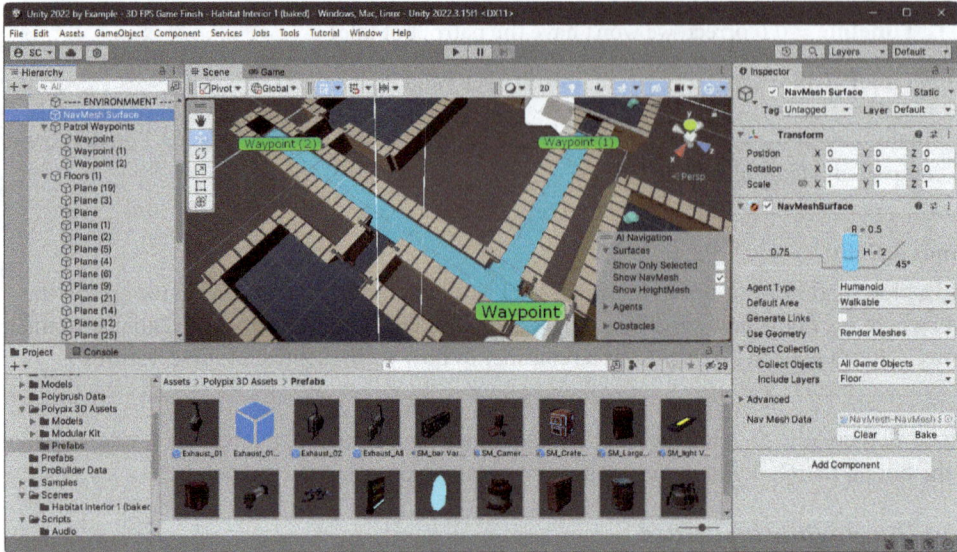

Figure 13.7 – NavMesh Surface baked and waypoints added

We'll first have to add a new layer to select the specific surfaces for NavMesh baking. Using the **Layers** drop-down list at the top-right part of the **Unity Editor** window, select **Edit Layers…** then enter Floor in the first available numbered **User Layer** fields.

Next, select the contiguous corridor floor sections in the **Scene** view while holding the *Ctrl/Cmd* key to add each to the current selection. Then, with the floor sections selected, assign the Floor layer by selecting it from the Inspector's **Layer** dropdown (to the right of the **Tag** dropdown). Done. On to adding the NavMeshSurface component.

Configuring the NavMesh Surface

After assigning floor sections to their designated layer, we're ready to add the NavMeshSurface component and bake the surfaces – it's like embedding invisible pathways that dictate the agent's movements. This baking process will empower the agents we set up to navigate simple and complex paths with speed and accuracy, ensuring seamless AI navigation for our patrolling enemy hover bot!

To bake the NavMesh floor surfaces now, follow these steps:

1. Add a new empty GameObject to the **Scene** Hierarchy and name it NavMesh Surface.

    You can place this just under an organizational object in the **Scene** Hierarchy with a title of "---- ENVIRONMENT ----".

> **Tip**
>
> Set the **EditorOnly** tag on organization-only GameObjects in the **Scene** Hierarchy so these objects aren't included in builds, saving some resources.

2. Add a `NavMeshSurface` component to the `NavMesh Surface` GameObject:

    - To ensure that only the floor sections we want the enemy bot to patrol get baked, in the **Object Collection | Include Layers** dropdown, first select **Nothing**, then select **Floor**.

    - Click the **Bake** button. The selected floor sections should now show a blue mesh that represents the baked navigation surface, as shown in *Figure 13.7*.

And that's really all there is to setting up the navigation surfaces for our NavMesh agents. Make sure to revisit the `NavMeshSurface` component whenever you make changes and re-bake. We'll now make one final adjustment to the navigation setup via **Agent Type** before adding waypoints and testing it all out.

Configuring the agent type

An additional consideration we'll make while baking the NavMesh Surface is to adjust the agent's size to ensure its capabilities align with our needs. I've arbitrarily decided that we don't want the bot to come too close to the sides of the corridor as it finds its way from waypoint to waypoint.

While referring to *Figure 13.8*, proceed to create a new navigation agent type by following these steps:

1. Open the **Navigation** window by going to **Window | AI | Navigation**.
2. Click the plus (+) icon at the bottom-right of the **Agent Types** list to add a new type.
3. In the **Name** field, rename from the default name to `Bot`.
4. In the **Radius** field, change the value to `1` to give the bot more of a margin from the walls of the corridors.
5. In the **Step Height** field, ensure the value is `0.75`.

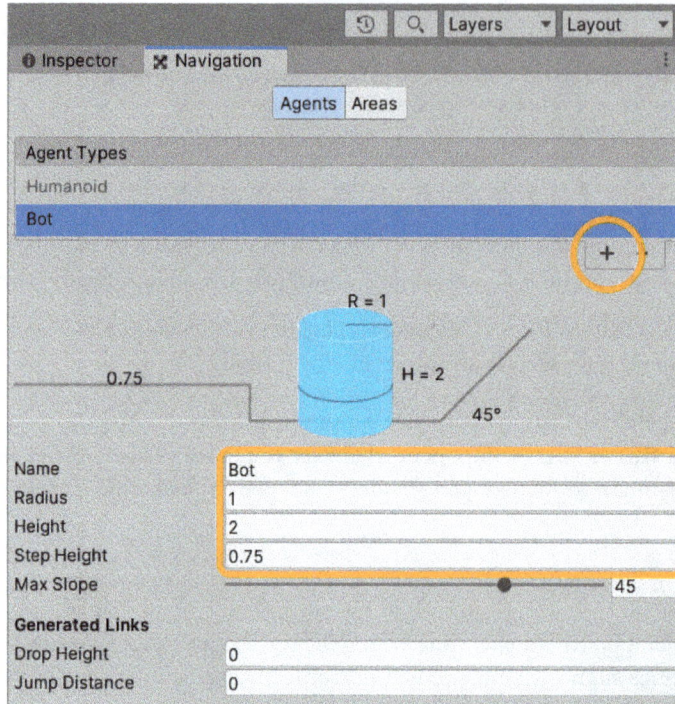

Figure 13.8 – Specifying a Bot agent type

> **Tip**
> You can also access the **Navigation** window from the `NavMeshSurface` component by selecting **Open Agent Settings…** in the **Agent Type** dropdown.

Now that we have a specific bot agent type to use with our navigation mesh surface, return to both the following components to set **Agent Type** to `Bot`:

- The `NavMeshSurface` component: On the `NavMesh Surface` object in the **Scene** Hierarchy.
- The `NavMeshAgent` component: On the `Enemy Hover Bot A` Prefab (which can be seen as (B) in *Figure 13.6*).

Having `NavMeshSurface` and `NavMeshAgent` use the same agent type is a requirement of the AI Navigation system – you'll receive odd *set destination*-related issues in the console if they don't, and baking just won't work!

With a different agent type selected, we'll have to click the **Bake** button again to calculate the new navigation mesh surface – make sure you return to re-bake as needed when adjustments are made to the **Agent Type** values.

> **We've only touched the surface**
>
> Sorry, I couldn't help the pun! The Unity AI Navigation system is capable of a *lot* more than what we've touched on here. I encourage you to explore the documentation and try implementing additional features into your game to level up the player experience even more:
>
> `https://docs.unity3d.com/Packages/com.unity.ai.navigation%401.1/` `manual/`.

Now that we have the navigation surface baked, we can finally add waypoints to get the bot performing its patrol duties while walking the beat!

Adding waypoints to the level and testing

NPC patrolling enhances gameplay by making environments feel alive, increasing player immersion and challenge. While having an NPC patrol a set of waypoints may not be considered an **emergent behavior** on its own, encounters can still appear unpredictable, keeping gameplay exciting and testing the player's adaptability.

> **Emergent behavior**
>
> Emergent gameplay refers to complex situations that could arise from the implementation of simple game mechanics in both video and tabletop board games.

We've already configured all the requirements for adding the ability for our enemy bot to patrol; we just need to define the specific patrol points now. So, follow these steps to add our waypoints to specific locations within the navigation mesh surface we made:

1. Add an empty GameObject named `Patrol Waypoints` to the **Scene** Hierarchy to parent (group) the waypoint objects under.

    Don't forget to reset the transform position to (`0, 0, 0`) after adding the object.

2. Add three GameObjects, one to each end of the corridors – whose transform position represents the actual waypoint location; see *Figure 13.7* as a location reference.

3. With a waypoint object selected, use the Inspector to assign an icon for quick identification in the **Scene** view – *Figure 13.7* shows a green label icon having been selected.

Now that we have waypoints in our scene, we can assign them to the `PatrolWaypoints` behavior component's **Waypoints** field so the enemy bot has their assigned patrolling. So, select the `Enemy Bot A` object in the **Scene** Hierarchy, then lock the Inspector. Now, select all three waypoint objects we just made and drag them to the **Waypoints** field to add them all at once. If we didn't lock the Inspector, it would have changed when we multi-selected the waypoint objects in the hierarchy, preventing the assignment. Alternatively, without locking the Inspector, you can drag each waypoint one at a time to assign them (yuck).

> **Obligatory saving reminder**
>
> You've been hitting *Ctrl/Cmd + S* to save your scene periodically, right? Consider this a reminder to do it now. The waypoints assignment to the `Enemy Bot A` Prefab is a scene-level save; they won't be saved with the file-based Prefab asset.

Test out navigating the waypoints. Enter **Play mode**, switch back to the **Scene** view, adjust the view, and watch the bot navigate a path to the waypoints in the order they appear in the list of waypoints – make any adjustments to the **Config** values for acceleration, speed, or the navigation agent type radius, and so on. Repeat playtesting as needed to get the initial enemy bot patrolling in a, well, good patrolling fashion.

In this section, we learned how to refactor a 2D patrolling waypoint behavior to its 3D equivalent counterpart using Unity's NavMesh system. Next, let's make the enemy bot have the ability to detect – or sense – the player and respond with appropriate behavior.

Dynamic enemies with sensors and behavior trees

In video games, sensors can play a crucial role in creating interactive and dynamic AI behaviors. They enable NPCs to *perceive* their surroundings, allowing them to react to not only the player but also other environmental things in a *more realistic* manner. For example, in stealth games such as **Metal Gear Solid**, enemies are often equipped with a **field of view** (**FOV**) sensor that detects the player when they enter their line of sight. Similarly, games such as **The Last of Us** incorporate audio sensors that enable enemies to detect the player based on noises they make. These types of sensors add depth to the gameplay by having players change the strategy of their movement and actions more carefully.

In short, sensors are abilities added to our objects that make them aware of their surroundings and other objects – especially true concerning an enemy NPC and the player!

In this section, we'll look at some code that can be used to implement the two types of sensors we just identified: detecting the player based on an FOV and the player based on audio – such as the player's footsteps.

Creating sensory behaviors

The first sensor code we'll look at is for detecting the player within an FOV of the patrolling enemy NPC. We'll make this an event-based class that does not inherit from `MonoBehaviour`. That way, we can use it directly in the implementing class, `EnemyController`, by simply creating a new instance. It will be very straightforward.

First, look at the code template for our sensor classes:

```
using UnityEngine;
using UnityEngine.Events;
```

```
public class SensorTemplate
{
    private readonly MonoBehaviour _context;
    public event UnityAction OnSensorDetected;

    public SensorTemplate(MonoBehaviour context)
    {
        _context = context;
    }

    public void Tick()
    {
        // Invoke only if detection occurred.
        OnSensorDetected?.Invoke();
    }
}
```

Let's break it down:

- class SensorTemplate: Note that we're not a derived class, and there is no inheritance from MonoBehaviour since we won't be using this class as a component in the Inspector. We'll create an instance of the sensor class in the implementing class using the new keyword (remember, too, that we can't use new with MonoBehaviour).

- MonoBehaviour _context: Speaking of MonoBehaviour... we may still need access to it for running a coroutine or getting the Transform values or GameObject of the implementing class. We'll use the class constructor to assign this variable.

> **Constructor (C#)**
>
> Every time an instance of a class or struct is created, its constructor is called. More information can be found here: https://learn.microsoft.com/en-us/dotnet/csharp/programming-guide/classes-and-structs/constructors.

- event UnityAction OnSensorDetected: I mentioned the sensor class will be event-based, so here it is. This is the event, UnityAction, that we'll invoke when the sensor detects, well, what we design it to detect. The event keyword enforces that only the declaring class can invoke it.

- SensorTemplate(MonoBehaviour context): We have our class's constructor here. The method that shares the same name as the class is the constructor. We've added a parameter to set the _context member variable when creating the class instance.

- void Tick(): This is where the magic happens. Typically, Tick() will be called from the implementing class's Update() method to perform the detection work required for each sensor's purpose. We'll invoke the OnSensorDetected event only when the detection occurs.

Now, before continuing onto the two sensor examples I've provided, let's see how the sensor class is used in the implementing class. We'll use EnemyController for this example, where we previously just used a simple member method, IsPlayerInRange(). You can see how we can instead use a sensor class as a way to construct an NPC's abilities in the following code:

```
public class EnemyController : MonoBehaviour
{
    private SensorTemplate _sensor;

    private void Start()
    {
        _sensor = new SensorTemplate(this);
        _sensor.OnSensorDetected += HandleSensor_Detected;
    }

    private void HandleSensor_Detected()
    {
        Debug.Log("Sensor triggered!");
        ChangeState(State.Detected);
    }

    private void Update()
    {
        _sensor.Tick();
    }
}
```

Here is another quick breakdown of the implementation:

- private SensorTemplate _sensor: This variable will hold the instance of the sensor class we create to refer to throughout its implementation in this class.

- _sensor = new SensorTemplate(this): Using the new keyword, we create a new instance of the sensor class for its use, passing in the this keyword, which represents the EnemyController class (derived from the MonoBehaviour base class).

> **this (C#)**
>
> The this keyword refers to the current instance of a class. More information can be found here: https://learn.microsoft.com/en-us/dotnet/csharp/language-reference/keywords/this.

- _sensor.OnSensorDetected += HandleSensor_Detected: Adding the handler method for listening when the sensor detects something.

In this template example, the `HandleSensor_Detected()` method simply changes to a different state where the controller will take action based on the sensor detection. Don't forget that we can also pass parameters to the handler method via the event; simply change the event type declaration to something like `UnityAction<float>`.

- `void Update()`: We'll run the `Update()` method every frame in the sensor class for it to perform its detecting workload.

That's all there is to implementing custom class instances for the purpose of adding sensing abilities to your NPCs. If you want to further solidify the sensor class's architecture, you could use a base abstract class or an interface. Consider this your challenge to refactor the sensor template code with either of these patterns in mind!

Abstract class or interface?

An abstract class can provide a starting point with some predefined ways to do things, whereas an interface simply states requirements that must be followed. Imagine you have a box of crayons with some colors already defined – that's the abstract class. And a coloring book with outlines of pictures where you have to stay within the lines – that's the interface.

In the following two subsections, I've provided two code examples for sensor classes that can be added to any controller class to detect a target within an FOV and detect any nearby player's audio source that is playing.

Detecting the player within an FOV sensor

Imagine having a pair of glasses that can assist you in finding your friends in a game of hide and seek. These glasses work similarly to the *player within the FOV sensor* in a video game, where you have a limited seeing distance and area of view and need to look in the correct direction to identify your friends. If the player is within the viewing angle and distance while not hidden behind anything, the sensor detects them and triggers an alert. By using this type of sensor in a video game, NPCs can detect, or sense, the player, similar to how the glasses help during the game of hide and seek.

I've provided the complete example `SensorTargetInFOV` class in the 3D FPS game project code available for download and its initial implementation in the `EnemyController` class. The script is located here: `Assets/Scripts/Sensors/SensorTargetInFOV.cs`.

Download the complete sensor code

The code for all the example sensors can be found on the GitHub repository here: `https://github.com/PacktPublishing/Unity-2022-by-Example/tree/main/ch13/Unity-Project/Assets/Scripts/Sensors/`.

The class constructor has additional parameters added for specifying the FOV angle and the visual range:

```
public SensorTargetInFOV(MonoBehaviour context,
    float fovAngle, float fovRange)
{
    _context = context;
    _fovAngle = fovAngle;
    _fovRange = fovRange;
}
```

These additional values are then used for the sensor's detection calculation within its `IsTargetInsideFOV()` method called every frame tick, which does the following:

1. Calculates the direction of the specified target object.

    The target object is set with a call to `SetTarget()` just before calling `Tick()`.

2. If the direction to the target angle is within the specified `fovAngle`, it then calculates the distance to the target.

3. If the distance to the target is within `fovRange`, we finally do a physics **raycast** to detect the target object as the player.

Additional reading | Unity documentation

In game development, a *raycast* is a common way of detecting objects. It involves projecting an invisible laser beam in a specified direction to report on any objects that it intersects with.

More information about `Physics.Raycast` can be found here: `https://docs.unity3d.com/2022.3/Documentation/ScriptReference/Physics.Raycast.html`.

---

4. If the hit object of the raycast is the player – determined by comparing its tag – we invoke the *target detected* event.

Explore and discover the sensor code and its implementation to further solidify your understanding of the different ways in which we can create code architecture. We'll now look at the second sensor example, the hearing sensor.

Detecting the player's audio sensor

In a video game, a hearing sensor gives your NPCs the ability to identify sounds from the surrounding environment with super-sensitive ears. This type of sensor can help the NPC locate and identify objects based on the sounds they make, even if the object is not visible to the NPC. My favorite example of this is the horrifying game **Alien: Isolation**, in which the alien NPC uses a sophisticated hearing sensor to not only hear noises made by the player but also recognize their severity – sound plays a crucial role in the survival-horror experience.

Using a hearing-type sensor in your game may or may not be so frightening, but it will elevate the gameplay experience nonetheless. I've provided the complete example `SensorHearing` class in the 3D FPS game project code available for download and its initial implementation in the `EnemyController` class. The script is located here: `Assets/Scripts/Sensors/SensorHearing.cs`.

The class constructor has additional parameters added for specifying the hearing range and update frequency (how often to check for audio sources):

```
public SensorHearing(MonoBehaviour context,
    float hearingRange, float updateFrequency)
{
    _context = context;
    _hearingRange = hearingRange;
    _updateFrequency = updateFrequency;

    _context.StartCoroutine(
        PeriodicallyUpdateAudioSources());
    ...
```

These additional values are then used for the sensor's detection calculation within its `IsAudioDetected()` method called every frame tick. Additionally, we're using the passed-in `MonoBehaviour` context to start a coroutine: coroutines require a `MonoBehaviour` object to run – to periodically update the list of audio sources in the scene (not particularly useful for our 3D FPS game but especially relevant for multiplayer games where players will drop in and out of gameplay).

The audio detection method, `IsAudioDetected()`, does the following:

1. We'll simply start by iterating through a list of audio sources found in the scene.

2. If an audio source is playing and we can hear the audio source, invoke the *target detected* event.

    `CanHearAudioSource()` is used to determine whether the currently playing audio source is within `hearingRange`, adjusted for its playing volume.

Player footsteps 3D audio source

Note that for our player footsteps audio, we implemented it as a 2D sound using the `AudioPlayerSFX` component. For the enemy NPC to sense the player's footstep sounds, with the `SensorHearing` class, we'd need to use the `AudioPlayerSFX3D` component instead. Refer to *Chapter 12* if you'd like to refactor the player footsteps to support audio-sensing behavior.

Similarly, for this sensor, explore and discover the sensor code and its implementation to further solidify your understanding of the different ways in which we can create code architecture.

In this section, we learned what a sensor is in game development terms and how to use a class that does not derive from `MonoBehaviour` to implement sensory abilities for our NPCs. Next, we'll see how a behavior tree can help to manage the complexity of AI for NPCs.

Wrangling behaviors with a behavior tree

A **behavior tree** (**BT**) is a powerful and flexible tool for implementing diverse AI-driven NPCs because it enables more complex hierarchical decision-making compared to a more traditional **finite state machine** (**FSM**) that would simply hold the current state. BTs work really well with predefined sensors, such as our *player within FOV sensor* and *hearing player audio* sensors we've discussed, since these sensors could be integrated into the BT as custom nodes in the graph. This sensor integration would allow NPCs to make decisions based on sensory input conditions, favorably enhancing enemy NPC behavior for the gameplay.

Because BTs are graph-based, with sequences, actions, and other nodes, they offer a more manageable way to visualize the relationship of the AI behavior being implemented. However, BTs are not visual scripting (although some BT frameworks do provide functionality that crosses over into what might be considered visual scripting). Conditional nodes are what allow the BT-driven NPC to react dynamically to the player's actions, such as changing the NPC's behavior from patrolling to investigating a noise or spotting the player.

Here is a simple example UML diagram of a BT for a patrolling NPC that can sense the player's position within its FOV or hear the player, with the child nodes representing the resulting state of the triggered sensors:

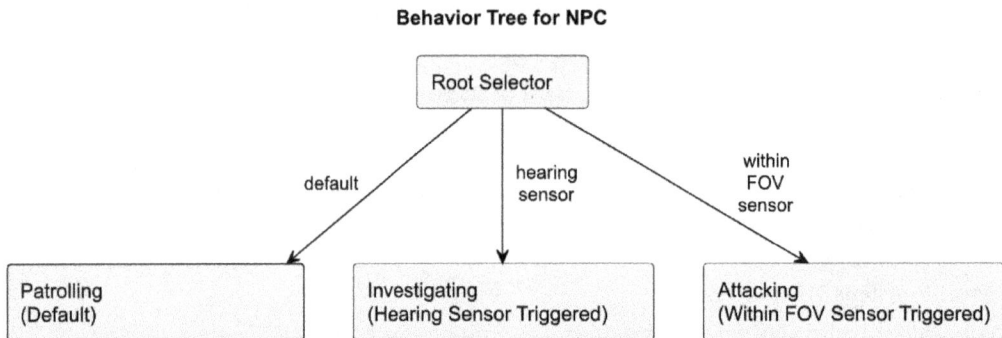

Figure 13.9 – A simple BT graph

As you can see, the hierarchical structure of BTs offers a clear and manageable way to design AI behaviors. Tools such as **Behavior Designer**, **Node Canvas**, and **Schema** in the Unity Asset Store provide frameworks for creating BTs, mostly without the need to write any code. They offer intuitive editors, many pre-built nodes, plus their own custom actions and actions specific to the Unity Scripting API, too.

BT tools are designed to make it easier for Unity developers to implement complex behaviors for their NPCs. They're also more accessible for integrating custom conditional nodes (e.g., sensors), allowing for rapid iteration of configurations to attain the desired behaviors.

Free behavior tree asset | Unity Asset Store

Schema is described as "*A fast, easy platform to build artificial intelligence with Behavior Trees. Create complex and intelligent behaviors for your game without writing a single line of code. Behavior trees are used extensively by AAA studios to bring lifelike behaviors into their AI.*"

Here is its Asset Store link: `https://assetstore.unity.com/packages/tools/behavior-ai/schema-200876`.

I've already gone ahead and added the Schema BT asset to the 3D FPS project files in the book's GitHub repo (link in the *Technical requirements* section) for you to explore and experiment with. You can use the Schema graph editor to create custom nodes for the sensors we made in the previous *Creating sensory behaviors* section and apply them to the enemy hover bot NPC.

Unity Muse AI tools – Behavior

Muse Behavior is a new Unity built-in graph-based BT tool with an intuitive flow graph and action node stories for AI designers. It is currently in pre-release but looks to be an up-and-coming full-featured BT solution for AI NPC design – including the ability to use text-input generative AI to create node actions rapidly.

Here is the link to a Muse Behavior tutorial project: `https://assetstore.unity.com/packages/templates/tutorials/muse-behavior-tutorial-project-269570`.

As a game developer, you'll gravitate toward certain technologies over others and favor either a code-only approach or adding some visual tools to your skill set over time. At the end of the day, remember, it's about overcoming challenges effectively to implement all the features in the scope of your project, and especially to finish the game!

In this section, we learned the value of creating sensors for adding detecting abilities to our NPCs and using a BT as a visual tool for creating, managing, and setting up dynamic AI-driven NPC behaviors. Next, we'll look at AI-driven behavior further by investigating Unity's ML tools.

Introducing ML with ML-Agents

Many parts of developing a video game touch on technologies that could fill entire books all on their own, and ML is definitely one of them! Unity has its own tooling for ML that is particularly suited for games, and it's called **ML-Agents**.

ML-Agents is an **AI toolkit** for Unity developers that helps them create advanced and complex behaviors in their games and simulations using ML techniques. Unlike BTs, which rely on hand-crafted sets of rules, ML-Agents leverages ML to enable NPCs to *learn* and adapt their behavior by interacting with their surroundings. This is accomplished by training an agent using ML techniques such as **reinforcement learning**, **imitation learning**, or other custom methods. The process enables trained agent NPCs to determine the best course of action, all on their own, constantly changing and surprising you with dynamic and, hopefully, unpredictable gameplay. I briefly noted **emergent gameplay** before, and this is it!

Unity ML-Agents applies to various scenarios, from simple games to much more complex simulations. Here is a summarized list of what the ML-Agents toolkit offers:

- **Different training environments**: You can create environments as simple or complex as the real world to train agents.

- **Variety of training algorithms**: Also relatable to the toolkit's flexibility, it supports various cutting-edge ML algorithms customizable for the agent training task at hand.

- **Unity integration**: ML-Agents is designed to work specifically with Unity, which makes it very accessible to Unity game developers.

Unity also provides a breadth of resources to learn ML-Agents, including the **ML-Agents: Hummingbirds** project, a comprehensive example project with approximately 10 hours of content!

> **Additional reading | Unity documentation**
>
> More information about ML-Agents can be found here: `https://unity.com/products/machine-learning-agents`.
>
> More information about the ML-Agents toolkit can be found here: `https://github.com/Unity-Technologies/ml-agents/tree/latest_release`.
>
> More information about ML-Agents: Hummingbirds can be found here: `https://learn.unity.com/course/ml-agents-hummingbirds/?tab=overview`.

There are three basic steps to how ML-Agents can work for Unity games:

1. **Integrating**: This step involves integrating the Unity ML-Agents package into a Unity project designed for training.

2. **Training agents**: This step involves connecting the Unity project to train the agents to learn the desired behavior.

3. **Embedding**: This final step involves embedding the trained agent model into your game project.

Let's keep this simplified, high-level, three-step process in mind and build on our previous work with the Unity AI Navigation package. In this way, we can use a NavMesh Agent to optimize the efficiency of the ML-Agents agent training by having it focus on behavior training and not navigation.

Navigating training efficiency with NavMesh

When developing a game using ML-Agents, we can use the `NavMeshAgent` component to handle the task of agent navigation while leaving the ML-Agents training to focus on the higher-level decision-making processes such as patrolling, investigating, and attacking the player character.

This approach of blending intelligent AI behavior with an established navigation setup results in a more streamlined and effective training result. The following figure shows a simple training scene with a NavMesh Surface setup:

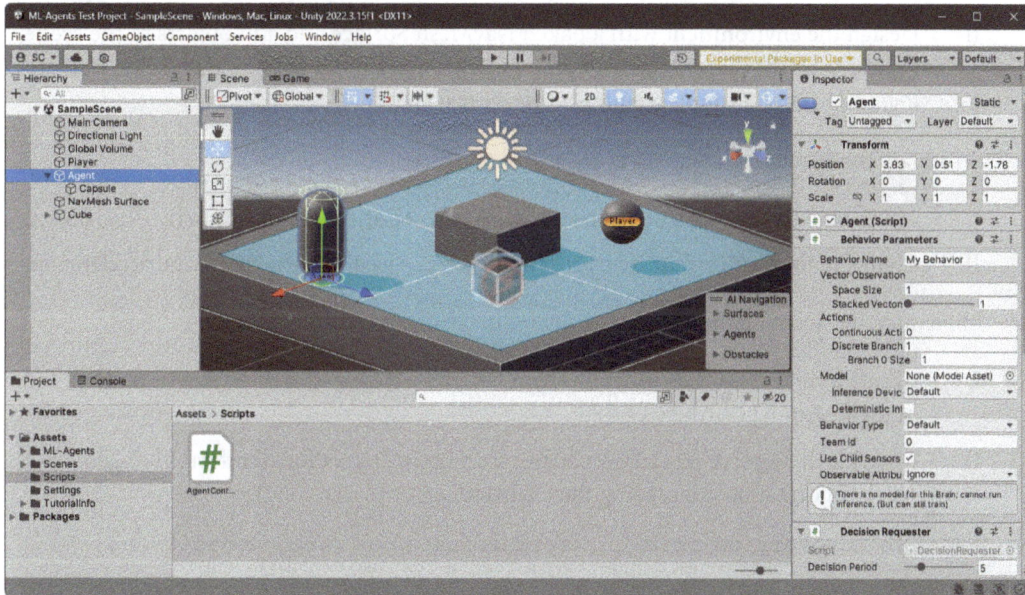

Figure 13.10 – Example ML-Agents training scene

Having noted this efficiency, let's see what a general ML-Agents training setup looks like next.

Examining an ML-Agents setup

Here is a more detailed breakdown of the steps we'd typically see setting up ML-Agents training and implementation within Unity:

1. The ML-Agents package and dependencies.

    The latest installation instructions can be found in the ML-Agents GitHub repo at `https://github.com/Unity-Technologies/ml-agents/blob/latest_release/docs/Installation.md`:

I. I recommend installing the latest version using the GitHub URL via the Package Manager: `git+https://github.com/Unity-Technologies/ml-agents.git?path=com.unity.ml-agents#release_21`.

II. Dependencies include ML-Agents extensions, **Python**, the `mlagents` Python package, **PyTorch** (Windows), **Visual C++ Redistributable**, and **Unity Sentis**.

2. Unity project setup:

I. Create a new Unity project, selecting a **3D Core** template.

II. Design the environment with a baked NavMesh Surface and add objects to the scene representing the player character and the NPC that should observe the player. Refer to *Figure 13.10* for a simple training scene representation.

3. Setting up the agents:

I. Attach both the `NavMeshAgent` and ML-Agents `Agent` components to the NPC object.

II. Configure the `NavMeshAgent` properties such as speed, angular speed, and acceleration.

4. Implement the agent methods:

I. Create a new class that inherits from `Unity.MLAgents.Agent`, named, for example, `AgentController`.

II. Override the `Agent.CollectObservations()` method to inform the agent with sensor information concerning the environment or player data:

```
public override void CollectObservations(VectorSensor sensor)
{
    // TODO: Implementation of sensor observations.
}
```

III. Override the `Agent.OnActionReceived()` method to apply NPC actions such as setting the destination for the NavMesh Agent or turning to face the player:

```
public override void OnActionReceived(ActionBuffers
actionBuffers)
{
    // TODO: Implementation of actions.
}
```

5. Define the rewards and training:

I. Define rewards within the `OnActionReceived()` method:

i. **Positive reward**: Call the `SetReward()` method and pass in a positive float value for successfully reaching a patrol point or facing the player, for example. Note that this function replaces any rewards given to the agent during the current step. You can also use `AddReward()` to incrementally change rewards instead of overriding.

ii. **Negative reward**: Call the `SetReward()` method and pass in a negative float value for undesirable behaviors such as running into walls or losing sight of the player:

```
public override void OnActionReceived(ActionBuffers
actionBuffers)
{
    ...
    SetReward(0.1f); // Positive reward.
    SetReward(-1);   // Negative reward.
}
```

II. Create a training configuration file (i.e., `trainer_config.yaml`) where the specified training algorithm will be defined (e.g., reinforcement learning algorithm) along with its configuration.

III. Configure both a `BehaviorParameters` and `DecisionRequestor` component on the player object agent for the observation and decision-making settings.

6. Run the training process:

I. Using Command Prompt or Terminal, navigate to your Unity project folder, run the training command (e.g., `mlagents-learn trainer_config.yaml --run-id=firstNPCPatrol` – here, `run-id` is any unique name for the session), and use **TensorBoard** to monitor the training process of the agent.

7. Integrate the training results:

I. After finishing training, the inference model is saved as a `.nn` file. Import it into a folder in your Unity project.

II. On the agent, assign the model to the `BehaviorParameters` component's **Model** property and set **Behavior Type** to **Inference Only** to test the behavior in your scene.

8. Playtest and iterate on the design:

I. In Unity, play the scene and observe the NPC agent's behavior (for example, as it patrols, senses, or detects the player).

II. Adjust rewards and rerun the training as needed to refine the NPC behavior incrementally.

By following these steps, you can set up a training environment that uses a NavMesh Agent for pathfinding. At the same time, ML-Agents is left to train only the complex NPC behavior, such as patrolling and player detection in our example.

> **ML-Agents sample environments**
>
> The **ML-Agents Toolkit** GitHub repository contains a set of example environments that demonstrate the various features of the toolkit. These environments can also be used as a starting basis for new environments or a predefined setup for testing new ML algorithms.
>
> Some example learning environments can be found here: `https://github.com/Unity-Technologies/ml-agents/blob/latest_release/docs/Learning-Environment-Examples.md`.

If you're looking to create a truly exceptional gaming experience for your players, consider using ML with ML-Agents. It's a powerful tool that can help you achieve remarkable results!

> **Unity Muse | Additional reading**
>
> During the course of the Unity 2022 tech stream releases, Unity has released new AI tools for developers under the moniker **Unity Muse**. The tools are currently offered as a subscription service, starting with a free trial period.
>
> More information about Unity Muse AI tools for developers can be found here: `https://unity.com/products/muse`.

In this section, we explored how Unity's ML-Agents toolkit can be used to train NPCs with ML algorithms. The result is an NPC with adaptable behaviors that go beyond what we can pre-script for an AI (for example, with a BT). We further understood the required knowledge to use ML-Agents in a Unity project to elevate the player experience with adaptive NPC behavior.

Summary

In this chapter, we refactored our game's enemy NPCs from 2D to 3D environments while still using waypoints but also leveraging Unity's NavMesh system for AI navigation. We also improved our NPCs' behavior complexity by enabling them to interact with the player and the environment more realistically by using sensors – the effect of which is to challenge players more engagingly.

We continued discussing dynamic enemy behavior by incorporating our sensors as conditions within BTs. We completed our advanced AI discussion with an introduction to ML by using Unity's ML-Agents, enabling NPCs that can learn and evolve, allowing us to integrate advanced AI-based gameplay into our games for remarkable player experiences.

In the next chapter, we'll finish things up with the 3D FPS game by creating a classic boss room battle in **mixed reality** (**MR**). Additionally, you will learn how to design a challenging boss room and its mechanics quickly by applying all the previous lessons learned for developing game systems.

14

Entering Mixed Reality with the XR Interaction Toolkit

In *Chapter 13*, we made some changes to our 3D FPS game's enemy NPCs. We upgraded them from 2D to 3D components while still using waypoints for navigation but utilized Unity's NavMesh system to rapidly implement the patrolling behavior. We also enhanced the complexity of the NPC behavior by adding sensors that allow them to interact with the player and the environment in a more realistic way.

We continued by discussing how to create dynamic enemy behavior using our sensors as conditions within behavior trees. We then completed our AI discussion with an introduction to **machine learning (ML)** using Unity's ML-Agents, which enables NPCs to learn and evolve. We can create remarkable experiences for our players by integrating advanced AI-based gameplay!

In this chapter, we'll finish the journey that started with the 3D FPS game by creating the final boss room encounter in **mixed reality (MR)**. We'll accomplish this by using the **Unity XR Interaction Toolkit** along with assets, reusable components, and systems accumulated from previous efforts, all coming together to create a battle to take place in your own room!

In this chapter, we're going to cover the following main topics:

- Introduction to MR and development frameworks

- Designing a boss room

- Working with AR planes (AR Foundation)

- Placing interactable objects in the world

- Implementing the boss room mechanics

By the end of this chapter, you'll be able to make an MR game or experience that incorporates the player's physical space – such as walls, the floor, and tables – to create a novel experience for players. You'll also learn how to create interactable objects and manage their instantiation, particularly with regard to detected surface planes that define the boundaries and objects of the physical space. The chapter completes the accumulation of knowledge required to rapidly build out features and behaviors when making games.

Technical requirements

To follow along with this chapter, you'll need a Meta Quest 2 or 3 headset and a USB-C cable to connect it to your computer. The cable lets you push the Unity project build to your device and test some functionality directly in the Unity Editor's play mode.

> **If you don't have an MR headset**
>
> You can still follow along in this chapter even without owning an MR headset – by using the Meta XR Simulator, available from the Unity Asset Store: `https://assetstore.unity.com/packages/tools/integration/meta-xr-simulator-266732`.

You can download the complete project from GitHub at `https://github.com/PacktPublishing/Unity-2022-by-Example`.

Introduction to MR and development frameworks

Mixed reality has only recently hit the forefront of what's possible with the latest **head-mounted displays** (**HMDs**) to create environments where the physical and virtual worlds are blended to have digital and physical objects co-exist and appear to interact with one another. An MR gaming, educational, healthcare, or industrial application combines aspects of both **virtual reality** (**VR**) and **augmented reality** (**AR**) to offer an immersive experience where virtual content is anchored in the real world.

One doesn't need to look much further than popular VR adaptations of popular PC games such as **Skyrim VR** or **Resident Evil VR** to understand that VR-based technology has a strong outlook for the future of virtual entertainment. Additionally, games such as **Minecraft VR** and **Roblox VR**, with their enormous and engaged player base no less, offer immersive experiences that turn otherwise static surroundings into dynamic worlds that allow interaction and exploration in unprecedented ways never experienced before.

The breakout success of the original VR title **Beat Saber** also shows the diverse potential of the platform, not only for entertainment but also for physically involved gameplay. The future of VR, AR, and MR will continue to interest us, so let's be sure we're armed with the tools to succeed in this space.

In this section, we'll review the technology we'll use to build our boss room game. The tech stack includes the Unity **XR Interaction Toolkit**, **AR Foundation** framework, and the **OpenXR** Meta package. These technologies on their own are powerful but combine one with another, and something new is created. They enable developers to create impressive MR experiences much more quickly when used in tandem.

Let's have a brief overview of each and see how they harmonize.

XR Interaction Toolkit (XRI)

Unity's XRI is a versatile interaction system for VR/AR that simplifies and streamlines cross-platform creation. It provides a common framework for various interactions such as poking, gazing (i.e., rays), and grabbing for controllers and hands. It also includes virtual hands, haptic feedback, and responses for selections using scaling, animation, or even blend shapes.

> **Additional reading | XR Interaction Toolkit (XRI)**
>
> XRI: `https://docs.unity3d.com/Packages/com.unity.xr.interaction.toolkit%402.5/manual/index.html`.
>
> XRI examples: `https://github.com/Unity-Technologies/XR-Interaction-Toolkit-Examples`.
>
> XR: `https://docs.unity3d.com/Manual/XR.html`.

The XRI toolkit dramatically simplifies the process of developing interactive VR and AR experiences by providing a comprehensive set of interaction components and systems, minimizing the barrier to entry for developers looking to enter this space. It allows for easy implementation of common functions such as head tracking, locomotion (i.e., movement), object interactions, and the UI within the virtual space. The toolkit is also flexible and modular, which provides an excellent foundation for creating an MR game.

Specifically for Unity 2022, Unity's cross-platform MR development tools for the Meta Quest HMDs have moved from the experimental preview state to fully supported in the 2022 LTS release!

XRI provides the interaction part; now, let's look at the environment part of these technologies.

AR Foundation

Unity's AR Foundation is a cross-platform framework that provides a unified API for simplifying building applications for mobile and head-worn AR/MR devices. The package is designed to work natively with XRI (and XR Hands), significantly reducing any hurdles for developers accessing the specific device features to support building AR applications.

> **Additional reading | AR Foundation**
>
> AR Foundation: `https://unity.com/unity/features/arfoundation`
>
> Unity documentation: `https://docs.unity3d.com/Packages/com.unity.xr.arfoundation%405.1/manual/index.html`

More specifically, AR Foundation is the layer that unifies **ARCore** (Google) and **ARKit** (Apple) APIs into a single higher-level API. This single API allows developers to write code once where the specific feature implementations of the underlying platforms are handled automagically.

AR Foundation simplifies building spatial awareness into applications, making digital objects appear interactable with the real world. This is crucial for creating MR experiences that seamlessly blend the virtual and the real world.

We'll be working specifically on the **Meta Quest HMD** platforms. Our boss room game will be compatible with Quest 2 and Quest 3 devices. The AR Foundation support for Meta Quest is built using a familiar industry-adopted standard interface for XR hardware and software, and that interface is called **OpenXR**.

AR Foundation provides the visual part; now, let's look at the platform support part of these technologies.

OpenXR: Meta package

OpenXR is an open, royalty-free standard that enables high-performance access via a unified interface across multiple AR and VR hardware and software platforms and devices, collectively known as XR.

Developing with OpenXR simplifies the development process by allowing developers to target any supporting OpenXR system without worrying about specific platform details. The Meta package (available since Unity 2022.3.11.f1) contains **Meta-specific OpenXR** extensions and Meta's **AR Foundation provider plugin** for its Quest devices – it ensures compatibility and interoperability between the software and hardware to support its specific input devices, head-mounted displays, and other peripherals.

> **Additional reading | OpenXR**
>
> Kronos Group: OpenXR: `https://www.khronos.org/openxr/`

To summarize, OpenXR is the glue that binds the interaction and visual systems to any supporting hardware devices – especially new devices that feature better graphics performance and sensors. The trio of technologies, when combined, enable developers to rapidly create prototypes and deploy production-ready MR games and experiences – XRI provides the foundation for interactive elements, AR Foundation builds on the ability to merge digital and physical-world visuals, and OpenXR ensures the experiences are accessible across different devices.

Unity XR Tech Stack

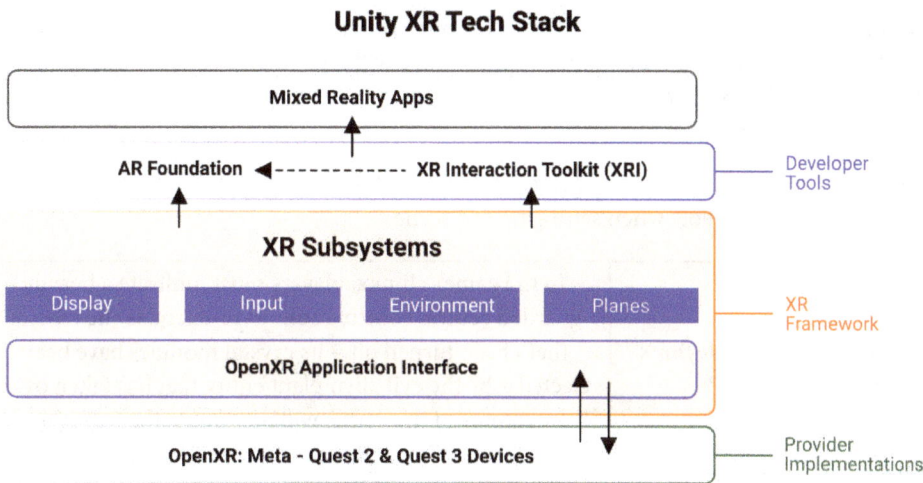

Figure 14.1 – Unity XR tech stack

In this section, we learned what Unity MR technologies are available to us and that this combination of MR-based technologies not only simplifies development but also enables the creation of complex, engaging MR applications to have broad end-user reach. This brings us right into the next section, where we'll get down to the business of designing our MR boss room.

Designing a boss room

Designing a boss room encounter is a critical part of game creation that combines aspects of narrative, mechanical, and environmental considerations to create an engaging and challenging experience for players.

There are several key areas to consider when designing a boss room encounter, and we'll take a shallow dive into a few:

- **Narrative element**: The encounter should feel like a natural progression or even the story's climax.

- **Boss mechanics**: The player's battle with the boss element should stand out as a unique experience, separate from the player mechanics mainly being used, requiring players to adapt a strategy to overcome attack patterns and other behaviors.

- **Environment design**: The layout of the boss room itself should complement the narrative and mechanics being implemented. This is a special consideration for MR because we'll use the player's own room (i.e., their physical space) to construct the gameplay environment and place the interactive elements, creating a novel challenge for each player.

- **Balancing**: Challenges presented by boss encounters should be demanding yet feel pretty balanced to avoid undue frustration while still providing a solvable challenge for the player.

By incorporating these elements into our boss room, we aim to offer players an enjoyable and unforgettable experience. Overcoming the *boss challenge* will give them a sense of satisfaction and accomplishment. Moreover, with MR included in our case, the experience becomes even more remarkable and rewarding.

Let's revisit our GDD for a moment to get a quick update for the boss room battle added that will provide the context we'll follow when setting up our scene.

What is the habitat interior's boss encounter?	In the game's climax, players must infiltrate a heavily guarded central control room to re-energize the reactor that's been turned off – its crystal modules have been ejected – by the evil alien plant entity that has taken over. The outcome of this battle will determine the future of the Kryk'zylx race on the planet.

Table 14.1 – GDD snippet setting the scene for the boss battle

Very nice. The context has been set, and we have some story background for the purpose of the boss battle. You are not just some kid from a trailer park; you are a Kryk'zylx scout! As such, you are armed with the most advanced energy-based weaponry, such as this laser pistol: pew-pew!

Figure 14.2 – XR interactable gun

Let's start by first defining our physical space, then move on to creating the Unity project and testing our MR setup.

Setting up the physical space

Properly defining the physical space set up on the device for an MR game is of utmost importance, as it directly impacts the possibilities available in the immersive experience. It seamlessly blends virtual content, such as the horizontal AR surface planes defined for walls, the floor, the ceiling, tables, and seats, in addition to vertical AR surface planes, such as doors and windows. Having these virtual surface objects in place for their real-world counterparts enhances gameplay, ensures safety, and maximizes the player's engagement. The physical space environment setup also serves as an interactive canvas for the game developer's storytelling and exploration.

For Meta Quest 3, the headset includes a depth sensor to scan your room surroundings and detect the floor, walls, and ceiling to establish a starting point for your physical space setup. Once you've finished the room scan, you can manually confirm the walls and add furniture.

For Meta Quest 2, you'll have to set up your physical space entirely manually.

> **Meta Quest room setup**
>
> For plane detection to function correctly on a Meta Quest device, you must first complete the new **Room Setup** feature found in **Settings | Physical Space | Space Setup** on the Quest headset before entering an MR game. To ensure optimal performance, it's also recommended to include at least one piece of furniture with a horizontal surface, such as a table.

Please note that the MR game we'll create relies on providing examples for the different surface planes established by either the physical scan or manual space setup. You must ensure your room includes at least four walls and a table.

You can perform the room setup any time before running an MR game or experience, so you can do this at your leisure. But for now, we'll move on to creating and setting up our Unity project to get started with our boss room battle.

Creating the Unity project

Install the latest Unity 2022.3 LTS release if you haven't already – this is so we have access to the new VR and MR templates in **Unity Hub**. We'll also require the **Android Build Support** module to be available, so ensure you have that, along with the **OpenJDK** and **Android SDK & NDK Tools** modules installed.

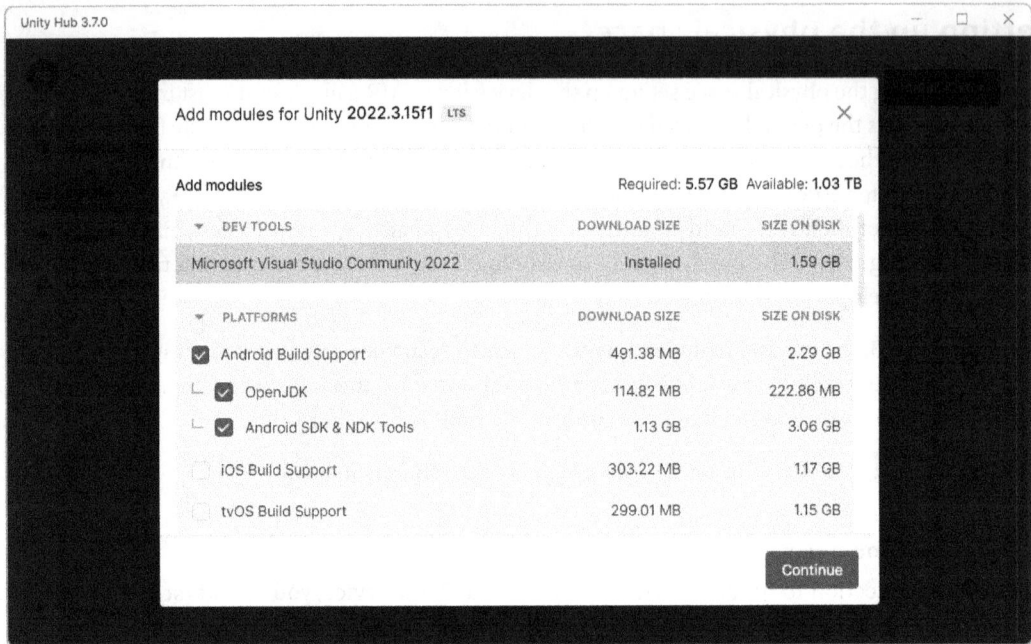

Figure 14.3 – Install the Android Build Support module

With the minimum required Unity Editor version and Android dependency modules now installed, we're ready to rapidly set up our MR boss room project using Unity's new **Mixed Reality** (Core) template. It's built upon the core MR technologies outlined in the *Introduction to MR and development frameworks* section. How very convenient!

The MR template project simplifies XR development by streamlining the implementation of advanced features such as plane detection, device passthrough, and spatial UI creation alongside designer-friendly XR interactable components. It comes pre-configured with essential packages such as XRI, AR Foundation, Unity OpenXR Meta, and XR Hands, making project setup and package management a breeze. This MR project template approach targets the needs of MR creators for richer content and reduces developer friction in accessing advanced MR features.

Additional reading | Unity MR template

Mixed Reality Template Quick Start Guide: `https://docs.unity3d.com/Packages/com.unity.template.mixed-reality%401.0/manual/index.html`

Explore cross-platform MR development on Meta Quest 3: `https://blog.unity.com/engine-platform/cross-platform-mixed-reality-development-on-meta-quest-3`

Open Unity Hub and click the **New Project** button (top-right corner) to create a new project. Then, referring to *Figure 14.4*, follow these steps.

1. Ensure the Unity 2022.3 LTS version previously installed is selected in the **Editor Version** dropdown at the top.

2. In the center template list, scroll down and select **Mixed Reality** (Core).

3. In the right pane, if you're prompted with a **Download template** button, click to download the template.

4. Once the template finishes downloading, provide these options:

   - **Project name**: Remember not to start with special characters. Use something like MR Boss Room.

   - **Location**: Select the folder path for where to store your project files.

   - **Unity Cloud Organization**: You must select the organization to which this project will belong. When you create a new Unity ID account, Unity generates an organization associated with your username and ID. The base feature a Unity organization provides is the ability to organize your projects, services, and licenses.

   - **Connect to Unity Cloud**: Only enable this if you wish to take advantage of gaming services for your project (generally, yes, you'll want this).

   - **Use Unity Version Control**: Enable this option if you want to have Unity Cloud's **version control system** (**VCS**) back up your project to the cloud and allow additional team members to collaborate on the project (we'll cover Unity Version Control in *Chapter 15*).

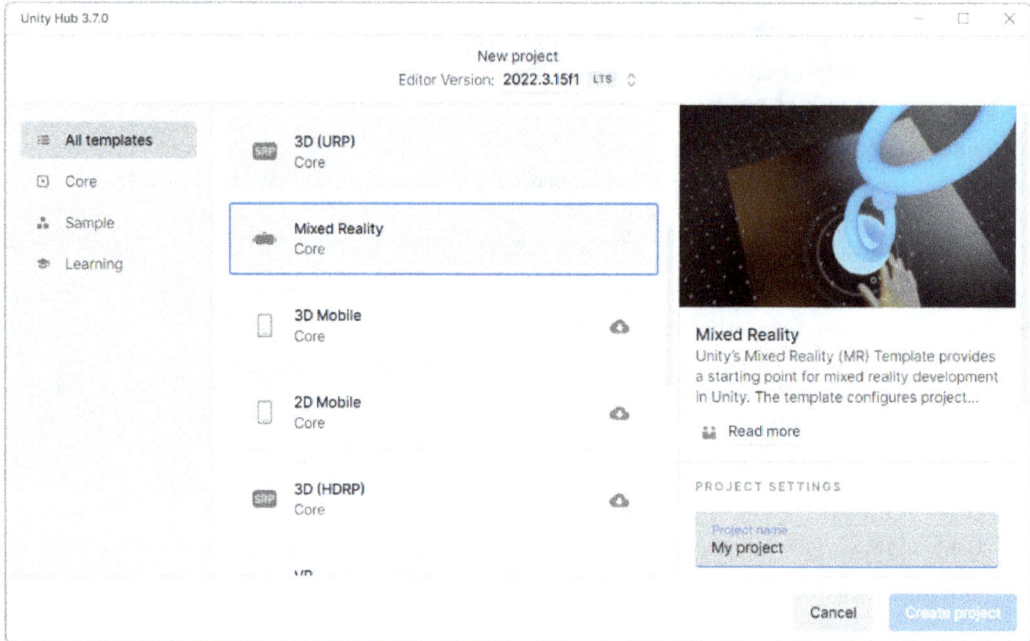

Figure 14.4 – New Mixed Reality project from a template

5. To start creating the project, click the **Create project** button and relax for a few minutes.

Let's finalize some essential setup steps once the project opens in the Unity Editor.

Open **File | Build Settings…** and follow these steps to configure the platform to support building to our Meta Quest device:

1. Select **Android** in the **Platform** list.

2. Select **ASTC** from the **Texture Compression** dropdown.

3. Click the **Switch Platform** button.

Done!

> **ASTC**
>
> **Adaptable Scalable Texture Compression (ASTC)** is a texture compression method that uses variable block sizes instead of a single fixed size and replaces older formats while also providing additional features.

You can now open the `SampleScene` scene located in the `Assets/Scenes` folder to examine the scene setup, including the following GameObjects responsible for managing the XR features, including controller and hand tracking, interaction with the UI and virtual objects, and AR features such as surface plane detection and passthrough: **MR Interaction Setup**, **UI**, and **Environment**.

Universal RP renderer settings

Please note that the Meta Quest headset is sensitive to the Unity renderer settings being correctly configured for the platform. Therefore, I suggest keeping the URP renderer and quality settings at the default values provided by the MR project template (unless you really know what you're doing).

Also note that standalone VR hardware, such as the Quest, requires additional consideration for performance optimization to sustain a minimum FPS (typically, not less than 72 FPS): this is to prevent vection (the visual perception of motion while the body is still), which can make people nauseous.

With your Quest headset attached to your system with the USB-C cable, you can verify the device is recognized by Unity by going to the **Build Settings** window with the **Android** platform selected, in the **Run Device** field, and clicking the dropdown that currently shows **Default device**. Your Meta/Oculus Quest 3 (or 2) device should be listed there.

Testing in Play Mode with Quest Link

To thoroughly test and play our game, we must build to the device because plane detection and passthrough are currently not supported with Quest Link when entering Play Mode in the Unity Editor. I would still recommend leveraging Quest Link for rapidly iterating on setting up object interactions independent of the gameplay and then building to the device for complete gameplay testing.

To use Quest Link, ensure you are connected to your system with the USB-C cable and have the Oculus app (Meta Quest Link) running, then click the Quest Link button on the Quest headset's **Quick Settings** menu. Once established, you can enter Play Mode in Unity to playtest your scene.

Now, still in the **Build Settings** window, click **Build And Run**, or press *Ctrl/Cmd + B* and pop on that headset!

Tip

It's vital to recenter or reset your orientation when you begin a MR environment. This ensures that virtual objects are placed correctly about your current position and facing direction. Doing this will enhance your experience, especially as device orientation detection technology continues to improve.

The MR template is configured to use either controllers or hands. Still, we'll focus on using the controllers for our game, so I'd recommend you play around with using the controllers as input in the sample project.

This verifies your Quest headset device and Unity MR project setup for XR development are ready. So, let's start building out the boss room!

Laying out the boss room scene

According to our GDD, we must clear the central control system from the evil plant entity infestation. So, to do that, we need to energize the sabotaged control console and restart the reactor (yes, believe me, that will do the trick). Therefore, the boss room's layout requires objects related to this context to be present in our scene.

Here are the objects we'll need for our boss room setup:

- **Control console**: Maintains the state of all the habitation station's primary systems, including the main power reactor. It has three crystal module slots for energizing the system.

- **Power reactor**: Provides power to the central systems, especially the ones in charge of environmental control and eradicating foreign entities.

- **Corridors**: The habitat station comprises several rooms and connecting corridors – this should already be familiar from the 3D FPS project.

These are the primary objects we need in our scene, which, again, is your room, to provide the setting for the boss encounter. We'll construct the layout to have the control console near the player and virtual corridors extending from the center of the room, providing a central focal point for the action.

With this layout in mind, in the following screenshot – from a Quest 3 with a passthrough visible – we can see the objects instantiated in this fashion in a real-world room.

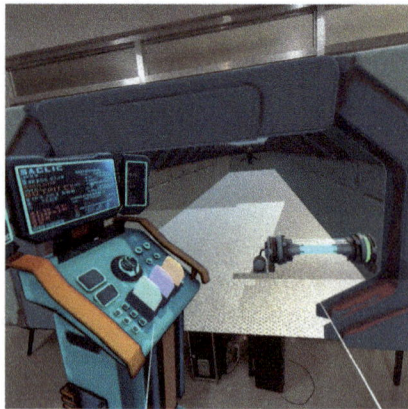

Figure 14.5 – Virtual objects spawned in a physical space

Let's first duplicate the provided MR template's sample scene to start setting up the scene:

1. Find the `SampleScene` scene in the `Assets/Scenes` folder.

2. Duplicate it by selecting and pressing *Crtl/Cmd + D*.

3. Rename it to `Boss Room`.

Now, open the scene, and we'll deactivate some example content provided in the scene Hierarchy that we won't use. Select and deactivate the following objects (using the checkbox at the top of the Inspector to the left of the object's name):

1. Disable these two child objects of `MR Interaction Setup`:

    *   `Goal Manager`

    *   `Object Spawner`

2. Disable these two child objects of `UI`:

    *   `Coaching UI`

    *   `Hand Menu Setup`

3. Save the scene.

We're done! We now have an empty scene with all the XR setup ready for us to create our MR game. Easy-peasy.

In this section, we covered the fundamental principles of designing a boss room and the steps required to set up our physical space with our Quest devices. Additionally, we learned how to create a basic starting project using Unity's MR template and configuring it for our use.

When creating the virtual objects that construct our boss room, we use AR surface planes representing real-world objects in our physical space to spawn them dynamically. Let's explore how to spawn virtual objects using detected AR planes next.

Working with AR planes (AR Foundation)

AR planes are virtual representations of flat planar surfaces, both horizontal and vertical, represented by dimensions and boundary points and detected by the AR Foundation technology. The planes provide a foundation for accurately placing digital objects and interacting with the surfaces.

As previously mentioned, these planes represent the walls, floor and ceiling, tables, and so on, and we'll use the walls, floor, and table specifically in this example boss room to blend the gameplay seamlessly with the player's physical surroundings.

> **Tip**
>
> AR Plane Manager allows you to specify a prefab for plane visualization. The AR Plane Prefab, provided by the MR template, uses a shader that occludes objects assigned a material with transparency, so if you want objects that are meant to be seen past the AR planes, ensure you don't use a transparent material.

Now, let's start working with our first horizontal plane type, the table, to see how we can detect the plane type and use its properties to spawn an object.

Spawning using planes with AR Plane Manager

The AR Plane Manager component, located on the XR Origin (XR Rig) object as a child of the MR Interaction Setup root object, is responsible for the detection of the horizontal and vertical surfaces in the physical space and creates the virtual plane objects (AR Plane Prefab) that our virtual content can be placed and interacted with.

> **Optimization note**
>
> With AR Plane Manager, in addition to specifying the AR Plane Prefab, you can choose between horizontal, vertical, or both for **Detection Mode**. Turning off vertical plane detection is recommended if you only need to detect horizontal planes.

One of the first things we'll have to do is enable the AR Plane Manager component because it is deactivated by default within the MR sample scene. We'll do that with our first script, the game manager.

Create a new script named GameManager in the Assets/Scripts folder, and start with the following code:

```
using UnityEngine.XR.ARFoundation;

public class GameManager : MonoBehaviour
{
    [SerializeField]
    private ARPlaneManager _planeManager;

    private IEnumerator Start()
    {
        yield return new WaitForSeconds(2f);
        EnablePlaneManager();
    }

    public void EnablePlaneManager()
```

```
        => _planeManager.enabled = true;
}
```

Now, create a GameObject in the scene and attach this script to it. As you can see, we've changed the method signature for the `Start()` method to be an `IEnumerator` (yes, you can do that), and we've delayed the execution of the `EnablePlaneManager()` method call for 2 seconds (to give XR components time to initialize).

Assign `_planeManager` in the Inspector by dragging in the `XR Origin (XR Rig)` object to the field on the `GameManager` component. We're enabling the component in the `EnablePlaneManager()` method like so:

```
_planeManager.enabled = true;
```

You can now playtest the boss room scene by first ensuring the scene is added to **Scenes in Build** in the **Build Settings** window, connecting your Quest device to your system via the USB-C cable, and clicking **Build And Run** (*Ctrl/Cmd + B*, i.e., *build and run*).

You should see something similar to the following figure, where the detected planes for the walls, floor and ceiling, and any horizontal surfaces such as tables have a fading dotted material. Note that I've manually added the magenta lines for better visibility of the planar surfaces (which include a table, the walls, and the floor).

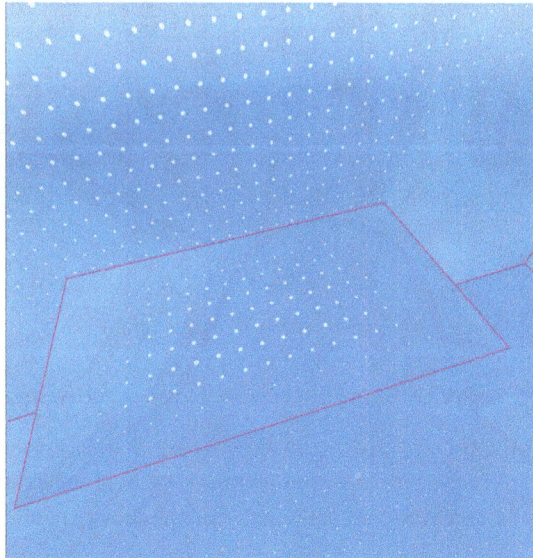

Figure 14.6 – Detected surface planes in the room

Now that we've verified we have planes to work with, let's start spawning in our virtual objects for the boss room.

Instantiating on a table plane

Whoa, wait a second… we need the objects we'll be spawning into the room. We can thank **Polypix Studios** again for providing 3D art for these assets: console, module, reactor, corridor, gun, and hover bot.

Boss room virtual objects

You can download `VirtualObjects-start.zip` from the book's GitHub repo here: `https://github.com/PacktPublishing/Unity-2022-by-Example/tree/main/ch14/Art-Assets`.

Unzip the file to get the `'.unitypackage'` file and import it into your project – you can do that by dragging and dropping the file from your system's file manager to Unity's **Project** window.

The first asset we'll work with is the reactor model imported as the `Reactor` Prefab in the `Assets/Prefabs` folder, as seen in the following screenshot:

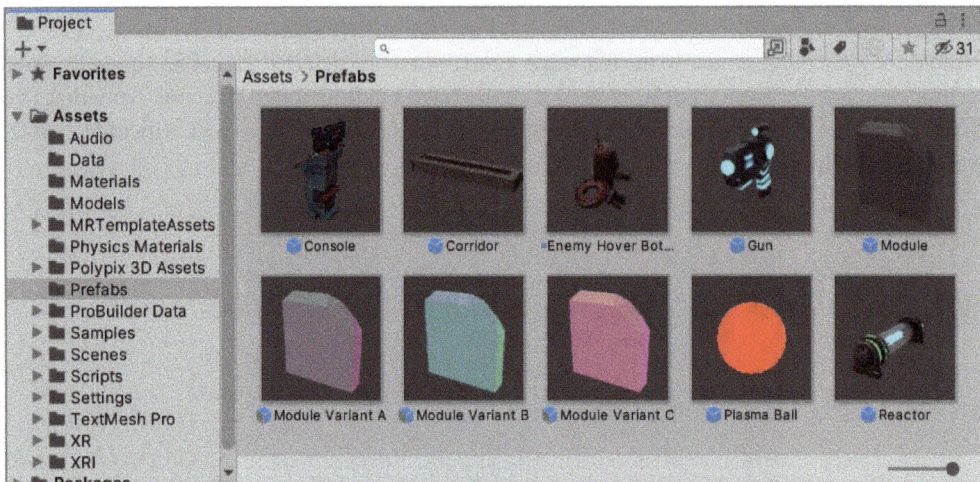

Figure 14.7 – Boss room 3D asset Prefabs

And we'll spawn the `Reactor` Prefab on the first table detected in the room.

`AR Plane Manager` provides a `planesChanged` event that we'll subscribe to, and when a `PlaneClassification` is of the `Table` type, we know we should spawn the reactor. So, let's add the listener in our `GameManager` class and the handler method for spawning the object:

```
public void EnablePlaneManager()
{
    _planeManager.enabled = true;
    _planeManager.planesChanged += OnPlanesChanged;
}
```

Don't forget to unsubscribe from the listener in `OnDisable()` or `OnDestroy()`.

Now, add the handler method:

```
private void OnPlanesChanged(ARPlanesChangedEventArgs args)
{
    foreach (var plane in args.added)
    {
        switch (plane.classification)
        {
            case PlaneClassification.Table:

                if (!_hasSpawnedPrefab_Reactor)
                {
                    SpawnPrefab(plane, _prefabReactor);
                    _hasSpawnedPrefab_Reactor = true;
                }
                break;
        }
    }
}
```

Here, you can see we loop through all the detected planes (`ARPlane` type) provided by the handler's `args` parameter with a `foreach` statement. Then, the `switch` statement allows us to work with the specific plane classification for our needs, which, again, is `Table`. We use a helper `SpawnPrefab()` method, which performs the instantiation – passing in the specific plane and Prefab:

```
private void SpawnPrefab(ARPlane plane, GameObject prefab)
{
    Instantiate(prefab,
        plane.transform.position,
        plane.transform.rotation);
}
```

You can see we're using the regular `Instantiate()` method and the plane's `position` and `rotation` values for the point of instantiation.

To ensure we're only spawning one reactor in the scene, if the physical space has multiple table surfaces defined, we'll use the `_hasSpawnedPrefab_Reactor` Boolean to limit it to one and call the `SpawnPrefab()` method, specifying the `ARPlane` and `Reactor` Prefab as the parameters. Setting `_hasSpawnedPrefab_Reactor` to `true` after calling the spawn method ensures only one Prefab is spawned.

You must ensure you've declared the variables to assign the `Reactor` Prefab and *has spawned* bool. Then, assign the `Reactor` Prefab from the **Project** window to the `GameManager` field in the Inspector.

If you build and run the project on your device now, you should see the reactor appear on your table!

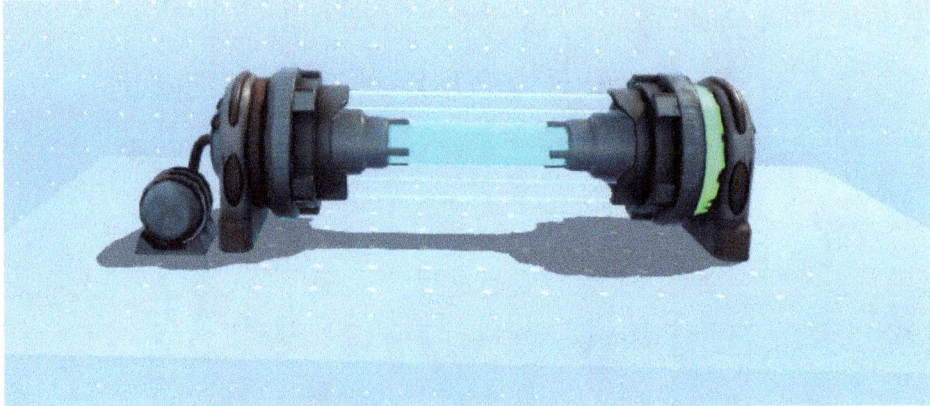

Figure 14.8 – The reactor Prefab spawned on the table plane

> **Enabling passthrough**
>
> If you want to temporarily skip ahead to see the virtual objects sitting in your real-world surroundings, we'll need to toggle the passthrough to be visible. In the *Toggling MR visuals with XR Input section*, we'll add the ability to toggle passthrough visibility.

That takes care of demonstrating how to instantiate an object on the table plane. Let's move on to spawning on the floor plane next.

Instantiating using the floor plane

Unlike the reactor's placement, we'll prioritize the orientation of the controller console object relative to the player's forward direction (when the planes are detected) to ensure immediate interaction availability for the player.

The process is relatively the same, except for said orientation. So, let's first add to our `switch (plane.classification)` block to handle the `Floor` classification:

```
case PlaneClassification.Floor:
    if (!_hasSpawnedPrefab_Console)
    {
        SpawnPrefab(plane,
            _prefabConsole,
            new Vector3(-1f, 0f, 0f));
        _hasSpawnedPrefab_Console = true;
    }
    break;
```

We have a similar Boolean check to ensure we don't have multiple control consoles spawned into the room with the _hasSpawnedPrefab_Console variable and a method overload for SpawnPrefab(). The new spawn method signature takes an additional parameter for an offset – and we'll assume that the offset is from the player's location.

The new spawner method with the offset looks like the following – add it to the GameManager class:

```
private void SpawnPrefab(
    ARPlane plane, GameObject prefab, Vector3 playerOffset)
{
    var playerTransform = Camera.main.transform;
    playerTransform.position = new Vector3(
        playerTransform.position.x,
        plane.transform.position.y,
        playerTransform.position.z);

    var worldOffset =
        playerTransform.TransformDirection(playerOffset);
    var spawnPosition = playerTransform.position
        + worldOffset;
    var directionToPlayer =
        (playerTransform.position
            - spawnPosition).normalized;
    var spawnRotation =
        Quaternion.LookRotation(-directionToPlayer,
            Vector3.up);
    Instantiate(prefab, spawnPosition, spawnRotation);
}
```

The points of interest in the new spawn method are as follows:

- playerTransform: We get the player's current position in the world from the main camera, which is attached to XR Origin (XR Rig) to represent the player's head.

- playerTransform.position: We apply a new Vector3 Y-value to playerTransform to anchor the instantiated model to the floor plane (its Y-value).

- worldOffset: We take the player offset value provided and use the TransformDirection() method to ensure the player offset will be applied in the appropriate world space coordinates – there is no way we could know the world coordinates to pass into the spawn method since it is relative to the player's current position.

- spawnRotation: We want to ensure the console faces the player when instantiated, so we use Quaternion.LookRotation() to accomplish that.

We finish by just calling the `Instantiate()` method, like before. Add the required variables and make the Inspector assignments. Then, save your changes and go ahead and do another *build and run* to see the console in your room. An example of the console placement can be seen in the virtual objects spawned into the actual room in the *Figure 14.5*.

Okay, we're making great progress on the boss room layout! The last environmental object to get into our room is the corridor, which will virtually extend the reality of the room and set the stage for our enemy hover bots to engage the player.

Instantiating with wall planes

Our last AR-detected plane instantiation example will still be relatively the same as the previous two, except we now have to account for spawning an object relative to a vertical surface. This will require a bit of additional care during positioning because the plane anchors are located in the center of the plane's surface object.

Fortunately, we have access to all the basic `Bounds` properties, such as `extents`, but we still need the surface extent and orientation. So, let's first add to our `switch (plane.classification)` block to handle the `Wall` classification:

```
case PlaneClassification.Wall:
    if (!_hasSpawnedPrefab_Corridor)
    {
        SpawnPrefabAtWallBase(
            plane, _prefabCorridorDoorway);
        _hasSpawnedPrefab_Corridor = true;
    }
    break;
```

You can see the same pattern here as before, using the Boolean to determine whether we've spawned the corridor Prefab already and calling a spawn Prefab method, passing in just the plane and the Prefab again this time.

The `SpawnPrefabAtWallBase()` method looks like this:

```
private void SpawnPrefabAtWallBase(
    ARPlane plane, GameObject prefab)
{
    var planeCenter = plane.transform.position;
    var heightOffset = plane.extents.y;
    var basePosition =
        new Vector3(planeCenter.x, planeCenter.y
            - heightOffset, planeCenter.z);
```

```
    var prefabRotation =
        Quaternion.LookRotation(-plane.normal, Vector3.up);

    Instantiate(prefab, basePosition, prefabRotation);
}
```

Just a bit more calculation is required here to ensure we are anchoring the Prefab's instantiation point to the vertical bottom of the surface plane – at the same Y-value as the floor – by using `plane.extents` and subtracting from the plane's transform position (at the center of the plane).

For the rotation of the spawned Prefab, we'll again use `LookRotation()`, but this time, instead of using the *direction to player* vector, we'll use the plane's surface normal vector. The plane's surface normal is pointing away from the center of the room, so we want to invert it for the instantiation of the corridor Prefab that has its forward direction looking down the corridor (for your own 3D models, you can invert either the normal vector or rotate the pivot's forward direction for the correct orientation).

Again, add the required script variables, save the script, make the corridor Prefab assignment to the field on `GameManager`, save your scene, and build and run the app to test the placement of the corridor addition to the boss room.

Now that we have the layout of the boss room completed, with all the elements required for the battle, let's see how we can work with the MR visuals to set a proper gameplay experience without AR planes being visible and passthrough enabled.

Toggling MR visuals with XR Input

You know how much of a fan I am of reusable components to build out functionality that is also designer-friendly in our games. So, let's approach input from our XR controllers similarly by adding an *on button press* component that relies on an `InputAction` input signal (courtesy of the new Input System) to identify the button presses.

You can get the `OnButtonPress` script file from the GitHub repo here: `https://github.com/PacktPublishing/Unity-2022-by-Example/tree/main/ch14/Code-Assets`, then import it into your project in the `Assets/Scripts/Interaction` folder.

We're going to wire up buttons for the following in-game actions:

- Right controller, primary button (**A**) → Toggle passthrough visibility.
- Right controller, secondary button (**B**) → Toggle AR plane surfaces.
- Left controller, primary button (**X**) → Start game.

Let's see what the buttons for the Oculus controllers with Unity's **XR Input** mappings look like in the following figure.

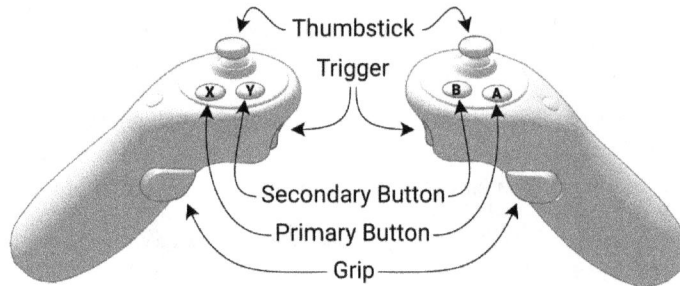

Figure 14.9 – Oculus XR controller button mapping

> **Additional reading | XR Input**
>
> Unity XR Input mappings: `https://docs.unity3d.com/Manual/xr_input.html#XRInputMappings`

Okay, simple enough. Let's start by wiring up the passthrough visibility toggle for when the **A** button is pressed.

Toggling passthrough

The MR template showcases the ability to switch between the virtual environment and device passthrough. This is achieved by using a simple fade transition on an environment mesh. The mesh employs a vertex color **ShaderGraph**, which has an alpha property that can be smoothly transitioned.

We don't even have to write a script to perform the fade transition. The MR template's `Environment` Prefab already includes a `FadeMaterial` component with an exposed public method for fading!

So, let's hook into it quickly to toggle the fade, starting with making a new controller script named `SceneController` in the `Assets/Scripts` folder with the following code:

```
using UnityEngine.Events;
using UnityEngine.XR.ARFoundation;

public class SceneController : MonoBehaviour
{
    [Header("Triggered Events")]
    public UnityEvent<bool> OnTogglePassthrough;
}
```

Save the script and add it to the `GameManager` object. As you can see, we'll use a `UnityEvent` to assign the reference to the `FadeMaterial` function, passing a Boolean parameter representing the visible state to fade to.

Let's assign the `OnTogglePassthrough(Boolean)` callback in the Inspector and then finish by adding the toggle logic code afterward. So, start by clicking the + icon to add a new event callback entry. Then, using *Figure 14.10* as a reference, find the **UI | Environment** object in the scene Hierarchy and drag it to the **Object** field.

Now, select the `FadeMaterial.FadeSkybox` function at the top in the **Dynamic bool** section – we want to dynamically pass the `UnityEvent`'s `bool` parameter when the event is invoked.

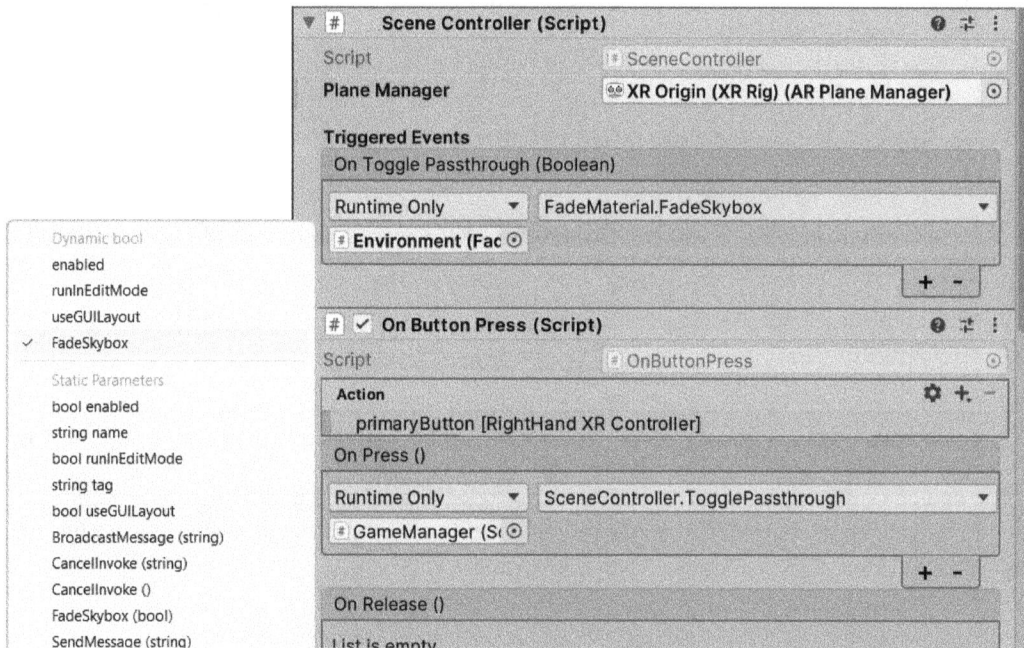

Figure 14.10 – SceneController component setup

All that's left to do is invoke the `OnTogglePassthrough` event when the player presses the right controller's primary button (**A**). Let's add the following toggle code to the `SceneController` class for that:

```
public void TogglePassthrough()
{
    _isPassthroughVisible = !_isPassthroughVisible;
    SetPassthroughVisible(_isPassthroughVisible);
}
```

```
public void SetPassthroughVisible(bool visible)
{
    _isPassthroughVisible = visible;
    OnTogglePassthrough.Invoke(_isPassthroughVisible);
}
```

Straightforward, single-responsibility, and well-named methods. Add the private member variable _isPassthroughVisible for keeping track of the current toggle state, which defaults to false and is the correct default passthrough state.

Lastly, referring to *Figure 14.10* for these steps, to receive the player input when the controller's button is pressed, let's use the OnButtonPress component with the input action configured for the **A** button:

1. Add the OnButtonPress component to the **GameManager** object (just below SceneController).

2. Click the + dropdown, then **Add binding**.

3. Double-click **<No Binding>**.

4. In the **Path** dropdown, select **XR Controller | XR Controller (Right Hand) | Optional Controls | primaryButton**.

    Alternatively, enter the following text (click the **T** button first): <XRController>{RightHand}/ primaryButton.

5. Lastly, assign the OnPress() UnityEvent function to SceneController. TogglePassthrough.

Doing a *build and run* to your device now, you can toggle the passthrough visibility by pressing the **A** button on your right controller. This will be your first time seeing the boss room's digital objects spawned within your real-world space – pretty awesome, right?

Camera setup to support passthrough

The MR template's main camera comes preconfigured to enable device passthrough, but it's worth mentioning the setup. The camera's background type is Solid Color with the background color set to black, with 0 alpha. The AR Camera Manager component is also explicitly included to control passthrough on Meta Quest devices.

Passthrough, check! Now, let's see about toggling the AR plane visibility when the **B** button is pressed.

Toggling AR plane visibility

It shouldn't be any surprise that we'll mimic the passthrough toggle setup. We've already seen how to reference and use the AR Plane Manager to spawn the virtual objects for the different plane classifications. Well, we'll be using it again here to access the current set of **trackables**.

A trackable is a component that represents AR objects detected in the real world. Examples include planes (you're already familiar with these), point clouds, anchors, environment probes, faces, bodies, images, and 3D objects.

> Trackables (AR Foundation)
>
> Trackables and trackable managers: `https://docs.unity3d.com/Packages/com.unity.xr.arfoundation%405.1/manual/architecture/managers.html#trackables-and-trackable-managers`

Let's start by adding the following toggle code to the `SceneController` class for plane visibility:

```
[SerializeField]
private ARPlaneManager _planeManager;

public void TogglePlaneVisibility()
{
    _arePlanesVisible = !_arePlanesVisible;
    SetPlaneVisible(_arePlanesVisible);
}

public void SetPlaneVisible(bool visible)
{
    _arePlanesVisible = visible;
    foreach (var plane in _planeManager.trackables)
    {
        if (plane.gameObject.
            TryGetComponent<FadePlaneMaterial>(
                out var planeFader))
        {
            planeFader.FadePlane(_arePlanesVisible);
        }
    }
}
```

The `foreach` loop in the `SetPlaneVisible()` method is responsible for implementing the fade on the planes found while iterating the trackable collection based on finding a `FadePlaneMaterial` component. If found, we simply call that plane's `FadePlane()` method. The `FadePlaneMaterial` component provides us with the MR template's `AR Plane` Prefab. Easy-peasy.

Let's hook it up now to the controller button press – which will be the right controller's secondary button (**B**):

1. Add another `OnButtonPress` component to the `GameManager` object (just below the previous `OnButtonPress` component).

2. Click the + dropdown, then **Add binding**.

3. Double-click <**No Binding**>.

4. In the **Path** dropdown, select **XR Controller | XR Controller (Right Hand) | Optional Controls | secondaryButton**.

    Alternatively, enter the following text (click the **T** button first): <XRController>{RightHand}/ secondaryButton.

5. Lastly, assign the `OnPress()` UnityEvent function to `SceneController.TogglePlaneVisibility`.

Doing a *build and run* to your device now, you can toggle the AR plane visibility by pressing the **B** button on your right controller. Toggling the plane visibility will mostly serve the purpose of debugging; in case objects spawn in unexpected ways, you can verify the plane detected in the physical space (any that do might indicate you need to revisit your Quest headset's **Room Setup** configuration).

Mixed Reality template script fix!

At the time of writing, the `/Assets/MRTemplateAssets/Scripts/FadePlaneMaterial.cs` script has an error with the `FadePlane()` method that must be corrected for the planes to fade in and out correctly after first being called in its `Awake()` method.

On *line 91*, within the `FadeAlpha()` method, replace the `k_DotViewRadius` variable with `k_Alpha`. The line should now read: `rend.material.SetFloat(k_Alpha, alphaValue);`.

Note that we'll turn off the plane visibility when we start the game. Speaking of... let's wire up starting the game now.

Starting the game

Offering players a choice to start a game via a button or menu selection creates a sense of control and anticipation. In contrast, for MR and even VR, an automatic game start can be disorienting or, worse, jarring. Well, that is, unless you want to be hostile and throw players directly into unforgiving action without warning (ahem, *Dark Souls*, I see you).

As mentioned, we won't be so cruel and will provide a button press for the player to enter the MR environment to start the game. With that in mind, we need to make some additions to the GameManager class. Let's add the following code:

```
private SceneController _sceneController;

private void Awake()
    => _sceneController = GetComponent<SceneController>();
```

Here, we're just getting a reference to the SceneController sibling component. Go ahead and add a [RequireComponent] attribute for the SceneController component, too.

Here's the StartGame() method we'll call from the button press:

```
public void StartGame()
{
    EnablePlaneManager();
    StartCoroutine(DelayStartGame());

    IEnumerator DelayStartGame()
    {
        yield return new WaitForSeconds(1.5f);

        _sceneController.SetPlaneVisible(false);
        _sceneController.SetPassthroughVisible(true);
    }
}
```

The first thing we'll do with the EnablePlaneManager() call is to enable the plane manager to spawn the virtual objects that make up our boss room game. We'll then use a coroutine to delay calling the local DelayStartGame() function by 1.5 seconds, setting planes invisible and the passthrough visible – this will ensure we see the virtual objects unobstructed in our real-world space.

Remove the IEnumerator Start() method

Don't forget, we'll have to remove the Start() method we started with to set up the object spawning in the earlier *Spawning using planes with AR Plane Manager* section. We're now going to wait until the player presses the button to start the game to enable spawning the virtual objects.

Okay, let's finish things up by adding the **X** button press to start the game by following these steps again – but simplified this time (you got this):

1. Add another `OnButtonPress` component to the `GameManager` object.
2. Assign input action binding for the XR left-hand controller's primary button (**X**), like so: `<XRController>{LeftHand}/primaryButton`.
3. Assign the `OnPress()` function to `GameManager.StartGame`.

This time, when you *build and run*, you'll see yourself in the empty virtual environment until you press the **X** button to start the game and enter the boss room battle!

In this section, we learned how to work with the generated AR planes to spawn objects in our rooms about their transform position and rotation. We also learned how to access the AR systems and pre-made components (provided by the MR template) to toggle the visual state of passthrough and AR planes.

We have the boss room environment taken care of now, and we have the game starting, but there's still nothing for us to do or interact with. We'll solve that problem now by adding XRI interactable objects to the room.

Placing interactable objects in the world

In MR game design, interactive objects are essential to bridging the virtual and real worlds. Interactable objects are designed to respond to user input, even as basic as hand (or controller) movements that allow natural and intuitive interactions like pushing, grabbing, throwing, or even complex multi-hand manipulation (for example, rotating and scaling the object). They really help to sell the reality of the environment, and as a result, they significantly enhance the player's engagement and overall gameplay experience.

For our game's purposes, we'll have examples of a simple grab and placement interaction and, with the gun, a secondary interactable event action for shooting. Note that while many MR games and experiences are built for use with hands (hand tracking), our boss room example game will use controllers.

Let's start by configuring the modules for grabbing – these will then be configured to be inserted into the slots on the control console (refer to the GDD in the *Designing a boss room* section).

Making objects XR interactables

The first grabbable object we'll work with is the crystal module. The player must be able to grab the module and insert it into the control console, so we'll open up the provided `Module Prefab` asset in Prefab Mode (double-click on it in the **Project** window) and add an `XR Grab Interactable` component to the root.

As seen in the following screenshot, grabbable objects should have a transform positioned and appropriately rotated for grabbing the item with the correct orientation for proper usage – here, we see both the `Module` and the `Gun` assets with their `Attach` object positioned and rotated for a good grab.

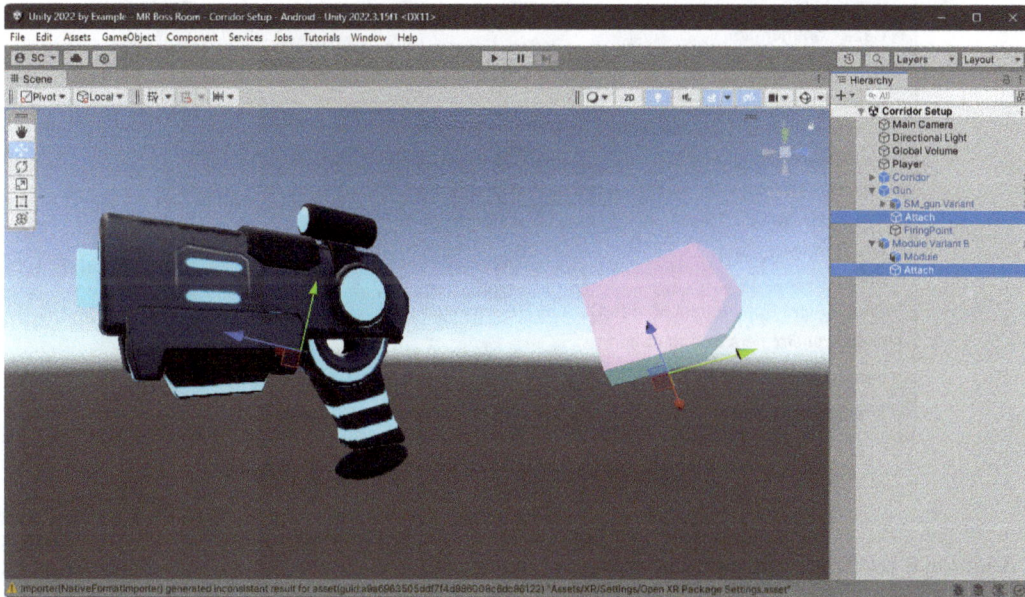

Figure 14.11 – Configuring the XR grab attach transforms

Note from the screenshot that the forward direction (Z-axis, blue arrow) of the `Attach` transform is pointing away from the player holding the object. Some experimentation may be done to attain the desired grab position.

Now, we just need to assign the `Attach` object to the `XR Grab Interactable` **Attach Transform** field to ensure it gets properly attached to the player's controller. You can find the **Attach Transform** field hidden within the many options the interactable component provides.

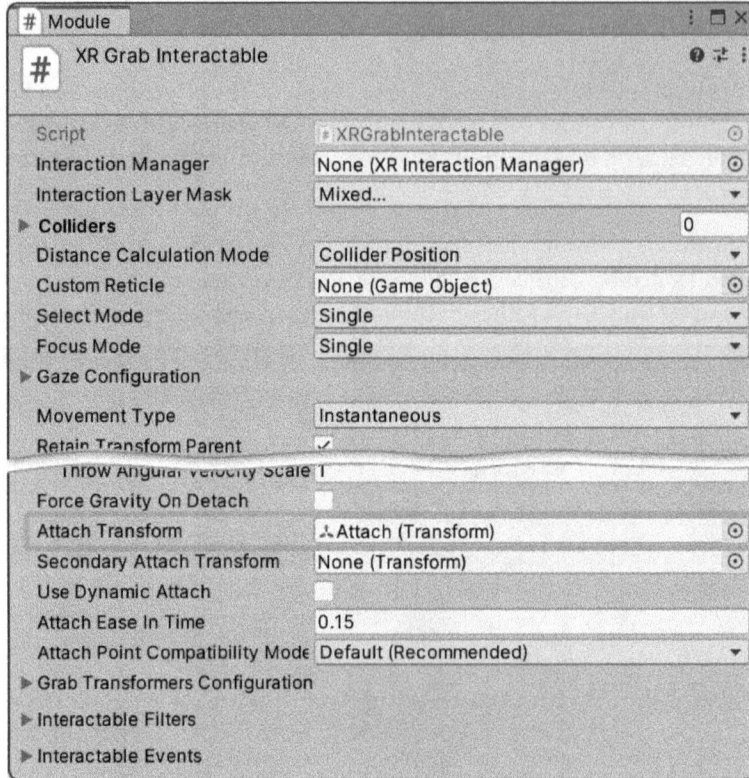

Figure 14.12 – XR Grab Interactable Attach Transform assignment

Additional reading | Affordance system

The XRI affordance system gives visual color and audio feedback cues when interacting with objects, especially when haptics are unavailable while using hands, using an XR Interactable Affordance State Provider component with the interactable source. Samples are provided in the XRI example project.

Affordance system: `https://docs.unity3d.com/Packages/com.unity.xr.interaction.toolkit%402.5/manual/affordance-system.html`

Save the module and temporarily add it to your Boss Room scene near the MR Interaction Setup object. Enter Play Mode and test grabbing the module and moving it around with the controller (by using the grip button on the side of the controller, with your middle finger). Notice I said *enter play mode* this time, not *build and run*. That's because we want to iterate changes like grab point attachment positions more quickly. For details, refer to the Quest Link callout in the *Creating the Unity project* section.

> **Completed interactable objects**
>
> All of the completed XR interactable objects are provided in the completed Unity project files for this chapter in the book's GitHub repository here: `https://github.com/PacktPublishing/Unity-2022-by-Example/tree/main/ch14/XR-Assets`.

That's all there is to making an object interactable in XR. XRI makes it very easy to get the minimum required interactions, such as grabbing in place for games and experiences. We saw how to dynamically place other digital objects in the world; let's do the same for the modules, but with a twist.

Placing the modules in the room

In our boss room battle, our primary objective, besides just staying alive, is to collect the crystal modules to restore the functionality of the control console and energize the reactor to expel the evil plant entity. So, we have a sort of collection game here again! However, let's add to the challenge of collecting and managing the modules.

Collecting objects in an MR game can be made more engaging by having these objects move around the room. The player will need to rely on their spatial awareness and timing skills, which introduces a more dynamic and novel challenge requiring them to explore the room. With the objects reacting to not only the player's actions but the physicality of the space, it also deepens the immersion of the MR gameplay experience.

If the collectible objects were to float somehow, the modules could also contribute to the game's narrative or aesthetic theme. As it happens, crystal modules have a strange other-worldly property – gravity does not affect them, but forces do. ¯_(ツ)_/¯

With that context set, let's first create the three required modules, then spawn them into the room when the game starts.

Creating unique module variants

The three `Module` Prefab variants we'll create can be seen in *Figure 14.7*, and each will have a unique identifier – the ID of the module will come into play when we configure the control console slots.

There are multiple ways to create a Prefab variant, but this time, we'll use the following steps to create each unique module:

1. Make a Prefab variant of `Module` by right-clicking on it in the **Project** window and selecting **Create | Prefab Variant**.
2. Name it `Module Variant A` (the proceeding variants will be B and C).
3. Double-click on `Module Variant A` to open it in Prefab Mode.
4. In the **Module ID** field of the `Module` component, set it to A (followed by B and C). (The `Module` script is provided as part of the imported base assets.)

5. From the Assets/Materials folder, assign the Module_A material. You can easily do this by dragging the material from the **Project** window onto the model visible in the **Scene** view.

6. **Use Gravity** = false on the Rigidbody component.

Repeat these steps to create variants for modules B and C, respectively. Remember, any edits you make to a Prefab variant, such as modified property values or added/removed components, become overrides of the base Prefab, so you don't want to apply these overrides, or you'll be applying them to the base Prefab asset, and we don't want that!

> **XR interactable required component**
>
> Adding an XR Grab Interactable component to our objects will automatically add a Rigidbody component with its default values.

Three unique modules, check! We can now add the necessary code to our game manager to spawn the modules when the game starts.

Spawning the modules to get things moving

There is no sense in reinventing the wheel to spawn another Prefab into the scene; we can rely on the work we've already coded (as we should generally do). We will, however, change the spawning up just a bit because we don't want to instantiate the objects about a plane object. We want a more arbitrary position in the world, but still in relation to the player position.

Open up the GameManager script for editing. Let's first create the serialized private member variable where we can assign all the module variants in the Inspector that need to be spawned into the scene:

```
[SerializeField]
private GameObject[] _prefabModules;
```

Now, we can create another method overload for the SpawnPrefab() method:

```
private void SpawnPrefab(GameObject[] prefabs,
    Vector3 playerOffset,
    Vector3 forceDirection, float force)
{
    var playerTransform = Camera.main.transform;
    var spawnPosition = new Vector3(
        playerTransform.position.x
            + (playerTransform.right * playerOffset.x).x,
        playerTransform.position.y
            + (playerTransform.up * playerOffset.y).y,
        playerTransform.position.z
            + (playerTransform.forward * playerOffset.z).z
```

```
    );

    foreach (var item in prefabs)
    {
        var module = Instantiate(item,
            spawnPosition, Quaternion.identity);
    }
}
```

In the method signature, we've made the prefabs parameter an array, GameObject[] prefabs, to accept any number of Prefabs to spawn, then added forceDirection and force parameters, which we'll use to apply a force to the objects after instantiation.

The primary difference with this Prefab spawning method is that we're using a foreach statement to iterate the array of Prefabs to ensure each one is instantiated.

Now, we can add the call to SpawnPrefab() to do the module spawning. For simplicity's sake, we'll just tag it onto the console spawning. Add the following call to SpawnPrefab() in the switch statement's floor plane classification case statement:

```
case PlaneClassification.Floor:
    if (!_hasSpawnedPrefab_Console)
    {
        ...
        SpawnPrefab(_prefabModules,
            new Vector3(0f, 1.5f, 0.8f),
            Vector3.up, 0.05f);
    }
    break;
```

A new vector position is passed in as the offset from the player's position (world space), the Vector3.up is the direction force, and 0.05f is the force applied to the modules when they are instantiated. Simple.

Okay, we've talked about adding a force to the crystal modules so that they float about the room... now's the time to implement it! Add the following lines to this iteration of the SpawnPrefab() method:

```
// Existing line in foreach body.
var module = Instantiate(item,
    spawnPosition, Quaternion.identity);

// Added lines.
if (forceDirection != Vector3.zero || force != 0)
{
    if (module.TryGetComponent<Rigidbody>(out var rb))
    {
```

```
                    ApplyForce(rb);
        }
    }
```

If we have a force direction and amount passed as parameters to the `SpawnPrefab()` call that are not zero, we attempt to get the `Rigidbody` component of the instantiated Prefab. If the `Rigidbody` component reference is successfully retrieved, we call `ApplyForce()` and pass it in.

All that remains is to add the `ApplyForce()` method as a local function to work its physics magic:

```
void ApplyForce(Rigidbody rb)
{
    rb.AddForce(forceDirection * force, ForceMode.Impulse);
    var torqueMultiplier = 3f;
    var randomRotation = new Vector3(
        Random.Range(-1f, 1f),
        Random.Range(-1f, 1f),
        Random.Range(-1f, 1f)).normalized
            * (force * torqueMultiplier);
    rb.AddTorque(randomRotation, ForceMode.Impulse);
}
```

The physics API methods we're taking advantage of here are `rb.AddForce()` and `rb.AddTorque()` to apply forces using an `Impulse` force mode.

Additional reading | Unity documentation

`Rigidbody.AddForce:` `https://docs.unity3d.com/2022.3/Documentation/ScriptReference/Rigidbody.AddForce.html`

`Rigidbody.AddTorque:` `https://docs.unity3d.com/2022.3/Documentation/ScriptReference/Rigidbody.AddTorque.html`

Save the script and assign all the crystal module Prefab variants to the `GameManager`'s **Prefab Modules** field. Playtest and adjust the spawn position of the modules to your liking. Have fun chasing them down!

Applying impact force

The provided `Module` Prefab comes with an `ImpactApplyForce` script added to it that will apply an opposite force to the module when it collides with any other object with a collider. Combined with a very bouncy physics material assigned to the collider, this attempts to keep the modules moving about the room constantly.

In this section, we got the crystal modules floating about the room, adding the first challenge to the boss room battle mechanics. The second half of the challenge with the modules has to do with inserting them correctly into the slots of the control console. In the next section, we'll perform the XR interactable configuration necessary for this interaction.

Making the module slots interactable

To have objects that can work together to create an intuitive system that mimics how things work in the real world, we use an XR Grab Interactable object and an XR Socket Interactor object – we have an *interactable* and an *interactor*. The grab interactor allows players to pick up and interact with objects, while the socket interactor provides the designated spots to place them. This handshake between the two components makes it easier for users to interact with objects and provides a more seamless and immersive experience in virtual or MR environments.

This means we'll be configuring each control console slot with a socket interactor. Go ahead and open up the Console Prefab in Prefab Mode from the Assets/Prefabs folder. Add the XR Socket Interactor component for the Slot A, Slot B, and Slot C objects parented to the ConsoleSlots object. The object hierarchy can be seen in the following screenshot:

Figure 14.13 – Console slot configuration

An *attach* transform object can also be seen in the preceding screenshot; each slot has an object parented to it, and named Socket Attach. For each socket interactor added to the slot objects, assign the *attach* object to the interactor's **Attach Transform** field (just like we did for the grab interactables).

We also want to ensure that only modules are inserted into the slots on the control console; we can do something about that. We can use the `Interaction Layer Mask` property of both `XR Grab Interactable` and `XR Socket Interactor`.

It doesn't matter which one you start with, but it's essential first to add a `Module` interaction layer. You can do that from any **Interaction Layer Mask** field by clicking the dropdown and selecting **Add layer…** (at the bottom), then going back to the component and selecting **Nothing**, then **Module** for each.

> **Setting the interactive layer with the asset**
>
> Alternatively, find the interactive layer asset at `Assets/XRI/Settings/Resources/InteractionLayerSettings`, add the `Module` layer, then return to the components and set the layer.

The last part of the slot configuration is that the slots are configured with a `ConsoleSlot` component already, similar to how we configured the module's `Module` component; ensure **Slot ID** for each of the slots has their ID assigned: A, B, and C again.

Speaking of the `ConsoleSlot` component, let's take a closer look at the code. It's more than just a slot ID – it can detect when a module is inserted or removed. This allows it to tell the parent console controller when the specific slot is interacted with, which can then respond accordingly:

```
public class ConsoleSlot : MonoBehaviour
{
    [SerializeField] private char _slotID;
    private ConsoleController _controller;
    private XRSocketInteractor _socketInteractor;

    private void Awake()
    {
        _controller =
            GetComponentInParent<ConsoleController>();
        _socketInteractor = GetComponent<XRSocketInteractor>();

        _socketInteractor.selectEntered.
            AddListener(HandleModuleInserted);
        _socketInteractor.selectExited.
            AddListener(HandleModuleRemoved);
    }
}
```

We declare our variables, then, in Awake(), once we have the references to the required components, we register the listeners for responding to the socket interactor selectEntered and selectExited events for handling inserting and removing modules, respectively.

Here are the handler method declarations:

```
private char _moduleID;

private void HandleModuleInserted(SelectEnterEventArgs arg)
{
    _moduleID = arg.interactableObject.transform.
        GetComponent<Module>().ModuleID;

    if (!char.IsWhiteSpace(_moduleID))
    {
        _controller.InsertModule(_slotID, _moduleID);
    }
}

private void HandleModuleRemoved(SelectExitEventArgs arg)
    => _controller.ResetSlots();
```

The first thing we do is get the ID of the inserted module (remember, only modules can be inserted due to the interaction's layer mask assignment). We then call a method of the ConsoleController instance for either the module being inserted, InsertModule(), or simply resetting the slots, ResetSlots(), when the module is removed.

You might be considering having ConsoleController subscribe to a ConsoleSlot exposed event. Since there are three slots, it is more efficient to have each slot handle its own interactions (objects should be responsible for their own state) and notify the controller (by passing its ID and the module's ID). This is a more simplified approach.

> **Bonus activity**
> Feel free to *flip the script* and experiment with the console controller listening to events on all three slots to compare the required code differences.

You should now be able to playtest the console slot interactions by grabbing a crystal module and placing it in any slot. Fun!

There's more fun to be had… let's get that laser pistol configured to provide us some protection against infiltrated hover bots.

Configuring the laser gun

The configuration for the interactable gun object is pretty much the same as the crystal module; we already saw how to configure an attach transform in *Figure 14.11*. Except now, we'll add a secondary action for shooting when the trigger is pulled.

Implementing shooting with XR Interactable Events

We only want shooting triggered when we're actually grabbing the gun, so we won't be relying on the reusable OnButtonPress component this time. Instead, we'll use the **Interactable Events** of the XR Grab Interactable component, specifically, Activated. Activated is called when the interactor selecting the interactable sends a command to activate the interactable – precisely what we need.

> **Additional reading | Grab interactables**
>
> Both basic and advanced examples of grab interactions are available in the XRI examples: https://github.com/Unity-Technologies/XR-Interaction-Toolkit-Examples/blob/main/Documentation/GrabInteractables.md.

To set up the Gun Prefab, take the following steps:

1. Either modify the provided Gun Prefab directly or make a Prefab variant to work with.

2. Open the Prefab in **Prefab Mode**.

3. Add an XR Grab Interactable component to the root.

    I.   Assign the Attach object to the **Attach Transform** field.

    II. For **Interactable Events** | Activated, assign the Gun.Shoot function.

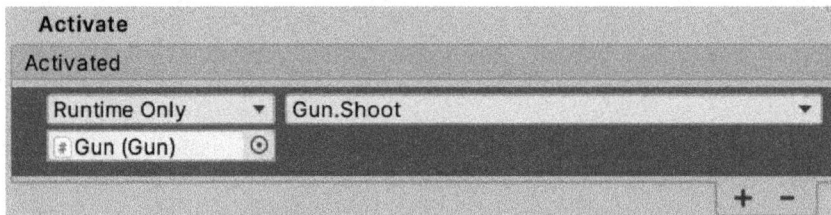

Figure 14.14 – XR Grab Interactable event Activated assignment

4. For the Rigidbody component, use the following property values (the gun will stay floating in the air right where the player releases their grip; Kryk'zylx military tech is truly advanced!):

    *   **Use Gravity** = false

    *   **IsKinematic** = true

And that's all that's required to configure the Gun Prefab to make it an interactable object that players can pick up and shoot. Pew-pew!

Gun sound FX

We also have sound FX added for the shooting, courtesy of `AudioManager` and the `AudioPlayerSFX3D` audio-playing component. So, add the audio manager to the boss room scene, create an audio mixer and the required mixer groups, and then assign the mixer groups to the audio manager. For a refresher, visit *Chapter 12*.

All the code responsible for making the gun shoot a laser beam when the `Shoot()` method is called is contained entirely within the Gun class. It's single-responsibility for its specific use case in this game, and the code is simple and straightforward, so I didn't feel the need to overcomplicate the architecture here.

Code architecture philosophy

"When you have a hammer, everything looks like a nail" is a metaphor we can apply to a common pitfall in software development. People may use their favorite approaches to solve every problem they encounter, unintentionally leading to overcomplicated and inefficient code. Choosing the most appropriate solution for each problem or situation is important, rather than relying solely on a software doctrine.

Sometimes, you just need to embrace simplicity. Knowing when to – or not to – is called experience.

When you examine the Gun script, you'll see that we're simply using `Physics.Raycast()` and `LineRenderer` with the two points for drawing the line set to the firing point and the end of the gun's firing range, or the point at which the ray hits a damageable object (filtered by use of a **layer mask**).

Tip

Unity provides a specialized `XRLineRenderer` component for producing an XR-optimized line render compared to the regular `LineRenderer` component. It's also capable of producing very inexpensive glow effects, which is fantastic for laser beams!

XR Line Renderer: `https://github.com/Unity-Technologies/XRLineRenderer`

If the raycast hits a damageable object, we pass the damage amount specified in `_damageAmount` in a call to `TakeDamage()`. This is how we'll work within our health system, from *Chapter 8* (yes, reusable system for the win!), to cause damage to objects that have health (i.e., a `HealthSystem` component added).

Now that we have a functional self-defense weapon, let's get it into the player's hands.

Spawning the gun position

Alright, this will be a piece of cake! We're already pros at spawning virtual objects into the room. We'll reuse most of what we already have in place for spawning objects because we'll spawn the gun near the player, on their right-hand side (sorry, left-handers).

First things first, open up the `GameManager` script and add a declaration for a serialized private variable, `_prefabGun`, to hold the reference to the Gun Prefab:

```
[SerializeField] private GameObject _prefabGun;
```

We're already using the `Console` Prefab spawning section to spawn other objects, so let's tag the gun instantiation onto it:

```
case PlaneClassification.Floor:
    if (!_hasSpawnedPrefab_Console)
    {
        ...
        SpawnPrefab(_prefabGun,
            new Vector3(0.5f, 1.2f, 0.15f));
    }
    break;
```

Notice this time, when we call `SpawnPrefab()`, we have another new method signature. This is very much like the method overload we used to spawn the modules, except we're going to spawn a single Prefab and won't apply any physics force in a specified direction to it.

In this version, let's create a new method overload for spawning a single Prefab. This method will simply pass values to our previous `SpawnPrefab()` method, which requires an array of Prefabs. So, we just need to add the single Prefab to a *single item array* first:

```
private void SpawnPrefab(GameObject prefab, Vector3 playerOffset)
    => SpawnPrefab(new GameObject[] { prefab },
        playerOffset, Vector3.zero, 0f);
```

Notice we preset the parameter values for `forceDirection` and `force` to zeros to ensure no physics forces will be applied to the spawned object.

Save the script, assign Gun to the GameManager's **Prefab Gun** field, save the scene, and playtest with all the elements in place for the start of our game.

In this section, we learned how to create interactable Prefab variants for the player and collect and place modules into slots on the control console, enhancing player engagement within the environment. We also learned how to implement shooting for the gun as a secondary activate action for objects held by the player. Now, with the added ability to shoot, let's see how we bring everything together with the gameplay mechanics.

Implementing the boss room mechanics

In our climactic boss room battle, the player will collect the crystal module puzzle pieces while defending against virtual enemies – the patrolling hover bots the evil plant entity has so rudely infected. With the unique blend of a physical space for exploration and interactive gameplay with digital objects the MR technology provides, our players will be challenged to think strategically while physically exerting themselves. This innovative and novel approach to boss room mechanics pushes the boundaries of traditional video game design, and I'm very excited to see this technology continue to mature and break more boundaries!

This chapter is dedicated to introducing the Unity technologies that enable game developers and creators to rapidly produce compelling and immersive MR experiences for players to consume and enjoy enthusiastically. As such, the concepts for the boss room mechanics will be discussed in a broader sense and we'll only dive into the details where additional clarification is required.

In this section, we'll finalize the puzzle mechanic by implementing the logic required to solve it and energize the console. We'll also set up the enemy bots to spawn and move toward the player, as well as their shooting behavior. Finally, we will complete the game loop by updating the game state.

So, first, concerning the crystal modules, let's work with the problem of solving the control console puzzle.

Solving the crystal modules puzzle

As previously discussed, the crystal modules must be collected and placed back into the control console. Each console slot and module have corresponding IDs, but the order in which they must be placed is not apparent to us – the console just shows some garbled characters. Let's have a look at the console controller script to set the correct combination for the modules and determine when they've been inserted successfully.

To complete the task of restoring the console and reenergizing the reactor successfully, the modules must be inserted in the exact order, starting with the first one – you cannot just randomly place them to end up in the correct order (that's just how this tech works; I don't think you should blame me for this). This makes the puzzle more challenging for the player because you must fend off the enemy hover bots while figuring this out!

Inspecting the `ConsoleController` script, we first see the solution code as a serialized private `string` variable, `_consoleCode`, so we can inspect and set it in the Inspector at any time:

```
public class ConsoleController : MonoBehaviour
{
    [SerializeField] private string _consoleCode = "CBA";
    ...
    public void InsertModule(char slotID, char moduleID)
    {
        // TODO: Solve module sequence logic.
```

```
        ConsoleEnergized();
    }
}
```

The `InsertModule()` code will efficiently handle modules inserted in a specific order, provide feedback for incorrect placement, and signal success for the correct sequence. The module insertion and solving logic should be carried out like so:

1.   Using the current slot index, starting at zero, check the slot order against the inserted module.

2.   Increment the slot index when the next correct module ID has been provided; otherwise, reset the slots (index set back to zero).

3.   Check for code completion and update UI or trigger events accordingly.

Bonus challenge

In the `ConsoleController` class, based on the above steps, code the puzzle-solving logic for the correct order of modules inserted into the console with the `InsertModule(_slotID, _moduleID)` method yourself first.

You can get just the completed console puzzle code in the `ConsoleController` script from the GitHub repo here: `https://github.com/PacktPublishing/Unity-2022-by-Example/tree/main/ch14/Code-Assets`.

These steps are designed to ensure that the modules are inserted in the correct order and that a module's ID matches the expected ID for the current slot index. The console is activated if all modules are correctly inserted, while progress is displayed on the console screen (UI).

World space UI

For the console screen (UI) mentioned above, in step 3's *update UI*, the control console Prefab includes a world space **uGUI Canvas**. A world space UI is a user interface that appears within the game's 3D world instead of as a screen overlay. It is rendered on a canvas that can be positioned, rotated, and scaled just like any other 3D object in the scene. Developers use them to create interactive elements within the game world, such as control panels, information displays, or interactive menus.

Previously, we worked with a screen space UI, as first introduced in *Chapter 3*. We use the `SetText()` method of `TMP_Text` in the following `ConsoleEnergize()` code snippet, but I previously showed setting the text using `TextMeshProUGUI` and `.text = "string";` either approach is acceptable (i.e., developer style).

When the player has restored the console, we can notify the reactor to energize it via an event. And what better event implementation could we use than our very own global event system? Refer to *Chapter 9* for a refresher on the event system's setup and usage (just make sure to add the EventSystem component to the scene somewhere). But you can see how we trigger this event in the ConsoleEnergized() method here:

```
public void ConsoleEnergized()
{
    _consoleScreen.SetText(MSG_SOLVED);
    EventSystem.Instance.TriggerEvent(
        EventConstants.OnConsoleEnergized, true);
}
```

Then, in the Reactor script, we respond to the event by simply swapping the mesh renderer's material to one with an emissive property to visually indicate that it has been energized (this can be much more than just a simple material change; think of a particle system, **VFX Graph**, or a custom **Shader Graph**):

```
public class Reactor : MonoBehaviour
{
    ...
    private void OnEnable()
        => EventSystem.Instance.AddListener<bool>(
            EventConstants.OnConsoleEnergized, Energize);

    public void Energize(bool energize)
        => _renderer.material = _matEnergized;
}
```

With the problem now solved to implement the puzzle mechanics, let's move on to spawning the waves of hover bot enemies… since they are supposed to get in our way and make solving the puzzle even more challenging!

Spawning enemies

We have a challenge. Now, let's make it even more difficult for the player! Spawning waves of enemies in a boss room encounter dramatically enhances the player's challenge and boosts the intensity of the gameplay. The approach of spawning an unrelenting number of adversaries not only heightens the excitement and satisfaction of overcoming the challenge but also deepens the player's engagement with the mechanics of the battle.

The Corridor Prefab provided with the boss room starter assets includes game objects for the locations of the spawner and the target position at the doorway end of the hallway for where the hover bot will travel. So, let's complete the setup by adding the EnemySpawner component and configuring its properties.

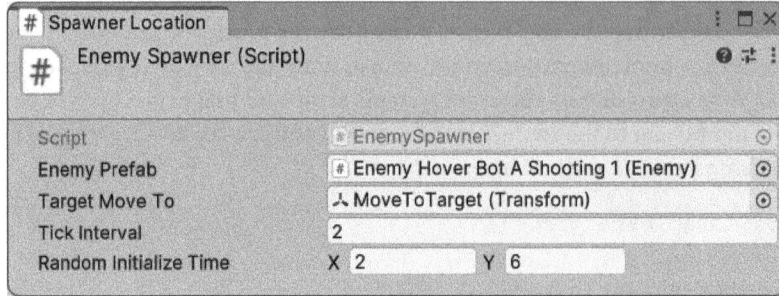

Figure 14.15 – The enemy spawner component values

> **Note on the starter assets provided**
>
> The assets included in the provided starter assets package already imported have been created using the techniques in previous chapters. You will find that the design patterns, code architecture, and all the components used will be familiar. Therefore, we won't be covering everything again. However, I recommend taking some time to examine the components used for configuring these Prefabs, especially `Enemy Hover Bot A Shooting 1`, since it offers the most significant example.

You may very well already be able to handle adding the spawner component and configuring it. If so, congrats! Still, for reference, here are the steps we can follow to configure it now:

1. Open the `Corridor` Prefab in the `Assets/Prefabs` folder in Prefab Mode.

2. Add the `EnemySpawner` script to the `SpawnerLocation` child object. We add the script here because we'll use the object's transform forward as the direction of travel for the spawned hover bot.

3. Assign the component's field values:

   - **Enemy Prefab** = **Enemy Hover Bot A Shooting 1**.

   - **Target Move To** = **MoveToTarget**.

   - **Tick Interval** = 2 (starting value – this is the time between checking if the currently instantiated bot has been destroyed).

   - **Random Initialize Time** = (2, 6) (the time between instantiating the bot and sending it on its way down the corridor).

4. Exit Prefab Mode and save the changes.

Now, when we do the *build and run* drill, we'll have enemy hover bots repeatedly spawning when destroyed, moving toward us in their menacing fashion, and when they get within range, they'll start shooting (using our pooled shooting setup, of course).

We'll have to ensure a few things are configured still because importing assets from `'.unitypackage'` is not the same as having a starting Unity project. Things such as **Layers** and **Tags**, **AI Navigation** settings, and **Build Settings** are not brought in with the imported assets.

Ensure the following layers are added to the project (**Layers | Edit Layers…**): `Projectile, Damageable`

Now, within the `Assets/Prefabs` folder, make the following Prefab objects' assignments:

- `PlasmaBall` Prefab: Select it and, in the Inspector, using the **Layer** dropdown, assign its layer to `Projectile`.

  Also, use **Layer Collision Matrix** at the bottom of the **Edit | Project Settings… | Physics** settings page to disable collisions between **Placeable Surface** and **Projectile** – we don't want the two to have a physics interaction.

- Gun Prefab: Select it and, in the Inspector, for the Gun component, set **Damage Mask** to `Damageable`.

- `Enemy Hover Bot A Shooting 1`: Select it and, in the Inspector, set **Layer** to `Damageable`. You can probably recognize how this layer assignment correlates to the damage mask assignment for the preceding gun.

> **Complete code**
>
> For reference, the complete code and project setup for this section can be found in the Unity project files provided for this chapter in the book's GitHub repository here: `https://github.com/PacktPublishing/Unity-2022-by-Example/tree/main/ch14/Unity-Project`.

This time, playing the game will have the proper interactions between the interactable objects and the virtual room surfaces where the gun damages and destroys the hover bots, and the bot's projectile weaponry, the plasma ball, approaches the player – you – fully. All that's left to do is complete the game loop for win-and-lose conditions… making it an actual game challenge.

Completing the game loop

Completing a game loop with clear conditions for winning or losing is fundamental in game design. Winning conditions often culminate in overcoming a final challenge like our boss room and reward players with an enormous sense of accomplishment for winning, while losing conditions such as depleting health increases the game's challenge. The fragile balance between these conditions is critical to ensure the experience is engaging, rewarding, and fair in terms of the time players invest in your game.

I'm not going to make the claim that I have accomplished perfect, or even near perfect, game balance in the boss room battle we've created with the default values provided in this chapter. It's simply the foundation for what's possible when creating an immersive, engaging, and fun MR game experience.

Let's finish up the boss room MR game by looking at how we tie in the winning and losing conditions from the gameplay created, starting with losing.

Losing the battle

When it comes to playing video games, losing is just as common as winning – anyone who's played a video game has experienced losing. For our MR boss room battle, with enemy hover bots shooting at us, the apparent game-losing scenario is we run out of health to continue playing. So, that's just what we're going to do.

To complete the setup in our scene to support losing by player health depletion, we first need to ensure we have a `Player` object in the scene – a GameObject tagged `Player` with the `Player` script added. Follow these steps to complete the player setup:

1. Add the `Player` Prefab as a child of `MR Interaction Setup`.

2. Ensure it's tagged as `Player`. Add the `Player` tag for assigning it now if it's not already there.

3. Ensure the `Player` object's layer is set to `Damageable`.

    `Player` implements the `IHaveHealth` interface for the health system, so also ensure the enemy's `PlasmaBall` Prefab's `ProjectileDamage` component has `Damageable` set for **Damage Mask**.

Within the `Player` class, we have an event system event triggered when the player's health has fully diminished after taking too many plasma ball hits from the enemy hover bots:

```
public class Player : MonoBehaviour, IHaveHealth
{

    ...

    public void Died()
        => EventSystem.Instance.TriggerEvent(
            EventConstants.OnPlayerDied, true);

}
```

Our GameManager class will listen to the event and respond by setting the `_isConditionMetLose` condition variable for losing and responding accordingly. With `true` being passed as the Boolean value for the event, we use it to set the condition:

```
private void OnEnable()
{

    EventSystem.Instance.AddListener<bool>(
        EventConstants.OnPlayerDied, SetLoseCondition);}
}
private void SetLoseCondition(bool value)
    => _isConditionMetLose = value;
```

We can use the individual event to update the UI, turn the passthrough off, or fade to black. However, the **finite state machine (FSM)** we have implemented in GameManager will respond to the condition change by ending the game:

```
private void Update()
{
    switch (_currentState)
    {
        case State.Playing:
            if (_isConditionMetLose || _isConditionMetWin)
                ChangeState(State.GameOver);
            break;
    }
}
```

> **FSM refactor**
>
> The best solution to keep track of our game states is, of course, a state pattern. We previously used a simple **finite state machine (FSM)** in our EnemyController class; you can also find it here. A refactor of the enum-based FSM was beyond the scope of what I had planned for the book, so I've left the refactor to you, but with a provided example using the **UnityHFSM (Unity Hierarchical Finite State Machine)** package available on GitHub here: https://github. com/Inspiaaa/UnityHFSM?tab=readme-ov-file#simple-state-machine.
>
> In the project files provided by the book's GitHub repo, inspect and evaluate the refactored Assets/ Scripts/Refactored/GameManager_HFSM script for an example implementation compared to the enum-based switch statements in GameManager and implement it in your project.

Losing the battle is not fun, but if you don't succeed... what's the saying *"try, try again?"* I'm sure if you do, you will win. Let's see how the win condition is wired; it's really just like the lose condition – the global event system really makes this easy.

Winning the battle

For our MR boss room battle, with enemy hover bots as the primary adversaries, the winning scenario is clearly defeating all waves of enemies, right? No. That's not what we've laid the groundwork for here; as you already know, we win the game when we solve the puzzle of the control console slots and reenergize the reactor.

Since everything is already in place for reenergizing the reactor with a global event system event, we're just going to add another listener to the OnConsoleEnergized event in GameManager to set the win condition variable, exactly like we did for losing:

```
private void OnEnable()
{
```

```
    ...
    EventSystem.Instance.AddListener<bool>(
        EventConstants.OnConsoleEnergized, SetWinCondition);
}

private void SetWinCondition(bool value)
    => _isConditionMetWin = value;
```

And that's it! Setting the game state to `State.GameOver` handles the rest!

By carefully setting up the game states and defining a clear path to victory, we create a rewarding gameplay loop that challenges players to develop strategies to overcome the game's challenges and ultimately achieve success in their MR boss room battle.

In this section, we explored a basic set of boss room mechanics for our MR game. We learned how to integrate and solve a puzzle mechanic that players must figure out under pressure and introduced how to spawn waves of enemies equipped with projectile weaponry. Learning about the player's laser pistol configuration allows for damaging and destroying the hover bots. Then, we further understood how to connect the win and lose conditions in the game manager state machine to complete the game design, emphasizing the strategic balance between solving puzzles and surviving.

Summary

In this chapter, we explored the Unity MR technologies that streamline development and empower developers to create immersive MR experiences. We took a dive into the design principles of crafting a captivating boss room while setting up our physical space with Quest devices and configuring Unity's MR template to suit our needs. Furthermore, we gained insights into leveraging AR planes and components to manipulate AR visual elements and spawn virtual objects dynamically within our environment.

Additionally, we honed our skills in creating interactive Prefab variants and integrating shooting mechanics to support core gameplay. By enabling players to collect and place modules into control console slots, we deepened the XR interactive potential of our game world. Bringing together all the various MR elements, we crafted a compelling MR gaming experience.

In the next chapter, we'll see what it means to operate a published game by exploring Games as a Service, including Unity DevOps and LiveOps, look at safeguarding the investment you make in your projects through robust source code management and strategies for engaging players through in-game economies, and touch on some essentials of platform distribution. This overview will equip you with the tools and knowledge to effectively manage, maintain, expand, and distribute your finished games.

15

Finishing Games with Commercial Viability

In *Chapter 14*, we wrapped up the development process that started with us creating the 3D **first-person shooter** (**FPS**) game and ended with us designing a truly immersive boss room encounter – in our own room – using **mixed reality** (**MR**). To achieve this, we took advantage of the power and flexibility of the **Unity XR Interaction Toolkit** (**XRI**) and combined 3D assets with our reusable components and systems from our previous work.

We also explored additional Unity-specific MR technologies such as AR Foundation, which we used to create immersive experiences using detected planes from the player's physical surroundings. We used walls, floors, and tables to spawn virtual objects and create a virtual game environment. We also designed the boss room, set up Quest devices, and customized Unity's Mixed Reality Template. Additionally, we crafted the interactive module Prefab Variants, along with puzzle and shooting mechanics, to deepen the player experience.

In this chapter, we'll explore **Games as a Service** (**GaaS**), Unity DevOps and LiveOps resources, and source code management via a **version control system** (**VCS**). We'll also discuss in-game economies for commercialization and finish by exploring different storefront platforms for distributing our games, including an example of implementing **Unity Gaming Services** (**UGS**) and publishing.

In this chapter, we're going to cover the following main topics:

- Introducing GaaS – UGS
- Safeguarding your investment! Source code management with Unity Version Control
- Engaging players with an in-game economy
- Getting your game out there! Platform distribution
- Implementing UGS and publishing

By the end of this chapter, you'll have gained confidence in your ability to manage and secure your project development life cycle during both the production and release phases. You'll also know how to distribute a commercially viable finished game effectively.

Technical requirements

To follow along in this chapter, you must have Unity Hub and a 2022 version of Unity Editor installed. You'll also need to be able to sign in to Unity Cloud using your Unity ID account.

You can download the complete project from GitHub at `https://github.com/PacktPublishing/Unity-2022-by-Example`.

Introducing GaaS – UGS

The most important lesson is to finish your games. You have no potential to graduate as a hobbyist or have a self-sustaining career as an indie game developer unless you finish your games. A finished game also serves as proof of finely-tuned skills as a game developer and a showcase for studios, should you be looking to get hired by one.

Therefore, finishing games boils down to treating the game-making process more like a business. This is **GaaS**, blending development and operating strategies to create dynamic, remarkable, commercial, and continuously evolving game experiences for players in your *business* of making games.

When developing a game using a GaaS model, the focus shifts more toward long-term engagement and monetization strategies in the later stages of production. If we aim to maximize commercial viability with our game's release – for a sustained revenue stream – we'll need to employ continuous content delivery, social interaction, and data analysis to guide our decision-making processes.

Unity not only provides a game engine for the foundation of your game's development but also cloud services you should leverage to increase your chance for success. Namely, **Unity Cloud** provides a group of efficient and time-saving tools and workflows that substantially assist in securing your projects and releasing commercially viable games.

> **Additional reading | Unity Cloud**
> Unity Cloud: `https://cloud.unity.com/`

Unity Cloud provides access to **UGS**, which they describe as "*a complete service ecosystem for live games.*" Considering the breadth of service offerings and ease of integration into our game projects, it should be no surprise why we would benefit from adopting one or more services.

Here's a quick rundown of what UGS provides:

- **Foundational game systems**: Systems such as authentication and player accounts enable cloud saves and platform crossplay, and content management enables deployment and dynamic updating of game content. Meanwhile, multiplayer game hosting can scale your game to millions of players (as games such as **Among Us** and **Apex Legends** did). Rounding things out are version control and build automation, which comprise Unity's **DevOps** features.

- **Player engagement**: Features such as analytics enable data-driven decisions to maximize the player experience. Engagement tools with A/B testing allow features with the most player enjoyment to surface, while communications such as voice and text chat build communities. And don't forget the ever-so-important crash reporting to know where your game crashes so that you can respond quickly with fixes. This grouping of services is Unity's **LiveOps** features.

- **Game growth**: Grow your game with user acquisition and drive monetization revenue with in-game ads, ad mediation, and in-app purchase features to establish your game's economy.

Additional reading | UGS

From DevOps to LiveOps in one platform: `https://unity.com/solutions/gaming-services`

The aforementioned UGS topics can fill a whole book, so we'll focus on just the core services that are crucial for game development production and are intended for commercial release in this chapter. That means we'll focus on **Unity DevOps** and **Unity LiveOps** in this section, as well as the basics of a game economy and publishing in the following sections.

The success of GaaS depends on the complementary roles played by DevOps and LiveOps. DevOps is responsible for updating the game efficiently. At the same time, LiveOps decides which updates to include and how to engage players (determining what to buff or nerf in the game based on player engagement is a fun challenge!). Both tools work together to create a cycle of continuous improvement (that is, **Kaizen**) that sustains and improves the game with updates over time.

Kaizen (改善)

Kaizen is the Sino-Japanese word for "improvement." It's all about continuous improvement. In game development, this could mean making small changes weekly or even daily to mechanics, balance, storylines, or even coding practices. It can help identify issues so that we can make adjustments early before they become more significant problems.

Let's explore these core services in more detail, starting with DevOps.

Introducing Unity DevOps

DevOps, derived from the fusion of **development (Dev)** and **operations (Ops)** practices, plays a vital role in software development and is particularly valuable in game development, especially in the application of GaaS. Unity DevOps tools facilitate accelerated development and deployment cycles – throughout the entire production life cycle – and help ensure reliable game releases.

Examining the problems that can be solved using Unity DevOps tools can help us better understand the benefits. The service offerings tackle essential challenges in game development by addressing critical aspects such as the following:

- **Managing assets (complex and/or large)**: You can trust that your assets will be maintained and archived in your project repository, no matter the size or the complexity of their structure. This applies to working solo or in collaboration with a team.

- **Managing code base changes**: In addition to managing assets, tracking changes via version history will ensure quality is maintained and progress on work is never lost.

- **Streamlining build and deployment**: Automating the build process and deployment for your game's executables significantly accelerates the distribution of updates – getting updates in the hands of your players quicker, whether that's to squash bugs or improve player engagement.

Additional reading | Unity DevOps

Unity DevOps: `https://unity.com/products/unity-devops`

There are two specific services Unity DevOps offers:

- **Unity Version Control**: Game development teams and solo developers can overcome project continuance challenges by using a specialized VCS that's also tightly integrated with Unity Editor. Unity Version Control offers local and private cloud code repositories that provide a scalable collaboration platform for programmers, artists, and creators.

Important note

You may be wondering why I suggest using Unity Version Control for our game development projects even though this book's projects are shared with you through a link to GitHub at the beginning of each chapter. Well, the reason is straightforward: GitHub offers public repositories, which are commonly used for open source projects, while Unity Version Control does not provide this feature.

- **Unity Build Automation**: A customizable build pipeline that seamlessly integrates with Unity Version Control or third-party repositories (for example, **GitHub**, **GitLab**, and **Bitbucket**). It provides robust platform build (for example, on Windows, Mac, Linux, Android, and iOS) and testing capabilities that streamline the process of building and deploying game builds.

The Unity DevOps services are much more than the brief introductions provided here, so let's explore each in more detail. We'll start with Unity Version Control.

Unity Version Control

Unity Version Control (formerly **Plastic SCM**) is a modern VCS that can be used seamlessly with Unity. It provides Unity-tailored version control capabilities and mitigates many of the risks associated with game development. For the record, you can use Unity Version Control with software platforms and projects other than Unity.

> **Additional reading | Unity Version Control**
>
> Unity Version Control: `https://unity.com/solutions/version-control`

So, what is version control? Version control is a system that tracks file changes over time. It enables multiple people to collaborate on the same set of files and tracks the different file versions. It also allows you to revert to previous file versions if needed.

As mentioned repeatedly, due to its importance, it's crucial to use a VCS to avoid losing work while working on game development projects. Version control, a central pillar of DevOps, also enables team collaboration while more easily resolving conflicts that can occur with multiple people working on the same code base and assets.

As such, Unity Version Control offers the following essential VCS features:

- **Efficient management of large assets**: The Unity Version Control workspace optimizes handling large binary files in your projects (such as texture images and 3D model assets). Unlike other VCS offerings, which rely on problematic large file support add-ons, Unity Version Control ensures efficient storage and retrieval of large assets without any decrease in performance.

- **Enhanced collaboration with branching and merging**: Unity Version Control provides robust capabilities for branching implemented features into isolated environments and merging changes when two people have worked on the same file. This makes for a smoother collaboration experience across the project and teams by managing parallel or asynchronous work effectively to mitigate conflicts.

- **Cloud archiving for project continuance**: Most VCSs offer cloud archiving, as does Unity Version Control, which prevents losses from local system failures – which can and do occur! If your project source doesn't exist in more than one location, it doesn't exist. Yikes!

- **Integration with DevOps pipelines**: Unity Version Control works with any game engine or software development environment, not just Unity. There are integrations into DevOps toolchains, issue-tracking applications, team communication platforms, IDEs, and more.

Now, let's explore Unity Build Automation, which picks up where Unity Version Control leaves off and is the second half of Unity DevOps.

Unity Build Automation

Using **Unity Build Automation** tools, you can automate the process of creating and deploying builds across platforms, where frequent and reliable game updates are essential for GaaS in a DevOps environment – players can be brutal with developer expectations for rapid responses and high quality. As competitive as the games market can sometimes be, you generally get only one chance to hit a player's expectations; otherwise, they'll bail hard on your game.

Build automation complements version control by addressing developer challenges related to an individual and manual build creation process, as well as consistency and efficiency across build distributions. A winning combination of version control and build automation achieves a comprehensive solution for managing game development workflows, from code management to build distribution.

> **Additional reading | Unity Build Automation**
>
> Unity Build Automation: `https://unity.com/solutions/ci-cd`

As such, Unity Build Automation offers the following essential build features:

- **Streamlined central build process**: Building projects is a time-consuming endeavor, and keeping a developer's system pegged at 100% CPU utilization during a build on their development machine crashes their productivity. Offloading the build and deployment processes from the local system to a cloud-based pipeline eliminates bottlenecks and provides a standardized process, especially when targeting multiple platforms.

- **Streamlined build deployment process**: Automating the deployment process significantly saves time and eliminates the – error-prone – manual process of copying and uploading game builds to distribution servers across multiple platforms, ensuring your game binaries are always ready for release to your testing team (for solo Indie devs, this is your friends and family) or players.

- **Real-time integration and feedback**: Since your QA testers and players have already expeditiously received the latest game update, there should be no delay in receiving feedback and bug reports. This is vital to promptly address your players' needs and efficiently close the loop on iterating changes, maintaining the quality of the game in the process.

> **Additional reading | CI/CD**
>
> **CI/CD** stands for **continuous integration** and **continuous deployment** (or **delivery**). It encompasses best practices for development teams using automated code integration and delivery.
>
> What is CI/CD?: `https://www.infoworld.com/article/3271126/what-is-cicd-continuous-integration-and-continuous-delivery-explained.html`

By leveraging Unity DevOps, game developers can streamline their workflows, accelerate development cycles, and deliver higher-quality gaming experiences to their players. DevOps takes care of the production phase of your game development journey. Now, let's review Unity LiveOps, which takes care of the release phase.

Introducing Unity LiveOps

LiveOps, derived from the fusion of, well, **live (Live)** and **operations (Ops)** practices, also plays an important role in game development, especially in applying GaaS during the post-release phase. Unity LiveOps focuses on the activities required to run live games, such as content updates, community management, and analytics, all of which help keep your game relevant and engaging for players.

Examining the problems Unity LiveOps tools can solve will help us better understand the benefits. The service offerings tackle essential challenges for operating a live game by addressing critical aspects such as the following:

- **Optimizing player engagement**: To enhance player engagement and retention, we can employ LiveOps strategies that allow us to update game content dynamically, offer personalized player experiences, and respond to real-time analytics data to identify and reduce churn rate (that is, the number of players who leave the game and never return)

- **Efficient game management**: You can streamline game operations with automated event management, enable direct player communication, and rapidly iterate with A/B testing to align better with player expectations

- **Monetization strategies**: Unity LiveOps provides tools to analyze in-app purchases and manage in-game ads, allowing us to optimize revenue with promotions and pricing while keeping our players happy

> **Additional reading | Unity LiveOps**
>
> Use LiveOps to get the insights you need for a better player experience: `https://unity.com/solutions/gaming-services/player-insights`.
>
> Power up your LiveOps strategy for better player retention: `https://unity.com/solutions/gaming-services/continuous-game-improvements`.

Many services are included under the Unity LiveOps umbrella. These services collectively empower game developers to extend the life cycle of their games, so I encourage you to review the Unity LiveOps links in the preceding callout.

For our purposes, a few services stand out as particularly crucial for the success and longevity of a live published game. So, we'll only look at a few essential services for maintaining engagement and ensuring a high-quality player experience. These services can be found on Unity Cloud under the **Products** section, as shown in the following screenshot:

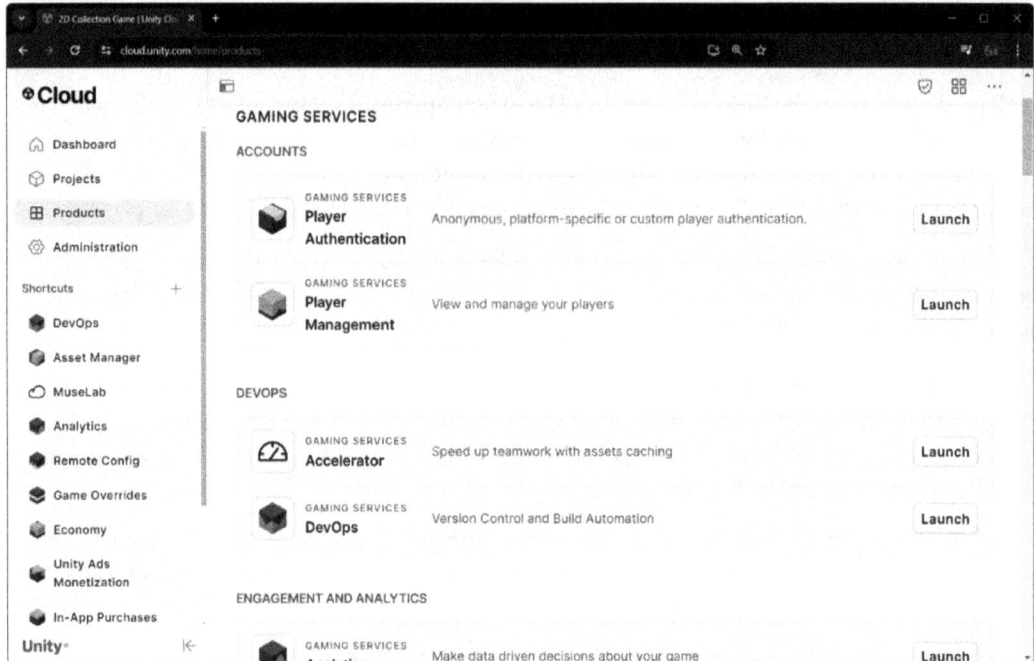

Figure 15.1 – Unity Cloud products list

I'll list the services in the same categories Unity displays in the Unity Cloud dashboard:

- **GAMING SERVICES | ACCOUNTS**:

  - **Player Authentication**: Enables secure and consistent player authentication, including anonymous, platform-specific, or custom options, and offers cross-device support and **Cloud Save**

- **GAMING SERVICES | ENGAGEMENT AND ANALYTICS**:

  - **Analytics**:

    - **Analyze**: A comprehensive set of analytics features, including **Game performance**, **Retention**, **Revenue**, and **User acquisition**, to make focused data-driven decisions. Custom dashboards and data explorers are available for your deep dives into player metrics, and they use funnels to identify your player journeys.

- **Manage**: You can both manage and analyze in-game events (predefined and custom) with the **Event Manager** and **Event Browser** features. Well-placed events allow you to adjust gameplay strategies based on your players' engagement.

- **Game Overrides**: Utilizing **Remote Config** and **A/B Testing** allows for dynamic content and game-logic adjustments to target your player's preferences (without the need to deploy a new game build), resulting in more engagement and more fun!

- **GAMING SERVICES | CRASH REPORTING**:

  - **Cloud Diagnostics**: Utilize crash reporting tools to promptly detect and resolve issues that impact game stability and performance, resulting in a seamless and bug-free gaming experience for players

---

Unity Cloud pricing

Unity DevOps and LiveOps consist of a pay-as-you-go consumption model. All developers start with a generous free plan: `https://unity.com/solutions/gaming-services/pricing`.

Note that **Authentication** and **Game Overrides** are 100% free.

By implementing these essential LiveOps services into our games, we can strive for increased player engagement and satisfaction, both of which contribute to our game's longevity. Completing our first published commercial game is only the start of our journey as career game developers. We want to enable our next game, so we should leverage all the tools at our disposal to help us get there.

In this section, we learned how combining Unity DevOps and Unity LiveOps technologies, such as Unity Version Control and Analytics, equips game developers with the tools required to form the backbone of a successful GaaS project. Next, we'll learn how to configure Unity Version Control to secure our game project code and assets.

Safeguarding your investment! Source code management with Unity Version Control

Unity Version Control provides a customized version control solution for Unity projects with tight editor integration. This ensures our work is protected and that for team environments, a smooth collaboration experience is provided for team members. As previously stated, in software development, including game development, version control is essential to prevent data loss and effectively resolve conflicts.

In this section, we'll explore the steps to implement Unity Version Control for our projects. First, we'll break down Unity Version Control's approach to catering the VCS experience that's specific to programmers and artists. We'll start with programmers.

Catering VCS for programmers

Unity Version Control provides customized workflows for artists and programmers in one repository. It offers developers the flexibility to work in centralized or distributed environments with comprehensive branching and merging capabilities. Artists and designers can use a simpler, user-friendly workspace interface called **Gluon** to take advantage of intuitive file-based asset workflows that enhance their creativity.

Figure 15.2 shows the developer-centric workspace interface, which provides programmers with every feature on offer, such as **Pending Changes**, **Changesets**, **Branches**, and **Branch Explorer**:

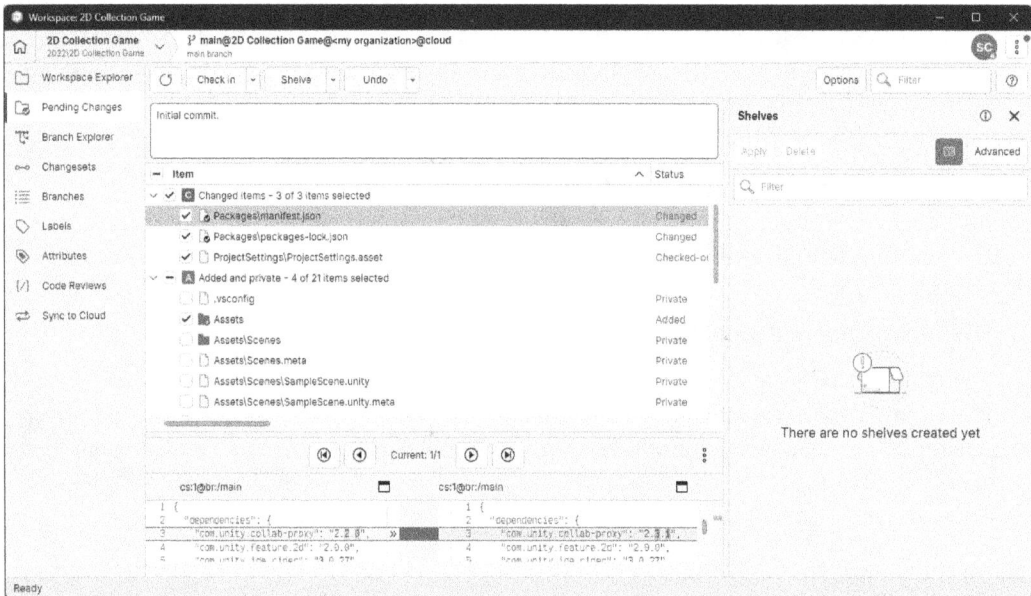

Figure 15.2 – Unity Version Control workspace

Compare this to the simpler, more straightforward, artist-centric workspace interface in *Figure 15.3* in the *Catering VCS for artists* section.

Next, we'll delve into each interface's features as we walk through setting up a new Unity project using Unity Version Control. But before that, we'll look at the artist-centric workflow.

Catering VCS for artists

Gluon is recommended for artists who are using Unity Version Control to streamline art production workflows. Gluon provides an easy-to-use file management interface that allows you to select specific files to work on without the need to download and manage the entire project. Additionally, Gluon enables users to lock files (across branches), ensuring exclusive access to art assets so that no one else can modify the same file simultaneously and then seamlessly submit changes back to the repository.

The ability to lock and unlock assets as needed during work enables smoother collaboration within a game production team.

Figure 15.3 shows the simplified artist-centric Gluon workspace interface, which provides artists and designers with the **Workspace Explorer**, **Checkin Changes**, **Incoming Changes**, and **Changesets** features:

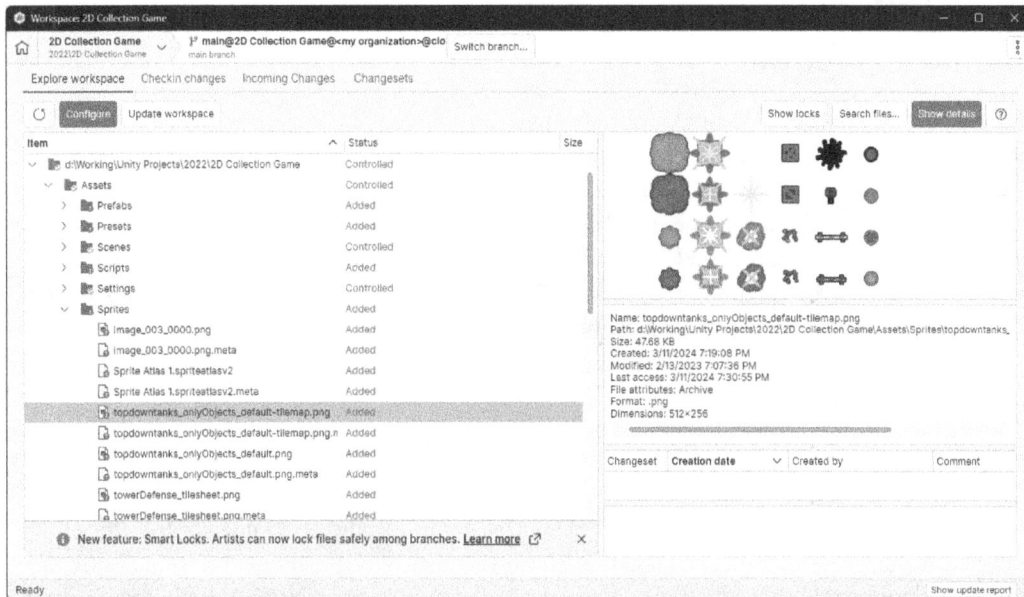

Figure 15.3 – Unity Version Control Gluon workspace

Additional reading | Gluon

Version control for artists: `https://unity.com/solutions/version-control-artists`

Gluon also offers a unique feature that benefits artists directly: it has a built-in image viewer diff tool! With it, you can compare two versions of the same file from the change history so that there's never a question of what changes were made, when, and by whom to the art assets in our projects. A Gluon workspace alone can be used as a robust asset management platform. Gluon works well with standard image formats such as PNG and JPG but can support additional formats with preview generators (for example, by adding **ImageMagick** as an external preview tool to add support for over 100 additional formats).

Gluon's image diff tool

With Gluon's image diff tool, you can view images side by side, use an "onion skin" preview, calculate differences, do a "swipe," and even compare image properties in text format.

Not only does Unity Version Control provide all the VCS tooling for daily operations such as checking in changes (that is, committing and pushing) and uploading to a cloud repository, handling incoming changes (that is, pulling), and merging and resolving file change conflicts within its workspace, it's also fully compatible with the popular Git **distributed version control system (DVCS)**. Let's quickly look at how Unity Version Control operates as a Git client.

Catering VCS for Git

For Git users, yes, Unity Version Control also has you covered – it also speaks the Git network protocol. It can push and pull changes directly with any remote Git server (such as **GitHub**, **GitLab**, or **Bitbucket**). The **Git Sync** feature, which is implemented in **Branch Explorer** (seen in *Figure 15.7*) with a simple right-click menu, where you can choose **Push/Pull | Sync with Git…**, immediately turns Unity Version Control into a VCS that's fully compatible with Git bidirectional synchronization. The advantage is that you can use Unity Version Control for all your DVCS client needs, either with a Unity Version Control workspace or a Git project.

> **Additional reading | Git**
> Git is a popular free and open source DVCS: `https://git-scm.com/`.

However, Unity Version Control is designed to manage game-specific assets quickly and efficiently, even when dealing with huge files and binaries. This makes it an effective alternative to Git, especially Git **large file support (LFS)**, which is known for its tendency to have issues. Whether your game assets are large or small, Unity Version Control offers a fast and efficient way to manage them all in your game projects.

Okay, enough on practices and workflows – let's set up our first Unity Version Control cloud workspace for a new Unity project.

Setting up Unity Version Control

Unity has made adding version control to your projects as frictionless as possible by adding an option directly in the **Unity Hub** interface when creating a new project. You simply need to select a checkbox, as seen in *Figure 15.4*. Use Unity Version Control? Check – yes, please!

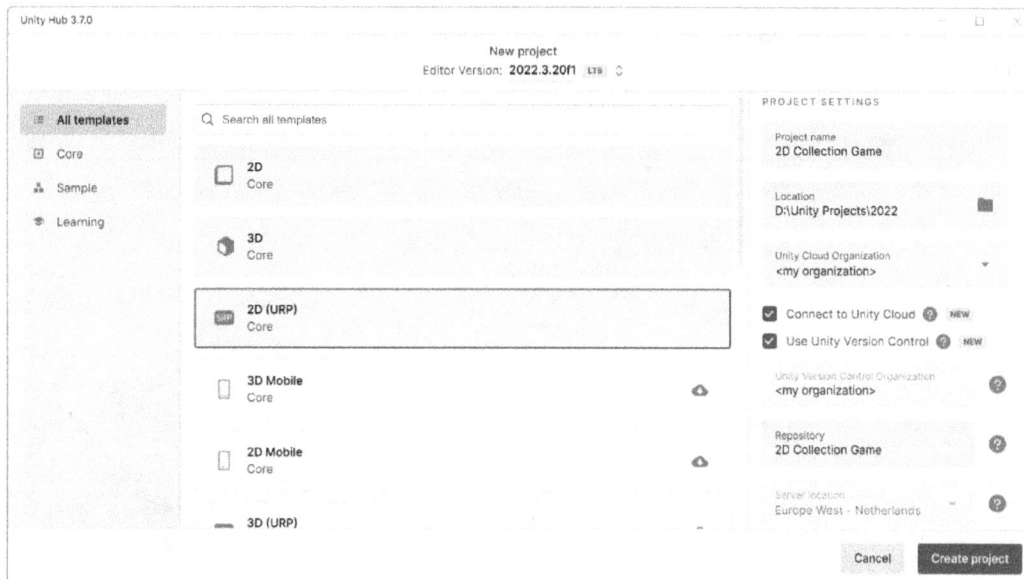

Figure 15.4 – Unity Hub showing a new version-controlled project

When creating a project in Unity Hub, referring to *Figure 15.4*, we must create and link a version control workspace by following these steps:

1. Open **Unity Hub** and click the **New Project** button (top-right of the window).

2. Select a template on which to base your new project (download the template if required).

3. Now, under the **PROJECT SETTINGS** section (right-hand side of the window), fill out the required fields:

   - **Project name**: The name you'll assign this project (something like "my awesome game" – no special characters).

   - **Location**: Where you'll store the project files on your local system drive (use a short path and no special characters).

   - **Unity Cloud Organization**: You must select an organization for your project.

   - **Connect to Unity Cloud**: Optional; enable this option if you intend to use UGS with this project (considering we already discussed all the benefits of UGS, we'd be doing ourselves a disservice if we don't enable it).

   - **Use Unity Version Control**: Finally, here we are. This is optional, but we should surely enable this option for our "my awesome game" project so that we don't lose any of our development work!

4. Click **Create project**.

Let Unity do its thing, creating the project and Unity Version Control workspace locally and in Unity Cloud. When Unity Editor opens, we'll be greeted by the **Unity Version Control** window. As we can see, we already have some pending file changes to check in:

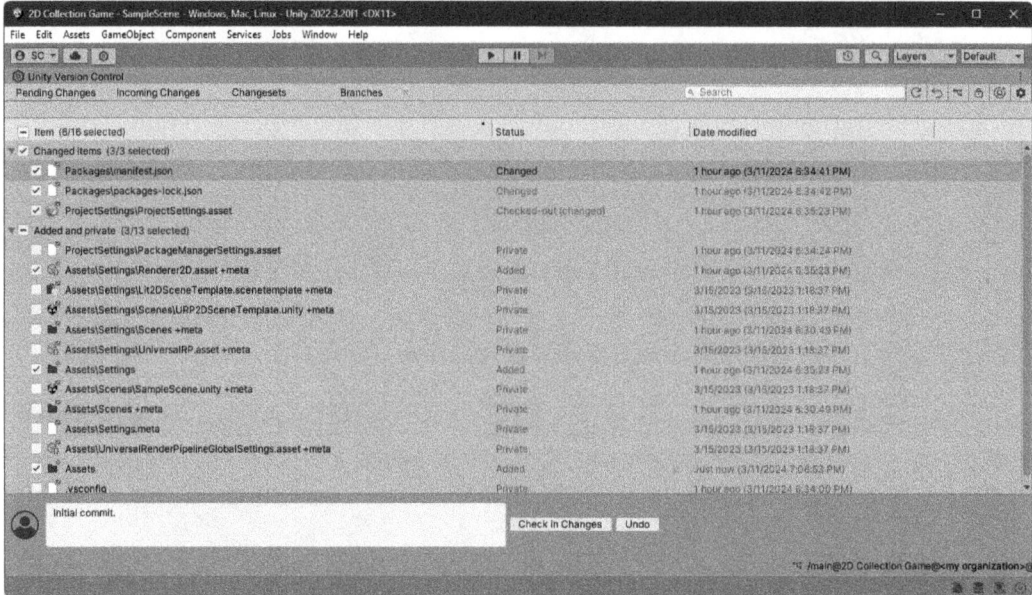

Figure 15.5 – Unity Editor's Unity Version Control window

At any time, if you need the **Unity Version Control** window because it's not currently showing, go to **Window | Unity Version Control** or click the button to the right of the *Manage services* (cloud icon) button (top left, directly under the file menu).

> **Tip**
> You can also quickly use the standalone Unity Version Control desktop application, shown in *Figure 15.5*, by clicking the gear icon and selecting **Open in Desktop App**.

As seen in *Figure 15.5*, these files are the default assets Unity adds to a new project. They only exist locally in our project folder, so we need to inform the Unity Version Control workspace that we have these changes to track. We can do that by using the **Pending Changes** tab, confirming the files we want selected, and checking in the changes (or, in Git terms, **commit**).

Initial commit is a common first commit message for a changeset, meaning this is the beginning of our workspace, or repository, history, so enter that in the text box provided at the bottom. Check the check box next to **Added and private** because we want all these files, too, then click **Check in Changes**. This will create a changeset and a file version history for all the files in the changeset. Yay! You've just made your first VCS commit!

> **Committing best practices**
>
> Commit small, commit often. Or, at the very least, commit at the end of your workday to ensure no work progress is lost. It's also good to test your work before committing and use a clear and concise commit message (for example, include why the change was made, not just what was changed).

The Unity Cloud dashboard also provides a browser-based interface for managing Unity Version Control repositories. It allows users to visualize and interface with their project's version history, branches, changesets, code reviews, and file locks. Seats and user groups are also available for team management, and usage reporting is available for tracking billing.

You can view your game project's cloud repository by opening Unity Cloud in your browser (`http://cloud.unity.com`) and going to **DevOps | Repositories**:

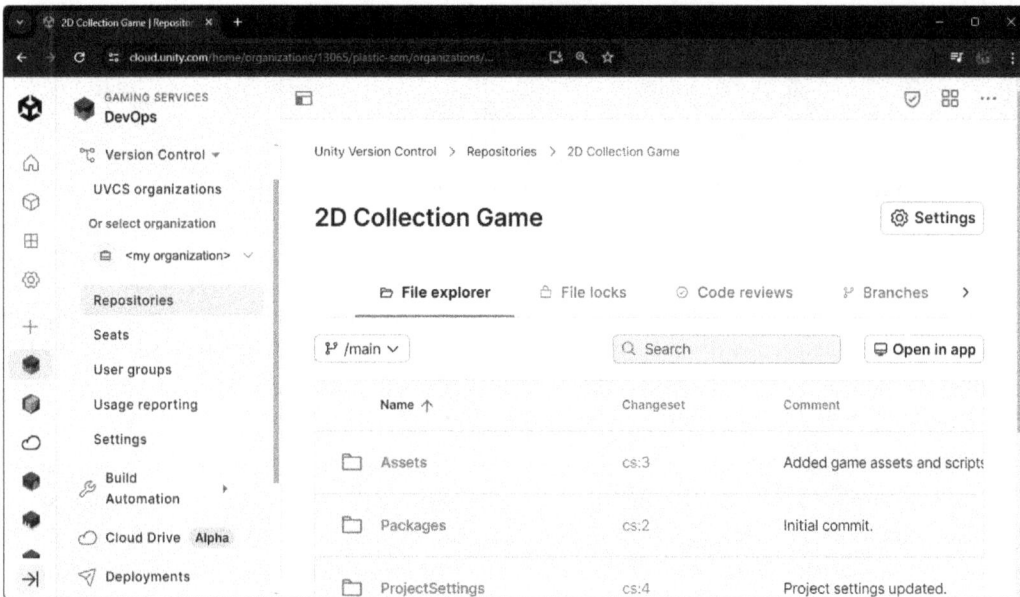

Figure 15.6 – Unity Cloud repositories dashboard

The VCS workflow is very simple and straightforward for a solo developer. You are the only person making changes, and you'll rarely, if ever, conflict with yourself while performing daily work in your solo-member workspace.

> **Pulling best practices**
>
> A common best practice for project teams is to regularly update their working copy by pulling the latest changes from the cloud repository to work from before starting new work or making further changes. That way, the team always builds upon moving the project code base forward.

You may, however, want to revert some changes to an earlier version. So, for those situations and if you're working with a team of developers in the same workspace, we must consider some best practices for setting up a Unity project for collaboration to minimize file merge conflicts.

Setting up a collaborative project structure

When setting up a Unity project for version control, adopting a project architecture designed for better collaboration is essential to help prevent workflow conflicts between team members. A well-organized project structure and clear guidelines for scene organization and prefab workflows can significantly reduce the likelihood of conflicts and problematic merge issues that are often encountered with Unity scene and Prefabs assets and any binary files that cannot merge.

> **Project organization and VCS best practices eBook (Unity)**
>
> Version control and project organization best practices for game developers: `https://unity.com/resources/version-control-project-organization-best-practices-ebook`

The two aspects for better project structure that we touched on previously, namely Prefabs and scenes, deserve a bit more context, so let's have a quick look at them:

- **Prefab workflow**: Prefab workflows are crucial for efficient collaboration in Unity projects. Prefabs allow for modular game design, where game objects are pre-made objects that form the Scene Hierarchy and are put together to create all the necessary functionality. Prefab changes are carefully managed and communicated to minimize conflicts and ensure consistency. Still, they are updated separately from scene changes to prevent unintentional simultaneous scene updates by multiple team members.

- **Additive scene workflow**: When it comes to managing scenes in Unity, it's important to have collaborative workflows and organization strategies in place. Using an additive scene approach allows multiple developers to work on different aspects of the same level simultaneously by sectioning a level into main scenes and additive scenes (such as lighting, gameplay elements, or UI components); teams can work on different scenes in parallel and bring them together at runtime without conflicts. Additionally, ensure that Unity's scene serialization is used in text mode (the default for version control) to make it easier for VCS to track changes and use merge tools such as **UnityYAMLMerge**.

In addition to these Unity project-specific collaborative workflows, we can also use VCS team workflows, such as *branch per feature* or even *branch per task*.

When developing new features or working on specific project tasks, it's often helpful for teams to work in isolated branches rather than use a single main (or master) branch. This approach facilitates collaboration by allowing each feature to progress on its own timeline, reducing conflicts and enabling parallel development among the team. As a result, team workflow is smoother, and project management becomes more organized.

> **Git Flow**
>
> In Git, **Git Flow** is a workflow that employs different branches for features, bug fixes, and releases.

In *Figure 15.7*, we can see how Unity Version Control visualizes the workspace branches. Additional branches will show up as forked from the changeset of the source branch:

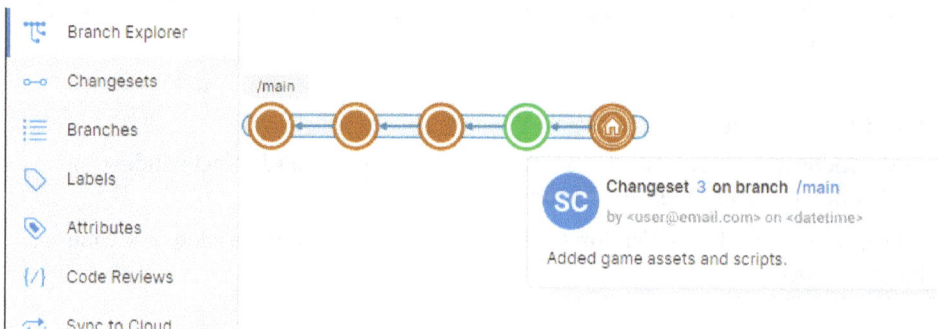

Figure 15.7 – Unity Version Control branch explorer with changesets

> **Pull request (PR) | code review**
>
> After completing work on a feature branch, it's a good practice to create a **pull request** (**PR**) to merge the changes into the dev/develop branch (before merging into the main branch for distribution). The team leader or a designated senior developer on the DevOps team should review the changes before accepting and merging the request.
>
> In Unity Version Control, we don't have PRs per se (it's a Git concept). Instead, we have a whole code review system: `https://docs.plasticscm.com/code-review`.

Once work has been completed on a features branch, using a PR or code review process is good practice to merge the work into the Dev (develop) branch. It's the responsibility of the team leader or designated senior developer on the DevOps team to review before accepting and merging the request.

At the end of the day, what matters most is that you – or your development team – feel at ease with the VCS workflow. However, it's not always possible to avoid merge conflicts, no matter how strictly we adhere to the best workflow practices and collaboration strategies. So, when faced with the inevitable task of resolving merge conflicts, we rely on merge tools to help us through.

Merging conflicting file changes

Unity Version Control comes with a powerful merge tool and merge configuration settings designed specifically for handling merges in Unity projects. The merge configuration allows us to specify asset-specific merge tools such as **UnityYAMLMerge** for Unity scene files and **Semantic Merge** for most other text-based file assets:

- **UnityYAMLMerge**: A merge tool that's specific for merging scene and prefab files in YAML format. It understands the structure of Unity-specific assets, making it more intelligent than text-based merge tools. It is an invaluable asset for teams working on Unity projects, and Unity Version Control's default configuration includes it.

- **Semantic Merge**: An intelligent C# language-dependent merge tool that can resolve code conflicts efficiently, mostly automatically, because it understands code structure beyond simple text differences.

> **Viewing file history and differences**
>
> You can right-click on any script or asset at any time in the Unity Editor's **Project** window and select **Unity Version Control** | **View file history**.
>
> Alternatively, you can select **Unity Version Control** | **Diff with previous revision** to see file-specific changes between the two.

I hope I have convinced you of the value of using the Unity version control system in your game project – or any VCS. We learned how to set up Unity Version Control for a new Unity project, the basics of checking in our work, and some best practices for organizing our project for team collaboration.

> **Unity Version Control pricing**
>
> Unity Version Control enables you, or your team, to safely and securely store and work with your game project assets and data in parallel: `https://unity.com/products/unity-devops#pricing`.
>
> Unity provides 1 to 3 seats and 5 GB of storage for free. You only pay for additional seats (per seat per month for additional team members) and anything above 5 GB of storage (per GB per month).

I relied on Unity Version Control throughout the entire process of creating the projects for this book. You didn't think I would want to risk losing my work at any point during the project's progress, did you?

In the next section, we'll continue our discussion for preparing our games for commercial release with an overview of an in-game economy.

Engaging players with an in-game economy

In-game economies are critical for commercial success and warrant our attention when designing our games. Different monetization mechanisms, when designing our game mechanics, play a significant role in our ability to generate revenue within the game, which is distinct from commercialization efforts such as marketing and distribution.

Economy mechanisms can cover a wide range of strategies and differ, sometimes drastically, between game platforms and genres. These strategies can include microtransactions, ads, virtual currencies, subscription models, or premium game purchases. By understanding a few basic nuances between these mechanisms, game producers and developers can gain an understanding of effective commercial strategies for their games.

> **Commercialization versus monetization**
>
> The distinction between commercialization and monetization lies in their respective focuses. Commercialization is primarily concerned with bringing a game to market and making it available to players, while monetization deals with how the game generates revenue once it's in players' hands.

Producers have two primary platforms to choose from when launching commercial games, and how game economies are structured varies significantly between them. These are **mobile** and **PC** – we could even say "free" and premium.

So, let's quickly look at what each platform's economic strategy is designed to do, starting with mobile.

Economies for mobile games

Mobile games often follow a "freemium" or **free-to-play** (**F2P**) model, which means they are available for players to download and play without any up-front payment. However, the monetization strategy includes buying virtual currency for in-game items, watching ads, or paying microtransactions to obtain different gameplay advantages. Unsurprisingly, this model is designed to attract a wide range of players to download and play the game, meaning your game has to be remarkable and fun to be successful since most mobile games only earn from a small fraction of their players (churn is real – players need to find the fun fast!).

To get the most out of a commercial game release on mobile, here are some key strategies you can focus on in your game design:

- **Encourage regular player engagement**: Mobile game developers commonly use daily login rewards, quests, events, timers, speed-ups, and even weekly or monthly challenges to encourage players to return to the game regularly with the ultimate goal of getting players to buy.

- **Rewarded ad watch**: One way of generating ad revenue is by offering rewards to players in exchange for watching the entire ad. Rewards can speed up gameplay progress, double virtual currency rewards, or even save the player's progress. You get the ad revenue but also lead the player down a path to buying later on.

- **In-game purchase offers**: Cosmetic items to gameplay advantages are on offer here. However, it's crucial to balance gameplay advantages to avoid "pay-to-win" scenarios, which can tend to drive players away.

- **Subscriptions and season passes**: Providing unique content for a limited time – evoking a sense of urgency and excitement – can motivate players to keep coming back and engage them to keep spending on your game.

It's not an extensive list by any means, and implementing these strategies in your games' monetization mechanics requires a lot of subtlety. Still, it would be best to have a general sense of what's needed to introduce an in-game economy for your mobile game's commercial potential.

Next, we'll examine similar strategies for premium games.

Economies for premium games

Premium games, which are purchased upfront and only offer a *free* play experience if the publisher provides a demo, focus more on the player's initial investment but may include additional incentives for revenue longevity. To contradict a bit of what I just said, though, some titles are F2P but monetize purely through in-game purchases and expansion packs. As such, the premium game economy differs significantly from mobile games.

Here are some key strategies you can focus on in your game design to get the most out of a commercial game release on PC:

- **Expansion packs and downloadable content (DLC)**: To extend the life cycle and longevity of premium games, new stories and content can be purchased by players after their initial release to generate additional revenue. It's best to account for DLC during the development phase because the game architecture must support it.

- **In-game purchase offers**: This one crosses over from mobile since we also include cosmetic items and gameplay advantages here. You don't need to look any further than the free-to-play **Fortnite** to know how much revenue can be generated from cosmetic items! Again, here, it's important to balance gameplay advantages to avoid "pay-to-win" scenarios – I believe the PC crowd is more inclined to be driven away here.

- **The game's initial quality and depth = player trust**: Premium games should offer gameplay experiences that justify the price tag to the player (that is, the feeling that what they paid for the game is worthwhile). So, in this case, the game's economy should focus on enriching the player experience and not so much on encouraging more player spending, which is a difficult balance to strike when we try to ensure our investment in making the game is positive from a revenue perspective.

This is also not an extensive list, with the obvious subtleties required to implement these strategies being left to your creative mind while we wear our economist hat. However, it's still a good sense of what's needed to bring in-game economies for a premium game's commercial potential.

Let's finish our introduction to the in-game economy by examining the contrasting strategies and mechanisms of the mobile and PC platforms.

Contrasting economic strategies

The platforms we've discussed regarding the in-game economy each have distinct approaches to engaging players and generating the resulting revenue (hopefully). To better understand the differences in the strategies laid out for each one, we can simply and clearly compare the key strategies.

The following matrix table shows the primary differences between each platform's in-game economy strategy and associated player engagement:

Strategy	Mobile Games	PC (Premium) Games
Engagement versus investment	Focus on constant player engagement and frequent but small incentivized transactions	Focus on the initial player investment and later offer expanded content with DLCs
Accessibility versus depth	Games must be easy to pick up and play, be addictively fun, and attract a broad audience of players	Games must offer deeper and more complex gameplay experiences to hook long-term players
Monetization frequency	Relies on continuous monetization opportunities employing several strategies for player spending	Relies on monetizing through fewer but more significant player spending, if any, beyond the initial game purchase

Table 15.1 – In-game economy contrast between mobile and PC games

Creating an in-game economy is a delicate task that requires many disciplines, including a mashup of gameplay design and fair monetization strategies. For mobile games, the challenge lies in attracting and then engaging a small percentage of players to convince them to spend real money on an otherwise free experience. And for premium PC games, the challenge lies initially in offering enough upfront value to players that they'll gladly justify purchasing your game. Then, it's up to you to engage them to want to spend more for additional content. Both economies require a thorough understanding of your target player demographic and your ability to avoid introducing undesirable monetization approaches.

In this section, you learned the basics of an in-game economy for mobile and PC platforms. Of course, a lot more can be learned here, so I encourage you to do additional reading on the topics that appeal most to your game's design. As it would happen, Unity provides best practices and guides on the subject too!

> **What is an in-game economy | Unity**
>
> The first guide in a series about the in-game economy: `https://unity.com/how-to/what-is-in-game-economy-guide-part-1`.

In the next section, we'll explore the platform distribution aspect of a game production's life cycle.

Getting your game out there! Platform distribution

In the previous section, we talked about the mobile and PC platforms concerning the in-game economies. However, we did not go into detail about distributing to these platforms. Well, now's the time.

In this section, we'll examine the options available for distributing our games on mobile and PC platforms. This will be a high-level review of the processes, with a focus geared more toward commercial releases. I'll also provide links for each platform for the – in most cases – step-by-step details of preparing our game for distribution on that platform.

> **Console distribution**
>
> For console distribution platforms such as **PlayStation**, **Xbox**, and **Nintendo Switch**, a game developer – who is not associated with a large AAA studio – will face additional challenges regarding these platforms. The additional set of requirements for getting our games on consoles generally involves such things as a rigorous approval process, purchase of a costly developer's kit, a lengthy and complicated code porting process, and conforming to specific standards and performance requirements.

Starting with mobile, let's discover what's available to get our game into players' hands.

Distributing on mobile platforms

In the world of mobile game distribution, there are two key players to be aware of – let's face it, most people old enough to buy a smartphone or tablet device know what they are – the **Google Play Store** and the **Apple App Store**.

These mobile-centric platforms, or storefronts, offer developers access to a vast global audience without much friction in accessing the platform and building our games – especially when using Unity. However, developers are challenged with standing out in a crowded marketplace and competing with large studios (and large marketing budgets!), so it's critical to optimize for discoverability with not only a remarkable game but also learning the ins and outs of **app store optimization** (**ASO**).

The free-to-play model is king on mobile storefronts since it's what the storefronts provide significant support for and is generally what players still expect. As mentioned in the previous section, *Engaging players with an in-game economy*, and aligning with player expectations, we'll want to consider the platform's built-in payment systems for in-app purchases and ads for revenue.

Let's start with an overview of the publishing requirements for the Google Play Store.

Publishing on the Google Play Store

Developers looking to publish their games on the Google Play Store must understand both the technical and financial aspects of the platform – this is true not only for the Google Play Store but also for any of the distribution platforms available to us.

Considering the technical aspects, publishing to the Google Play Store means building our games for the **Android** platform. Unity offers great support building for Android and provides many settings for customizing the game builds (such as a graphics API, texture compression, quality settings, and a scripting backend). In addition, Unity provides scripting API access to mobile-specific device features such as touch input, camera, vibration, gyroscope and accelerometer, location services, notifications, and AR support (**ARCore**).

> **Android screen resolutions**
>
> Adapting games for Android devices poses a significant challenge to mobile developers because of the eye-widening number of screen resolutions, aspect ratios, display notches and cutouts, and now foldables out in the wild west of device manufacturers. This requires the game developer to put on their UI designer hat and implement responsive and flexible UI design strategies to ensure an optimal viewing experience within the screens' safe areas across devices.

Each distribution platform generally offers a unique set of value-added services specific to its platform that developers can incorporate into their game designs. Google offers its **Google Play Games Services**, which include sign-in, saved games, achievements, leaderboards, friends, player stats, and events.

> **Google Play Store | distribution information**
>
> Release with confidence: `https://play.google.com/console/about/guides/releasewithconfidence/`

Now, let's review the publishing requirements for the Apple App Store.

Publishing on the Apple App Store

Developers looking to publish their games on the Apple App Store must understand that it has its own technical and financial aspects for distributing games on the platform. Again, this is something true for all platforms.

When looking to publish games to the Apple App Store, the technical aspects involve building our games for the **iOS** platform. Unity has us covered here, too, and offers excellent support for building iOS **Xcode** projects. Correct – I didn't say game build. Unity generates an Xcode project to build iOS apps and games, so if you want to build a game locally, you must have Xcode installed – it's only available on **macOS** systems. However, Unity Build Automation, which we already know from the *Introducing Unity DevOps* section, can build and deploy iOS games for you, making it possible to develop an iOS application on non-macOS systems.

> **Additional reading | Unity documentation**
>
> Build for iOS: `https://docs.unity3d.com/2022.3/Documentation/Manual/iphone-BuildProcess.html`

Unsurprisingly, Unity also provides scripting API access for iOS device-specific features such as touch input, camera, vibration, gyroscope and accelerometer, location services, notifications, and AR support (**ARKit**).

> **iOS screen resolutions**
>
> In contrast to Android, the iOS device ecosystem offers a more consistent range of screen resolutions, aspect ratios, and "notched" safe areas to display game content and UI within. Apple's well-controlled hardware designs also mean there are fewer device models, each with its well-documented screen specifications. This uniformity simplifies game development as we wear our UI-designer hats to ensure our games are presented more seamlessly across Apple's mobile devices.

Apple also offers a range of value-added services specific to its platforms that developers can incorporate into their game designs. Apple offers **Game Center**, which can be utilized for games targeting **Apple Arcade**. The following services are available: player identity, saved games, leaderboards, achievements, and turn-based and multiplayer games.

> **Apple App Store | distribution information**
>
> Submit your iOS apps to the App Store: `https://developer.apple.com/ios/submit/`.

Additional third-party app stores are available for Android and target smaller but more direct markets, such as **Amazon Appstore**, **Samsung Galaxy Store**, **Huawei AppGallery**, and **APPTUTTi**.

Unity Distribution Portal (`https://api-udp.unity.com/`) attempts to simplify the process of releasing and managing our Android game releases to multiple global app stores, so that may be of interest. But there are (currently) no such third-party app store markets for iOS.

Let's finish up the mobile distribution platforms with a quick breakdown of the financial considerations.

Publishing financials

For the financial aspect, I've compiled the essential publishing information for developers interested in the Google Play Store and Apple App Store platforms in the following table:

	Registration Fee	Revenue Cut	Platform
Google Play Store	$25 one-time	70/30% split* 30% to Google	Wide audience reach, rich games market
Apple App Store	$99 annual	70/30% split* 30% to Apple	Apple device users only, Apple Arcade

Table 15.2 – Mobile store publishing details

> **Note regarding Table 15.2**
>
> *I've greatly simplified this statement because it's constantly evolving; please consult the platform documentation for current information. In some situations, the revenue fee is lowered to 15%.

That covers mobile-specific distribution platforms, so let's move on to PC-based platforms and explore some of their specific requirements.

Distributing on PC platforms

PC games have a wider range of distribution options for both commercial sale of and non-commercial sharing of our games. For mobile games, the distribution is purely digital delivery, whereas PC games can have physical distribution channels, too, but we'll focus only on digital storefronts.

Steam and **Epic Games Store** are the two primary parties involved in the PC-based game digital storefronts. The primary commercialization strategy for these storefronts is premium. Generally speaking, PC-based players usually seek more immersive gameplay experiences than mobile offers, which aligns more with the premium model's focus on upfront purchases or subscriptions.

However, just as much as with mobile storefronts, developers launching on PC game distribution platforms must maximize visibility, navigate channels, and leverage platform-specific features to entice players, enhance their engagement, and, of course, drive sales.

> **Direct sales**
>
> Game developers can offer their games directly to players from their websites as an alternative to PC game storefronts. This approach provides higher revenue per sale and complete control over the sales process but does require significant effort and investment in marketing, customer service, and infrastructure. Do you want to make games and get them in players' hands or manage an eCommerce website?

Let's dive into each of these primary PC game storefronts now, with similar coverage to the mobile storefronts, and finish with **Itch.io**, which caters more to smaller independent game developers for both premium, donation-based, or completely free game distribution.

Publishing on Steam

It should be no surprise that developers looking to publish their PC games on Steam must also navigate and understand this platform's technical and financial aspects.

Considering the technical aspects, publishing to Steam means building our games for standalone PC platforms, but primarily Windows (Mac and Linux are also supported). Before being able to publish our Unity games on Steam, we first need to integrate the **Steamworks software developer kit** (**SDK**). The SDK provides all the necessary tools and resources to ensure compatibility with the Steam distribution platform (including such essential things as setting our Steam **App ID**). Games must also meet specific technical requirements set by the platform. These requirements include such things as minimum system specifications, controller support, and game performance.

Like what mobile platforms provide, PC platforms also have unique services that add value to our games! For Steam, we have a full breadth of services available to integrate into our games, including achievements, game statistics, leaderboards, OpenID, game notifications, voice chat and commands, input, player-created content, inventory, microtransactions, game servers, and matchmaking and lobbies for multiplayer.

> **Steam | distribution information**
>
> See what Steamworks has to offer: `https://partner.steamgames.com/`.

Now, let's look at the other side of the coin for PC game distribution platforms, the Epic Games Store.

Publishing on the Epic Games Store

Developers who want to distribute their PC games on the **Epic Games Store** (**Epic**) face the same familiar platform technical and financial aspects. Let's review them.

The technical aspects are up first; publishing to Epic means building our games for standalone PC platforms, which Epic supports for Windows and Mac. Unlike Steam, Epic does not require us to integrate any specific platform SDKs, but games still must meet the particular technical requirements set by the platform. The requirements are designed to provide the best gameplay experiences and, surprisingly, not lock players into a single storefront. The requirements include supporting crossplay for multiplayer games across any PC storefront, implementing achievements for Epic if implemented in other stores' distributions, and ensuring games download, install, launch, and function consistently – while also being of sufficient quality.

That's not to say that Epic doesn't offer gaming services similar to Steam – they provide a dizzying array of online services. **Epic Online Services** (**EOS**) has built-in support for all kinds of platforms: Windows, Mac, Linux, PlayStation, Xbox, Nintendo Switch, Android, and iOS. The services on offer for both account and game services include achievements, gameplay stats, leaderboards, login, player accounts and data storage, title storage, player management, friends, presence, game invites, voice chat, analytics, anti-cheat, and multiplayer services with crossplay, peer-to-peer, matchmaking, and lobbies. Whew! Mind you, they all are entirely free to use.

> **Why is Epic providing free online game services?**
>
> Scaling a game and strengthening player communities requires various backend services and infrastructure; finishing our games is only half the battle. Epic offers help for game developers to succeed with their games by providing services – initially made for Fortnite and currently operating the Epic Games Store – for free for all developers. The sentiment from Epic is that they succeed when, through widespread adoption of their services, the participating developers succeed.

Epic has also recently started allowing indie game developers to self-publish on their store through the **Epic Developer Portal**, stating that the store is open to all developers and game publishers, so long as their games meet the store's requirements.

> **Epic Games | distribution information**
>
> Start distributing PC games on the Epic Games Store: `https://store.epicgames.com/en-US/distribution`.

In addition to self-publishing as a differentiating factor, Epic Games Store is also known for exclusivity deals, regular free game giveaways, grant awards, and support for blockchain-based games (wading into a more controversial side of game development).

> **Web3 (blockchain) games**
>
> The Epic Games Store is one of the only leading storefronts to embrace blockchain technology games. Blockchain, NFT, and cryptocurrency products can be published but only if they follow specific policies: `https://dev.epicgames.com/docs/epic-games-store/requirements-guidelines/distribution-requirements/blockchain`.

Transitioning from the leading PC distribution platform storefronts to a more indie-friendly distribution platform is a pivot away from purely commercial revenue generation aspirations. Let's have a look at how more indie-friendly platforms, such as Itch.io and **Game Jolt**, prioritize creative freedom and a more experimental space for game developers to share their game creations with the world.

Publishing on Itch.io

In the landscape of digital distribution platform storefronts, Itch.io stands apart. The platform distinctly caters to indie game developers and creatives. That's not to say it's not home to some well-known and popular titles such as *Celeste*, *Night in the Woods*, and *Doki Doki Literature Club!* (for you visual novel fans). The platform offers a unique opportunity for game publishers to start small and scale up as their games grow.

The technical aspects of the platform are pretty open because the type of platforms supported varies, depending on the kind of game being distributed via Unity or literally any other game engine or coding framework. The platform supports playing **WebGL** games directly within the web browser (very well supported with Unity). Also, it supports game downloads for any platform, such as Windows, Mac, and Linux executables, Android APK files for sideloading to devices, and any number of other platforms in compressed archive format (for example, `*.zip` and `*.rar` files), as well as media files (that is, image, audio, and video files).

> **Itch.io | Distribution information**
>
> Create a custom page to instantly distribute or sell your indie games: `https://itch.io/developers`.

Regarding the financial aspect, Itch.io allows sellers to determine the revenue percentage share to Itch.io, ranging from 0% to 100%, with a default rate of 10% if you don't make any changes to it. Yay!

> **Itch.io revenue sharing**
>
> Introducing open revenue sharing: `https://itch.io/updates/introducing-open-revenue-sharing`.

As a side note, for Unity developers just looking to quickly share their games (usually early prototypes or development in progress for testing), the **SIMMER.io** website provides drag-and-drop simplicity for quickly sharing WebGL game builds. I believe a "tip jar" is in the works.

SIMMER.io

A place for Unity developers to share WebGL games fast and easy (and free): `https://simmer.io/`.

Let's finish up the PC distribution platforms with a quick breakdown of the financial considerations.

Publishing financials

For the financial aspect, the essential publishing information for developers interested in releasing on Steam, Epic Games Store, or Itch.io is shown in the following table:

	Registration Fee	Revenue Cut	Platform
Steam	$100 fee for each game Reimbursed with $1,000 earned revenue	70/30% split* 30% to Steam	Massive community, Steamworks tools for developers
Epic Games Store	$100 fee for each game (self-publish)	88/12% split* 12% to Epic	Exclusivity deals, game funding, free gaming services (EOS), makers of Unreal Engine, home of Fortnite
Itch.io	None	Open revenue sharing 0% to 100% to Itch.io*	Large indie game dev community, creative freedom, flexible pricing, donations

Table 15.3 – Steam publishing details

Note regarding Table 15.3

*I've greatly simplified this statement because it's constantly evolving; please consult the platform documentation for current information.

In this section, we learned about the larger game distribution platforms for mobile and PC. We explored the technical and financial aspects of publishing games on these storefronts and looked at what distinct services they offer for integrating into our game's design.

Next, we'll walk through an example implementation of UGS and publish the game to a PC-based store platform.

Implementing UGS and publishing

Now that we have a foundational knowledge of Unity's cloud services for DevOps and LiveOps, let's walk through an example of adding base LiveOps analytics and crash reporting to a project and implementing a basic player engagement strategy by updating game content dynamically. We'll finish up by automating our build process with DevOps.

We'll start by adding the LiveOps services to our game project. For this example, and as seen in this chapter's screenshots, I'll use the 2D collection game we started with in *Chapter 2*. To follow along, you'll either need to have made a game based on that project, or you can download the Unity project from this book's GitHub repository here: `https://github.com/PacktPublishing/Unity-2022-by-Example/tree/main/ch2/Unity%20Project`.

Adding LiveOps services

The two services we'll be adding to our game project are **Analytics** and **Cloud Diagnostics**. These services are the most basic activities required to run live games successfully, and I wouldn't publish a game without them! We can easily understand how our game is performing with our players, and Unity's providing these services out-of-the-box is a definite plus.

Both services are available to add and enable directly within the Unity Editor. Let's start by adding the Analytics service.

Adding Analytics

The Unity Analytics service collects essential data about your game and players. Core game events and user properties are automatically collected when a player opens and runs your game. Additionally, you can define and track your custom in-game events. All the data that's collected from the player activity is aggregated and displayed in the Unity Cloud dashboard, which allows you, or any team member, to analyze the data and gain insights related to player engagement. Analytics is invaluable in determining how to optimize your game for increased player retention and satisfaction.

> **Additional reading | Unity documentation**
>
> Deep data insights about your game: `https://unity.com/products/unity-analytics`.
>
> Get started with Analytics: `https://docs.unity.com/ugs/en-us/manual/analytics/manual/get-started`.

Luckily, it's easy to quickly add to our games by following these steps:

1. Open **Window | Package Manager**.
2. In the **Packages** dropdown, change the context to **Unity Registry**.

3. Search for `analytics`.

4. Select **Analytics** and click the **Install** button.

5. Once installed, click the **Configure** button.

6. Click the **Go to Dashboard** link to open the Unity Cloud dashboard (in your default web browser) to configure and monitor events and get an overview of your game's activity in **Game performance**, drill down into critical metrics with **Data Explorer**, and see how events change over time with **Event Browser.**

Initializing the Analytics service

To start collecting player data to be shown and analyzed in the Unity Cloud dashboard, we'll have to do two things for the Analytics service. First, we need to call `UnityServices.InitializeAsync()` when the game starts. More importantly, due to data privacy regulations (such as GDPR, CCPA, or PIPL), we'll also need to have the player opt into having data collected so that we can call `StartDataCollection()`.

You can find detailed information on how to perform this initial setup here: `https://docs.unity.com/ugs/en-us/manual/analytics/manual/sdk-guide`.

I've provided a basic analytics initialization script example for you to get started with here: `https://github.com/PacktPublishing/Unity-2022-by-Example/tree/main/ch15/Game-Assets/AnalyticsInitialization.zip`.

Make opting in a positive experience for your players by using a statement like the following: "*Help us create a better game experience for you! Opting into analytics lets us learn how you play the game and what we can improve. We take your privacy seriously; all data is collected anonymously and used solely to enhance gameplay and features. Thank you for your support!*" (Disclaimer: I am not a lawyer; you should consult accordingly when it comes to regulations.)

Adding our next LiveOps service will be even easier.

Adding Cloud Diagnostics

The most essential service that's included under Cloud Diagnostics is, by far, **Crash and Exception Reporting**. When your game is out in the wild with players banging on it, it is incredibly challenging to know what's going on with its operational stability. You can get real-time data on crashes and exceptions in your games by simply adding and enabling the service in your project from Unity Editor. Then, you can periodically view any occurrences with their crash and exception details (including stack trace) to take action on.

To add Cloud Diagnostics to your project, follow these steps:

1. Open **Window | Package Manager**.

2. In the **Packages** dropdown, change the context to **Unity Registry**.

3. Search for `diagnostics`.

4. Select **Cloud Diagnostics** and click the **Install** button.

5. Once installed, click the **Configure** button (or, at any time, go to **Services | Cloud Diagnostics | Configure**).

6. Click the slider in the top-right corner of the **Project Settings** window to enable Cloud Diagnostics.

7. Click the **Go to Dashboard** link to open the Unity Cloud dashboard (in your default web browser) to monitor and review any crash occurrences in Crash and Exception Reporting.

Additional reading | Unity documentation

Setting up Cloud Diagnostics: `https://unity.com/products/cloud-diagnostics`

Setting up Crash and Exception Reporting: `https://docs.unity.com/ugs/ manual/cloud-diagnostics/manual/CrashandExceptionReporting/ SettingupCrashandExceptionReporting`

To test our configuration of the Cloud Diagnostics service, we can use a `Debug.LogException()` statement to view an exception message within the Console window in the **Editor** area and the Unity Cloud dashboard | **Cloud Diagnostics | Crash and Exception Reporting**. To do this, add the following line of code – temporarily – to the `Start()` method of any script added to a GameObject in your scene and enter **Play Mode**:

```
Debug.LogException(
    new System.Exception(("Cloud diagnostics test!")));
```

To finish our LiveOps example, let's see how we can dynamically change game content by adding a seasonal holiday theme to our ladybug player character for Halloween.

Updating game content dynamically

The autumn season in the Northeastern part of the United States has always been my favorite time of the year, both as an adult and while growing up. I particularly enjoy the Halloween season during this time. With that being said, let's begin the dynamic content update by taking inspiration from this season and seeing how I've added to the artwork for our ladybug player character from the 2D collection game:

Figure 15.8 – Halloween-themed player character

Scary, right? As seen in the **Hierarchy** window in *Figure 15.8,* I created a second set of graphics parented to the `Player` Prefab root named `Graphics_Halloween`. In the **Inspector** window on the right, you can see I've added a `GameObject` array with a header of `Theme Graphics`, where I've assigned both the original graphics, renamed to `Graphics_Default`, and the new holiday graphics.

> **To follow along…**
>
> You can either update your ladybug player character with holiday graphics and add the `_graphics` array to the `PlayerController` script yourself (refer to the *Updating the player controller script* section that follows) or find the updated player character with both of these already completed in this book's GitHub repository: `https://github.com/ PacktPublishing/Unity-2022-by-Example/tree/main/ch15/Game-Assets/ SeasonalPlayer.zip`.

Let's see how we'll dynamically switch to the Halloween graphics by setting up and coding a **Remote Config** integration.

Adding Remote Config

With Remote Config, we can fine-tune game adjustments for difficulty or time values, make personalization changes, or run timed events without having to distribute a new game build – providing we account for these features during our design and development phases, that is.

For our example, we'll use a simple Boolean value for our configuration key value to turn on or off the display of the Halloween graphics. First things first, however, we need to install and configure Remote Config by following these steps:

1.   Open **Window | Package Manager**.

2.   In the **Packages** dropdown, change the context to **Unity Registry**.

3.   Search for `remote`.

4.   Select **Remote Config** and click the **Install** button:

      I.   Note that Remote Config requires **Authentication** and will also install it.

5.   Once installed, go to **Window | Remote Config** (no, I don't know why it's not also under **Services**):

Figure 15.9 – Remote Config key value settings in the Editor area

As seen in *Figure 15.9*, by default, our environment will be **production**, but just know that you have the option of creating any number of additional environments for your needs – the first logical one would be **development** for working within the **Editor** area and testing before pushing to production (that is, your live players).

As shown in *Figure 15.9*, you can quickly access the Unity Cloud dashboard using the **View in Dashboard** button. The web dashboard allows you to add and change key values without firing up Unity Editor and is available from any computer or mobile device with a web browser; just go to `https://cloud.unity.com/`.

Let's add the config key for controlling the visibility of our holiday graphics:

1.   With the **Remote Config** window open, click the **View in Dashboard** button to open the Unity Cloud dashboard:

      I.   Yes, we can simply add a new key-value pair directly in the **Remote Config** window right within Unity Editor, and you can certainly do that now by clicking the **Add Setting** button. Still, we're going to leverage the dashboard to cheat – ah, I mean, generate some starter code for this first setting.

2. If the first screen you see does not say **Setup Guide**, you can find the link in the column to the left of the main window titled **GAMING SERVICES Remote Config**. There are a few options at the bottom, and one is **Setup guide** – click it.

3. Review the options for checking your environments and installing the package by confirming or simply clicking the defaults for **Next** and **Finish** since we won't need to change anything here.

4. Now, click **Create key** and use the following values:

    - **Name**: Theme_Holiday

    - **Type**: **Boolean**

5. Click **Next**, set **Value** to **true**, and click **Finish**.

6. Now, we want to click the **Implement code** button, which will generate a C# script that forms the basis for integrating Remote Config settings in our project (nice cheat, huh?):

    I.   Copy the code, then, back in Unity, create a new script named RemoteConfigSettings in a new Assets/Scripts/Services folder.

    II.   Open the script in your IDE, delete all the existing template code, and paste the copied code.

    III. Rename the public class declaration from ExampleSample to RemoteConfigSettings and save (*Ctrl/Cmd + S*).

Okay; we have started the first phase of our integration, but we'll want to make two minor additions to make it compatible with our other scripts:

- First, we'll go ahead and make it a singleton instance so that we can access it from our PlayerController component

- Second, we'll add an event to listen for when the remote configuration has completed being fetched from the server so that we can update our graphics based on the current published value

---

Automating dynamic content | UGS use cases

Note that for this example, we'll manually toggle the value for displaying the holiday theme graphics at the time we decide to. Still, Unity provides another way to automate this on a calendar schedule, which is by using **Game Overrides**. This use case can be found in the Unity documentation here: https://docs.unity.com/ugs/en-us/manual/game-overrides/manual/use-cases.

Many additional use cases for implementing UGS can be found here: https://docs.unity.com/ugs/en-us/solutions/manual/Welcome.

Update the `RemoteConfigSettings` script with the following singleton pattern and `OnSettingsChanged` event:

```
public class RemoteConfigSettings : MonoBehaviour
{
    public static RemoteConfigSettings Instance
        { get; private set; }

    private void Awake()
    {
        if (Instance == null)
            Instance = this;
        else
            Destroy(gameObject);
        DontDestroyOnLoad(gameObject);
    }

    public event UnityAction<RuntimeConfig> OnSettingsChanged;
    ...
}
```

Now, to be able to invoke the event when the server configuration values have been retrieved, add the following `OnSettingsChanged` invocation to the `ApplyRemoteSettings()` method:

```
private void ApplyRemoteSettings(ConfigResponse configResponse)
{
    ...
    OnSettingsChanged?.Invoke(
        RemoteConfigService.Instance.appConfig);
}
```

As you can see, we're passing in the `RemoteConfigService` class's `appConfig` object so that the event listeners can get their required config settings. With that, we can now add the Remote Config service to our game. Create a new GameObject in the scene named `RemoteConfig` and add the `RemoteConfigSettings` script. With that, we're done – easy-peasy.

> **Architecture tip**
>
> For brevity and to keep things simple for this dynamic content example, I've decided to add the graphics changing directly to the `PlayerController` script. However, you may want to consider making it a separate component for a more robust and reusable *dynamic graphics swapping system* component.

Now, all that's left to do is add to the capability of our `PlayerController` component so that we can get the configuration setting value to determine whether to show the holiday graphic.

Updating the player controller script

Open the `PlayerController` script for editing and add the following declarations for holding `GameObject` references for the graphics (both the default and holiday versions) and a constant for the Remote Config settings key we previously defined:

```
[SerializeField] private GameObject[] _graphics;
private const string THEME_HOLIDAY = "Theme_Holiday";
```

So that we can respond to the completion of the remote settings being fetched, let's add a listener to the `OnSettingsChanged` event on the `RemoteConfigSettings` singleton instance:

```
private void OnEnable()
    => RemoteConfigSettings.Instance.
        OnSettingsChanged += ConfigSettingsChanged;

private void OnDisable()
    => RemoteConfigSettings.Instance.
        OnSettingsChanged -= ConfigSettingsChanged;
```

We've also made sure we add a removal for the listener, like the good programmers we are.

Finally, we can handle the `OnSettingsChanged` event being triggered by adding the following methods:

```
private void ConfigSettingsChanged(RuntimeConfig config)
{
    var isThemeEnabled = config.GetBool(THEME_HOLIDAY, false);
    if (!isThemeEnabled)
        return;

    ShowThemeGraphics(1);
}

private void ShowThemeGraphics(int value)
{
    foreach (var g in _graphics)
    {
        g.SetActive(false);
    }
    _graphics[value].SetActive(true);
}
```

Alright, let's do this one last time. A breakdown of the methods looks like this:

- `ConfigSettingsChanged()`: This method is the `OnSettingsChanged` event handler. Using the `RuntimeConfig` object passed in the event, we use its `GetBool()` method with the string constant for the key name, `THEME_HOLIDAY`, to retrieve the current value. `false` is used as the default value that's returned should the key not be found:

  - `If (!isThemeEnabled)`: If the value that's returned from `GetBool()` is `false`, we'll short-circuit the method with `return`, leaving the default graphics as the currently visible graphics on the player character:

    - Otherwise, `ShowThemeGraphics(1)` is called and passed in an index value of `1`, meaning the second item that's assigned in the `_graphics` array – where we assigned the Halloween graphics in the **Inspector** area. Arrays are 0-based, so item #1 is index 0, item #2 is index 1, and so on.

- `ShowThemeGraphics()`: Here, we simply iterate through all the GameObjects in the `_graphics` array using a `foreach` statement, setting each object's active state to `false`. We immediately follow up by setting the object with the passed-in index value active, ensuring the graphics that are assigned to that index will be the graphics displayed with `_graphics[value].SetActive(true)`.

Save the script and try it out! After entering **Play Mode**, you should see the graphics on your ladybug player character change to the holiday graphics created for the theme. Nice!

Bonus activity

Update the `OnSettingsChanged` event architecture so that it uses an implementation of the global event system we created in *Chapter 9*.

After you build and publish your game, you can use the Unity Cloud dashboard to change the value of `Theme_Holiday` at any time to show/hide the holiday graphics without having to rebuild and redistribute a new game build.

Complete project code

You can download the completed 2D collection game project code for this chapter from this book's GitHub repository: `https://github.com/PacktPublishing/Unity-2022-by-Example/tree/main/ch15/Unity-Project`.

And speaking of building and distributing your game build, that's something you're going to have to do lots and lots of times throughout its life cycle. If only there were a way to simplify and automate that process… oh wait, there is!

Publishing with Unity Build Automation

Moving the build process of our local system to dedicated build servers – cloud-based systems, in this case – in our game development workflow offers several benefits but primarily improved productivity because our machines won't be locked up during the build time. Builds generated on dedicated servers can also help to ensure consistency and reliability across all our game versions – significantly reducing the "*but it works on my machine*" problem.

To help identify issues and catch bugs for our players early on, we can also integrate automated testing and **quality assurance** (**QA**) into our automated build process. Additionally, we can automate the distribution of completed builds to various teams for further testing and, if required, publishing.

It all starts with understanding the build process, so let's have a look at building locally.

Building your game

In *Chapter 14*, we saw how we can use **Build Settings** to set the build target platform to **Android** and build to our head-mounted device for testing our MR game. Here, we can do the same thing but build our 2D collection game for standalone platforms, such as Windows and Mac PCs, by performing the following steps:

1. Open **Build Settings** from **File | Build Settings**.
2. Ensure that **Platform** is set to **Windows, Mac, Linux** (also known as *standalone*).
3. Select the desired platform in the **Target** dropdown (for example, Windows).
4. Click **Build**, or **Build And Run**, to automatically launch the game when the build finishes, and select a folder on your system to store the build.
5. Wait until the build finishes.

You don't have any choice; you have to sit there and wait… staring at the progress indicator minute by minute until it finishes. Luckily, with our small 2D game, this doesn't take long at all, but that won't always be the case as our games grow in size and complexity. Prepare to get up from your desk, go make a cup of coffee, and come back only to see that the progress bar has barely moved!

When the build finishes, you'll have the files necessary to run your game in the folder you selected for the build. Open the folder in your system's file explorer and run the EXE file to play the game. To distribute your game – share it with your team, friends, and family – zip up the contents of the folder (minus any folders ending in DoNotShip) and share it via an online cloud storage service (such as Google Drive, Dropbox, OneDrive, or Box).

Still, we can do better. Let's offload the build process to Unity Build Automation.

Automating the build pipeline

We've already seen how to add Unity Version Control to our projects' DevOps strategy. Well, automating a build is the second half of this DevOps story – specifically **CI/CD** – since we'll use our project's cloud workspace as the source for the cloud build configuration. This is how the cloud service gains access to our project files to perform the build.

First, let's add **Build Automation** to our project by following these steps:

1. Open **Window | Package Manager**.

2. In the **Packages** dropdown, change the context to **Unity Registry**.

3. Search for `build`.

4. Select **Build Automation** and click the **Install** button.

5. Once installed, click the **Configure** button (or, at any time, go to **Services | Build Automation | Configure**).

6. In the top-right corner of the **Project Settings** window, click the slider to enable Build Automation.

7. Click the **Manage Build Targets** button to open the Unity Cloud dashboard (in your default web browser). This is where we'll add our first build target:

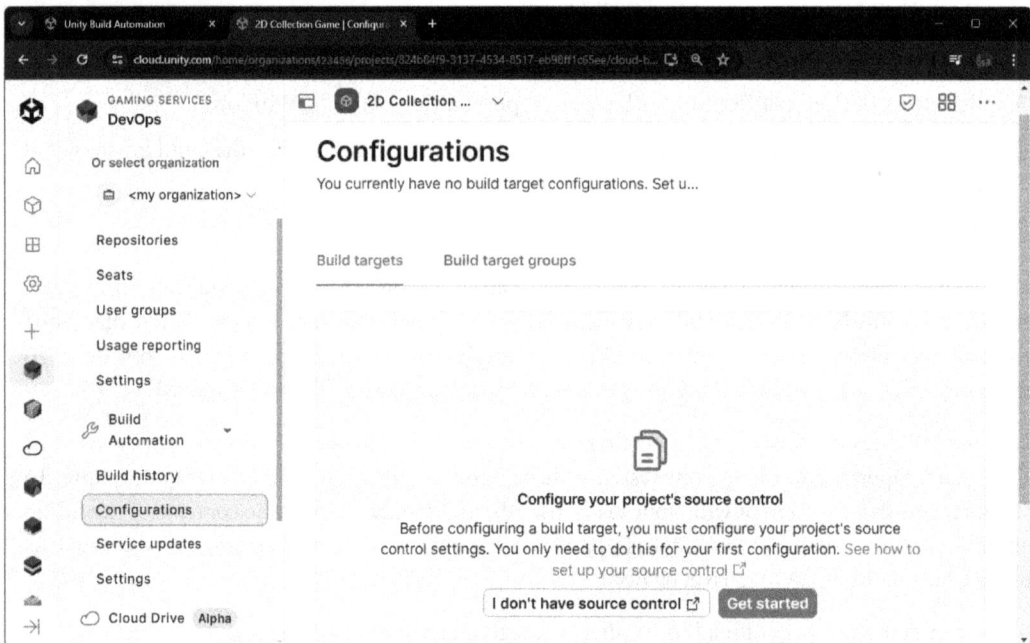

Figure 15.10 – Unity Cloud Build Automation configurations

Now, click the **Get Started** button to connect the source control provider and repository (that is, the workspace):

1. Select **Unity Version Control** in the **Source control provider / SCM type** dropdown.

2. Select your Unity organization in the **UVCS organization Server URL** dropdown.

3. Leave **Authenticate with Unity ID** selected.

4. Click **Save** (at the top).

Once the source control settings have been saved, proceed to **Build Automation | Configurations** and follow these steps to add our build target:

1. Click the **Quick target setup** button.

2. Select **Windows desktop 64-bit** in the **Select a platform to build for:** dialogue.

3. Confirm that the **Repo** field has been populated with your game:

    I. If you have more than one branch in your workspace, select the branch for building in the **Branch** dropdown.

4. Under **Versioning**, select **Always Use Latest 2022.3** in the **Unity version** dropdown.

5. Select **Windows 11** in the **Builder Operating System and Version** dropdown.

6. Click **Next**.

7. On the **Builder configuration** screen, ensure **STANDARD** is selected and click **Next**.

8. On the **Scheduling** screen, for this example, we're going to manually kick off the build process from within Unity Editor, so I prefer to enable **Auto-cancel**.

> **Important tip**
> You can change the scheduling of your builds at any time. The most popular strategy is to dedicate a branch in your workspace to automatically start a build when the branch has been updated.

9. Click the **Save configuration** button.

Now, return to Unity Editor and reopen the Build Automation settings. When it refreshes, you'll see we now have a build target that's been added: **Default Windows desktop 64-bit**.

First, ensure your Unity Version Control workspace is up to date by checking in any pending changes for your project, and then click the **Build** button. A console entry will be created with a status of **Build #1 Default Windows desktop 64-bit added to queue** to notify you that the build has been queued.

You can check the status of your cloud builds by clicking the **Build History** button or opening a web browser on any device in the Unity Cloud dashboard under **Build Automation | Build History**.

> **Build Automation pricing**
>
> You can build faster for multiple platforms, including Windows, Mac, Linux, Android, and WebGL, and on multiple machines simultaneously: `https://unity.com/products/unity-devops#pricing`.
>
> Windows build includes 200 minutes for free per month. You pay for additional build minutes, Mac build minutes, additional concurrent build machines, and storage above 5 GB (per GB per month).

When the build finishes, if you're still within Unity Editor, you'll see a console message appear: **Build #1 Default Windows desktop 64-bit success**.

You'll also receive an email with the subject **Built '2D Collection Game' for Windows x86_64**, and the message **'2D Collection Game' (Default Windows desktop 64-bit) #1 has been built for Windows x86_64!** The email will also list any warnings or errors encountered during the build process and links to the cloud dashboard configuration.

Most importantly, an **INSTALL** link will be provided to download the build artifact (that is, files your team uses to deploy or test your application). Clicking the link will bring you to the Unity Cloud dashboard's **Build Details** page for this build. You can download the game build by clicking the **Download .ZIP file** button. You can also easily share this build with anyone by clicking the **Share** button – you'll be provided a share link and QR code that will remain valid for 14 days.

> **Automatic build sharing**
>
> You can enable automatically creating the build share link on the **Build Automation Settings** page. When enabled, the share link will be included in the email notification you receive when the build succeeds (preventing a trip to the Unity Cloud dashboard to grab it).
>
> In addition to enabling automatic build sharing, Unity allows integrations with popular developer tools such as Discord, Slack, Jira, and Trello to create automated notifications in these spaces for both Build Automation and Cloud Diagnostics. You can configure your integrations in the Unity Cloud dashboard under **Administration | Project integrations** (`https://docs.unity3d.com/Manual/UnityIntegrations.html`).

Now that we have built our first standalone PC game, let's publish it on **Itch.io** and share it with the world!

Publishing your game

For this publishing example, we'll publish our game to **Itch.io**. It's the perfect platform for our little 2D collection game to start getting our indie title into players' hands for some valuable playtesting, feedback, and bug squashing (Ack! But not our ladybug!).

Open Itch.io and register on the site; ensure you check off **I'm interested in distributing content on itch.io**: `https://itch.io/register`.

Once registered, go to the **Creator Dashboard** area (that is, **Dashboard**) and click the **Create new project** button. This will bring you to the following screen, where you can start filling out all the required details and upload a screenshot for a cover image to display your game listing on the storefront:

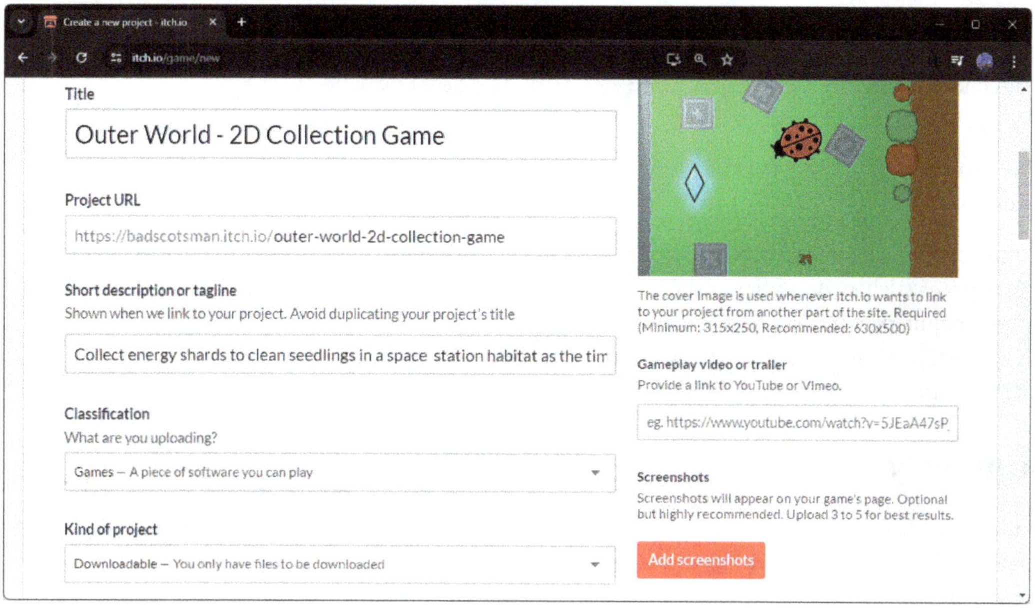

Figure 15.11 – Creating a new project on Itch.io

The field we need to pay attention to when uploading our game build to the site is **Kind of project**. Ensure it's set to **Downloadable**. Then, in the **Uploads** section below, click the **Upload files** button and select the ZIP file we created to distribute our game build in the *Building your game* section or from the build automation process.

Building for WebGL | Unity documentation

Note that for Itch.io, you have the option of publishing a WebGL build of the game that will run directly within the player's web browser instead of having to be downloaded. Before committing to a WebGL version of your game, you must ensure your game is playable and performant in a web browser to provide a good player experience.

Build your WebGL application: `https://docs.unity3d.com/2022.3/Documentation/Manual/webgl-building.html`.

Fill out the remainder of the fields, including all the fields under the **Details** section, which includes **Description**, **Genre**, **Community interaction**, and **Visibility & access**. Also, don't forget to add **Screenshots** and a **Gameplay video or trailer** to excite and encourage visitors to your page to download and play your game.

> **Itch.io Devlogs**
>
> Have you heard of a Devlog? It's a forum post where game developers share updates on their ongoing projects. Anyone can leave comments and thoughts, which is an excellent way for players to stay informed about the latest developments in games they are interested in. Additionally, developers can use the forum to generate excitement and increase project engagement.
>
> Itch.io Community Devlogs: `https://itch.io/devlogs`.

When you're happy with what you've entered, click **Save & view page** to finish publishing your game on Itch.io. Yay!

In this section, we learned how to add Unity LiveOps services to our 2D collection game, including core services such as Analytics and Cloud Diagnostics, and updating in-game content dynamically without requiring distributing a new game build with Remote Config. We also learned how to set up Unity DevOps to automate our game build process while offloading it from our local machine. We finished with a quick game publishing example.

Summary

In this chapter, we introduced and explored the concepts and strategies for operating and publishing GaaS to help achieve our commercial viability goals and, ultimately, success. By adopting GaaS and implementing the tools and technologies of both DevOps and LiveOps, we can build a solid foundation to support the ongoing release of games and content updates. This helps keep our players engaged and can provide longer lifetime value for our game releases.

Additionally, we learned how to safeguard all the hard work we put into our game projects by introducing source code management with Unity Version Control. We also learned how to set up and use version control within Unity Editor for a cloud-based DVCS solution and better structure our projects to minimize conflicts for team collaboration.

We gained a fundamental understanding of game economies and distribution channels while discussing how to better leverage player engagement to convert mobile free-to-play games into revenue through in-game purchases, ads, subscriptions, and premium purchases for PC games. Then, we provided a technical and financial feature breakdown of the top game distribution platforms for both free and premium games.

We concluded this chapter with examples of implementations of Unity LiveOps and DevOps. We achieved this by adding core cloud services, dynamically updating in-game content, building and automating the build process, and finally publishing our game.

Final words!

Throughout these chapters, from the creative processes of designing games and working with art assets to the detailed implementations of programming mechanics and systems, we explored fundamental principles, solved problems, and celebrated overcoming challenges together. Whether you're a seasoned professional, an indie dev, a student, or an aspiring hobbyist, I hope you've been provided with inspiration, insights, moments of revelation, and the occasional laugh.

As this chapter comes to a close, I want to express my heartfelt appreciation to each of you who has embarked on this adventure with me. Thank you. I am honored to have the opportunity to share my knowledge and experiences with such an incredible developer community.

All stories have an ending, and as such, this one has come to an end now, too. Still, it's just one of many phases in our journey as game developers as we continue to work toward finishing and releasing our remarkable games.

Until next time… I have a game to finish now (and you do, too). Have fun!

Index

‹packt›

packtpub.com

Subscribe to our online digital library for full access to over 7,000 books and videos, as well as industry leading tools to help you plan your personal development and advance your career. For more information, please visit our website.

Why subscribe?

- Spend less time learning and more time coding with practical eBooks and Videos from over 4,000 industry professionals

- Improve your learning with Skill Plans built especially for you

- Get a free eBook or video every month

- Fully searchable for easy access to vital information

- Copy and paste, print, and bookmark content

Did you know that Packt offers eBook versions of every book published, with PDF and ePub files available? You can upgrade to the eBook version at packtpub.com and as a print book customer, you are entitled to a discount on the eBook copy. Get in touch with us at customercare@packtpub.com for more details.

At www.packtpub.com, you can also read a collection of free technical articles, sign up for a range of free newsletters, and receive exclusive discounts and offers on Packt books and eBooks.

Other Books You May Enjoy

If you enjoyed this book, you may be interested in these other books by Packt:

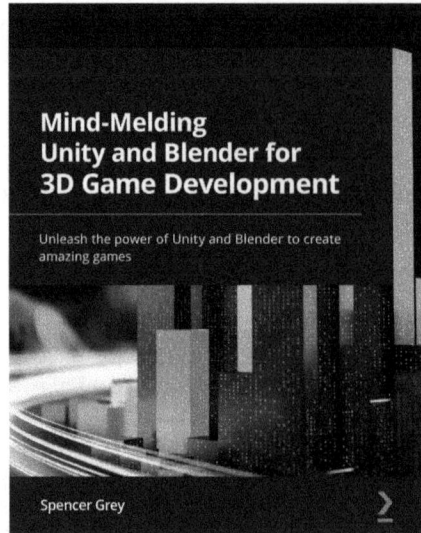

Mind-Melding Unity and Blender for 3D Game Development

Spencer Grey

ISBN: 978-1-80107-155-0

- Transform your imagination into 3D scenery, props, and characters using Blender.
- Get to grips with UV unwrapping and texture models in Blender.
- Understand how to rig and animate models in Blender.
- Animate and script models in Unity for top-down, FPS, and other types of games.
- Find out how you can roundtrip custom assets from Blender to Unity and back.
- Become familiar with the basics of ProBuilder, Timeline, and Cinemachine in Unity

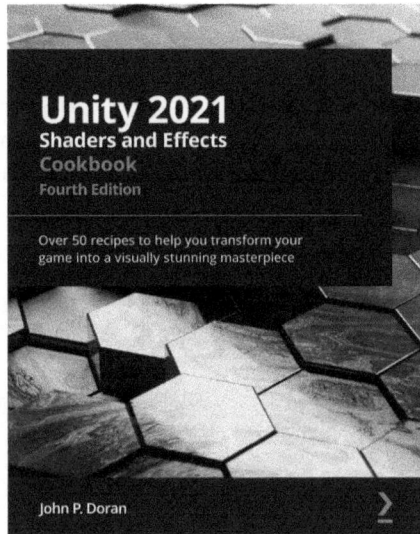

Unity 2021 Shaders and Effects Cookbook

John P. Doran

ISBN: 978-1-83921-862-0

- Use physically based rendering to fit the aesthetic of your game.
- Create spectacular effects for your games by testing the limits of what shaders can do.
- Explore advanced shader techniques for your games with AAA quality.
- Use Shader Graph to create 2D and 3D elements for your games without writing code.
- Master the math and algorithms behind the commonly used lighting models.
- Get to grips with the Post-Processing Stack to tweak the appearance of your game.

Packt is searching for authors like you

If you're interested in becoming an author for Packt, please visit `authors.packtpub.com` and apply today. We have worked with thousands of developers and tech professionals, just like you, to help them share their insight with the global tech community. You can make a general application, apply for a specific hot topic that we are recruiting an author for, or submit your own idea.

Share Your Thoughts

Now you've finished *Unity 2022 by Example*, we'd love to hear your thoughts! Scan the QR code below to go straight to the Amazon review page for this book and share your feedback or leave a review on the site that you purchased it from.

`https://packt.link/r/1-803-23459-8`

Your review is important to us and the tech community and will help us make sure we're delivering excellent quality content.

Download a free PDF copy of this book

Thanks for purchasing this book!

Do you like to read on the go but are unable to carry your print books everywhere?

Is your e-book purchase not compatible with the device of your choice?

Don't worry!, Now with every Packt book, you get a DRM-free PDF version of that book at no cost.

Read anywhere, any place, on any device. Search, copy, and paste code from your favorite technical books directly into your application.

The perks don't stop there, you can get exclusive access to discounts, newsletters, and great free content in your inbox daily

Follow these simple steps to get the benefits:

1. Scan the QR code or visit the following link:

https://packt.link/free-ebook/9781803234595

2. Submit your proof of purchase.
3. That's it! We'll send your free PDF and other benefits to your email directly.

www.ingramcontent.com/pod-product-compliance
Lightning Source LLC
Chambersburg PA
CBHW072005230326
41598CB00082B/6769